Key Transitions
In
Animal Evolution

Key Transitions
In
Animal Evolution

Editors

Rob DeSalle

American Museum of Natural History
Division of Invertebrate Zoology
New York
USA

Bernd Schierwater

ITZ, Ecology and Evolution
TiHo Hannover
Germany

CRC Press

Taylor & Francis Group
Boca Raton London New York

CRC Press is an imprint of the
Taylor & Francis Group, an **informa** business

CRC Press
Taylor & Francis Group
6000 Broken Sound Parkway NW, Suite 300
Boca Raton, FL 33487-2742

First issued in paperback 2019

ISBN-13: 978-1-57808-695-5 (hbk)
ISBN-13: 978-0-367-38319-0 (pbk)

Library of Congress Cataloging-in-Publication Data

Key transitions in animal evolution / [edited by] Rob Desalle,
Bernd Schierwater.
       p. cm.
   Includes bibliographical references and index.
   ISBN 978-1-57808-695-5 (hardback)
1. Phylogeny. 2. Evolution (Biology) 3. Genetic transcription.   I. DeSalle, Rob II. Schierwater, B. (Bernd),
1958-
   QH367.5.K49 2010
   591.3'8--dc22

                                                    2010033462

Visit the Taylor & Francis Web site at
http://www.taylorandfrancis.com

and the CRC Press Web site at
http://www.crcpress.com

Introduction
A Phylogenomic Journey Through the Animal Tree of Life: Key Innovations in the Evolution of Metazoa

Bernd Schierwater [1,2] and *Rob DeSalle* [2]

> *"All animals are equal, but some animals are more equal than others."*
> —George Orwell

> *"You can observe a lot just by watchin'."*
> —Yogi Berra

The origin of multicellular animals is one of those difficult and delicate biological problems that makes careers and produces enemies. Many ideas have been proffered and many hypotheses have been tested with respect to the subject. In order to address this subject further and to assess progress in the examination of animal origins and transitions, an international group of scientists was convened at the Society for Comparative Biology meeting in 2007 in Phoenix, Arizona. Since that meeting the organizers have worked to put together a volume of essays stemming from the talks at the meetings. While the meeting occurred a few years back, we have updated and expanded all of the papers in the symposium and have added several chapters that did not occur in the original symposium. This book seeks to bring together experts in many areas of research on the origin of multicellular animals from a variety of biological perspectives. An extra dimension of this book is the addition of chapter ending questions for

[1] ITZ, Ecology & Evolution, TiHo Hannover, Bünteweg 17d, D-30559 Hannover, Germany.
E-mail: bernd.schierwater@ecolevol.de
[2] American Museum of Natural History, New York, Division of Invertebrate Zoology 79 St. at Central Park West, New York, NY 10024, USA.
E-mail: desalle@amnh.org

discussion that we hope students reading the volume will find useful in strengthening their knowledge of the individual subjects covered by the chapters. The book is divided into three sections. The first section entitled "Tangled Roots in the Animal Tree of Life" examines the phylogenetic relationships of animals from the single celled close outgroups of animals to fine tuned arthropod phylogeny. The second section entitled "The Earliest Animals: from Genes to Transitions" takes groups of living animals that represent the earliest body plans for animals and examines their evolution and development. The final section of the book entitled "Pattern and Process at the Base of the Metazoan Tree of Life" examines the life history of basal metazoans and the major transitions such as feeding, germ line evolution and adaptation to the environment, these organisms experienced in evolution.

Animals and bauplans: The background on body plans for Metazoa is interesting. As DeSalle and Schierwater (2007) describe:

Multicellular animals (metazoans) almost certainly derive from a cum grano salis "single" celled protozoan ancestor. Metazoa—by definition—harbor more than one somatic cell type and can be classified into basal (or diploblastic) and derived (or triploblastic) groups. These groups differ by the existence of three (triploblasts) or two germ layers only (diploblasts). It is generally accepted that triploblasts derive from a diploblastic ancestor, implying that multicellular animals evolved from simple to more complex (anagenetic evolution). The basal (diploblastic) metazoans consist of sponges, placozoans, cnidarians, and ctenophores. The higher (derived triploblastic) metazoan groups are represented by two lines which differ by the fate of the blastoporus. In the Deuterostomia lineage the blastoporus becomes the anus of the adult, in the Protostomia the blastoporus becomes the mouth of the adult. Because of the definition of ventral and dorsal according to the position of the final mouth, Protostomia and Deuterostomia represent opposite dorsal-ventral bauplans and endo- versus exogenuous skeleton systems. The above transitions from single celled to multicellular, from diploblastic to triploblastic, and from protostomic to deuterostomic are—despite substantial efforts—not yet understood.

While the anatomical transitions between groups is well known, the jury on the phylogenetic relationships of the major groups of animals at the base of the metazoan tree of life is still out. Within just 12 months three papers appeared in the literature with three rather different hypotheses concerning the topology of the tree of life in this area. Dunn et al. (2008) using EST data on over 70 taxa, suggested that Ctenophores were the most basal animal, with Bilateria sister to sponges AND Cnidaria. Schierwater et al. (2009) followed with a detailed analysis of anatomy, molecules and molecular structure to come to a similar conclusion of Dunn et al. (2008)

that Bilateria are sister to Cnidaria AND Porifera. However, by adding the Placozoa to their analysis, Schierwater et al., hypothesized that there was an early Diploblast (Placozoa, Cnidaria, Ctenophora and Porifera)— Triploblast (Bilateria) split, with Placozoa as the most basal diploblast (cf. Dellaporta et al. 2006). The final study in this trio, came from Philippe et al. (2009) who suggest that the tree of life in this area is more like traditional understanding, where Porifera are the most basal metazoan, followed by Placozoa, and then Cnidaria and Bilateria. A clue to the diversity of hypotheses for the topology in this part of the tree of life can be found in Siddall (2009) who reanalyzed all three of these data sets. Siddall points out which relationships in the three studies are robust and which are not and it is clear that much more work needs to be done on this problem.

To set the stage for what the student might need to know about metazoan phylogenetics we present in this volume, several papers that discuss the metazoan part of the tree of life. Vazquez et al. start off this first section of the book entitled "Tangled Roots in the Animal Tree of Life", with a description of phylogeny of single celled eukaryotes and how these relate to the relationships of metazoa. The outgroup status for the Metazoa is followed by two chapters that address the relationships of the major animal groups; specifically the chapter by Kocot et al. focuses on broad based nuclear gene studies and the chapter by Lavrov utilizes mtDNA whole genomes to examine the relationships of the major animal groups. To demonstrate the flexibility of phylogenetic analysis in the Metazoa, we also include a summary of the relationships of arthropod taxa by Deutsch.

Comparative biology of early metazoans: The rapidly increasing number of whole genome projects has increased the number of model systems for evolutionary developmental research. As DeSalle and Schierwater (2007) suggest:

Genomic features require a comparative framework to be understood, and thus the choice of the right model organisms becomes particularly crucial (Martindale 2005). Genes require a comparative classification and when animal genomes are sequenced, homologs need be assigned to various classes with some certainty (Copley et al. 2004). The interpretation of "gene homologs" has been a hot subject of debate and different researchers use crucial terms like "anterior/posterior and oral/aboral in different, and interpretive ways (e.g. Finnerty 2003, Abouheif 1999), leading to conflicting views sometimes on the same (or similar) data set (Kamm et al. 2006, Chourout et al. 2006, Ryan et al. 2007). Great expectations derived from the study of Antp-class genes to unravel the evolution of axis formation, and thus the basic bauplan setup for all major metazoan groups (e.g. Deutsch & Mouchel-Vielh 2003).

Since the initial application of in situ techniques and antibody staining approaches to evolutionary questions in the 1990's, a proliferation of studies has appeared in the literature. These studies opened the way for a better understanding of the genes involved in specific morphologies of plants and animals. The combination of genome level gene content analysis and where genes are expressed in animals have led to a real transition in how scientists look at key transitions in animal evolution. The first two chapters in this section entitled "The Earliest Animals: from Genes to Transitions", look closely at the evolution of the nervous system. Nickel examines the biology and genomics of sponges to develop a framework for a pre-nervous system in animals. Galliot et al. examine neural genes and neurogenesis in cnidaria. Boero and Piraino use the existing ideas about development to hypothesize a one step transition from cnidaria to bilateria. This interesting hypothesis is an important example of transitional thinking. The next chapter by Jacobs et al. looks at the evolution of sensory structures and mechanisms in basal animals from an evo-devo perspective. While the nervous system and sensory apparatus are two excellent examples of transitional structure in animal evolution, the basic body plan change to bilateral symmetry has also been a major target for evo-devo research. The chapter by Ball et al. examines many of the developmental aspects of the change to a bilateral body plan. Finally, developmental mechanisms control other more specific anatomical aspects of animals. To cover all of these would be an immense undertaking so we give the student a "taste" of the kinds of studies that can be accomplished with genome level information and evo-devo approaches. Oakely and Plachetzki focus on phototransduction and how this important transition evolved in metazoa. Hox genes have been the "poster genes" for evo-devo since its inception. In order to represent the far reaching work in the area of Hox gene evolution for this volume, Kappen presents a summary of the role of Hox genes in the reproductive biology of mammals. While the subject matter of this section is vast, we hope we give the student a good background and a broad view of the kind of research being conducted.

Major transitional problems at the origin of animals: Scientists can focus on specific aspects of development. Their relevance to how animal live evolved in very direct fashion as a result of genomics and the developmental approaches available. However, larger more quantitative transitions such as feeding and whole scale changes in developmental strategies such as evolution of the germ line, are also a fertile area of examination for the base of the animal tree of life. In addition, the transition from a very simple metazoan form to more complex forms can best be understood after a strong background for the simple animals have been presented. The first two papers in this section entitled "Pattern and Process at the Base of the Metazoan Tree of Life", address the life history and biology of Placozoa.

These chapters by Pearse and Voigt and Schierwater et al. should clearly be the reference point for understanding the biology of this little understood animal and should be an important stepping off point for anyone interested in animal evolution regardless of the phylogenetic placement of this enigmatic animal. Because the feeding apparatus of animals vary greatly across phylogeny, unraveling the events that have occurred is an important and compelling subject. Hence, Blackstone examines the evolution of feeding in animals. An understanding of how animals evolved is also dependent on the environmental conditions under which they evolved so the interrelationship of animal evolution with biogeochemical events is examined in the next chapter by Gaidos. Extavour finishes the section by examining the key events and transitions involved in the evolution of germ lines.

Conclusion. 'Watchin': Understanding the changes that occurred at the base of the metazoan branch of the tree of life is as Yogi Bera, the famous American baseball player and philosopher once said, all about 'watchin'; but "watchin" what? Readers of this volume will be directed to "watch" a lot of things and we hope that this book on Key Transitions in Animal Evolution contributes to the observational richness on the metazoan branch of the tree of life. It is clear that many of the questions relevant to this part of the tree of life have not been answered and some have not even been addressed. However, the more we observe (watchin') about this question, the richer our understanding becomes and more importantly, the more well defined our questions and hypotheses become.

Acknowledgments: We thank all speakers and particularly all authors for their intriguing contributions. The symposium was supported by a grant to BS and RD from the NSF (DEB-0706893) and general support from the SICB.

References

Abouheif, E. 1999. Establishing homology criteria for regulatory gene networks: prospects and challenges. Novartis Found. Symp. 222: 207–225.

Chourrout, D. and F. Delsuc, P. Chourrout, R.B. Edvardsen, F. Rentzsch, E. Renfer, M.F. Jensen, B. Zhu, P. de Jong, R.E. Steele, and U. Technau. 2006. Minimal ProtoHox cluster inferred from bilaterian and cnidarian Hox complements. Nature 442: 684–672.

Copley, R.R. and P. Aloy, R.B. Russell, and M.J. Telford. 2004. Systematic searches for molecular synapomorphies in model metazoan genomes give some support for Ecdysozoa after accounting for the idiosyncrasies of Caenorhabditis elegans. Evol. Dev. 6: 164–169.

Dellaporta, S.L, and A. Xu, S. Sagasser, W. Jakob, M.A. Moreno, L.W. Buss, and B. Schierwater. 2006. Mitochondrial genome of *Trichoplax adhaerens* supports Placozoa as the basal lower metazoan phylum. Proc. Natl. Acad. Sci. USA 103: 8751–8756.

DeSalle, R. and B. Schierwater .2007. Key transitions in animal evolution. Integr. Comp. Biol. 47: 667–669.

Deutsch, J.S. and E. Mouchel-Vielh. 2003. Hox genes and the crustacean body plan. Bioessays 25: 878–987.

Dunn, C.W. and A. Hejnol, D.Q. Matus, K. Pang, W.E. Browne, S.A. Smith, E. Seaver, G.W. Rouse, M. Obst, G.D. Edgecombe, M.V. Sørensen, S.H. Haddock, A. Schmidt-Rhaesa A. Okusu, R.M. Kristensen, W.C. Wheeler, M.Q. Martindale, and G. Giribet. 2008. Broad phylogenomic sampling improves resolution of the animal tree of life. Nature 452:745–749.

Finnerty, J.R. 2003. The origins of axial patterning in the metazoa: how old is bilateral symmetry? Int. J. Dev. Biol. 47: 523–529.

Kamm. K. and B. Schierwater, W. Jakob, and D. Miller. 2006. Axial patterning and diversification in the Cnidaria predate the Hox system. Curr. Biol. 16: 920–926.

Martindale, M.Q. 2005. The evolution of metazoan axial properties. Nat. Rev. Genet. 6: 917–927.

Philippe, H. and R. Derelle, P. Lopez, K. Pick, C. Borchiellini, N. Boury-Esnault, J. Vacelet, E. Renard, E. Houliston, E. Quéinnec, C. Da Silva, P. Wincker, H. Le Guyader, S. Leys, D.J. Jackson, F. Schreiber, D. Erpenbeck, B. Morgenstern, G. Wörheide, and M. Manuel. 2009. Phylogenomics Revives Traditional Views on Deep Animal Relationships. Curr. Biol. 19: 706–712.

Ryan, J.F. and M.E. Mazza, K. Pang, D.Q. Matus, A.D. Baxevanis, M.Q. Martindale, and J.R. Finnerty. 2007. Pre-bilaterian origins of the Hox cluster and the Hox code: evidence from the sea anemone, Nematostella vectensis. PLoS ONE 2: 1–23.

Schierwater, B. and M. Eitel, W. Jakob, H.-J. Osigus, H. Hadrys, S. Dellaporta, S.O. Kolokotronis, and R. DeSalle. 2009. Concatenated Molecular and Morphological Analysis Sheds Light on Early Metazoan Evolution and Fuels a Modern "Urmetazoon" Hypothesis. PlosBiology 7: e1000020.

Siddall, M.E. 2009. Unringing a bell: metazoan phylogenomics and the partition bootstrap. Cladistics 25: 1–9.

Contents

Section 1

Tangled Roots in the Animal Tree of Life

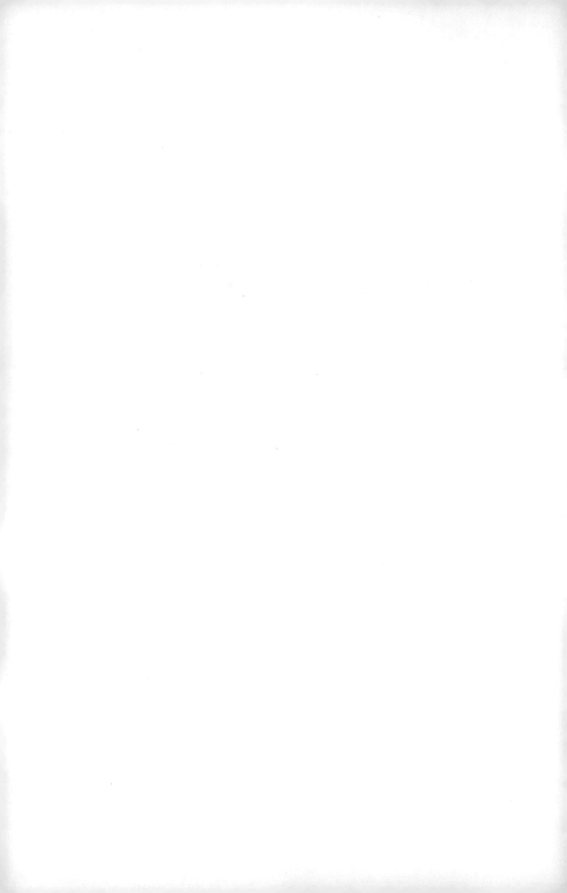

Chapter 1

Putting Animals in their Place within a Context of Eukaryotic Innovations

Danielle Vazquez,[1] Laura Wegener Parfrey[2] and Laura A. Katz[1,2]

Animals represent just one of an estimated 70+ lineages of eukaryotes (Patterson 1999), and the bulk of the remaining lineages are microbial. During the past decade, perspectives of the organization of eukaryotic diversity have undergone a major shift to a system that recognizes five or six high-level groupings (Adl et al. 2005, Tekle et al. 2009). While deep nodes on the eukaryotic tree of life remain elusive (reviewed in Parfrey et al. 2006), the position of animals emerging from within microbial lineages has been robustly resolved by several lines of evidence (King 2004). Animals are placed within the supergroup Opisthokonta, along with fungi and numerous microbial relatives (Adl et al. 2005, Steenkamp et al. 2006).

In this chapter we place animals in their phylogenetic context by focusing on lineages of eukaryotes defined by one or more ultrastructural identities (i.e., unique combinations of subcellular structures). Many of these clades are marked by major innovations in cell and/or genome structure at their base. We argue that this contrasts with animals, which lack clear innovations at the base of the lineage despite many within (e.g., triploblasty, true gut). We focus on eight exemplar clades: foraminifera, diatoms, ciliates, apicomplexa, dinoflagellates, euglenids, parabasalids

[1]Department of Biological Sciences, Smith College, Northampton, MA, USA.
[2]Program in Organismic and Evolutionary Biology, University of Massachusetts, Amherst, MA, USA.
E-mail: lkatz@email.smith.edu

and kinetoplastids. Many of the eukaryotic clades originally defined by morphological and/or ultrastructural characteristics are now grouped into larger clades—the Stramenopiles, Alveolates, and Euglenozoa—that are well-supported in molecular analyses.

Here, we aim to accomplish two goals: (1) putting animals into an evolutionary context by exploring major transitions at the base of diverse microbial clades; and (2) using examples from diverse eukaryotic lineages to highlight the complex evolutionary patterns of features that are often believed to be fixed across the eukaryotic tree or unique to animals (as well as plants and fungi). In both sections, we aim to be exemplary rather than exhaustive and our intent is to provide a broader context for interpreting major transitions within animals, the subject of the remainder of this book.

Part one: Understanding Innovations Across the Eukaryotic Tree of Life

We have chosen a limited set of clades to demonstrate the types of transitions that have occurred at the base of many microbial lineages. These examples were chosen partially due to the availability of data and partially based on the authors' areas of expertise.

Diatoms are among the most diverse clades of eukaryotes, with an estimated 100,000 extant species, and fall within the **stramenopiles** (Fig. 1). The stramenopiles are a well-supported clade that includes kelp (Phaeophyceae), watermolds (oomycetes), slime nets (labyrinthulids), and many other lineages (Andersen 2004). Synapomorphies for diatoms include the presence of a silica shell, which gives diatoms their distinctive appearance (Fig. 2a). These shells are thought to have evolved initially as a byproduct of silica metabolism, and appear to play a protective role (Schmid 2003).

Three of our exemplar lineages—ciliates, dinoflagellates, and apicomplexa—are members of a single clade, the **alveolates** (Fig. 1). Alveolates are united by the presence of alveoli, a system of abutting sacs that form a layer immediately underneath the plasma membrane that is supported by microtubules (Patterson 1999). These alveolar sacs are thought to add rigidity to the cells.

Ciliates (Fig. 2b) are marked by numerous innovations. Morphologically, all ciliates contain rows of cilia (flagella) on the cell surface that are supported by a complex series of microtubules (Lynn 1996). The diverse patterns of ciliary organization have given this clade tremendous morphological complexity. The organization of cilia also provides numerous characters by which ciliates are described, classified, and monographed (e.g. Corliss

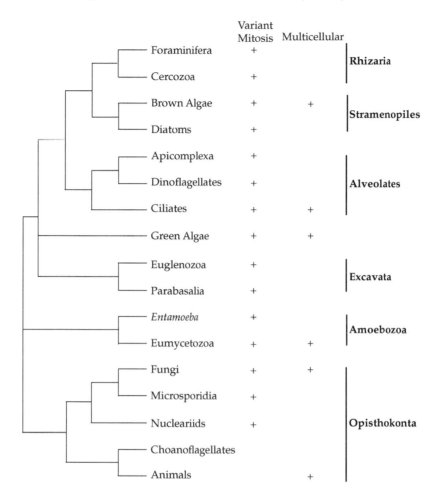

Fig. 1. Distribution of multicellularity and alternative types of mitosis across the eukaryotic tree of life. Variant mitosis: + denotes lineages in which mitosis deviates from the open orthomitosis of at least some members (Fig. 3), the "classic" animal and plant model. Multicellar: Clades with one or more members that have multicellular life cycle stages. Tree modified from Tekle et al. 2009.

1979). Ciliates are also defined by the presence of two distinct types of nuclei within each cell—the germline micronucleus and the somatic macronucleus. The genomic implications of this feature are discussed below.

Apicomplexa, a second major lineage within the Alveolates, include the malarial parasites *Plasmodium falciparum* (Fig. 2c) and *P. vivax*. Apicomplexa are defined by the presence of an 'apical complex' at the anterior end of the cells (Morrison 2009). This structure consists of a ring of specialized microtubules and an underlying system of modified vesicles, which secrete chemicals to break down host cell walls. The apical complex enables these parasites to invade a variety of tissues and cells, contributing to the success of this diverse clade (Morrison 2009).

Dinoflagellates (Fig. 2d), the third major alveolate lineage, are marked by innovations including a very unusual nucleus, called a dinokaryon. In most eukaryotes, nuclear DNA wraps around histone proteins, forming a nucleosome that can be further packaged depending on the degree of chromosome condensation. Most nuclei have condensed (heterochromatin) and decondensed (euchromatin) chromosome regions, the proportions of which vary according to transcriptional activity and life cycle stage. In contrast, the dinoflagellate nuclei are devoid of canonical histones— though histone-like proteins have been identified (Wong et al. 2003)—and their chromosomes are permanently condensed (Hackett et al. 2004). The dinoflagellate chromosome structure is proposed to be a cholesteric liquid crystal (Rizzo 2003, Costas and Goyanes 2005), but the implications of these features are not known. Dinoflagellate nuclei divide by closed mitosis (see below), with the ability to create tunnels for microtubules (Hackett et al.2004).

Foraminifera (Fig. 1, Fig. 2e) are defined by their granular reticulopodia (Bowser and Travis 2002), a dynamic network of branching and anastomosing pseudopodia that supports bidirectional movement of particles both within and on the surface of the pseudopodia. Reticulopodia are supported by an infrastructure of microtubule bundles that enable the dynamic motility of the network (Bowser and Travis 2002). The reticulopodia and attached particles move several orders of magnitude faster than the pseudopods of other amoeboid organisms. Unique microtubule assembly and disassembly underlain by divergent tubulins is hypothesized to enable such speed (Habura et al. 2005). Ultrastructural studies of the reticulopodia show that microtubules disassemble into helical filaments, unique tubulin polymorphs that can be stored as paracrystals and shipped out to rapidly growing points of the network (Bowser and Travis 2002). Foraminifera interact with their environment through the reticulopodia by capturing food, building shells (also called tests), and moving. The innovation of the reticulopodia is thought to have enabled the diversification and ecological success within the clade.

The supergroup **Excavata** (Fig. 1), as proposed by Cavalier-Smith (2002), seeks to unite a number of protist lineages, including the euglenids, kinetoplastids, and parabasalids. All members are either excavate taxa (they possess an excavated ventral feeding groove) or are thought to be descended from excavates, though the monophyly of the supergroup remains disputed (Simpson 2003). We describe a few of these lineages below.

Together, the **kinetoplastids** and euglenids comprise the bulk of the Euglenozoa, a taxon whose monophyly is robust (Simpson 2003, Parfrey et al. 2006). The kinetoplastids include marine flagellates such as *Rhynchomonas nasuta* (Fig. 2f) as well as the trypanosomes, parasites responsible for serious human illnesses such as sleeping sickness (*Trypanosoma brucei*) and leishmaniasis (e.g. *Leishmania major*) (Simpson et al. 2002). They are distinguished from other protist lineages by the presence of a single mitochondrion and a unique organization of the mitochondrial genome. Kinetoplastids have a condensed body of mitochondrial DNA called the kinetoplast that is large enough to be detected by light microscopy and is the source of group's name (Simpson et al. 2002). In trypanosomes, the kinetoplast DNA (kDNA) is arranged into interlocking maxi- and minicircles. Maxicircles are few in number and encode incomplete copies of typical mitochondrial genes, whereas the minicircles can number in the thousands (Simpson et al. 2002). Each minicircle encodes 1 to 4 small guide RNAs that mediate the extensive editing of maxicircle transcripts. Also gene organization along the nuclear chromosomes of kinetoplastids follows an unusual pattern termed absolute strand polarity, in which the genes are arranged in large clusters along just one strand of the DNA helix (McGrath and Katz 2004). In contrast, the genes of most other organisms are encoded along both strands.

The **euglenids** (Fig. 2g) are mostly free-living freshwater flagellates characterized by a unique pellicle made up of proteinaceous strips located beneath the cell membrane (Leedale 1967). The pellicle is further supported by a network of microtubules (Leedale 1967). While some move only using their flagella, in many euglenids the pellicle strips can slip past each other, thereby producing a distinctive inching motion called metaboly.

Parabasalids are a flagellate lineage composed of roughly 80 amitochondriate genera (Brugerolle and Müller 2000) such as the well-studied *Trichomonas vaginalis* and *Tritrichomonas muris* (Fig. 2h). Most members are either parasites or endosymbionts of other animals, notably termites and humans. The taxon is distinguished by their prominent Golgi bodies, which are linked to the basal bodies by parabasal fibers (Brugerolle and Müller 2000).

Animals are members of the **Opisthokonta**, a robust clade defined by both a morphological and molecular synapomorphy. The morphological synapomorphy is the presence of a single posterior flagellum, which is found in animal sperm cells as well as choanoflagellates, the closest relatives of animals. The similarity of cell architecture and flagellar position between sponges and choanoflagellates was first noticed more than a century ago (King 2004). This character has been lost in some Opisthokont groups including the fungi. A molecular synapomorphy for this group is the presence of an insertion in the Elongation factor 1a (EF1a) gene (Baldauf and Palmer 1993), for those taxa that have the canonical EF1a (Keeling and Inagaki 2004, Gile et al. 2006).

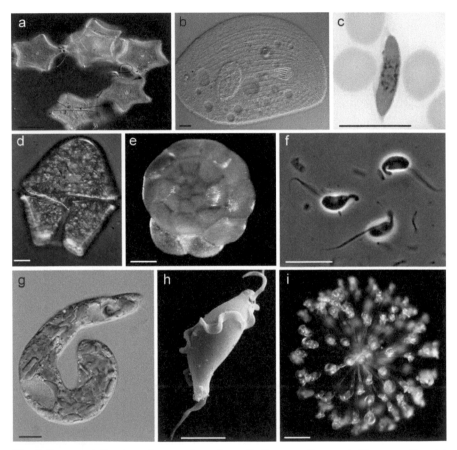

Fig. 2. Representative taxa from major eukaryotic lineages discussed. Stramenopiles: **(a)** *Triceratium pentacrinus*, a diatom; Alveolates (b-d): **(b)** *Trithigmostoma cucullulus*, a ciliate, **(c)** *Plasmodium falciparum*, an apicomplexan, and **(d)** *Akashiwo sanguinea*, a dinoflagellate; **(e)** *Ammonia* sp., a foraminiferan; Excavata: **(f)** *Rhynchomonas nasuta*, a kinetoplastid and **(g)** *Euglena mutabilis*, a euglenid; **(h)** *Tritrichomonas muris*, a parabasalid; Opisthokonta: **(i)** *Conochilus* sp., a colonial rotifer. Scale bars: b-d,f-h = 10 micrometers; a, e = 100 micrometers; i = 200 micrometers. All images are provided by micro*scope (http: //starcentral.mbl.edu/microscope/portal.php) except c, which is used under PLoS Biology's Open Access rules and can be found in Lacroix et al. (2005), credit Dr. Mae Melvin.

Color image of this figure appears in the color plate section at the end of the book.

Several lineages of eukaryotes are emerging from molecular studies that do not have morphological synapomorphies. These lineages, including Cercozoa, Amoebozoa, and Rhizaria, unite a surprising diversity of morphological forms. For example, molecular studies have placed more than 15 lineages of amoebae, parasites, flagellates, and heliozoa with ultrastructural identities into the Cercozoa. These include cercomonads (gliding flagellates), and chlorarachniophytes (filose amoebae with green algal plastids), and desmothoracids (heliozoa). The lack of a discernable morphological innovation uniting these groups may be a result of the ancient age of these clades.

Part two: Microbial Lineages Defy Many of the Long-Held Assumptions About Eukaryotes

Multicellularity

Although multicellularity has often been thought of as the exclusive territory of plants, fungi, and animals, there are, in fact, a number of microbial eukaryotic lineages that exhibit the trait. The largest protists are found among the brown algae, with giant kelps such as *Macrocystis* reaching lengths of up to 60 meters (Bonner 1998). A number of brown algae also undergo extensive cell differentiation, with some producing not only stem- and leaf-like parts, but also gas-filled bladders. Furthermore, the majority of giant kelps have photosynthetic filaments at the very edges of their stalks and blades, an example of tissue differentiation. Other lineages achieve muticellularity through mechanisms that are quite different from what one finds in animals, plants, and fungi. Some of them are more colonial rather than strictly multicellular: in the ciliate genus *Zoothamnium*, individual cells come together to form branched colonies. The cells are held together such that should one or a few cells contract, the rest of the colony will follow suit, presumably to avoid danger (Bonner 1998). The green algae genus *Volvox*, on the other hand, produces striking spherical colonies made up of a collection of individual cells held together by a gelatinous glycoprotein. Colonial species are generally not considered to be truly multicellular since they do not undergo functional tissue differentiation. However, the volvocine species that form larger colonies (up to 50,000 cells) actually do experience some functional differentiation, though it is usually limited to the formation of somatic and germ cells (Herron and Michod 2008).

Perhaps the most radically different approach to multicellularity is via aggregation, whereby individual cells come together (usually when conditions are poor). Unlike macrobes, though, protists that exhibit aggregative multicellularity nevertheless retain the ability to revert to their unicellular state. *Dictyostelium*, a genus of cellular slime mold, is one of the more extensively studied of the aggregative protist lineages. These amoebae have a sexual life cycle during which they feed as independent cells, coming together into a multicellular slug only after their food supply has been depleted (Bonner 1998). Cells then collectively form a fruiting body containing environmentally resistant spores. Such aggregation is characteristic of dictyostelid amoebae as well as the acrasids (the acellular "slime molds"), but it has also been found among the ciliates. *Sorogena stoianovitchae*, which lives in soil and feeds on other ciliates, is a unique ciliate species in that it forms aerial fruiting bodies in the face of nutrient scarcity (Lasek-Nesselquist et al. 2001).

Dynamic genomes

Much of our understanding of the structure and function of genomes is derived from model animals, fungi, and plants. However, this macrobial bias overlooks the tremendous diversity of genome structure and content present in microbial eukaryotes.

Data from microbial eukaryotes are challenging the assumption that genomes are largely static within species. In contrast to the textbook view, there is indeed intraspecific variation in genome architecture—the content and organization of nuclear DNA—within individuals during the life cycle and among individuals in populations (Parfrey et al. 2008). In many lineages, genome content changes during the life cycle through genome processing. Genome processing is the fragmentation, amplification, and/ or elimination of chromosomes or portions of chromosomes. Examples include: (1) ribosomal DNA amplification in numerous eukaryotic lineages (McGrath and Katz 2004, Zufall et al. 2005) and (2) whole genome rearrangements that occur in the somatic nuclei of ciliates and some animals (Prescott 1994, Jahn and Klobutcher 2002).

As mentioned above, every **ciliate** has two types of nuclei within a single cell: a germline micronucleus and somatic macronucleus. Micronucleuei go through canonical mitosis and meiosis, but are transcriptionally silent. The transcriptionally active macronucleus develops from the micronucleus through extensive genome fragmentation and amplification. In some types of ciliates this process results in a macronucleus with ~25 million gene-sized chromosomes (Prescott 1994). Genome processing has also been shown to affect patterns of molecular evolution in ciliates, with rates of protein evolution correlated with the extent of chromosomal processing. In other words, ciliates with extensively processed chromosomes have more divergent proteins (Katz et al. 2004, Zufall et al. 2006).

The genomes of **Entamoeba** vary from a ploidy of 4N to 40N in a single population (Lohia 2003). Populations can be synchronized to 4N, but the distribution of ploidy levels is restored within hours of normal growth

(Lohia 2003). *Entamoebae* also process their genomes by extensively amplifying their ribosomal DNA. They do not have any chromosomal copies of rDNA; instead, it exists only as a plasmid that varies in copy number (McGrath and Katz 2004).

Foraminifera and some related lineages go through a "nuclear cleansing" process prior to reproduction that indicates genome processing occurred during the development of the nucleus. During this process, termed Zerfall, nuclear contents including DNA are condensed, ejected from the nucleus, and degraded (Føyn 1936). The remaining nuclear material divides into hundreds to thousands of gametic nuclei (Goldstein 1997).

Variations in mitosis

Mitosis in eukaryotes assumes a diversity of forms, in contrast to the tidy, static process familiar from textbooks. In the "classic" form of mitosis—called eumitosis or open orthomitosis—the nuclear envelope breaks down completely and chromosomes are segregated by polar spindles into two daughter cells. While open orthomitosis is characteristic of animals and plants, it actually represents only one of the many variations of mitotic division found across the eukaryotic tree of life (Fig. 3 (Heath 1980, Raikov 1982)). There are a number of ways to distinguish different types of mitosis, including variations in microtubule quantity and chromatin condensation and organization (Heath 1980) as well as nuclear pore assembly (De Souza and Osmani 2007). Nevertheless, the major forms of mitosis (Fig. 3) are defined mainly by the behavior of the nuclear envelope and the

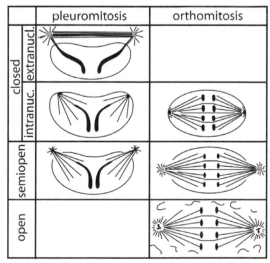

Fig. 3. The major forms of mitosis, as defined by the degree of nuclear envelope disintegration and spindle symmetry. Redrawn from Raikov 1994.

symmetry of the spindles (Raikov 1994). In this section we will briefly discuss differences in both.

The terms "open," "semiopen," and "closed" describe the degree to which the nuclear envelope has disintegrated, while "orthomitosis" and "pleuromitosis" refer to the symmetry (or lack thereof) of the mitotic spindles. Thus, in both semiopen orthomitosis and semiopen pleuromitosis, the nuclear envelope only disintegrates at the polar zones, leaving just enough space through which microtubules can enter the nucleus (Raikov 1982). In some cases, though, the nuclear envelope reseals itself soon after the microtubules have passed through (Heath 1980). Semiopen pleuromitosis differs from semiopen orthomitosis in that while the latter's microtubules produce a bipolar, symmetrical spindle—the definition of orthomitosis —, there are two independent spindles present in pleuromitosis (Raikov 1982). The spindles start out adjacent to one another but later move away at an angle, which initiates anaphase: there is neither a metaphase nor an equatorial plate in pleuromitosis (Raikov 1982). Semiopen pleuromitosis is typical of many apicomplexans (Raikov 1982), including the malarial parasites of the genus *Plasmodium*. Semiopen orthomitosis, meanwhile, is typical of the green flagellates Volvocales and Chloromonadida, as well as several gregarines and heliozoans (Raikov 1982).

One rare variant of "classic" open mitosis has only been observed in the gregarine genus *Stylocephalus*, in which the nuclear envelope completely disintegrates simply to reform around each individual chromosome (Heath 1980). All of the other major types of mitosis are closed—that is, the nuclear envelope does not disintegrate at all. In both closed intranuclear pleuromitosis and orthomitosis, the spindle remains entirely within the nucleus, though its positioning is different (Raikov 1982). Closed intranuclear pleuromitosis is typical of kinetoplastids, oxymonadids, foraminifera, radiolarians, as well as some green flagellates. Closed orthomitosis, on the other hand, is most typical of amoebae, ciliates, and some microsporidians (Raikov 1982). With extranuclear pleuromitosis, the spindle apparatus is found in the cytoplasm and contacts chromosomes through the nuclear envelope (Heath 1980, Ribeiro et al. 2000). Extranuclear pleuromitosis is typical of the parabasalids, such as the trichomonads (Raikov 1982) and dinoflagellates.

Summary

Here we have placed the evolutionary transitions leading to animals in the context of their microbial relatives on the eukaryotic tree of life. Microbial lineages make up the breadth of eukaryotic diversity, and many are marked by innovations in cell and/or genome structure at their base, which we believe contrasts with the origin of animals. An understanding of the diversity of microbial eukaryotes provides perspective in interpreting the evolution of eukaryotic features that are typified by animals, including multicellularity, genome structure, and mitosis, as discussed here. These features highlight the variation and diversity within eukaryotes.

References

Adl, S.M. and A.G.B. Simpson, M.A. Farmer, R.A. Andersen, O.R. Anderson, J.R. Barta, S.S. Bowser, G. Brugerolle, R.A. Fensome, S. Fredericq, T.Y. James, S. Karpov, P. Kugrens, J. Krug, C.E. Lane, L.A. Lewis, J. Lodge, D.H. Lynn, D.G. Mann, R.M. McCourt, L. Mendoza, O. Moestrup, S.E. Mozley-Standridge, T.A. Nerad, C.A. Shearer, A.V. Smirnov, F.W. Spiegel, and M. Taylor. 2005. The new higher level classification of eukaryotes with emphasis on the taxonomy of protists. J. Eukaryot. Microbiol. 52: 399–451.

Andersen, R.A. 2004. Biology and systematics of heterokont and haptophyte algae. Am J. Bot. 91: 1508–1522.

Baldauf, S.L. and J.D. Palmer. 1993. Animals and fungi are each other's closest relatives: congruent evidence from multiple proteins. Proc. Natl. Acad. Sci. USA 90: 11558–11562.

Bonner, J.T. 1998. The origins of multicellularity. Integr. Biol. 1: 27–36.

Bowser, S.S. and J.L. Travis. 2002a. Reticulopodia: Structural and behavioral basis for the suprageneric placement of Granuloreticulosan protists. J. Foram Res. 32: 440–447.

Brugerolle, G. and M. Müller. Amitochondriate flagellates. pp. 166–189. *In:* J.C. Green and B.S.C. Leadbeater. [eds.] 2000. Flagellates: Unity, Diversity and Evolution. Taylor and Francis, London.

Cavalier-Smith, T. 2002. The phagotrophic origin of eukaryotes and phylogenetic classification of Protozoa. Int. J. Syst. Evol. Microbiol. 52a: 297–354.

Corliss, J.O. 1979. The Ciliated Protozoa: Characterization, Classification and Guide to the Literature. Pergamon Press, Oxford.

Costas, E. and V. Goyanes. 2005. Architecture and evolution of dinoflagellate chromosomes: an enigmatic origin. Cytogenet. Genome Res. 109: 268–275.

De Souza, C.P.C. and S.A. Osmani. 2007. Mitosis, not just open or closed. Eukaryot Cell 6: 1521–1527.

Føyn, B. 1936. Über die Kernverhaltnisse der Foraminifere *Myxotheca arenilega* Schaudinn. Arch. Protistenkd 87: 272–295.

Gile, G.H. and N.J. Patron, and P.J. Keeling. 2006. EFL GTPase in cryptomonads and the distribution of EFL and EF-1alpha in chromalveolates. Protist. 157: 435–444.

Goldstein, S.T. 1997. Gametogenesis and the antiquity of reproductive pattern in the Foraminiferida. J. Foram Res. 27: 319–328.

Habura, A. and L. Wegener, J.L. Travis, and S.S. Bowser. 2005. Structural and functional implications of an unusual Foraminiferal beta-tubulin. Mol. Biol. Evol. 22: 2000–2009.

Hackett, J.D. and H.S. Yoon, M.B. Soares, M.F. Bonaldo, T. Casavant, T.E. Scheetz, T. Nosenko, and D. Bhattacharya. 2004. Migration of the plastid genome to the nucleus in a pridinin dinoflagellate. Curr. Biol. 14: 213–218.

Heath, B. 1980. Variant mitosis in lower eukaryotes: indicators of the evolution of mitosis? Int. Rev. Cytol. 64: 1–80.

Herron, M.D. and R.E. Michod. 2008. Evolution of complexity in the volvocine algae: Transitions in individuality through Darwin's eye. Evolution 62: 436–451.

Jahn, C.L. and L.A. Klobutcher. 2002. Genome remodeling in ciliated protozoa. Annu Rev Microbiol 56: 489–520.

Katz, L.A. and J.G. Bornstein, E. Lasek-Nesselquist, and S.V. Muse. 2004. Dramatic diversity of ciliate histone H4 genes revealed by comparisons of patterns of substitutions and paralog divergences among eukaryotes. Mol. Biol. Evol. 21: 555–562.

Keeling, P.J. and Y. Inagaki. 2004. A class of eukaryotic GTPase with a punctate distribution suggesting multiple functional replacements of translation elongation factor-1 alpha. Proc. Natl. Acad. Sci. USA 101: 15380–15385.

King, N. 2004. The unicellular ancestry of animal development. Dev. Cell 7: 313–325.

Lacroix, R. and W.R. Mukabana, L.C. Gouagna, and J.C. Koella. 2005. Malaria infection increases attractiveness of humans to mosquitoes. PLoS Biol. 3: 1590–1593.

Lasek-Nesselquist, E. and L.A. Katz. 2001. Phylogenetic position of *Sorogena stoianovitchae* and relationships within the class Colpodea (Ciliophora) based on SSU rDNA sequences. J. Eukaryot. Microbiol. 48: 604–607.

Leedale, G.F. 1967. Euglenoid Flagellates. Prentice-Hall, Inc Englewood Cliffs N.J., pp. 96–107.

Lohia, A. 2003. The cell cycle of *Entamoeba histolytica*. Mol Cell Biochem 253: 217–222.

Lynn, D.H. Systematics of ciliates. *In:* K. Hausmann and P.C. Bradbury. [eds.] 1996. Ciliates: Cells as Organisms. Gustav Fischer Verlag, Stuttgart.

McGrath, C.L. and L.A. Katz. 2004. Genome diversity in microbial eukaryotes. Trends Ecol. Evol. 19: 32–38.

Morrison, D.A. 2009. Evolution of the Apicomplexa: where are we now? Trends in Parasitol 25: 375–382.

Parfrey, L.W. and E. Barbero, E. Lasser, M. Dunthorn, D. Bhattacharya, D.J. Patterson, and L.A. Katz. 2006. Evaluating support for the current classification of eukaryotic diversity. PLoS Genet. 2: 2062–2073.

Parfrey, L.W. and D.J.G. Lahr, and L.A. Katz. 2008. The dynamic nature of eukaryotic genomes. Mol Biol Evol 25: 787–794.

Patterson, D.J. 1999. The diversity of eukaryotes. Amer. Nat. 154: S96–S124.

Prescott, D.M. 1994. The DNA of ciliated protozoa. Microbiol. Rev. 58: 233–267.

Raikov, I.B. 1982. The Protozoan Nucleus: Morphology and Evolution. Springer-Verlag, Wien.

Raikov, I.B. 1994. The Diversity of Forms of Mitosis in Protozoa: A Comparative Review. European J. Protistol. 30: 253–269.

Ribeiro, K.C. and L.H. Monteiro-Leal, and M. Benchimol. 2000. Contributions of the axostyle and flagella to closed mitosis in the protists *Tritrichomonas foetus* and *Trichomonas vaginalis*. J. Eukaryot. Microbiol. 47: 481–492.

Rizzo, P. 2003. Those amazing dinoflagellate chromosomes. Cell Res 13: 215–217.

Schmid, A.-M.M. 2003. The evolution of the silicified diatom cell wall—revisited. Diatom. Res. 17: 345–351.

Simpson, A.G.B. and J. Lukes, and A.J. Roger. 2002. The evolutionary history of kinetoplastids and their kinetoplasts. Mol. Biol. Evol. 19: 2071–2083.

Simpson, A.G.B. 2003. Cytoskeletal organization, phylogenetic affinities and systematics in the contentious taxon Excavata (Eukaryota). Int J Syst Evol Microbiol 53: 1759–1777.

Steenkamp, E.T. and J. Wright, and S.L. Baldauf. 2006. The protistan origins of animals and fungi. Mol. Biol. Evol. 23: 93–106.

Tekle, Y.I. and L.W. Parfrey, and L.A. Katz. 2009. Molecular data are transforming hypotheses on the origin and diversification of eukaryotes. Bioscience 59: 471–481.

Wong, J.T.Y. and D.C. New, J.C.W. Wong, and V.K.L. Hung. 2003. Histone-like proteins of the dinoflagellate *Crypthecodinium cohnii* have homologies to bacterial DNA-binding proteins. Eukaryot. Cell 2: 646–650.

Zufall, R.A. and T. Robinson, and L.A. Katz. 2005. Evolution of developmentally regulated genome rearrangements in eukaryotes. Journal of Experimental Zoology Part B-Molecular and Developmental Evolution 304B: 448–455.

Zufall, R.A. and C.L. McGrath, S.V. Muse, and L.A. Katz. 2006. Genome architecture drives protein evolution in ciliates. Molecular Biology and Evolution 23: 1681–1687.

Chapter 2

Elucidating Animal Phylogeny: Advances in Knowledge and Forthcoming Challenges

Kevin M. Kocot, Johanna T. Cannon, and *Kenneth. M. Halanych*

Introduction

Understanding key transitions in animal evolution is dependent in on an accurate evolutionary framework or phylogeny. Molecular phylogenetics greatly informed the study of animal evolutionary relationships during the late 1980s and 1990s and brought about several major changes to the animal tree of life (e.g. Lophotrochozoa, Ecdysozoa). More recently, our understanding of higher-level animal evolutionary relationships has been further resolved, although many questions remain unanswered. Major recent changes to our view of the animal tree of life include the placement of Acoela and Nemertodermatida at the base of Bilateria, the inclusion of sipunculans within Annelida, the placement of Onychophora as the sister taxon of Arthropoda, and the surprising find of a clade composed of Craniata and Tunicata to the exclusion of Cephalochordata. Despite these advances, a number of issues remain unresolved; including which of the basal metazoan phyla (Porifera, Placozoa, Cnidaria, or Ctenophora) is basal-most, the branching order of the three major lineages of Arthropoda, the relationships within Lophotrochozoa, and the placement of several so-called minor groups such as Chaetognatha. Here, we review the current understanding of animal phylogeny with emphasis on the most recent work and illustrate remaining gaps in our knowledge.

Department of Biological Sciences, 101 Rouse Life Science Building, Auburn University, Auburn, AL USA 36849.
E-mail: kmkocot@auburn.edu, cannojt@auburn.edu, ken@auburn.edu

One of the goals of this volume is to examine major transitions in form or function that allowed animals to become more evolutionarily successful (as judged by either number of species, diversity, or dominance). If such transitions occur in stem or ancestral lineages and are present in all or most of the descendents, we typically call them major, or key, transitions. Understanding of such transitions is dependent upon an accurate assessment of evolutionary history and in particular, the phylogenetic relationships of the organisms being examined. Without an accurate phylogenetic framework, we have little hope of assessing a number of evolutionary issues relating to such transitions, for example: Which is the most derived taxon that shows the pre-transition state? Was the transition a single major event or a graded series of events? Has the transition in question occurred once or multiple times? Does the transition represent a *de novo* evolutionary invention or just a rearrangement of parts?

The goal of this chapter is to address the current understanding of animal phylogeny so that key transitions can be placed in an appropriate evolutionary context. Although our understanding of animal phylogeny is far from complete, the past several years have seen significant advances in resolving animal phylogeny at all hierarchical levels. The advent of molecular approaches in the late 1980s to 1990s radically reshaped traditional hypotheses of animal phylogeny (e.g. Hyman 1959, Willmer 1990, Nielsen 2001). Several reviews (e.g. Halanych 1999, Eernisse and Peterson 2004, Halanych 2004, Giribet et al. 2007) have outlined advances based on molecular data. Halanych (2004) provided a summary of major findings from phylogenetic studies up until that time, most of which were molecular studies based on one to three genes. Due to space limitations, this contribution will essentially continue from 2004. Since then a number of multiple-gene studies have been conducted which have provided considerable insight into animal phylogenetics while raising new questions. Herein, we will consider some advantages and disadvantages of studies utilizing genome or transcriptome-level data (a.k.a. phylogenomics), provide an overview of the current understanding of metazoan phylogenetics, and discuss unresolved issues.

Molecular Approaches

Several of the major changes in our understanding of animal phylogeny in the late twentieth century were initially hypothesized on the basis of the nuclear ribosomal small subunit (SSU or 18S) rDNA gene (e.g. Field et al., 1988, Halanych et al. 1995, Aquinaldo et al. 1997, Ruiz-Trillo et al. 1999). The advantages of this gene include the presence of both variable and conserved regions, the latter of which facilitate oligonucleotide primer design for polymerase chain reaction (PCR). This marker also has some potential pitfalls, the most serious of which are rate heterogeneity across taxa and its susceptibility to long-branch attraction (Felsenstein 1988) that can mislead phylogenetic interpretation (Eernisse 1997, Abouheif et al. 1998). As the field matured, other markers, such as mitochondrial genomes, Hox genes, the large nuclear ribosomal subunit (28S rDNA) gene, the elongation factor 1α gene, and the myosin II gene, began to be used. Additionally, the number of taxa employed in phylogenetic analyses of Metazoa increased significantly from early studies with less than 10 taxa (e.g. Halanych 1995, Bromham and Degnan 1999) to studies with over 100 taxa (e.g. Giribet et al. 2000, Eernisse and Peterson 2001).

Studies that employ a target-gene approach are dependent on polymerase chain reaction (PCR) to amplify specific DNA markers for sequencing. Until recently, the use of single-copy nuclear protein-coding genes for metazoan phylogeny has been limited because of variation in intron boundaries, challenges involved in working with RNA, and the need for preexisting sequence data for primer design. As more genomic resources have become available for a number of taxa, and molecular methods have developed, phylogenetic investigations targeting multiple nuclear protein-coding genes have become more feasible (e.g. Peterson et al. 2004, Rokas et al. 2005, Bourlat et al. 2008, Paps et al. 2009). Also, high-throughput genomic approaches allowing for molecular phylogenetic data to be obtained in a PCR-independent manner have become more accessible and widespread. Studies using high-throughput sequencing have produced large amounts of data for phylogenetic analyses on growing numbers of taxa. Such approaches are typically referred to as phylogenomic approaches (Delsuc et al. 2005, Telford 2008).

Most recent phylogenomic studies have made use of expressed sequence tag (EST) data. By way of a brief description, EST data are generated by extracting mRNA, reverse transcribing it to cDNA, and then sequencing a randomly selected subset of the cDNA, producing partial sequences or 'tags.' Data collection for most phylogenomic studies has been conducted using capillary sequencers, and by necessity has included bacterial cloning of the cDNA. As of the time of writing this manuscript, most published EST studies used capillary sequencers and relied on data produced from 1000–5000 clones. However, the development of pryosequencing technologies (e.g. 454, Illumina) has made it possible to collect very large amounts of transcriptome data at a low cost per base pair relative to capillary sequencing. The phylogenomic approach is a powerful one for studies of deep animal relationships because many of the sequences obtained belong to 'housekeeping' genes which are vital to the function of any given animal cell. Because such genes are usually constitutively expressed, they are likely to be recovered in a typical EST survey regardless of the source of the material. Furthermore, because of their functional importance, housekeeping genes tend to be highly evolutionarily conserved, furnishing phylogenetic signal for the study of deep relationships.

The first major paper using the phylogenomic approach for deep animal phylogeny was Philippe et al. (2005) in which 146 genes from 35 species were analyzed. This manuscript was followed by a series of manuscripts (e.g. Bourlat et al. 2006, Delsuc et al. 2006, Matus et al. 2006, Webster et al. 2006, Philippe et al. 2007) that added new taxa but employed nearly the same set of genes. However, other studies (e.g. Marlétaz et al. 2006, Dunn et al. 2008) have independently examined primary sequence data and have generated alternative gene sets. Before we detail some of the findings from such studies (below), we wish to outline some issues of the phylogenomic approach as currently employed.

The appeal of phylogenomics stems from the fact that large amounts of genomic data corresponding to multiple genes can be brought to bear on phylogenetic issues. Additionally, the approach allows access to genetic data that would be otherwise difficult to obtain using a PCR-based approach as discussed above. However, phylogeneticists are still learning how to efficiently analyze such large data sets. Insights from previous theoretical discussions during the past 20 years focusing on, for example, effects of missing data, taxon sampling issues, signal in the data, and when to combine data have not been thoroughly addressed in a phylogenomic context. Most EST-based data sets have large amounts of missing data for any given taxon. For example, there was ~25% missing data in the matrix of Philippe et al. (2004) and ~50% missing data in the matrix of Dunn et al. (2008). While some have advocated this is not a serious problem (e.g. Philippe et al. 2004, Wiens 2006), the issue has not been adequately explored for large multigene data sets (but see Hartmann and Vision 2008). Also, some studies have had poor taxon sampling and therefore recovered artifactual results. For example, the Cephalochordata + Echinodermata clade recovered by Delsuc et al. (2006) has subsequently been shown to be an artifact of insufficient taxon sampling (Bourlat et al. 2006, further discussed below). The issue of conflicting signal in different gene partitions within a data set has received little attention in metazoan phylogenomics (but see Gatesy et al. 2007 and Wiens et al. 2008). Such analysis is commonplace for analyses of data generated using a PCR-based approach (e.g. Danforth et al. 2005, Paps et al. 2009). Lastly, one of the most represented groups of genes examined in EST-based studies is ribosomal protein genes. These genes can comprise the majority of the usable data produced by surveys of a few thousand clones but their utility has recently been questioned (e.g. Yeang and Haussler 2007, Bleidorn et al. 2009a). Because of these issues, we encourage careful consideration of the data matrix employed in phylogenomic analyses. As revealed by discussions below, phylogenomic analyses have substantial utility and we look forward to further insights provided by such studies.

Overview of Animal Phylogeny

Despite holes in our understanding of animal evolutionary relationships, several hypotheses regarding higher-level metazoan relationships are now widely accepted. At the base of the animal tree are Porifera, Placozoa, Ctenophora, and Cnidaria. The relationships among these basal metazoans remain very controversial; particularly, which group has the distinction of being the basal-most animal. The existence of the clade Bilateria has been strongly supported by virtually all studies of morphology and molecules. Within Bilateria, most animals fall into three major lineages: Deuterostomia (e.g. echinoderms and chordates), Lophotrochozoa (e.g. molluscs and annelids), and Ecdysozoa (e.g. arthropods and nematodes). Exceptions include Acoela and Nemertodermatida which appear to be basal to the rest of the Bilateria and Chaetognatha which may be basal to Protostomia (Lophotrochozoa + Ecdysozoa; Marlétaz et al. 2006, Marlétaz and Le Parco 2008, but see Helmkampf et al. 2008a, Matus et al. 2006, 2007). Also, the placement of several so-called minor phyla including Orthonectida, Dicyemida, and Myxozoa on the animal tree of life is currently unclear. In the following sections, we have attempted to concisely synthesize the most recent work on animal phylogenetics and illustrate gaps in our current knowledge.

Basal metazoans

Relationships among the basal metazoans have received a great deal of attention recently, but despite a growing number of datasets with multiple sources of information, a consensus has not been reached. The branching order of Placozoa, Porifera (as either a monophyletic group or a paraphyletic assemblage), Ctenophora, Cnidaria, and Bilateria has been the subject of intense debate (see Cartwright and Collins 2007, Martindale 2005, Schierwater et al. 2009a).

Placozoa, a phylum which has multiple lineages despite only one formally described species, *Trichoplax adherens* (Voigt et al. 2004, Pearse and Voigt 2007, Signorovitch et al. 2007), has been suggested to occupy every imaginable position at the base of the metazoan tree (reviewed in Schierwater et al. 2009a,b). Although placozoan mitochondrial genomes (Dellaporta et al. 2006, Signorovitch et al. 2007) and the nuclear genome of *Trichoplax adherens* have been sequenced (Srivastava et al. 2008), the phylogenetic affinities of Placozoa remain unresolved. Using 104 nuclear protein-coding genes from whole genomes, Srivastava et al. (2008) found

Trichoplax as sister to a clade of Cnidaria + Bilateria. Because this analysis was restricted to species with complete genomes, only nine taxa were sampled including one sponge, two cnidarians, and no ctenophores. In a more recent phylogenomic study with enhanced taxon sampling of basal metazoans, placozoans were found to be sister to a clade comprised of Bilateria and monophyletic Coelenterata (Cnidaria + Ctenophora) based on phylogenomic analyses (Philippe et al. 2009). Schierwater et al. (2009b) concatenated a dataset of morphological characters, mitochondrial and nuclear ribosomal gene sequences, molecular morphology, and nuclear protein-coding genes. Their analyses recovered a monophyletic Diploblasta comprised of Placozoa sister to a clade of Porifera + Coelenterata.

Early molecular studies on sponges using LSU and SSU indicated that Porifera is a paraphyletic grade, with Calcerea more closely related to Eumetazoa than Demospongia (e.g. Collins 1998, Borchiellini et al. 2001, Medina et al. 2001, Cavalier-Smith and Chao 2003). This finding has been upheld using a small set of nuclear protein-coding genes (Peterson and Butterfield 2005, Sperling et al. 2007). Phylogenomic studies, however, suggest Porifera is monophyletic (Dunn et al. 2008, Philippe et al. 2009). Insights gained from poriferan mitogenomics are discussed elsewhere in this volume by Lavrov. Monophyletic or not, the phylogenetic position of Porifera has fluctuated in recent years. Investigations regarding sponge relationships were recently reviewed by Erpenbeck and Wörheide (2007).

Historically, Cnidaria and Ctenophora were allied as Coelenterata but this relationship was refuted by a number of early molecular studies (see Halanych 2004). Unlike sponges and placozoans, cnidarians and ctenophores have nerve cells, true epithelial tissues, and germ layers which have been interpreted as shared derived features with Bilateria by some. Surprisingly, Dunn et al. (2008) found Ctenophora to be the basal-most metazoan phylum while Cnidaria and Porifera formed a clade basal to Bilateria. Placozoa was not included in the analysis. A more recent phylogenomic analysis by Philippe et al. (2009), however, had greatly expanded taxon sampling at the base of the metazoan tree and recovered a more traditional branching order—Porifera as the basal-most clade in Metazoa, followed by Placozoa, followed by a monophyletic Coelenterata (Ctenophora + Cnidaria). Schierwater et al.'s (2009b) combined dataset including morphology, ribosomal secondary structure, and sequence data found Placozoa, Porifera, Cnidaria, and Ctenophora to form a clade, Diploblasta, as the sister to Bilateria. Nuclear protein-coding genes suggest the myxozoan *Buddenbrockia plumatellae* is sister to Medusozoa within Cnidaria (Jiménez-Guri et al. 2007). Cnidarian systematics was recently reviewed by Daly et al. (2007).

Acoela and Nemertodermatida

In recent years, evidence that the acoelomorph flatworms (Acoela and Nemertodermatida) are not members of Platyhelminthes, but rather are basal bilaterians, has continued to accumulate (e.g. Jondelius et al. 2002, Cook et al. 2004, Wallberg et al. 2007, Baguñà et al. 2008, Wheeler et al. 2009,). There is also a growing body of evidence suggesting that Acoelomorpha is not monophyletic; SSU/LSU + nuclear protein coding gene phylogenies have shown Acoela and Nemertodermatida as distinct lineages at the base of Bilateria with Acoela basal (Ruiz-Trillo et al. 2002, Wallberg et al. 2007, Paps et al. 2009). To date, acoels have been unstable in phylogenomic studies (Philippe et al. 2007, Dunn et al. 2008). Acoela clustered with deuterostomes in the phylogenomic analyses by Philippe et al. (2007) which was interpreted as an artifact of long branch attraction (LBA). Notably, a relationship between Acoela and *Xenoturbella* has been proposed based on similarities in the ultrastructure of epidermal locomotory ciliary rootlets (Lundin 1998). EST data from nemertodermatids and more slowly-evolving acoels will hopefully aid resolution.

Deuterostomia

Deuterostome relationships are among the best known of all animal groups (Halanych 2004, Swalla and Smith 2008). However, recent studies utilizing genomic data, particularly multigene analyses utilizing ESTs, have called into question the evolutionary history of this well-studied clade, particularly chordates. Until recently, the most commonly accepted phylogenetic hypothesis concerning chordates consisted of Tunicata as sister to Euchordata (Cephalochordata + Craniata). Phylogenomic studies have renewed interest in an alternative hypothesis in which cephalochordates are sister to a clade termed Olfactores, consisting of Tunicata and Craniata. In the EST phylogenetic study of Philippe et al. (2005), tunicates were recovered as sister to craniates, to the exclusion of cephalochordates. This study included one representative each of tunicates, cephalochordates, and echinoderms, and two vertebrate taxa. Delsuc et al. (2006) included four additional tunicate species and seven additional vertebrate species, and found tunicates clustered with craniates. Unexpectedly, the cephalochordate *Branchiostoma* clustered with *Strongylocentrotus purpuratus*, the sole echinoderm included in the analyses. This result was later shown to be due to poor taxon sampling (Bourlat et al. 2006). Olfactores, on the other hand, has been recovered in a number of phylogenomic studies utilizing different EST screening protocols as well as models of evolution (Delsuc et al. 2008, Dunn et al. 2008, Lartillot and Philippe 2008, Philippe et al. 2009).

Independent support for Olfactores arrangement has come in the form of morphology (Ruppert 2005), combined rRNA and morphological data (Zrzavý et al. 1998), cadherin gene structure (Oda et al. 2004), chordate fibrillar collagen genes (Wada et al. 2006), and microRNA distribution (Heimberg et al. 2008). On the other hand, analyses based on mitochondrial data (Yokobori et al. 2005, Bourlat et al. 2006) and rRNA (Mallat and Winchell 2007) recover Euchordata (Craniata + Cephalochordata) with varying degrees of support. The potential sister relationship between craniates and tunicates suggests that similarities between chordates and cephalochordates may be chordate symplesiomorphies. Tunicates are unusual in their highly derived morphologies and modified genomes and have been difficult to place phylogenetically due to long branches (Winchell et al. 2002, Zeng and Swalla, 2005, Mallatt and Winchell 2007). Whether or not the Tunicata + Craniata relationship continues to be upheld with increased taxon sampling remains to be seen.

Molecular phylogenetic studies have continued to provide robust support for Ambulacraria, a clade comprised of Hemichordata and Echinodermata, first proposed by Metchnikoff in 1881 based on similarities in organization of the larval coeloms in these two groups, and formalized as a node-based name based on SSU data (Halanych 1995). Recent support for this clade has inlcuded phylogenomics (Bourlat et al. 2006, Dunn et al. 2008, Lartillot and Philippe 2008, Philippe et al. 2009), morphology (Cameron 2005, Ruppert 2005), microRNAs (Wheeler et al. 2009), and a combined analysis of ribosomal, mitochondrial and nuclear protein-coding genes (Bourlat et al. 2008). The Ambulacraria hypothesis suggests that shared morphological features between hemichordates and chordates, such as pharyngeal gill slits and the post-anal tail, were likely to have been present in the deuterostome ancestor. Recent efforts have been made to resolve relationships within hemichorates and echinoderms (Cannon and Rychel et al. 2009, Mallatt and Winchell 2007, respectively).

The newly erected deuterostome phylum Xenoturbellida comprises *Xenoturbella*, a genus of small (< 4cm), morphologically non-descript animals (Bourlat et al. 2006). Although most analyses have found *Xenoturbella* sister to Ambulacraria (Bourlat et al. 2003, Bourlat et al. 2006, Bourlat et al. 2008, Dunn et al. 2008, Lartillot and Philippe 2008), phylogenetic analyses based on Hox genes (Fritzsch et al. 2008), and mitochondrial genomes (Perseke et al. 2007) have placed Xenoturbellida at the base of the deuterostomes. However, Philippe et al.'s (2007) phylogenomic study found the acoel *Convoluta pulchra* grouping with deuterostomes in 90% of bootstrap replicates, and despite the extremely high rate of evolution in the acoel, the authors suggest the possibility of a Xenoturbellida-Acoel relationship.

Lophotrochozoa

Lophotrochozoa was defined by Halanych et al. (1995) on the basis of an analysis of 18S rDNA as the last common ancestor of the three traditional lophophorate taxa (Brachiopoda, Phoronida, and Bryozoa [=Ectoprocta]), molluscs, annelids, and all of the descendants of that common ancestor. Monophyly of Lophotrochozoa has since been supported by numerous molecular phylogenetic investigations (e.g. Anderson et al. 2004, Philippe et al. 2005, Hausdorf et al. 2007, Dunn et al. 2008, Helmkampf et al. 2008a,b). However, the interrelationships among the phyla that constitute this taxon are largely unclear. Broadly, most lophotrochozoans appear to fall into one of two major clades: Trochozoa and Platyzoa.

Trochozoa (*sensu* Peterson and Eernisse 2001) includes Mollusca, Annelida, Nemertea, Brachiopoda, Phoronida and Entoprocta; phyla in which the presence of a trochophore larval stage is thought to be a plesiomorphy (although it has been lost or modified in some). Recent studies have greatly improved our understanding of Trochozoan phylogeny. In their phylogenomic analyses, Dunn et al. (2008) recovered a monophyletic Eutrochozoa in which Mollusca was sister to a clade comprised of Annelida, Brachiopoda, Phoronida, and Nemertea. Annelida was sister to a clade comprised of a Brachiopoda/Phoronida clade, and Nemertea. A similar topology for Eutrochozoa was recovered in a ML analysis of ribosomal proteins derived from EST data by Helmkampf et al. (2008b). Paps et al. (2009) analyzed 13 nuclear protein-coding genes and also recovered a monophyletic Trochozoa (albeit with weak support).

Molluscs are an ancient group of morphologically diverse trochozoans whose class-level relationships have proven difficult to reconstruct. To date, investigations of molluscan relationships have been unable to resolve most deep branches. The most significant results of these studies include the rejection of the Diasoma hypothesis (which allies scaphopods and bivalves; Passamaneck et al. 2004) and evidence for a clade of cephalopods and scaphopods (Steiner and Dreyer 2002, 2003). The phylogenetic position of Monoplacophora was recently addressed by Wilson et al. (2009). Molluscan phylogenetics were recently reviewed in *Phylogeny and Evolution of the Mollusca* (Ponder and Lindberg, eds. 2008).

Recent molecular phylogenetic investigations have radically altered our understanding of Annelida by demonstrating that Clitellata is within 'Polychaeta' and several taxa formerly thought to be distinct phyla (Pogonophora [now Siboglinidae], Echiura, and Sipuncula) and are actually within the annelid radiation (McHugh 1997, Colgan et al. 2006, Passamaneck and Halanych 2006, Rousset et al. 2007, Struck et al. 2007). The annelid affinities of Myzostomida, a group of unusual marine parasites, are still being debated. Recent studies of myzostomid anatomy (Eeckhaut and Lanterbecq 2005, Lanterbecq et al. 2008), mitochondrial genome information, nuclear rDNA + protein-coding genes (Bleidorn et al. 2007, 2009a), and Hox genes (Bleidorn et al. 2009b) support a close relationship between Myzostomida and Annelida. See Zrzavý et al. (2009) for the most recent phylogenetic investigation of Annelida.

Brachiopods, phoronids, and bryozoans (= ectoprocts) have traditionally been allied as the Lophophorata (Nielsen 1985, 2001) and while they all possess a horseshoe-shaped ring of hollow tentacles termed a lophophore, there are structural differences (Halanych 1996). However, recent work has mostly supported the monophyly of Brachiopoda and Phoronida to the exclusion of Bryozoa (e.g. Passamaneck and Halanych 2006, Anderson et al. 2004, Dunn et al. 2008, Helmkampf et al. 2008b). The phylogenetic affinities of Bryozoa are presently unclear. Phylogenomic analyses by Hausdorf et al. (2007) and Helmkampf et al. (2008b) recovered a well-supported sister relationship between the ectoproct *Flustra* and the entoproct (= kamptozoan) *Barentsia* but this is surprising given morphology (Wanninger 2009). Other studies place Entoprocta as the sister taxon of Cycliophora (Passamaneck and Halanych 2006). See Fuchs et al. (2009) for the most up-to-date phylogeny of Bryozoa.

Platyzoa comprises most of the remainder of Lophotrochozoa and is usually defined to include Platyhelminthes (excluding Acoelomorpha), Gastrotricha, Rotifera (including Acanthocephala), Gnathostomulida, and Micrognathozoa (Garey et al. 1996, Garey and Schmidt-Rhaesa 1998, Giribet et al. 2000, Passamaneck and Halanych et al. 2006). Virtually nothing is known about the relationships among the taxa that comprise Platyzoa and it was even suggested by Dunn et al. (2008) that Platyzoa is an artifact of long-branch attraction.

Ecdysozoa

Ecdysozoa comprises eight phyla which are united by the molting of their cuticle. Aguinaldo et al. (1997) using SSU rDNA provided the first molecular support for this grouping. Since then, several lines of evidence have supported the existence of this clade (reviewed by Halanych 2004, Telford et al. 2008). Within Ecdysozoa, three major clades are typically recognized: Scalidophora (Priapulida, Kinorhyncha, and Loricifera), Nematoida (Nematoda and Nematomorpha), and Panarthropoda (Arthropoda, Tardigrada, and Onychophora). The evolution of this clade has recently been reviewed by Telford et al. (2008), Budd and Telford (2009) and Edgecombe (2009).

The clade Nematoida (Nematoda + Nematomorpha) has been supported by SSU rDNA and morphology (Peterson and Eernisse 2001), SSU rDNA and LSU rDNA (e.g. Petrov and Vladychenskaya 2005, Mallatt and Giribet 2006), concatenated nuclear, ribosomal and mitochondrial genes (Bourlat et al. 2008), and phylogenomics (Dunn et al. 2008). Some studies have found other taxa pairing with nematoids, including Loricifera and Tardigrada, with varying degrees of support (see further discussion below), but the nematomorph-nematode relationship seems to be robust.

Most molecular studies have supported a Priapulid + Kinorhynch clade (Petrov and Vladychenskaya 2005, Mallat and Giribet 2006, Dunn et al. 2008), but the inclusion of Loricifera in Scalidophora has been called into question by a molecular phylogeny by Sørensen et al. (2008), whose results suggested a possible relationship between Loricifera and Nematomorpha. Embryological studies have been inconclusive as to the relationships within Scalidophora—the developmental patterns of Priapulids (Wennberg et al. 2009), Kinorhynchs (Kozloff 2007), and Loriciferans (Gad 2005) do not appear to closely resemble each other.

Scalidophora and Nematoida have been united in a clade called Cyloneuralia or Introverta, with the names referring to their circum-oral brain and eversible anterior introvert, respectively. Molecular analyses based on SSU rDNA and LSU rDNA have not recovered this clade (Mallat and Giribet 2006, Sørensen et al. 2008), but Dunn et al. (2008) did recover Cycloneuralia (Loricifera was not included in the analysis).

Panarthropoda was hypothesized on the basis of morphological characters shared among Arthropoda, Onychophora, and Tardigrada. Evolution of this group, including information on several recently discovered stem-group fossil taxa, was recently reviewed by Budd and Telford (2009) and Edgecombe (2009). The immediate sister taxon of Arthropoda has been a long-standing question, but two recent phylogenomic studies have supported a sister taxon relationship between Arthropoda and Onychophora (Roeding et al. 2007, Dunn et al. 2008). The phylogenetic position of Tardigrada, however, remains unclear. Roeding et al. (2007) found tardigrades to be more closely related to Nematoda than Arthropoda or Onychophora. However, in the 64-taxon analysis by Dunn et al., tardigrades were recovered as sister to Arthropoda + Onychophora, consistent with the Panarthropoda hypothesis. Although numerous morphological features suggest a close relationship between tardigrades and arthropods, several cycloneuralian-like features of tardigrades may be in need of re-evaluation (reviewed by Edgecombe 2009).

Arthropoda has traditionally been divided into four major lineages: Chelicerata, Myriapoda, Crustacea, and Hexapoda. Classical hypotheses regarding arthropod phylogeny are reviewed by Budd and Telford (2009). More recently, evidence suggesting that Hexapoda is nested within Crustacea has come from molecular phylogenetic studies (Richter 2002, Regier et al. 2005, Regier et al. 2008) as well as morphological studies (e.g. Ungerer and Scholtz 2008). This clade of crustaceans and hexapods has been termed Pancrustacea or Tetraconata. The phylogenetic relationships among Pancrustacea, Myriapoda and Chelicerata remain uncertain. There is evidence for a clade of Pancrustacea + Myriapoda (=Mandibulata; Edgecombe et al. 2003, Scholtz and Edgecombe 2006, Mallatt and Giribet 2006) or, alternatively, a clade of Chelicerata + Myriapoda (=Paradoxopoda or Myriochelata; Negrisolo et al. 2004, Pisani et al. 2004). Support for both hypotheses has varied; recent studies including Dunn et al. (2008) found support for Paradoxopoda while the others such as Bourlat et al. (2008) and Regier et al. (2008) were more consistent with Mandibulata.

Conclusions

Despite many recent advances in the field of animal phylogenetics, a number of daunting questions remain unanswered, including the relationships among the basal metazoans, the branching order of the three major lineages of Arthropoda, relationships within Lophotrochozoa, and the placement of many so-called minor groups such as Chaetognatha. The phylogenetic tree in Fig. 1 presents a conservative view of our current understanding of animal evolutionary relationships. Polytomies

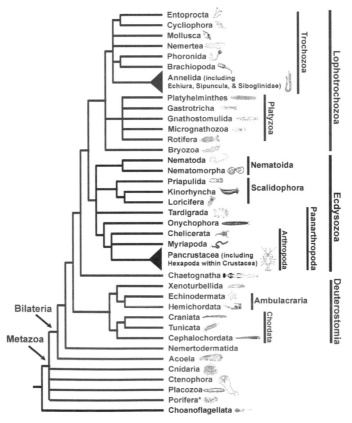

Fig. 1. Current understanding of animal phylogeny. This tree is conservatively drawn to illustrate major concepts inferred from recent studies as described in the text. The tree is color-coded: 'basal metazoans' are purple, deuterostomes are red, ecdysozoans are blue, and lohpotrochozoans are green. 'The monophyly of Porifera has been questioned, but for the sake of simplicity it is drawn as a single lineage here (see text).

Color image of this figure appears in the color plate section at the end of the book.

(unresolved nodes) within Arthropoda, Lophotrochozoa, and at the base of the tree represent forthcoming challenges to phylogeneticists. Recent developments in the fields of phylogenomics, gene expression analysis, pyrosequencing, microscopy and 3D visualization, and paleontology will undoubtedly continue to help refine our understanding of the animal tree of life, thereby broadly informing studies of metazoan evolution.

References

Abouheif, E. and R. Zardoya, and A. Meyer. 1998. Limitations of metazoan 18S rRNA sequence data: implications for reconstructing a phylogeny of the animal kingdom and inferring the reality of the Cambrian explosion. J. Mol. Evol. 47: 394–405.

Aguinaldo, A.M.A. and J.M. Turbeville, L.S. Linford, M.C. Rivera, J.R. Garey, R. Raff, and J.A. Lake. 1997. Evidence for a clade of nematodes, arthropods and other moulting animals. Nature 387: 489–493.

Anderson, F.E. and A.J. Cordoba, and M. Thollesson. 2004. Bilaterian phylogeny based on analyses of a region of the sodium-potassium ATPase alpha subunit gene. Mol. Evol. 58: 252–268.

Baguñà, J. and P. Martinez, J. Paps, and M. Riutort. 2008. Back in time: a new systematic proposal for the Bilateria. Phil. Trans. R. Soc. B 2008 363: 1481–1491.

Bleidorn, C. and I. Eeckhaut, L. Podsiadlowski, N. Schult, D. McHugh, K.M. Halanych, M.C. Milinkovitch, and R. Tiedemann. 2007. Mitochondrial genome and nuclear sequence data support Myzostomida as part of the annelid radiation. Mol. Biol. Evol. 24: 1690–1701.

Bleidorn, C. and L. Podsiadlowski, M. Zhong, I. Eeckhaut, S. Hartmann, K.M. Halanych, and R. Tiedemann. 2009a. On the phylogenetic position of Myzostomida: can 77 genes get it wrong? BMC Evol. Biol. 9: 150.

Bleidorn, C. and D. Lanterbecq, I. Eeckhaut, and R. Tiedemann. 2009b. A PCR survey of Hox genes in the myzostomid *Myzostoma cirriferum*. Dev. Genes Evol. 219: 211–216.

Borchiellini, C. and M. Manuel, E. Alivon, N. Boury-Esnault, J. Vacelet, and Y. Le Parco. 2001. Sponge paraphyly and the origins of metazoa. J. Evol. Biol. 14: 171–179.

Bourlat, S.J. and C. Nielsen, A.E. Lockyer, D.T.J. Littlewood and M.J. Telford, 2003. *Xenoturbella* is a deuterostome that eats molluscs. Nature 424: 925–928.

Bourlat, S.J. and T. Juliusdottir, C.J. Lowe, R. Freeman, J. Aronowicz, M. Kirschner, E.S. Lander, M. Thorndyke, H. Nakano, and A.B. Kohn. 2006. Deuterostome phylogeny reveals monophyletic chordates and the new phylum Xenoturbellida. Nature 444: 85–88.

Bourlat, S.J. and C. Nielsen, A.D. Economou, and M.J. Telford. 2008. Testing the new animal phylogeny: A phylum level molecular analysis of the animal kingdom. Mol. Phylogenet. Evol. 49: 23–31.

Bromham, L.D. and B.M. Degnan. 1999. Hemichordates and deuterostome evolution: robust molecular phylogenetic support for a hemichordate + echinoderm clade. Evol. Dev. 1: 168–171.

Budd, B. and M.J. Telford. 2009. The origin and evolution of arthropods. Nature 452: 812–817.

Cameron, C.B. 2005. A phylogeny of the hemichordates based on morphological characters. Can. J. Zool. 83: 196–215.

Cannon, J.T. and A.L. Rychel, H. Eccleston, K.M. Halanych, and B.J. Swalla. 2009. Molecular phylogeny of Hemichordata, with updated status of deep-sea enteropneusts. Mol. Phylogenet. Evol. 52: 17–24.

Cartwright, P. and A. Collins. 2007. Fossils and phylogenies: Integrating multiple lines of evidence to investigate the origin of early major metazoan lineages. Int. Comp. Biol. 47: 744–751.

Cavalier-Smith T. and E.E. Chao. 2003. Phylogeny of Choanozoa, Apusozoa, and other Protozoa and early eukaryote megaevolution. J. Mol. Evol. 56: 540–563.

Colgan, D.J. and P.A. Hutchings and M. Braune. 2006. A multigene framework for polychaete phylogenetic studies. Org. Div. and Evol. 2: 220–235.

Collins, A.G. 1998. Evaluating multiple alternative hypotheses for the origin of Bilateria: An analysis of 18S rRNA molecular evidence. Proc. Natl. Acad. Sci. USA 95: 15458–15463.

Cook, C.E. and E. Jiménez, M. Akam, and E. Saló. 2004. The Hox gene compliment of acoel flatworms, a basal bilaterian clade. Evol. Dev. 6: 154–163.

Daly, M. and M.R. Brugler, P. Cartwright, A.G. Collins, M.N. Dawson, D.G. Fautin, S.C. France, C.S. McFadden, D.M. Opresko, E. Rodriguez, S.L. Romano, and J.L. Stake. 2007. The phylum Cnidaria: A review of phylogenetic patterns and diversity 300 years after Linnaeus. Zootaxa 1668: 127–182.

Danforth, B.N. and C.P. Lin, and J. Fang. 2005. How do insect nuclear ribosomal genes compare to protein-coding genes in phylogenetic utility and nucleotide substitution patterns? Syst. Entomol. 30: 549–562.

Dellaporta, S.L. and A. Xu, S. Sagasser, W. Jakob, M.A. Moreno, L.W. Buss, and B. Schierwater. 2006. Mitochondrial genome of *Trichoplax adhaerens* supports Placozoa as the basal lower metazoan phylum. Proc. Natl. Acad. Sci. USA 103: 8751–8756.

Delsuc, F. and H. Brinkmann, and H. Philippe. 2005. Phylogenomics and the reconstruction of the tree of life. Nature Rev. Genet. 6: 361–375.

Delsuc, F. and H. Brinkmann, D. Chourrout, and H. Philippe. 2006. Tunicates and not cephalochordates are the closest living relatives of vertebrates. Nature 439: 965–968.

Delsuc, F. and G. Tsagkorgeorga, N. Lartillot, and H. Philippe. 2008. Additional molecular support for the new chordate phylogeny. Genesis 46: 592–604.

Dunn, C.W. and A. Hejnol, D.Q. Matus, K. Pang, W.E. Browne, S.A. Smith, E. Seaver, G.W. Rouse, M. Obst, G.D. Edgecombe, M.V. Sørensen, S.H.D. Haddock, A. Schmidt-Rhaesa, A. Okusu, R.M. Kristensen, W.C. Wheeler, M.Q. Martindale, and G. Giribet. 2008. Broad phylogenomic sampling improves resolution of the animal tree of life. Nature 452: 745–749.

Edgecombe, G.D. and S. Richter, and G.D.F. Wilson. 2003. The mandibular gnathal edges: homologous structures throughout Mandibulata. J. Afr. Invertebr. 44: 115–135.

Edgecombe, G.D. 2009. Palaeontological and molecular evidence linking arthropods, onychophorans, and other Ecdysozoa. Evol. Edu. Outreach 2: 178–190.

Eeckhaut, I. and D. Lanterbecq. 2005. Myzostomida: A review of the phylogeny and ultrastructure. Hydrobiologia 535: 253–275.

Eernisse, D.J. Arthropod and annelid relationships re-examined. pp. 43–56. *In:* R.A. Fortey and R.H. Thomas. [eds.] 1997. Arthropod Relationships. Chapman and Hall, London.

Eernisse, D.J. and K.J. Peterson. 2001. Animal phylogeny and the ancestry of bilaterians: inferences from morphology and 18s rDNA gene sequences. Evol. Dev. 3: 170–205.

Eernisse, D.J. and K.J. Peterson. 2004. The history of animals. pp. 197–208. *In:* J. Cracraft and M.J. Donoghue. [eds.] 2004. Assembling the Tree of Life. Oxford Univ. Press, New York.

Erpenbeck, D. and G. Wörheide. 2007. On the molecular phylogeny of sponges. Zootaxa 1668: 107–126.

Felsenstein, J. 1988. Phylogenies from molecular sequences: inference and reliability. Annu. Rev. Genet. 22: 521–565.

Field, K.G. and G.J. Olsen, D.J. Lane, S.J. Giovannoni, and M.T. Ghiselin, et al. 1988. Molecular phylogeny of the animal kingdom. Science 239: 748–753.

Fritzsch, G. and M.U. Böhme, M. Thorndyke, H. Nakano, O. Israelsson, T. Stach, M. Schlegel, T. Hankeln, and P.F. Stadler. 2008. PCR Survey of *Xenoturbella bocki* Hox genes. J. Exp. Zool. Mol. Dev. Evol. 310B: 278–284.

Fuchs, J. and M. Obst, and P. Sundberg. 2009. The first comprehensive molecular phylogeny of Bryozoa (Ectoprocta) based on combined analyses of nuclear and mitochondrial genes. Mol. Phylogenet. Evol. 52(1): 225–233..

Gad, G. 2005. A parthenogenetic, simplified adult in the life cycle of *Pliciloricus pedicularis* sp. n. (Loricifera) from the deep sea of the Angola Basin (Atlantic). Organisms Diversity and Evol. 5: 77–103.

Garey, J.R. and T.J. Near, M.R. Nonnemacher, and S.A. Nadler. 1996. Molecular evidence for Acanthocephala as a subtaxon of Rotifera. J. Mol. Evol. 43: 287–292.

Garey, J.R. and A. Schmidt-Rhaesa. 1998. The essential role of "minor" phyla in molecular studies of animal evolution. Am. Zool. 38: 907–917.

Gatesy, J. and R. DeSalle, and N. Wahlberg. 2007. How many genes should a systematist sample? Conflicting insights from a phylogenomic matrix characterized by replicated incongruence. Syst. Biol. 56: 355–363.

Giribet, G. and D.L. Distel, M. Polz, W. Sterrer, and W.C. Wheeler. 2000. Triploblastic relationships with emphasis on the acoelomates and the position of Gnathostomulida, Cycliophora, Plathelminthes, and Chaetognatha: A combined approach of 18S rDNA sequences and morphology. Syst. Biol. 49: 539–562.

Giribet, G. and C.W. Dunn, G.D. Edgecombe, and G.W. Rouse. 2007. A modern look at the animal tree of life. Zootaxa 1668: 61–79.

Halanych, K.M. 1995. The phylogenetic position of pterobranch hemichordates based on 18s rDNA sequence data. Mol. Phylogenet. and Evol. 4: 72–76.

Halanych, K.M. and J.D. Bacheller, A.M.A. Aguinaldo, S.M. Liva, D.M. Hillis, and J.A. Lake. 1995. Evidence from 18S ribosomal DNA that the lophophorates are protostome animals. Science 267: 1641–1643.

Halanych, K.M. 1996. Convergence in the feeding apparatuses of lophophorates and pterobranch hemichordates revealed by 18S rDNA. Biol. Bull. 190: 1–5.

Halanych, K.M. 1999. Metazoan phylogeny and the shifting comparative framework. pp. 3–7. *In:* Roubos, Wendelaar, Bonga, Vaudry, and De Loof. [eds.] Recent Developments in Comparative Endocrinology and Neurobiology. Shaker Publishing. Maastricht, The Netherlands.

Halanych and K.M. 2004. The new view of animal phylogeny. Annu. Rev. Ecol. Evol. Syst. 35: 229–256.

Hartmann, S. and T.J. Vision. 2008. Using ESTs for phylogenomics: Can one accurately infer a phylogenetic tree from a gappy alignment? BMC Evol. Biol. 8: 95.

Hausdorf, B. and M. Helmkampf, A. Meyer, A. Witek, H. Herlyn, I. Bruchhaus, T. Hankeln, T.H. Struck, and B. Lieb. 2007. Spiralian phylogenomics supports the resurrection of Bryozoa comprising Ectoprocta and Entoprocta. Mol. Biol. and Evol. 24: 2723.

Heimberg, A.M. and L.F. Sempere, V.N. Moy, P.C.J. Donoghue, and K.J. Peterson. 2008. MicroRNAs and the advent of vertebrate morphological complexity. Proc. Natl. Acad. Sci. USA 105: 2946–2950.

Helmkampf, M. and I. Bruchhaus, and B. Hausdorf. 2008a. Multigene analysis of lophophorate and chaetognath phylogenetic relationships. Mol. Phylogenet. Evol. 46: 206–214.

Helmkampf, M. and I. Bruchhaus, and Hausforf, B. 2008b. Phylogenomic analyses of lophophorates (brachiopods, phoronids and bryozoans) confirm the Lophotrochozoa concept. Proc. R. Soc. B. 275: 1927–1933.

Hyman, L.H. 1959. The Invertebrates: Smaller Coelomate Groups. McGraw-Hill, New York.

Jondelius, U. and I. Ruiz-Trillo, J. Baguñà, and M. Riutort. 2002. The Nemertodermatida are basal bilaterians and not members of the Platyhelminthes. Zool. Script. 31: 201–215.

Kozloff, E.N. Stages of development, from first cleavage to hatching, of an Echinoderes (Phylum Kinorhyncha: Class Cyclorhagida). Cahiers de Biologie Marine 48: 199–206.

Lanterbecq, D. and C. Bleidorn, S. Michel, and I. Eeckhaut. 2008. Locomotion and find structure of parapodia in *Myzostoma cirriferum* (Myzostomida). Zoomorphology 127: 59–68.

Lartillot, N. and H. Philipe. 2004. A Bayesian mixture model for across-site heterogeneities in the amino-acid replacement process. Mol. Biol. Evol. 21: 1095–1109.

Lartillot, N. and H. Brinkman, and H. Philippe. 2007. Suppression of long-branch attraction artifacts in the animal phylogeny using a site-heterogeneous model. BMC Evol. Biol. 7 (Suppl.1): S4.

Lartillot, N. and H. Philippe. 2008. Improvement of molecular phylogenetic inference and the phylogeny of Bilateria. Phil. Trans. R. Soc. B. 363: 1463–1472.

Lundin, K. 1998. The epidermal ciliary rootlets of *Xenoturbella bocki* (Xenoturbellida) revisited: new support for a possible kinship with the Acoelomorpha (Platyhelminthes). Zool. Script. 27: 263–270.

Mallatt, J. and G. Giribet. 2006. Further use of nearly complete 28S and 18S rRNA genes to classify Ecdysozoa: 37 more arthropods and a kinorhynch. Mol. Phylogenet. Evol. 40: 772–794.

Mallat, J. and C.J. Winchell. 2007. Ribosomal RNA genes and deuterostome phylogeny revisted: More cyclostomes, elasmobranchs, reptiles, and a brittle star. Mol. Phylogenet. Evol. 43: 1005–1022.

Marlétaz, F. and E. Martin, Y. Perez, D. Papillon, X. Caubit, C.J. Lowe, B. Freeman, L. Fasano, C. Dossat, and P. Wincker. 2006. Chaetognath phylogenomics: a protostome with deuterostome-like development. Curr. Biol. 16: R577–578.

Marlétaz, F. and Y. Le Parco. 2008. Careful with understudied phyla: The case of chaetognath. BMC Evol. Biol. 8: 251.

Martindale, M.Q. 2005. The evolution of metazoan axial properties. Nature Rev. Genet. 6: 917–927.

Matus, D.Q. and R.R. Copley, C.W. Dunn, A. Hejnol, H. Eccleston, K.M. Halanych, M.Q. Martindale, and M.J. Telford. 2006. Broad taxon and gene sampling indicate that chaetognaths are protostomes. Curr. Biol. 16: R575–576.

Matus, D.Q. and K.M. Halanych, and M.Q. Martindale. 2007. The Hox gene complement of a pelagic chaetognath, *Flaccisagitta enflata*. Int. Comp. Biol. 47: 854–864.

McHugh, D. 1997. Evidence that echiurans and pogonophorans are derived annelids. Proc. Natl. Acad. Sci. USA 94: 8006–8009.

Medina, M. and A.G. Collina, J.D. Silberman, and M.L. Sogin. 2001. Evaluating hypotheses of basal animal phylogeny using complete sequences of large and small subunit rRNA. Proc. Natl. Acad. Sci. USA 98: 9707–9712.

Metschnikoff, V.E. 1881. Über die systematische Stellung von *Balanoglossus*. Zool. Anz. 4: 139–157.

Negrisolo, E. and A. Minelli, and G. Valle 2004. The mitochondrial genome of the house centipede Scutigera and the monophyly versus paraphyly of myriapods. Mol. Biol. Evol. 21: 770–780.

Nielsen, C. 1985. Animal phylogeny in the light of the trochaea theory. Biol. J. Linn. Soc. 25: 243–299.

Nielsen, C. 2001. Animal Evolution: Interrelationships of the Living Phyla. Oxford University Press, Oxford.

Oda, H. and Y. Akiyama-Oda, and S. Zhang. 2004. Two classic cadherin-related molecules with no cadherin extracellular repeats in the cephalochordate amphioxus: distinct adhesive specificities and possible involvements in the development of multicell-layered structures. J. Cell Sci. 117: 2757–2767.

Paps, J. and B. Baguñà, and M. Riutort. 2009. Bilaterian phylogeny: A broad sampling of 13 nuclear genes provides a new Lophotrochozoa phylogeny and supports a paraphyletic basal Acoelomorpha. Mol. Biol. Evol. 26: 2397–2406.

Passamaneck, Y.J. and C. Schander, and K.M. Halanych. 2004. Investigation of molluscan phylogeny using large-subunit and small-subunit nuclear rRNA sequences. Mol. Phylogenet. Evol. 32: 25–38.

Passamaneck, Y.J. and K.M. Halanych. 2006. Lophotrochozoan phylogeny assessed with LSU and SSU data: Evidence of lophophorate polyphyly. Mol. Phylogenet. Evol. 40: 20–28.

Pearse, V.B. and O. Voigt. 2007. Field biology of placozoans (Trichoplax): distribution, diversity, biotic interactions. Int. Comp. Biol. 47: 677–692.

Perseke, M. and T. Hankeln, B. Weich, G. Fritzsch, P.F. Stadler, O. Israelsson, D. Bernhard, and M. Schlegal. 2007. The mitochondrial DNA of *Xenoturbella bocki*: genomic architecture and phylogenetic analysis. Theory Biosci. 126: 35–42.

Peterson, K.J. and D.J. Eernisse. 2001. Animal phylogeny and the ancestry of bilaterians: inferences from morphology at 18s rDNA gene sequences. Evol. Dev. 3: 170–205.

Peterson, K.J. and J.B. Lyons, K.S. Nowak, C.M. Takacs, M.J. Wargo, and M.A. McPeek. 2004. Estimating metazoan divergence times with a molecular clock. Proc. Natl. Acad. Sci. USA 101: 6536–6541.

Peterson, K.J. and N.J. Butterfield. 2005. Origin of the Eumetazoa: testing ecological predictions of molecular clocks against the Proterozoic fossil record. Proc. Natl. Acad. Sci. USA 102: 9547–9552.

Petrov, N.B. and N.S. Vladychenskaya. 2005. Phylogeny of molting protostomes (Ecdysozoa) as inferred from 18S and 28S rRNA gene sequences. Mol. Biol. 39: 503–513.

Philippe, H. and E.A. Snell, E. Bapteste, P. Lopez, P.W.H. Holland, and D. Casane. 2004. Phylogenomics of eukaryotes: impact of missing data on large alignments. Mol. Biol. Evol. 21: 1740–1752.

Philippe, H. and N. Lartillot, and H. Brinkman. 2005. Multigene analyses of bilaterian animals corroborate the monophyly of Ecdysozoa, Lophotrochozoa, and Protostomia. Mol. Biol. Evol. 22: 1246–1253.

Philippe, H. and H. Brinkmann, P. Martinez, M. Riutort, and J. Baguña. 2007. Acoel flatworms are not platyhelminthes: Evidence from phylogenomics. PLoS ONE 2: e717.

Philippe, H. and R. Derelle, P. Lopez, K. Pick, C. Borchiellini, N. Boury-Esnault, J. Vacelet, E. Renard, E. Houliston, E. Quéinnec, C. Da Silva, P. Wincker, H. Le Guyader, S. Leys, D.J. Jackson, F. Schreiber, D. Erpenbeck, B. Morgenstern, G. Wörheie, and M. Manuel. 2009. Phylogenomics revives traditional views on deep animal relationships. Curr. Biol. 19: 706–712.

Pisani, D. and L.L. Poling, M. Lyons-Weiler, and S.B. Hedges. 2004. The colonization of land by animals: molecular phylogeny and divergence among arthropods. BMC Biol. 2: 1.

Ponder, W.F. and D.R. Lindberg. 2009. Phylogeny and Evolution of the Mollusca. University of California Press. Berkeley and Los Angeles, CA.

Regier, J.C. and J.W. Shultz, and R.E. Kambic. 2005. Pancrustacean phylogeny: Hexapods are terrestrial crustaceans and maxillopods are not monophyletic. Proc. R. Soc. B. 272: 395 401.

Regier, J.C. and J.W. Shultz, A.R.D. Ganley, A. Hussey, D. Shi, B. Ball, A. Zwick, J.E. Stajich, M.P. Cummings, J.W. Martin, and C.W. Cunningham. 2008. Resolving arthropod phylogeny: exploring phylogenetic signal within 41 kb of protein-coding nuclear gene sequence. Syst. Biol. 57: 920–938.

Richter, S. 2002. The Tetraconata concept: hexapod-crustacean relationships and the phylogeny of Crustacea. Org. Divers. Evol. 2: 217–237.

Roeding, F. and S. Hagner-Holler, H. Ruhberg, I. Ebersberger, A. von Haeseler, M. Kube, R. Reinhardt, and T. Burmester. 2007. EST sequencing of Onychophora and phylogenomic analysis of Metazoa. Mol. Phylogenet. Evol. 45: 942–951.

Rokas, A. and D. Krüger, and S.B. Carroll. 2005. Animal evolution and the molecular signature of radiations compressed in time. Science 310: 1933–1938.

Rousset, V. and F. Pleijel, G.W. Rouse, C. Erséus, and M.E. Siddall. 2007. A molecular phylogeny of annelids. Cladistics 23: 41–63.

Ruiz-Trillo, I. and M. Riutort, T.J. Littlewood, E.A. Herniou, and J. Baguña. 1999. Acoel flatworms: earliest extant bilaterian metazoans, not members of Platyhelminthes. Science 283: 1919–1923.

Ruiz-Trillo, I. and J. Paps, M. Loukota, C. Ribera, U. Jondelius, J. Baguña, and M. Riutort. 2002. A phylogenetic analysis of myosin heavy chain type II sequences corroborates that Acoela and Nemertodermatida are basal bilaterians. Proc. Natl. Acad. Sci. USA 99: 11246–11251.

32

Ruppert, E.E. 2005. Key characters uniting hemichordates and chordates: homologies or homoplasies? Can. J. Zool. 83: 8–23.

Schierwater, B. and D. de Jong, and R. DeSalle. 2009a. Placozoa and the evolution of metozoa and intrasomatic cell differentiation. Int. J. Biochem. Cell Biol. 41: 370–379.

Schierwater, B. and M. Eitel, W. Jakob, H. Osigus, H. Hadrys, S.L. Dellaporta, S. Kolokotronis, and R. DeSalle. 2009b. Concatenated analysis sheds light on early metazoan evolution and fuels a modern "Urmetazoon" hypothesis. PLoS Biol. 7: e1000020.

Signorovitch, A.Y. and L.W. Buss, and S.L. Dellaporta. 2007. Comparative genomics of large mitochondria in placozoans. PLoS Genet. 3: e13.

Sørensen, M.V. and M.B. Hebsgaard, I. Heiner, H. Glenner, E. Willerslev, and R.M. Kristensen. 2008. New data from an enigmatic phylum: evidence from molecular sequence data supports a sister-group relationship between Loricifera and Nematomorpha. J. Zool. Syst. Evol. Res. 46: 231–239.

Sperling, E.A. and D. Pisani, and K.J. Peterson. Poriferan paraphyly and its implications for Precambrian palaeobiology. *In:* P. Vickers-Rich and P. Komarower [eds.] 2007. The Rise and Fall of the Ediacaran Biota. The Geological Society of London, London.

Srivastava, M. and E. Begovic, J. Chapman, N.H. Putnam, U. Hellstein, T. Kawashima, A. Kuo, T. Mitros, A. Salamov, M.L. Carpenter, A.Y. Signorovitch, M.A. Moreno, K. Kamm, J. Grimwood, J. Schmutz, H. Shapiro, I.V. Grigoriev, L.W. Buss, B. Schierwater, S.L. Dellaporta, and D.S. Rokhsar. The *Trichoplax* genome and the nature of placozoans. Nature 454: 955–960.

Steiner, G. and H. Dreyer. 2002. Cephalopoda and Scaphopoda are sister taxa: an evolutionary scenario. Zoology 105: 95.

Steiner, G. and H. Dreyer. 2003. Molecular phylogeny of Scaphopoda (Mollusca) inferred from 18S rDNA sequences: support for a Scaphopoda-Cephalopoda clade. Zool. Scr. 32: 343–356.

Struck, T.H. and N. Schult, T. Kusen, E. Hickman, C. Bleidorn, D. McHugh, and K.M. Halanych. 2007. Annelid phylogeny and the status of Sipuncula and Echiura. BMC Evol. Biol. 7: 57.

Swalla, B.J. and A.B. Smith. 2008. Deciphering deuterostome phylogeny: molecular, morphological and palaeontological perspectives. Phil. Trans. R. Soc. B 363: 1557–1568.

Telford, M.J. 2008. Resolving animal phylogeny: A sledgehammer for a tough nut? Dev. Cell 14: 457–459.

Telford, M.J. and S.J. Bourlat, A. Economou, D. Papillon, and O. Rota-Stabelli. 2008. The evolution of the Ecdysozoa. Phil. Trans. R. Soc. B 363: 1529–1537.

Ungerer, P. and G. Scholtz. 2007. Filling the gap between identified neuroblasts and neurons in crustaceans adds new support for Tetraconata. Proc. R. Soc. B. 275: 369–376.

Voigt, O. and A.G. Collins, V. Buchsbaum Pearse, J.S. Pearse, A. Ender, H. Hadrys, and B. Schierwater. Placozoa—no longer a phylum of one. Curr. Biol. 14: R944–R945.

Wada, H. and M. Okuyama, N. Satoh, and S. Zhang. 2006. Molecular evolution of fibrillar collagen in chordates, with implications for the evolution of vertebrate skeletons and chordate phylogeny. Evol. Dev. 8: 370–377.

Wallberg, A. and M. Curini-Galletti, A. Ahmadzadeh, and U. Jondelius. 2007. Dismissal of Acoelomorpha: Acoela and Nemertodermatida are separate early bilaterian clades. Zool. Script. 36: 509–523.

Wanninger, A. 2009. Shaping the things to come: Ontogeny of lophotrochozoan neuromuscular systems and the Tetraneuralia concept. Biol. Bull. 216: 293–206.

Webster, B.L. and R.R. Copley, R.A. Jenner, J.A. Mackenzie-Dodds, S.J. Bourlat, O. Rota-Stabelli, D.T.J. Littlewood, and M.J. Telford. 2006. Mitogenomics and phylogenomics reveal priapulid worms as extant models of the ancestral Ecdysozoan. Evol. Dev. 8: 502–510.

Wennberg, S.A. and R. Janssen, and G.E. Budd. 2009. Hatching and earliest larval stages of the priapulid worm *Priapulus caudatus*. Invertebr. Biol. 128: 157–171.

Wheeler, B.M., A.M. Heimberg, V.N. Moy, E.A. Sperling, T.W. Holstein, S. Heber, and K.J. Peterson. 2009. The deep evolution of metazoan microRNAs. Evol. Dev. 11: 50–68.

Willmer, P. 1990. Invertebrate Relationships, Patterns in Animal Evolution. Cambridge Univ. Press, New York.

Wiens, J.J. 2006. Missing data and the design of phylogenetic analyses. Biomed. Infomat. 39: 34–42.

Wiens, J.J. and C.A. Kuczynski, S.A. Smith, D.G. Mulcahy, J.W. Sites Jr., T.M. Townsend, and T.W. Reeder. 2008. Branch lengths support and congruence testing the phylogenomic approach with 20 nuclear loci in snakes. Syst. Biol. 57: 420–431.

Winchell, C.J. and J. Sullivan, C.B. Cameron, B.J. Swalla, and J. Mallatt. 2002. Evaluating hypotheses of deuterostome phylogeny and chordate evolution with new LSU and SSU ribosomal DNA data. Mol. Biol. Evol. 19: 762–776.

Wilson, N.G. and G.W. Rouse, and G. Giribet. 2009. Assessing the molluscan hypothesis Serialia (Monoplacophora + Polyplacophora) using novel molecular data. Mol. Phylogenet. Evol. 54(1): 187–193..

Yeang, C. and D. Haussler. 2007. Detecting coevolution in and among protein domains. PLoS Comp. Biol. 3: e211.

Yokobori, S. and T. Oshima, and H. Wada. 2005. Complete nucleotide sequence of the mitochondrial genome of *Doliolum nationalis* with implications for evolution of urochordates. Mol. Phylogenet. Evol. 34: 273–283.

Zeng, L.Y. and B.J. Swalla. 2005. Molecular phylogeny of the protochordates: chordate evolution. Can. J. Zool. 83: 24–33.

Zrzavý, J. and S. Mihulka, P. Kepka, and A. Bezdek. 1998. Phylogeny of the metazoa based on morphological and 18S ribosomal DNA evidence. Cladistics 14: 249–285.

Zrzavý, J. and P. Říhal, L. Piálek, and J. Janouškovec. 2009. Phylogeny of Annelida (Lophotrochozoa): total-evidence analysis of morphology and six genes. BMC Evol. Biol.9: 189.

Note

While this work was in press, several studies of metazoan phylogeny were published. Most notably, Regier et al. (2010) investigated arthropod phylogeny using 62 nuclear protein-coding genes and recovered a well-supported Arthropod tree with Pancrustacea (hexapods within crustaceans) and Mandibulata. Hejnol et al. (2009) found support for acoelomorph monophyly, a basal bilaterian placement for *Xenoturbella*, and placement of Cycliophora as the sister taxon of Entoprocta. Also, Pick and Philippe et al.'s (2010) investigation of basal metazoan phylogeny recovered the following branching order: (P orifera(Ctenophora(Cnidaria(Placozoa, Bilateria)))).

Hejnol, A. and M. Obst, A. Stamatakis, M. Ott, G.W. Rouse, G.D. Edgecombe, P. Martinez, J. Baguña, X. Bally, U. Jondelius, M. Wiens, W.E.G. Müller, E. Seaver, W.C. Wheeler, M.Q. Martindale, G. Giribet, and C.W. Dunn. 2009. Assessing the root of bilaterian animals with scalable phylogenomic methods. Proc. Royal Soc. B. 276(1161): 4261–4270.

Regier, J.C. and J.W. Shultz, A. Zwick, A. Hussey, B. Ball, R. Wetzer, J.W. Martin, and C.W. Cunningham. 2010. Arthropod relationships revealed by phylogenomic analysis of nuclear protein-coding sequences. Nature 463: 1079–1083.

Pick, K.S. and H. Philippe, F. Schreiber, D. Erpenbeck, D.J. Jackson, P. Wrede, M. Wiens, A. Alié, B. Morgenstern, M. Manuel, and G. Wörheide. 2010. Improved phylogenomic taxon sampling noticeably affects non-bilaterian relationships. Molecular Biology and Evolution (in press).

Chapter 3

Key Transitions in Animal Evolution: a Mitochondrial DNA Perspective

Dennis V. Lavrov

Introduction

When the first complete mitochondrial DNA (mtDNA) sequence—that of humans—was determined (Anderson et al. 1981), it was fittingly described by the phrase "small is beautiful" (Borst and Grivell 1981). Mitochondrial genomes of bilaterian animals are indeed small (~16 kpb), not only because of their limited coding capacity (typically 37 genes), but also due to the remarkable economy of their genomic organization (Fig. 1). Genes encoded in bilaterian mtDNA are compactly arrayed, separated by no, or only a few, nucleotides and typically contain neither introns (but see Valles et al. 2008) nor regulatory sequences. Protein and transfer RNA genes are even often truncated and completed by either posttranscriptional polyadenylation (Yokobori and Pääbo 1997) or, in some cases, editing (Lavrov et al. 2000). Furthermore, changes in the genetic code allowed animals to reduce the set of mitochondrial tRNA genes to 22, several tRNAs fewer than the minimum number required for translation under the standard genetic code (Marck and Grosjean 2002). Encoded ribosomal RNAs are also reduced in size, lacking many secondary structures present in homologous molecules in other groups. In addition to small size, bilaterian mtDNA displays several unusual genetic and genomic

253 Bessey Hall, Department of Ecology, Evolution and Organismal Biology, Iowa State University, Ames, Iowa 50011.
E-mail: dlavrov@iastate.edu

features such as unorthodox translation initiation codons, highly modified structures of encoded transfer RNAs, a high rate of sequence evolution, a relatively low rate of gene rearrangements, and the presence of a single large non-coding "control" region (Wolstenholme 1992b).

Although the organization of mtDNA is remarkably uniform across different groups of bilaterian animals [but see Armstrong et al. (2000), Helfenbein et al. (2004), and Suga et al. (2008) for some exceptions], it is far from typical for other eukaryotic lineages. In fact, it was known early on that mitochondrial genomes vary greatly in size, gene content, and genome architecture across eukaryotic groups (Wallace 1982, Lang et al. 1999). Interestingly, even the mitochondrial genome of the choanoflagellate *Monosiga brevicollis*, a close relative of animals, is several-fold larger than "typical" animal mtDNA and harbors 1.5 times as many genes (Fig. 1). It also encodes bacteria-like transfer and ribosomal RNAs, and uses a minimally-derived genetic code with TGA(Trp) as the only deviation from the standard code (Burger et al. 2003). Thus major changes

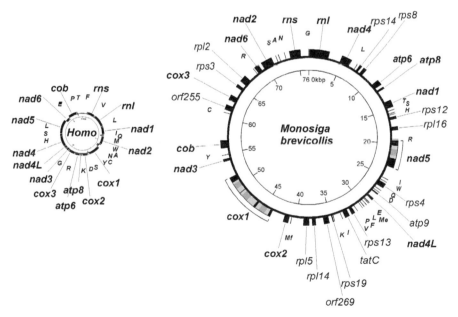

Fig. 1. mtDNA organization in bilaterian animals (represented by *Homo sapiens*) and the choanoflagellate *Monosiga brevicollis*. Protein and ribosomal genes are *atp6, atp8–9*: subunits 6, 8 and 9 of F_0 adenosine triphosphatase (ATP) synthase; *cob*: apocytochrome b; *cox1-3*: cytochrome c oxidase subunits 1–3; *nad1–6* and *nad4L*: NADH dehydrogenase subunits 1–6 and 4L; *rns* and *rnl*: small and large subunit rRNAs; *rps3–19* and *rpl2–16*: small and large subunit ribosomal proteins; *tatC*: twin-arginine translocase component C. tRNA genes are identified by the one-letter code for their corresponding amino acid. The drawings are proportional to genome size: *M. brevicollis* (76,568 bp) and *H. sapiens* (16,571 bp).

in mitochondrial genome architecture co-occurred with the evolution of multicellular animals. Studies of mtDNA in non-bilaterian animals can provide valuable insights into the progression of these changes and also supply data for phylogenetic analyses of animal relationships. Here I review recent progress in our understanding of non-bilaterian mtDNA and discuss the advantages and limitations of mitochondrial datasets for evolutionary and phylogenetic inferences.

The mitochondrial DNA of non-bilaterian animals

Complete, or nearly-complete, mitochondrial DNA sequences have been determined for >60 species of non-bilaterian animals, including 30 species of cnidarians, 27 species of sponges, and four placozoan strains (Fig. 2, Table 1). These mitochondrial genomes show many deviations from the typical bilaterian mtDNA described above. *First*, mtDNA is usually larger in size in non-bilaterian compared to bilaterian animals, averaging ~18.0 kbp in Cnidaria, ~20.0 in demosponges, and ~37.4 in placozoans. The larger size of non-bilaterian mtDNA is primarily due to the presence of larger non-coding intergenic regions, as well as more bacteria-like ribosomal and transfer RNA genes (Beagley et al. 1995, Signorovitch et al. 2007, Wang and Lavrov 2008). *Second*, mtDNA of non-bilaterian animals shows more variation in the gene content. Extra protein-coding genes are

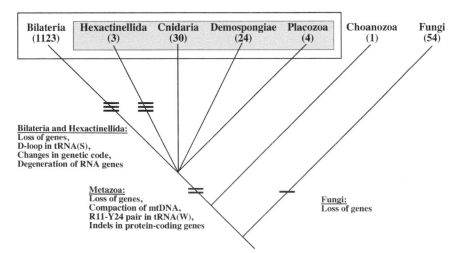

Fig. 2. mtDNA evolution in the Metazoa. Phylogenetic relationships among bilaterian and non-bilaterian animals are currently unresolved and are represented here by a polytomy (empty box). The present review is mostly limited to mtDNA in "non-bilaterian" animals (light-gray box). Only the taxa for which mtDNA data are available are shown. The numbers under the taxon names indicates the number of complete mitochondrial genomes available for each group.

Table 1. Mitochondrial genomes of non-bilaterian animals.

Species	GenBank #	Reference	Species	GenBank	Reference
Cnidaria			**Demospongiae**		
Acropora tenuis	NC_003522	(van Oppen et al. 2002)	*Agelas schmidti*	EU237475	(Lavrov et al. 2008)
Agaricia humilis	NC_008160	(Medina et al. 2006)	*Amphimedon compressa*	NC_010201	(Erpenbeck et al. 2007) (Lavrov et al. 2008)
Anacropora matthai	NC_006898	None	*Amphimedon queenslandica*	NC_008944	(Erpenbeck et al. 2007)
Astrangia sp. JVK-2006	NC_008161	(Medina et al. 2006)	*Aplysina fulva*	NC_010203	(Lavrov et al. 2008)
Aurelia aurita	NC_008446	(Shao et al. 2006)	*Axinella corrugata*	NC_006894	(Lavrov and Lang 2005b)
Briareum asbestinum	NC_008073	(Medina et al. 2006)	*Chondrilla* aff. *nucula*	NC_010208	(Lavrov et al. 2008)
Chrysopathes formosa	NC_008411	(Brugler and France 2007)	*Callyspongia plicifera*	NC_010206	(Lavrov et al. 2008)
Colpophyllia natans	NC_008162	(Medina et al. 2006)	*Cinachyrella kuekenthali*	EU237479	(Lavrov et al. 2008)
Discosoma sp. CASIZ 168915	NC_008071	(Medina et al. 2006)	*Ectyoplasia ferox*	EU237480	(Lavrov et al. 2008)
Discosoma sp. CASIZ 168916	NC_008072	(Medina et al. 2006)	*Ephydatia muelleri*	NC_010202	(Lavrov et al. 2008)
Hydra oligactis	EU237491	(Kayal and Lavrov 2008)	*Geodia neptuni*	NC_006990	(Lavrov et al. 2005)
Metridium senile	NC_000933	(Beagley et al. 1998)	*Halisarca dujardini*	NC_010212	(Lavrov et al. 2008)
Montastraea annularis	NC_007224	(Fukami and Knowlton 2005)	*Hippospongia lachne*	NC_010215	(Lavrov et al. 2008)
Montastraea franksi	NC_007225	(Fukami and Knowlton 2005)	*Igernella notabilis*	NC_010216	(Lavrov et al. 2008)

Table 1 contd...

Table 1 contd...

Montastraea faveolata	NC_007226	(Fukami and Knowlton 2005)	*Iotrochota birotulata*	NC_010207	(Lavrov et al. 2008)
Montipora cactus	NC_006902	None	*Negombata magnifica*	NC_010171	(Belinky et al. 2008)
Mussa angulosa	NC_008163	(Medina et al. 2006)	*Oscarella carmela*	NC_009090	(Wang and Lavrov 2007)
Nematostella sp. JVK-2006	NC_008164	(Medina et al. 2006)	*Plakortis angulospiculatus*	NC_010217	(Lavrov et al. 2008)
Pavona clavus	NC_008165	(Medina et al. 2006)	*Ptilocaulis walpersi*	EU237488	(Lavrov et al. 2008)
Pocillopora eydouxi	NC_009798	(Flot and Tillier 2007)	*Suberites domuncula*	NC_010496	
Pocillopora damicornis	NC_009797	(Flot and Tillier 2007)	*Tethya actinia*	NC_006991	(Lavrov et al. 2005)
Porites porites	NC_008166	(Medina et al. 2006)	*Topsentia ophiraphidites*	NC_010204	(Lavrov et al. 2008)
Pseudopterogorgia bipinnata	NC_008157	(Medina et al. 2006)	*Vaceletia* sp.	NC_010218	(Lavrov et al. 2008)
Rhodactis sp. CASIZ 171755	NC_008158	(Medina et al. 2006)	*Xestospongia muta*	NC_010211	(Lavrov et al. 2008)
Ricordea florida	NC_008159	(Medina et al. 2006)			
Sarcophyton glaucum	AF064823, AF063191	(Beaton et al. 1998)	**Hexactinellida**		
Savalia savaglia	NC_008827	(Sinniger et al. 2007)	*Aphrocallistes vastus,*	EU000309	(Rosengarten et al., 2008)
Seriatopora caliendrum	NC_010245	(Chen et al. 2008)	*Iphiteon panicea*	>19,045	(Haen et al. 2007)
Seriatopora hystrix	NC_010244	(Chen et al. 2008)	*Sympagella nux*	>16,293	(Haen et al. 2007)
Siderastrea radians	NC_008167	(Medina et al. 2006)			
Placozoa					
Trichoplax adhaerens	NC_008151	(Dellaporta et al. 2006)			

Species	GenBank #	Reference	Species	GenBank	Reference
BZ49	NC_008833	(Signorovitch et al. 2007)			
BZ10101	NC_008832	(Signorovitch et al. 2007)			
BZ2423	NC_008834	(Signorovitch et al. 2007)			

found in several lineages, including *atp9* for ATP synthase subunit 9 in most demosponges (Wang and Lavrov 2008) and glass sponges (Haen et al. 2007, Rosengarten et al. 2008), *tatC* for twin-arginine translocase subunit C in *Oscarella carmela* (Wang and Lavrov 2007), *mutS* for a putative mismatch repair protein in octocorals (Pont-Kingdon et al. 1995, 1998, Medina et al. 2006), and *polB* for the DNA-dependent DNA polymerase in the jellyfish *Aurelia aurita* and one strain of placozoans (Shao et al. 2006, Signorovitch et al. 2007). In addition, out of 22 tRNA genes typically present in mtDNA of bilaterian animals, at most two, *trnM* and *trnW*, are found in cnidarians (Beagley et al. 1995, Kayal and Lavrov 2008) and the G1 clade of demosponges (Wang and Lavrov 2008), only five are found in the demosponge *Plakortis angulospiculatus* (Wang and Lavrov 2008) and 17 are found in another demosponge *Amphimedon compressa* (Erpenbeck et al. 2007).[1] By contrast, additional mitochondrial tRNA genes are present in mtDNA of other demosponges and placozoans. *Third,* mitochondrial group I introns, with or without the homing endonucleases of the LAGLI-DADG type, are found in hexacorals (Beagley et al. 1996, van Oppen et al. 2002, Fukami and Knowlton 2005), placozoans (Signorovitch et al. 2007), and several species of demosponges (Rot et al. 2006, Wang and Lavrov 2008). *Fourth,* demosponges, cnidarians and placozoans use a minimally-modified genetic code for mitochondrial translation, with TGA=tryptophan as the only deviation. The same genetic code is used for mitochondrial translation in *Monosiga brevicollis* and in most (but not all) fungi, whereas most bilaterian animals have modified the specificity of at least ATA and AGR codons (Knight et al. 2001). *Finally,* many non-bilaterian animals display low rates of mitochondrial sequence evolution as revealed by intraspecific (Shearer et al. 2002, Duran et al. 2004, Hellberg 2006, Wörheide 2006) and interspecific (Lavrov et al. 2005, Medina et al. 2006) studies based on mtDNA sequences.

A Parallel Evolution of "Bilaterian-Like" mtDNA

Remarkably, one group of non-bilaterian animals—Hexactinellida, or glass sponges—evolved mitochondrial genomes very similar to those found in bilaterian animals. Mitochondrial genomes in glass sponges (Hexactinellida) and bilaterian animals share several characteristics, including a nearly identical gene content (37 genes, but with *atp9* instead of *atp8* in glass sponges), nucleotide composition of the coding strand, and the presence of a single large non-coding region (Haen et al. 2007, Rosengarten et al. 2008). Like those in bilaterian animals, glass sponge mt-

---

[1]Rosengarten et al. (Rosengarten et al. 2008) claim that three tRNA genes are also missing in the mitochondrial genome of the glass sponge *Aphrocallistes vastus*. However, we were able to identify the missing tRNAs in the non-coding regions of this genome.

tRNAs display a great degree of variation both in size and in nucleotide sequence of the DHU and TψC arms, including a highly variable sequence of the TψC loop and the lack of conserved guanine residues in the DHU loops. In addition, both groups share a reassignment of the mitochondrial AGR codons from arginine to serine and experienced similar changes in the secondary structure of the tRNA$^{Ser}_{UCU}$ that translates these codons (Haen et al. 2007). Many of these changes are rare and complex events that, so far, are known to occur only in the Bilateria and Hexactinellida.

However, a closer look at mitochondrial genomes from glass sponges and bilaterian animals also reveals some differences. First, although only one mitochondrial gene for a tRNA with the CAU anticodon has been found in both groups, the encoded tRNA in glass sponges does not contain the R11-Y24 pair, characteristic of initiator tRNAs in other organisms. Hence, it is unclear whether this gene codes for the initiator tRNA$^{fMet}_{CAU}$ or for the tRNA$^{Ile}_{LAU}$ that usually translates the isoleucine ATA codon in eubacteria and eubacteria-derived organelles, including mitochondria of demosponges and placozoans (see below). If the latter is the case, then glass sponges differ from several groups of bilaterian animals (hemichordates, echinoderms, and flat worms) where ATA also codes for isoleucine, but is translated by a modified tRNA$^{Ile}_{GAU}$ together with two other isoleucine codons, AUC and AUU, while the only encoded tRNA with CAU anticodon is clearly an initiator tRNA (Jacobs et al. 1988, Castresana et al. 1998). Second, an atypical R11-Y24 pair is present in tRNA$^{Pro}_{UGG/CGG}$ of glass sponges as well as all demosponges and placozoans but is not found in the outgroup species *Monosiga brevicollis* and *Amoebidium parasiticum* or in bilaterian animals. Because the R11-Y24 base pair is an important recognition element for the initiator tRNA, it is usually strongly counter-selected in elongator tRNAs (Marck and Grosjean 2002), and its presence in the proline tRNA of sponges and *T. adhaerens* spports the monophyly of non-bilaterian animals or at least the monophyly of the Porifera and Placozoa (this character is not available for Cnidaria, because they lack this and most other tRNA genes in their mtDNA). Finally, many genetic novelties of bilaterian animals, such as unusual initiation codons and incomplete termination codons, are rare or not found in glass sponges. These observations, together with phylogenetic inference based on mitochondrial coding sequences, indicate that shared mitochondrial features between glass sponges and bilaterian animals are the result of parallel evolution rather than a common ancestry.

Animal Mitochondrial Synapomorphies

Given that most of the traits previously thought to be characteristic of animal mtDNA now appear to have evolved within the Metazoa, one may wonder whether there are any derived mitochondrial features shared by all animals. Our analyses reveal two such putative mitochondrial synapomorphies. First, we found a set of Metazoa-specific insertions/ deletions (indels) in protein-coding genes (Fig. 3) that are well conserved across the Metazoa, but are absent in *M. brevicollis* and other non-animal species (Lavrov et al. 2005). Although the value of an individual indel for phylogenetic reconstruction is limited (Gribaldo and Philippe 2002), the presence of several such events in mitochondrial genes provides strong support for the monophyly of the Metazoa and can serve as a good indicator of metazoan affinity. Second, an atypical R11-Y24 pair is present in mitochondrial tRNA$_{UCA}^{Trp}$ of both bilaterian (Wolstenholme 1992a) and non-bilaterian animals, but is absent in non-metazoan outgroups. As explained above, the R11-Y24 pair is usually strongly counterselected in elongator tRNAs (Marck and Grosjean 2002). The lack of both of these synapomorphies in the mtDNA of *M. brevicollis* refutes the idea that choanoflagellates may be derived from sponges or other basal metazoans (e.g. King and Carroll 2001, Maldonado 2004).

Phylogenetic Inference Using Mitochondrial Sequences and Gene Orders

Mitochondrial genomes provide two primary datasets for phylogenetic inference: gene sequences and gene orders (Bruns et al. 1989, Boore and Brown 1998, Lavrov and Lang 2005a). Mitochondrial sequence data

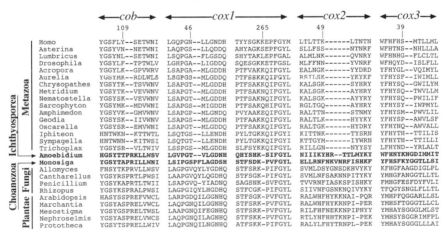

Fig. 3. Conserved indel events shared among multicellular animals. The four best-conserved genes (cob, cox1-3) were analyzed. Numbers above the alignment indicate positions in *H. sapiens* sequence.

are well suited for phylogenetic analysis within class-level lineages of non-bilaterian animals because of the low rate of sequence evolution and homogeneous nucleotide composition in most of them (Medina et al. 2006, Signorovitch et al. 2007, Wang and Lavrov 2007, Lavrov et al. 2008). For example, the phylogenetic analysis presented in Fig. 4 recovers phylogenetic relationships within the Demospongiae highly congruent with those based on 18S rRNA data and multiple nuclear proteins (see Lavrov et al. 2008 for details). The topology within Hexacorallia also presents a viable phylogenetic hypothesis, largely consistent with previous studies (Berntson et al. 1999, Daly et al. 2003, Collins et al. 2006). Notably, the use of an expanded dataset of anthozoan sequences helps to recover the monophyletic Scleractinia, forming a sister group to naked corals (order Corallimorpharia). This result contradicts the finding of the previous mtDNA-based study (Medina et al. 2006), which inferred the origin of the naked corals within the Scleractinia, but is largely congruent with analyses of other molecular and morphological datasets.

At the same time, the inference of global animal relationships based on mtDNA sequences is more problematic. The position of glass sponges is one example. The use of traditional empirical models of amino-acid evolution in maximum likelihood and Bayesian phylogenetic analyses results in a strong support (with high bootstrap and posterior probability numbers) for the sister group relationship between glass sponges and bilaterian animals (Haen et al. 2007). By contrast, Bayesian phylogenetic inference based on a CAT model that explicitly handles the heterogeneity of the substitution process across amino-acid positions (Lartillot and Philippe 2004) places glass sponges either with some fast-evolving cnidarians (Fig. 4) or within demosponges (Haen et al. 2007, Wang and Lavrov 2007). Another recurring result of studies based on mitochondrial protein sequences is the grouping of demosponges with some or all cnidarians (Lavrov et al. 2005, Shao et al. 2006, Haen et al. 2007, Kayal and Lavrov 2008, Lavrov et al. 2008). Although, this grouping may reflect a genuine phylogenetic signal [e.g. see Dunn et al. (2008) for a recent congruent result based on an alternative dataset], it can also be interpreted as an artifact of long-branch attraction (Felsenstein 1978, Hendy and Penny 1989) between rapidly evolving bilaterian animals and distantly-related outgroups (Lavrov et al. 2005). Similar uncertainty surrounds the interpretation of the phylogenetic position of Placozoa, which varies considerably in different analyses and is unresolved in the analysis conducted for this study (Fig. 4). Although one may argue that the missing mitochondrial data from Calcarea and Ctenophora will help to resolve the basal animal relationships, I think that the very absence of their mitochondrial sequences from public databases in this genomic age is symptomatic and indicates highly unusual and fast-evolving mitochondrial genomes in these two groups that will be little informative for phylogenetic reconstruction.

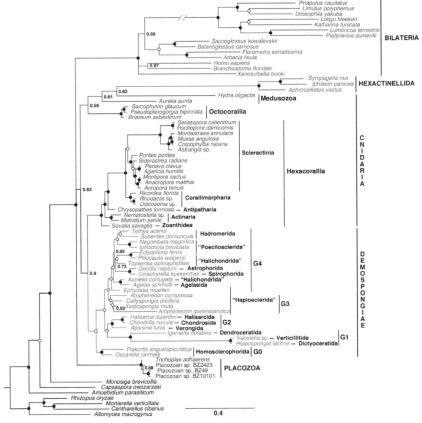

Fig. 4. Phylogenetic analysis of global animal relationships using mitochondrial sequence data. Posterior majority-rule consensus tree obtained from the analysis of 2,539 aligned amino acid positions under the CAT+F+Γ model in the PhyloBayes program is shown. We ran four independent chains for ~60,000 generations and sampled every 10th tree after the first 1000 burnin cycles. The convergence among the chains was monitored with the maxdiff statistics and the analysis was terminated after maxdiff became less than 0.3. The number / circle at each node represents the Bayesian posterior probability. Posterior probability = 1 is indicated by a filled circle; that > 0.95—by an open circle; smaller posterior probabilities are shown as numbers. Amino acid sequences for non-bilaterian animals were derived from the GenBank files listed in table 1. Amino-acid sequences for *Cantharellus cibarius* mtDNA were downloaded from http://megasun.bch.umontreal.ca/People/lang/FMGP/proteins.html; those for *Capsaspora owczarzaki* mtDNA were provided by Franz Lang (Université de Montréal). Other sequences were derived from the GenBank files: *Arbacia lixula* NC_001770, *Balanoglossus carnosus* NC_001887, *Branchiostoma floridae* NC_000834, *Drosophila yakuba* NC_001322, *Florometra serratissima* NC_001878, *Homo sapiens* NC_001807, *Katharina tunicata* NC_001636, *Limulus polyphemus* NC_003057, *Loligo bleekeri* NC_002507, *Lumbricus terrestris* NC_001673, *Platynereis dumerilii* NC_000931, *Priapulus caudatus* NC_008557, *Saccoglossus kowalevskii* NC_007438, *Xenoturbella bocki* NC_008556, *Amoebidium parasiticum* AF538042-AF538052, *Monosiga brevicollis* NC_004309, *Allomyces macrogynus* NC_001715, *Mortierella verticillata* NC_006838, *Rhizopus oryzae* NC_006836.

Color image of this figure appears in the color plate section at the end of the book.

While gene order comparisons commonly accompany descriptions of new mitochondrial genomes, a formal phylogenetic analyses on this dataset is rarely done. This is mainly due to the lack of user-friendly computer programs for such analyses but also to the algorithmic difficulties in analyzing gene order data (Moret et al. 2004). The paucity of gene order studies is rather unfortunate because they can provide important insights in cases where most other datasets fail, as has been shown on echinoderms (Smith et. al 1993), arthropods (Boore et. al 1995, Boore et al. 1998), and crustaceans (Lavrov et al. 2004, Morrison et al. 2002). Recently we used mitochondrial gene arrangements to infer phylogenetic relationships within the Demospongiae and recovered a tree largely congruent with that based on mitochondrial sequences, although with only a moderate support (Lavrov et al. 2008). However, the analysis of global animal phylogeny based on mitochondrial gene order data is again problematic and results in a phylogeny were most of the basal relationships are unresolved (Fig. 5). We note that there are several problems complicating mitochondrial gene-order based analysis in non-bilaterian animals. These include the lack of informative data from accepted outgroups, variation in the mitochondrial gene content across non-bilaterian animals, especially, an independent loss of mitochondrial tRNA genes in several lineages, and an unusual mitochondrial genomic organization in hexacorrals and placozoans, where some mitochondrial genes are located within introns of other genes.

Mitochondrial Size and Gene Content are Not Reliable Phylogenetic Indicators

Can mitochondrial genome size and gene content be used for phylogenetic reconstruction? At first glance there appears to be a trend in animal mtDNA evolution, with both genome size and gene content being reduced during transitions first to the Metazoa and then to the Bilateria (Fig. 2). Indeed, the mitochondrial genome of the choanoflagellate *Monosiga brevicollis* is 76.6 kbp in size and contains 55 genes, those of most demosponges are between 18 and 25 kbp and contain 40–44 genes, while bilaterian mtDNA is typically between 14 and 16 kpb and has 36–37 genes. Thus, it may be tempting to use mtDNA size and gene content as indicators of phylogenetic relationships (e.g. Dellaporta et al. 2006). It should be noted, however, that the reduction in mitochondrial DNA size and gene content are only overall trends in animal mitochondrial evolution and that both of these features can vary substantially within individual groups.

The largest animal mitochondrial genomes (up to 43 kbp) are found in the placozoan *Trichoplax adhaerens* (Dellaporta et al. 2006) as well as some very distantly related bilaterian animals, including three species of bark weevil *Pissodes* (Boyce et al. 1989), the deep-sea scallop *Placopecten magellanicus* (Snyder et al. 1987), and the nematode *Romanomermis culicivorax* (Powers et al. 1986), and have clearly evolved independently in these taxa. In addition, relatively closely related animals often display extensive variation in mtDNA size. For example, the size of the mitochondrial genome in *Drosophila melanogaster* is ~22% (3.5 kpb) larger than that in *Drosophila yakuba*, while the mtDNA length in four strains of Placozoa differs by as much as 35% (Signorovitch et al. 2007). The fact that mtDNA size can change rapidly and repeatedly within independent lineages makes it an unreliable character for phylogenetic reconstruction. Furthermore, it is likely that we underestimate the true range of length variation in bilaterian mtDNA because of the difficulties associated with PCR amplification and sequencing of the large non-coding region in this molecule (e.g. Lavrov and Brown 2001).

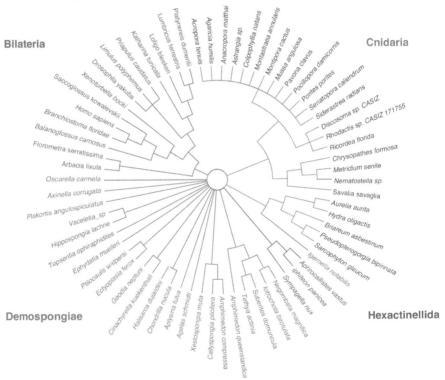

Fig. 5. Global animal relationships based on mitochondrial gene order data. Strict consensus tree is shown from the Maximum Parsimony analysis on Multistate Encodings (Boore et al. 1995, Wang et al. 2002). Gene orders were encoded as described previously (Lavrov and Lang 2005a) and analyzed using heuristic search with 100 random addition replicates in PAUP*4.0b10 and TBR branch swapping option.

Color image of this figure appears in the color plate section at the end of the book.

Mitochondrial gene content is also a poor indicator of phylogenetic relationships. In most cases when "extra" genes are present, they are inherited from a common ancestor and thus represent plesiomorphies not informative for phylogenetic reconstruction. By contrast, while the losses of genes from mtDNA are apomorphies, they are known to occur in parallel in different groups, resulting in homoplasious similarities in gene contents (Martin et al. 1998). As an example, a very similar mitochondrial gene content in animal and fungal mtDNA has likely evolved independently in the two groups given that the mitochondrial genome of the choanoflagellate *Monosiga brevicollis*, the sister group to animals, contains additional genes (Burger et al. 2003; Fig. 1). Another example comes from our recent study of mitochondrial genomes in demosponges. This study revealed that the G1 group of demosponges lost all but two tRNA genes from mtDNA, resulting in a gene content very similar to that in the phylum Cnidaria. Remarkably, in both of these groups the same two tRNA genes have been retained (*trnM(cau)* and *trnW(uca)*) supporting the inference of a special role of these tRNAs in animal mitochondria. As has been explained before (e.g. Shao et al. 2006), $tRNA_{CAU}^{Met}$ is used for the initiation of mitochondrial translation with formyl-methionine (Smith and Marcker 1968) while $tRNA_{UCA}^{Trp}$ must translate the TGA in addition to the TGG codons as tryptophan in animal mitochondria. Because of these special functions, the transfer of these tRNAs to the nucleus may be selectively disadvantageous.

Implications for mtDNA Evolution

Although mitochondrial genome size and gene content offer little information for resolving animal phylogenetic relationships, they do provide insights into mitochondrial genome evolution. Given that the transfer of nuclear genes to mitochondria is limited (in part by differences in genetic code), most additional genes found in demosponge mitochondrial DNA (*atp9, tatC, trnI(cau), trnR(ucu)*) were likely inherited from their common ancestor with animals. This is particularly the case for *trnI(cau)*. The maturation of $tRNA_{CAU}^{Ile}$ encoded by this gene involves a posttranscriptional modification of the cytosine at position 34 to lysidine (2-lysyl-cytidine) (Muramatsu et al. 1988, Weber et al. 1990). This modification is performed by the tRNA[Ile]-lysidine synthetase (Soma et al. 2003), an enzyme that is not involved in the maturation of cytoplasmic isoleucine tRNAs (Marck and Grosjean 2002). Because of the dispensable nature of this function for cytoplasmic protein synthesis, it is likely that in evolutionary terms, the loss of the metazoan mitochondrial $tRNA_{CAU}^{Ile}$ is followed by the loss of the nuclear lysidine synthetase gene, rendering a reacquisition of $tRNA_{CAU}^{Ile}$ practically impossible. The presence of *trnI(cau)* in the mtDNA of both demosponges and placozoans clearly indicates that the common animal ancestor used a less modified genetic code for mitochondrial translation than do most bilaterian animals. It is also likely that this common ancestor had more genes in its mitochondrial genome than extant metazoan taxa.

Signorovitch et al. (2007) suggest that the common ancestor of all animals possessed a large, noncompact mitochondrial genome. This inference is based on the observation that both placozoans and the choanoflagellate *Monosiga brevicollis* have large mtDNA and on the assumption that Placozoa forms the sister group to all other animals. There are two potential problems with this inference. First, phylogenetic analysis of global animal relationships based on mtDNA sequence data does not support the placement of Placozoa as the sister group to other animals (Signorovitch et al. 2007, Wang and Lavrov 2007). Second, even if Placozoa is the sister group to the rest of the animals, the most parsimonious reconstruction of genome size in the common animal ancestor can be grossly misleading (e.g. Cunningham et al. 1998) given the long branch leading to the Placozoa and rapid changes in the size of mtDNA observed among closely-related animals (see above).

A more interesting question, in my view, is about the evolutionary forces that maintain the compact nature of mtDNA in most modern animals. Studies of non-bilaterian animals can help to answer this question because these animals display a different combination of features in their mitochondrial organization. For example, these studies indicate that small size and gene content are not directly linked with an accelerated rate of evolution of mitochondrial sequences. As shown above, while the rate of nucleotide substitutions appears to be extremely low in some non-bilaterian animals, their genomes are still very compact and mostly intron-less (with the exception of Placozoa), challenging the idea that elevated mutation pressure is solely responsible for the evolution of these features in animal mtDNA (Lynch et al. 2006).

Implications for Animal Evolution

It is commonly believed that the study of non-bilaterian animals can also provide insights into metazoan morphological evolution. This is not because non-bilaterian animals are phylogenetically more "basal" or "lower"—these terms are misleading and should be avoided (Crisp and Cook 2005)—but because they evolved their bodyplans earlier in metazoan evolution (as revealed by the fossil record) and so may resemble the ancestors of other groups. Following this reasoning, sponges are often considered as living representatives of an intermediate stage in animal evolution that never reached the tissue-level grade of organization. Our recent study of mtDNA from the homoscleromorph *Oscarella carmela* challenges this idea (Wang and Lavrov 2007).

Homoscleromorpha is a small group of sponges that share? several features with "higher" animals such as the presence of type-IV collagen, acrosomes in spermatozoa and cross-striated rootlets in the flagellar basal apparatus of larval cells. More importantly, it has been shown that the epithelial cells in homoscleromorph larvae meet all the criteria of a true epithelium in higher animals: cell polarization, apical cell junctions, and a basement membrane (Boury-Esnault et al. 2003). Unless these shared cytological features evolved independently in Homoscleromorpha and Eumetazoa (an unlikely scenario), two alternative explanations are possible for these findings: either Homoscleromorpha is more closely related to other animals than to sponges or most sponges lost the aforementioned features. Our analysis of demosponge relationships based on mitochondrial genomic data did not find any support for the first of these hypotheses and instead provided strong support for the placement of the Homoscleromorpha with demosponges. This result suggests that the bodyplan of sponges might represent a secondary simplification in animal morphology, potentially due to their sedentary and water-filtering lifestyle. Similar conclusion should be made if sponges and cnidarians form a monophyletic group (Fig. 1) or if ctenophores form the sister group to the rest of the animals (Dunn et al. 2008).

Conclusions

The evolution of multicellular animals is associated with major changes in their mitochondrial genome architecture. Recent studies on mtDNA from non-bilaterian animals suggest that these changes occurred in two steps, roughly correlated with the origin of animal multicellularity and the origin of the bilaterality. The transition to multicellular animals is associated with the loss of multiple genes from mtDNA and a drastic reduction in the amount of non-coding DNA in the genome, resulting in its "small is beautiful" nature. The transition to bilaterian animals is correlated with multiple changes in the genetic code and associated losses of tRNA genes, the emergence of several genetic novelties, and a large increase in the rates of sequence evolution. Although we do not know whether the observed changes co-occurred with the morphological transitions or evolved independently in different lineages, the remarkable uniformity of animal mtDNA suggests an ongoing evolutionary pressure that maintains its unique organization. Studies of mtDNA in non-bilaterian animals are important because they provide insights not only into the history of mtDNA evolution, but also into the evolutionary factors that shape modern-day animal mtDNA. Although our sampling of mtDNA from non-bilaterian animals is still limited, substantial progress has been made in the last few years that revealed new information about phylogenetics, the evolution of mitrochondrial genomes, and even morphological evolution in animals.

Acknowledgements

I thank Bernd Schierwater, Stephen Dellaporta, and Rob DeSalle for organizing the symposium and its complementary session, Karri Haen, Ehsan Kayal and Xiujuan Wang for valuable comments on an earlier version of this manuscript, and the College of Liberal Arts and Sciences at Iowa State University for funding.

References

Anderson, S. and A.T. Bankier, B.G. Barrell, M.H.L. De Bruijn, A.R. Coulson, J. Drouin et al. 1981. Sequence, and organization of the human mitochondrial genome. Nature 290: 457–465.

Armstrong, M.R. and V.C. Block, and M.S. Phillips. 2000. A multipartite mitochondrial genome in the potato cyst nematode *Globodera pallida*. Genetics 154: 181–192.

Beagley, C.T. and J.L. Macfarlane, G.A. Pont-Kingdon, R. Okimoto, N. Okada, and D.R. Wolstenholme. 1995. Mitochondrial genomes of Anthozoa (Cnidaria). Palmieri F., Papa S., Saccone C., and Gadaleta N. (eds) Progress in cell research – Symposium on "Thirty Years of Progress in Mitochondrial Bioenergetics and Molecular Biology". Elsevier Science BV Amsterdam, The Netherlands, pp 149–153.

Beagley, C.T. and N.A. Okada, and D.R. Wolstenholme. 1996. Two mitochondrial group I introns in a metazoan, the sea anemone *Metridium senile*: one intron contains genes for subunits 1 and 3 of NADH dehydrogenase. Proc. Natl. Acad. Sci. USA 93: 5619–5623.

Beagley, C.T. and R. Okimoto, and D.R. Wolstenholme. 1998. The mitochondrial genome of the sea anemone *Metridium senile* (Cnidaria): introns, a paucity of tRNA genes, and a near-standard genetic code. Genetics 148: 1091–1108.

Beaton, M.J. and A.J. Roger, and T. Cavalier-Smith. 1998. Sequence analysis of the mitochondrial genome of *Sarcophyton glaucum*: conserved gene order among octocorals. J. Mol. Evol. 47: 697–708.

Belinky, F. and C. Rot, M. Ilan and D. Huchon. 2008. The complete mitochondrial genome of the demosponge *Negombata magnifica* (Poecilosclerida). Mol. Phylogenet. Evol. 47: 1238–1243.

Berntson, E.A. and S.C. France, and L.S. Mullineaux. 1999. Phylogenetic relationships within the class Anthozoa (phylum Cnidaria) based on nuclear 18S rDNA sequences. Mol. Phylogenet. Evol. 13: 417–433.

Boore, J.L. and T.M. Collins, D. Stanton, L.L. Daehler, and W.M. Brown. 1995. Deducing the pattern of arthropod phylogeny from mitochondrial DNA rearrangements. Nature 376: 163–165.

Boore, J.L. and W.M. Brown. 1998. Big trees from little genomes: mitochondrial gene order as a phylogenetic tool. Curr. Opin. Genet. Dev. 8: 668–674.

Borst, P. and L.A. Grivell. 1981. Small is beautiful—portrait of a mitochondrial genome. Nature 290: 443–444.

Boury-Esnault, N. and A. Ereskovsky, C. Bezac, and D. Tokina. 2003. Larval development in the Homoscleromorpha (Porifera, Demospongiae). Invertebr. Biol. 122: 187–202.

Boyce, T. M. and M.E. Zwick, and C.F. Aquadro. 1989. Mitochondrial DNA in the bark weevils: Size, structure and heteroplasmy. Genetics 123: 825–836.

Brugler, M.R. and S.C. France. 2007. The complete mitochondrial genome of the black coral *Chrysopathes formosa* (Cnidaria:Anthozoa:Antipatharia) supports classification of antipatharians within the subclass Hexacorallia. Mol. Phylogenet. Evol. 42: 776–788.

Key Transitions in Animal Evolution 51

Bruns, T.D. and R. Fogel, T.J. White, and J.D. Palmer. 1989. Accelerated evolution of a false-truffle from a mushroom ancestor. Nature 339: 140–142.

Burger, G. and L. Forget, Y. Zhu, M.W. Gray, and B.F. Lang. 2003. Unique mitochondrial genome architecture in unicellular relatives of animals. Proc. Natl. Acad. Sci. USA 100: 892–897.

Castresana, J. and G. Feldmaier-Fuchs, S. Yokobori, N. Satoh and S. Paabo. 1998. The mitochondrial genome of the hemichordate *Balanoglossus carnosus* and the evolution of deuterostome mitochondria. Genetics 150: 1115–1123.

Chen, C. and C.F. Dai, S. Plathong, C.Y. Chiou, and C.A. Chen. 2008. The complete mitochondrial genomes of needle corals, *Seriatopora* spp. (Scleractinia: Pocilloporidae): an idiosyncratic atp8, duplicated trnW gene, and hypervariable regions used to determine species phylogenies and recently diverged populations. Mol. Phylogenet. Evol. 46: 19–33.

Collins, A.G. and P. Schuchert, A.C. Marques, T. Jankowski, M. Medina, and B. Schierwater. 2006. Medusozoan phylogeny and character evolution clarified by new large and small subunit rDNA data and an assessment of the utility of phylogenetic mixture models. Syst. Biol. 55: 97–115.

Crisp, M.D. and L.G. Cook. 2005. Do early branching lineages signify ancestral traits? Trends Ecol. Evol. 20: 122–128.

Cunningham, C.W. and K.E. Omland, and T.H. Oakley. 1998. Reconstructing ancestral character states: a critical reappraisal. Trends Ecol. Evol. 13: 361–366.

Daly, M. and D.G. Fautin, and V.A. Cappola. 2003. Systematics of the Hexacorallia (Cnidaria : Anthozoa). Zool. J. Linn. Soc. Lond. 139: 419–437.

Dellaporta, S.L. and A. Xu, S. Sagasser, W. Jakob, M.A. Moreno, L.W. Buss, and B. Schierwater. 2006. Mitochondrial genome of *Trichoplax adhaerens* supports Placozoa as the basal lower metazoan phylum. Proc. Natl. Acad. Sci. USA 103: 8751–8756.

Dunn, C.W. and A. Hejnol, D.Q. Matus, K. Pang, W.E. Browne, S.A. Smith et al. 2008. Broad phylogenomic sampling improves resolution of the animal tree of life. Nature 452: 745–749.

Duran, S. and M. Pascual, and X. Turon. 2004. Low levels of genetic variation in mtDNA sequences over the western Mediterranean and Atlantic range of the sponge *Crambe crambe* (Poecilosclerida). Mar. Biol. 144: 31–35.

Erpenbeck, D. and O. Voigt, M. Adamski, M. Adamska, J. Hooper, G. Worheide, and B. Degnan. 2007. Mitochondrial diversity of early-branching Metazoa is revealed by the complete mt genome of a haplosclerid demosponge. Mol. Biol. Evol. 24: 19–22.

Felsenstein, J. 1978. Cases in which parsimony or compatibility methods will be positively misleading. Syst. Zool. 27: 401–410.

Flot, J.F. and S. Tillier. 2007. The mitochondrial genome of Pocillopora (Cnidaria: Scleractinia) contains two variable regions: the putative D-loop and a novel ORF of unknown function. Gene 401: 80–87.

Fukami, H. and N. Knowlton. 2005. Analysis of complete mitochondrial DNA sequences of three members of the *Montastraea annularis* coral species complex (Cnidaria, Anthozoa, Scleractinia). Coral Reefs 24: 410–417.

Gribaldo, S. and H. Philippe. 2002. Ancient phylogenetic relationships. Theor. Popul. Biol. 61: 391–408.

Haen, K.M. and B.F. Lang, S.A. Pomponi, and D.V. Lavrov. 2007. Glass sponges and bilaterian animals share derived mitochondrial genomic features: A common ancestry or parallel evolution? Mol. Biol. Evol. 24: 1518–1527.

Helfenbein, K.G. and H.M. Fourcade, R.G. Vanjani, and J.L. Boore. 2004. The mitochondrial genome of *Paraspadella gotoi* is highly reduced and reveals that chaetognaths are a sister group to protostomes. Proc. Natl. Acad. Sci. USA 101: 10639–10643.

Hellberg, M.E. 2006. No variation and low synonymous substitution rates in coral mtDNA despite high nuclear variation. BMC Evol. Biol. 6: 24.

Hendy, M.D. and D. Penny. 1989. A framework for the quantitative study of evolutionary trees. Syst. Zool. 38: 297–309.

Jacobs, H.T. and D.J. Elliott, V.B. Math, and A. Farquharson. 1988. Nucleotide sequence and gene organization of sea urchin mitochondrial DNA. J. Mol. Biol. 202: 185–217.

Kayal, E. and D.V. Lavrov. 2008. The mitochondrial genome of *Hydra oligactis* (Cnidaria, Hydrozoa) sheds new light on animal mtDNA evolution and cnidarian phylogeny. Gene 410: 177–186.

King, N. and S.B. Carroll. 2001. A receptor tyrosine kinase from choanoflagellates: molecular insights into early animal evolution. Proc. Natl. Acad. Sci. USA 98: 15032–15037.

Knight, R.D. and S.J. Freeland, and L.F. Landweber. 2001. Rewiring the keyboard: evolvability of the genetic code. Nat. Rev. Genet. 2: 49–58.

Lang, B.F. and M.W. Gray, and G. Burger. 1999. Mitochondrial genome evolution and the origin of eukaryotes. Annu. Rev. Genet. 33: 351–397.

Lartillot, N. and H. Philippe. 2004. A Bayesian mixture model for across-site heterogeneities in the amino-acid replacement process. Mol. Biol. Evol. 21: 1095–1109.

Lavrov, D.V. and W.M. Brown, and J.L. Boore. 2000. A novel type of RNA editing occurs in the mitochondrial tRNAs of the centipede *Lithobius forficatus*. Proc. Natl. Acad. Sci. USA 97: 13738–13742.

Lavrov, D.V. and W.M. Brown. 2001. *Trichinella spiralis* mtDNA: A nematode mitochondrial genome that encodes a putative ATP8, normally-structured tRNAs, and has a gene arrangement relatable to those of coelomate metazoans. Genetics 157: 621–637.

Lavrov, D.V. and L. Forget, M. Kelly, and B.F. Lang. 2005. Mitochondrial genomes of two demosponges provide insights into an early stage of animal evolution. Mol. Biol. Evol. 22: 1231–1239.

Lavrov, D.V. and B.F. Lang. 2005a. Poriferan mtDNA and animal phylogeny based on mitochondrial gene arrangements. Syst. Biol. 54: 651–659.

Lavrov, D.V. and B.F. Lang. 2005b. Transfer RNA gene recruitment in mitochondrial DNA. Trends Genet. 21: 129–133.

Lavrov, D.V. and M. Kelly, and X. Wang. 2008. Reconstructing ordinal relationships in the Demospongiae using mitochondrial genomic data. Mol. Phylogenet. Evol. 49: 111–124.

Lynch, M. and B. Koskella, and S. Schaack. 2006. Mutation pressure and the evolution of organelle genomic architecture. Science 311: 1727–1730.

Maldonado, M. 2004. Choanoflagellates, choanocytes, and animal multicellularity. Invertebr. Biol. 123: 1–22.

Marck, C. and H. Grosjean. 2002. tRNomics: analysis of tRNA genes from 50 genomes of Eukarya, Archaea, and Bacteria reveals anticodon-sparing strategies and domain-specific features. RNA 8: 1189–1232.

Martin, W. and B. Stoebe, V. Goremykin, S. Hapsmann, M. Hasegawa, and K.V. Kowallik. 1998. Gene transfer to the nucleus and the evolution of chloroplasts. Nature 393: 162–165.

Medina, M. and A.G. Collins, T.L. Takaoka, J.V. Kuehl, and J.L. Boore. 2006. Naked corals: skeleton loss in Scleractinia. Proc. Natl. Acad. Sci. USA 103: 9096–9100.

Moret, B.M. and J. Tang, and T. Warnow. Reconstructing phylogenies from gene-content and gene-order data. pp. 1–32. *In:* O. Gascuel. [ed.] 2004. Mathematics of Evolution and Phylogeny. Clarendon Press, Oxford.

Muramatsu, T. and K. Nishikawa, F. Nemoto, Y. Kuchino, S. Nishimura, T. Miyazawa, and S. Yokoyama. 1988. Codon and amino-acid specificities of a transfer RNA are both converted by a single post-transcriptional modification. Nature 336: 179–181.

Pont-Kingdon, G.A. and N.A. Okada, J.L. Macfarlane, C.T. Beagley, D.R. Wolstenholme, T. Cavalier-Smith, and G.D. Clark-Walker. 1995. A coral mitochondrial *mutS* gene. Nature 375: 109–111.

Pont-Kingdon, G. and N.A. Okada, J.L. Macfarlane, C.T. Beagley, C.D. Watkins-Sims, T. Cavalier-Smith, G.D. Clark-Walker, and D.R. Wolstenholme. 1998. Mitochondrial DNA of the coral *Sarcophyton glaucum* contains a gene for a homologue of bacterial MutS: a possible case of gene transfer from the nucleus to the mitochondrion. J. Mol. Evol. 46: 419–431.

Powers, T.O. and E.G. Platzer, and B.C. Hyman. 1986. Large mitochondrial genome and mitochondrial DNA size polymorphism in the mosquito parasite, *Romanomermis culicivorax*. Curr. Genet. 11: 71–77.

Rosengarten, R.D. and E.A. Sperling, M.A. Moreno, S.P. Leys, and S.L. Dellaporta. 2008. The mitochondrial genome of the hexactinellid sponge *Aphrocallistes vastus*: evidence for programmed translational frameshifting. BMC Genomics 9: 33.

Rot, C. and I. Goldfarb, M. Ilan, and D. Huchon. 2006. Putative cross-kingdom horizontal gene transfer in sponge (Porifera) mitochondria. BMC Evol. Biol. 6: 71.

Shao, Z. and S. Graf, O.Y. Chaga, and D.V. Lavrov. 2006. Mitochondrial genome of the moon jelly *Aurelia aurita* (Cnidaria, Scyphozoa): A linear DNA molecule encoding a putative DNA-dependent DNA polymerase. Gene 381: 92–101.

Shearer, T.L. and M.J. Van Oppen, S.L. Romano, and G. Worheide. 2002. Slow mitochondrial DNA sequence evolution in the Anthozoa (Cnidaria). Mol. Ecol. 11: 2475–2487.

Signorovitch, A.Y. and L.W. Buss, and S.L. Dellaporta. 2007. Comparative genomics of large mitochondria in placozoans. PLoS Genet. 3: e13.

Sinniger, F. and P. Chevaldonne, and J. Pawlowski. 2007. Mitochondrial genome of *Savalia savaglia* (Cnidaria, Hexacorallia) and early metazoan phylogeny. J. Mol. Evol. 64: 196–203.

Smith, A.E. and K.A. Marcker. 1968. N-formylmethionyl transfer RNA in mitochondria from yeast and rat liver. J. Mol. Biol. 38: 241–243.

Snyder, M. and A.R. Fraser, J. Laroche, K.E. Gartner-Kepkay, and E. Zouros. 1987. Atypical mitochondrial DNA from the deep-sea scallop *Placopecten magellanicus*. Proc. Natl. Acad. Sci. USA 84: 7595–7599.

Soma, A. and Y. Ikeuchi, S. Kanemasa, K. Kobayashi, N. Ogasawara, T. Ote, J. Kato, K. Watanabe, Y. Sekine, and T. Suzuki. 2003. An RNA-modifying enzyme that governs both the codon and amino acid specificities of isoleucine tRNA. Mol. Cell. 12: 689–698.

Suga, K. and D.B. Mark Welch, Y. Tanaka, Y. Sakakura, and A. Hagiwara. 2008. Two circular chromosomes of unequal copy number make up the mitochondrial genome of the rotifer *Brachionus plicatilis*. Mol. Biol. Evol. 25: 1129–1137.

Valles, Y. and K.M. Halanych, and J.L. Boore. 2008. Group II introns break new boundaries: presence in a bilaterian's genome. PLoS ONE 3: e1488.

van Oppen, M.J. and J. Catmull, B.J. McDonald, N.R. Hislop, P.J. Hagerman, and D.J. Miller. 2002. The mitochondrial genome of *Acropora tenuis* (Cnidaria; Scleractinia) contains a large group I intron and a candidate control region. J. Mol. Evol. 55: 1–13.

Wallace, D.C. 1982. Structure and evolution of organelle genomes. Microbiol. Rev. 46: 208–240.

Wang, L.S. and R.K. Jansen, B.M. Moret, L.A. Raubeson, and T. Warnow. 2002. Fast phylogenetic methods for the analysis of genome rearrangement data: an empirical study. Proc. 7th Pacific Symp. on Biocomputing (PSB'02) 524–535.

Wang, X. and D.V. Lavrov. 2007. Mitochondrial genome of the homoscleromorph *Oscarella carmela* (Porifera, Demospongiae) reveals unexpected complexity in the common ancestor of sponges and other animals. Mol. Biol. Evol. 24: 363–373.

Wang, X. and D.V. Lavrov. 2008. Seventeen new complete mtDNA sequences reveal extensive mitochondrial genome evolution within the Demospongiae. PLoS ONE 3: e2723.

Weber, F. and A. Dietrich, J.H. Weil, and L. Maréchal-Drouard. 1990. A potato mitochondrial isoleucine tRNA is coded for by a mitochondrial gene possessing a methionine anticodon. Nucleic Acids Res. 18: 5027–5030.

Wolstenholme, D.R. 1992a. Animal mitochondrial DNA: structure and evolution. Int. Rev. Cytol. 141: 173–216.

Wolstenholme, D.R. 1992b. Genetic novelties in mitochondrial genomes of multicellular animals. Curr. Opin. Genet. Dev. 2: 918–925.

Wörheide, G. 2006. Low variation in partial cytochrome oxidase subunit I (COI) mitochondrial sequences in the coralline demosponge *Astrosclera willeyana* across the Indo-Pacific. Mar. Biol. 148: 907–912.

Yokobori, S. and S. Pääbo. 1997. Polyadenylation creates the discriminator nucleotide of chicken mitochondrial tRNA(Tyr). J. Mol. Biol. 265: 95–99.

Chapter 4

Pending Issues in Development and Phylogeny of Arthropods

Jean S. Deutsch

Presentation

Arthropoda is by far the most numerous and most speciose phylum of the animal kingdom. From antiquity, arthropods have attracted the interest of naturalists. During the last century, an arthropod, the fly *Drosophila melanogaster* has been the model animal in many fields of biology, and can be viewed as the 'source' of modern genetics (Morgan 1928) and developmental genetics (Lewis 1978; Nüsslein-Volhard and Wieschaus 1980). Still, at the dawn of the 21st century, many questions remain unsolved about the phylogeny of arthropods, which most often are related to development.

In the present chapter, due to limited space, I will present only some of the pending issues, this choice being obviously personal and arbitrary. In some cases, in addition of presenting pros and cons, I will give my personal opinion, always debatable. I apologize to those who will not find discussion on their preferred group or topic. Because I will often focus on molecular and on developmental data, I will keep aside the fossil taxa, although I acknowledge the value of paleontological data for phylogeny.

UMR 7622 Biologie du Développement, CNRS et Université P. et M. Curie (Paris 6); case 24, 9 quai St-Bernard, 75252 Paris Cedex 05, France.
E-mail: jean.deutsch@snv.jussieu.fr

Homology, Evolution and Development

It is far beyond the scope of the present chapter to discuss exhaustively on what is 'homology', one of the most important, most difficult and debated concept in evolutionary biology. Hereafter, I will discuss homology between different species on a structural basis, by analysing the relationships between the various elements composing the structure of interest. This is valid at every level of biological organisation, from the whole body plan to organs, tissues, cells, molecules. This approach was first developed by Étienne Geoffroy Saint-Hilaire (1818) as *"the principle of connections"*; see Scholtz (2005) for an extended and modern view.

This structural approach is not contradictory with the Darwinian concept of homology as *"similarity that resulted from common descent"* (*"homogeny"* by Lankester 1870). Rather, it provides the best practical criterion for deriving hypotheses on homologous characters (de Beer 1938 p. 70). Indeed, any phylogenetic analysis requires constructing a set (data matrix) of homologous characters between the different taxa under study (primary homology). We all know how much molecular phylogenetic analyses depend on a correct alignment, which is nothing but application of Geoffroy's principle of connections to molecules.

Owen (1849) distinguished *"special homology"* that we can today equate to homology by descent (Amundson 2007), from *"serial homology"*, which is homology between parts within the same individual. Darwin's own definition of homology in the glossary of the 6th edition of the *Origin of Species* (1872) takes both types into account. Serial homology is not due to common descent, but derives from the same processes (including similar genetic networks) in constructing different organs. The same structural analysis can be used to characterize both types of homologues.

The Position of the Arthropoda Within the Animal Kingdom

Cuvier (1817) grouped the Arthropoda with the Annelida, into the 'Articulata', one of his four 'Embranchements' of the animal kingdom. The same idea is reflected in the etymology of the Greek word 'En-tomon' and Latin 'In-secta', or segmented animals.

In 1997, Aguinaldo et al., after the results of a molecular phylogenetic analysis of 18S RNA, coined the name 'Ecdysozoa' for a clade of moulting animals—'ecdysis' is the Greek word for 'moult'—grouping the Introverta (Nielsen 2001) (i.e. Nematoda, Nematomorpha, Priapulida, Kinoryncha and Loricifera)[1] with the Arthropoda and related phyla, Onychophora (velvet worms) (Ballard et al. 1992) and Tardigrada (water bears) (Giribet

---

[1]The phylogenetic affinities of the Gastrotricha are still much debated. They have been placed either without or within the Ecdysozoa or again as their sister-group. This controversy is out of the scope of the present discussion.

et al. 1996). This was the start of the so-called 'new phylogeny' (Adoutte et al. 2000) proposing a major revision of the previous phylogenetic picture of the animal kingdom. Although still debated [see Nielsen (2001, pp. 119, 510–519), Brusca and Brusca (2003, p. 499), Scholtz (2002, 2003), Nielsen 2003, and Jenner and Scholtz (2005) for extensive discussion] the Ecdysozoa grouping gained more and more support among the community of zoologists, on the basis of both molecular and morphological (Schmidt-Rhaesa et al. 1998, Peterson and Eernisse 2001) studies. Curiously, to accommodate the 'Articulata' grouping, Nielsen (2001, 123–137, 215–217) still includes the Arthropoda within the Spiralia, protostome animals whose embryo develops through a spiral cleavage, thus following Anderson (1973). Evidence for arthropods being spiralians has been discussed by Costello and Henley (1976) and evidence against has been reported by Scholtz (1997).

It is worth mentioning the work of de Rosa et al. (1999), who examined Hox genes, developmental genes known to have a critical role in the design of the body plan. The Hox genes' complement in a variety of bilaterian phyla support the dispatching of the Bilateria into three 'super-phyla', namely Deuterostomia, Lophotrochozoa and Ecdysozoa. As for the present purpose, the main outcome of the 'new phylogeny' is the dismissal of the 'Articulata'.

Monophyly of the Arthropoda: a Story of Appendages

'Arthropoda' means 'articulated legs'. It is thus an important diagnostic character. Monophyly of the Arthropoda has been disputed by Sidnie Manton and co-workers (see Fryer 1996, 1997), on the basis of differences in appendages: mainly branching or not of the limbs and articulation of the mandible. Crustacea, possessing limbs with two branches (biramous), would have been issued from an annelid-like stem by a convergent process of 'arthropodization' of a shoot different from that of Onychophora, Hexapoda and Myriapoda, together called 'Uniramia'. Under the polyphyletic hypothesis Chelicerata would be issued from another branch (Fig. 1A).

On the opposite side, homology of the appendages among arthropods has been supported by a structural analysis by Snodgrass (1935, Chap. 5): all arthropod legs are built on a similar pattern of 'podomeres' (leg segments). More recently, evo-devo approaches have shown that the construction of legs in all arthropod classes depends on the same network of interacting transcription factors and signalling molecules as the one which has been dissected in detail in *Drosophila* (Abu-Shaar and Mann 1998): Crustacea (Panganiban et al. 1995, Williams 1998, Nulsen and Nagy 1999, Abzhanov and Kaufman 2000, Williams et al. 2002, Hejnol and Scholtz 2004, Pripc

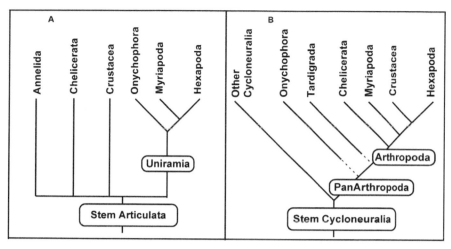

Fig. 1. Polyphyly vs. monophyly of the Arthropoda.
A. The polyphyletic hypothesis. **B.** The monophyletic hypothesis.

and Telford 2008, Sewell et al. 2008, Williams 2008); Myriapoda (Prpic and Tautz 2003, Prpic 2004); Chelicerata (Abzhanov and Kaufman 2000, Prpic et al. 2003). There are some variations in the expression of the corresponding genes along the proximal to distal axis (e.g. Prpic et al. 2003, Popadic 2005) but they do not overcome those observed within the Hexapoda (e.g. Jockusch et al. 2000, Jockusch et al. 2004). This common genetic system, including *wingless* (*wg/Wnt-1*) (Nulsen and Nagy 1999); *Distal-less* (*Dll*) (Panganiban et al. 1995, Williams 1998, Williams et al. 2002, Hejnol and Scholtz 2004, Williams 2008), *extradenticle* (*exd*) (Williams et al. 2002) and *nubbin* (*nub*) (Averof and Cohen 1997, Damen et al. 2002, Popadic 2005), possibly extends to the crustacean limbs whenever uniramous, biramous or phyllopodous; reviewed in Williams (1999).

In the 'Uniramia' concept, homology was assumed between the onychophoran jaw and the mandible of hexapods and myriapods, as opposed to that of the Crustacea (see below). It is now clear that the onychophoran jaw is not homologous with the mandible of myriapods, crustaceans and hexapods. The 'Uniramia' view has now generally been abandoned, see Shear (1992) for detailed discussion of morphological evidence. Most recent molecular phylogenies also support monophyly of the Arthropoda (e.g. Turbeville et al. 1991, Mallatt et al. 2004, Philippe et al. 2005, Dunn et al. 2008).

Panarthropoda: Arthropods Plus Related Groups

The clade grouping Tardigrada and Onychophora with Arthropoda has been called 'Pan-Arthropoda' (e.g. Nielsen 2001), meaning 'all-arthropods'.

However, neither tardigrades nor onychophorans have articulated limbs. A synapomorphy uniting these three phyla is the presence of claws at the distal end of their limbs, as reflected in the word 'Onychophora' (i.e. 'bearing nails'). Both Nielsen (2001) and Brusca and Brusca (2003) consider Tardigrada and Arthropoda as sister-groups on the basis of morphological synapomorphies, some of which, as articulated limbs with intrinsic muscles, are debatable (see Dewell et al. 1993, p. 144).

Contrary to previous claims (Garey et al. 1996, Giribet et al. 1996), recent molecular phylogenies (Philippe et al. 2005, Baurain et al. 2007, Roeding et al. 2007, Timmermans et al. 2008) and a morphological study (Zantke et al. 2008) suggest that the Tardigrada could be closer to the Cycloneuralia (= Gastrotricha + Introverta). Similarity between the brains of onychophorans and cycloneuralians has also been proposed (Eriksson and Budd 2000). Several morphological characters, a heart with ostia and a pericardial septum, formation of the coelom, metanephridia, support a closer relationship between Onychophora and Euarthropoda than between the latter and Tardigrada (Schmidt-Rhaesa 2001). In a recent phylogenomic study (Dunn et al. 2008), Onychophora are closer to the Arthropoda with high support, while the position of Tardigrada, within Cycloneuralia or Pan-arthropoda depends on the phylogenetic method. On the other side, mitochondrial genes' arrangements support the Tardigrada being closer to the Arthropoda than to the Nematoda (Ryu et al. 2007). At present, considering contradictory arguments, the relative position of Onychophora, Tardigrada and Arthropoda is not yet resolved.

It has been suggested that Cycloneuralia are paraphyletic, that is, a monophyletic Pan-arthropoda would be issued from a cycloneuralian ancestor (Garey 2001, Mallatt et al. 2004). In my opinion, this hypothesis is most congruent with both molecular and morphological data. Noteworthy, Cycloneuralia paraphyly implies that the common ancestor of the Arthropoda and of the remaining cycloneuralians presented neither a ganglionic ventral nerve cord nor segmentation, characters that might be correlated (Schmidt-Rhaesa 2001).

Are Pycnogonida the Sister-Group of the Euarthropoda?

'Euarthropoda', meaning 'true' arthropods, comprises Chelicerata, Myriapoda, Hexapoda and Crustacea plus all their fossils and their stem lineage fossils by opposition to some arthropod or arthropod-like fossils. They can be distinguished from the latter and from the Onychophora by loss of the appendage linked to the anterior-most part of the brain, the protocerebrum (Scholtz and Edgecombe 2006). The position of the Pycnogonida (= Pantopoda) (sea spiders) has been a matter of issue (see Dunlop and Arango 2005 for review). Indeed, Zrzavy et al (1997),

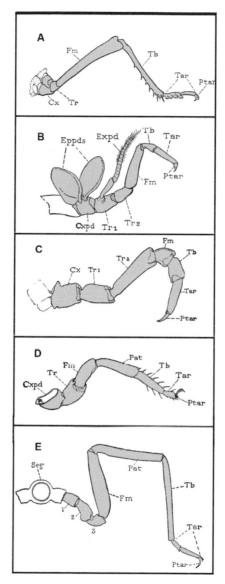

Fig. 2. Arthropod legs.
A. Second leg of a grass hopper (Orthoptera, Hexapoda); **B.** A crustacean biramous limb (*Anaspides*, Malacostraca, Crustacea); **C.** Leg of *Euryurus* (Diplopoda, Myriapoda); **D.** Leg of a solpugid (Arachnida, Chelicerata); **E.** Leg of a sea spider (Pycnogonida, Chelicerata).
(not to scale) Cx: coxa; Cxp: coxopodite; Eppds: epipodites; Expd: exopod; Fm; femur; Pat: patella; Ptar: pretarsus; Seg: body segment; Tar: tarsus; Tb: Tibia; Tr: trochanter; Tr1: first trochanter; Tr2: second trochanter. 1, 2, 3: proximal leg segments (in pycnogonid). All adapted from Snodgrass (1935).

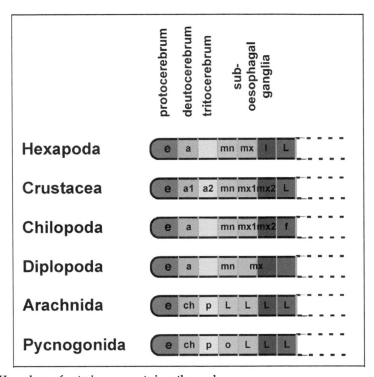

Fig. 3. Homology of anterior segments in arthropods.

Hexapoda: No appendage corresponding to the tritocerebral segment. Crustacea: In many species, the first legs of the thorax (or pereion) are transformed in a pair of maxillipedes, i.e. maxilla-like appendages. Chilopoda: No appendage corresponding to the tritocerebral segment; the first appendage of the trunk is transformed in a pair of forcipules (poison claws). Diplopoda: No appendage corresponding to the tritocerebral segment; the first and second pairs of maxillae are fused; no appendage in the first trunk segment. Arachnida: The deutocerebral segment bears a pair of chelicerae, the tritocerebral segment a pair of pedipalps; Arachnids have four pairs of walking legs. Pycnogonida: The first pair of legs is transformed in a pair of ovigers. Taking into account the ovigers as modified legs, pycnogonids have more legs than arachnids.

a: antennae; a1: first pair of antennae; a2: second pair of antennae; ch: chelicerae; e: eyes; f: forcipules; l: labium; L: legs; mn: mandibles; mx: maxillae; mx1: first pair of maxillae; mx2: second pair of maxillae; o: ovigers; p: pedipalps.

Color image of this figure appears in the color plate section at the end of the book.

Giribet et al. (2001) and Edgecombe (2004) placed them, although with weak support, as sister-group of all Euarthropoda on the basis of a combined molecular plus morphological phylogeny. This was supported by recent data from Maxmen et al. (2005) who claimed that the chelicerae (meaning 'horns in the form of pincers') of the pycnogonids belong to the protocerebral segment, thus excluding them from the Eu-arthropoda. Both morphological (discussed in Manuel et al. 2006) and molecular (Mallatt et al. 2004) data contradict this view. In addition, expression of Hox genes (Manuel et al. 2006, Jager et al. 2006) clearly show that pycnogonids' chelicerae are appendages belonging to the deutocerebral segment, homologous to those of 'true' chelicerates.

Most systematists distinguish the Pycnogonida from all other chelicerates: the group (Pycnogonida + or Euchelicerata) is then called 'Cheliceriformes'. In my opinion, the supposedly synapomorphies of the Euchelicerata, i.e. a prosomal carapace and genitals located in the first or second opisthosomal segment (Brusca and Brusca 2003, p. 697) can be interpreted as characters lost in pycnogonids in relation with their highly modified, and likely derived, body plan. Moreover, molecular phylogenies based on mitochondrial DNA retrieve Pycnogonida, not as a sister-group of all other chelicerates, i.e. Xiphosura (horseshoe crabs) and Arachnida, but instead nested within the Arachnida, as companions of Acari, Scorpiones, Aranae and others (Hassanin et al. 2005, Hassanin 2006, Park et al. 2007).

Monophyly vs. Polyphyly of the Myriapoda?

Monophyly of the Myriapoda has been debated (Dohle 1997, Strausfeld 1998, Kraus 2001). The Myriapoda comprise four classes, two 'major' ones, i.e. Chilopoda (centipedes) and Diplopoda (millipedes), and two 'minor' ones, comprising less species, also of smaller size, the Symphyla and the Pauropoda. The latter two are usually united with the Diplopoda into the Progoneata clade. The heart of the issue is that no convincing

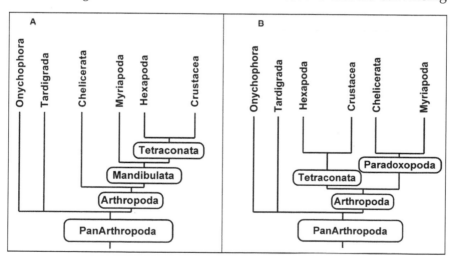

Fig. 4. Mandibulata vs. Paradoxopoda.
A. Phylogeny of the Arthropoda under the Mandibulata hypothesis; B. Phylogeny of the Arthropoda under the Paradoxopoda hypothesis.

morphological synapomorphy unites the Chilopoda with the other three myriapod classes (Dohle 1997). In addition, cladistic analyses of brain characters resulted in 'Myriapoda' being a biphyletic assemblage, with Diplopoda appearing as the earliest branch of the arthropod tree and Chilopoda as the sister-group of the other mandibulates (Strausfeld 1998, Loesel et al. 2002). In contrast, the pattern of serotonergic neurons of the Diplopoda displays a closer similarity to that of the Chilopoda than to that of any other arthropod group (Harzsch 2004, p. 209). From an extensive cladistic analysis of morphological and molecular characters, Edgecombe (2004) concluded in favour of the monophyly of Myriapoda, and underlined several synapomorphies, such as the structure of the tentorium (a piece of the cephalic skeleton), the secondary division of the gnathobasic mandible in two pieces, with corresponding muscles.

On the molecular side, most recent phylogenies with various molecules and methods consistently retrieve a myriapod clade comprising both millipedes and centipedes (Elongation factor 2: Regier and Shultz 2001; both 18S and 28S rRNA: Mallatt et al. 2004; EF1-alpha: Pisani 2004; 9 nuclear and 15 mitochondrial genes: Pisani et al. 2004; three nuclear protein genes: Regier et al. 2005; mitochondrial DNA: Hassanin 2006, *contra* Negrisolo et al. 2004); see discussion of both morphological and molecular phylogenies in Bitsch and Bitsch 2004).

Mandibles: Paradoxopoda vs. Mandibulata

As shown by their name, chelicerates bear chelicerae related to the deutocerebrum, whereas the other three sub-phyla bear a pair of antennae at the same location. In addition to a pair of antennae on the deutocerebral segment, all three groups share a pair of mandibles on the first post-oral (gnathal) segment, whereas the chelicerates exhibit their first pair of legs, hence the name 'Mandibulata' for the grouping of Myriapoda, Crustacea and Hexapoda. Homology of the anterior segments between the Chelicerata and Mandibulata is now conclusively supported by evidence for the presence of a deutocerebrum in chelicerates (Mittmann and Scholtz 2003), contrary to previous thoughts (e.g. Bullock and Horridge 1965, vol. II, chap. 16, p. 821, Wegerhoff and Breidbach 1995), and from the expression of Hox genes in chelicerates belonging to various groups (Telford and Thomas 1998, Damen et al. 1998, Jager et al. 2006). In addition, it is now clear that all cephalic appendages are serially homologous to the legs. This is based in particular on the use of a common network of developmental genes to build these appendages (Casares and Mann 1998, 2001, Ronco et al. 2008). Hence, a primitive leg-like appendage corresponding to the deuterocerebral segment (Mittmann and Scholtz 2003, Scholtz and Edgecombe 2006) has evolved as a chelicera in chelicerates (Abzhanov and Kaufman 2000, Prpic and Damen 2004) and as an antenna in the other arthropod sub-phyla. In addition, a primitive leg-like appendage has been transformed into a mandible altogether in myriapods, crustaceans and hexapods. Homology between mandibles of these three groups is not only supported by their segmental position, but also by their detailed structure (Edgecombe et al. 2003), which makes convergent evolution unlikely.

However, the clade 'Mandibulata' has recently been rejected by molecular phylogenies based on various molecules: hemocyanin (Kusche and Burmester 2001); 28S and 18S ribosomal RNA (Mallatt et al. 2004); mitochondrial DNA (Hwang et al. 2001, Hassanin et al. 2005, Hassanin 2006), Hox genes (Cook et al. 2001), various nuclear genes (Pisani et al. 2004). Indeed, Myriapoda, instead of grouping with Crustacea and Hexapoda, were found as sister-group of the Chelicerata. This clade was coined 'Paradoxopoda' by Mallatt et al. (2004), also called 'Myriochelata' (Pisani et al. 2004). To my knowledge, only a few developmental-morphological similarities have been brought in support to the Paradoxopoda. In both Myriapoda and Chelicerata neurogenesis does not proceed with the formation of single neuroblasts as in Hexapoda and Crustacea (Gerberding 1997, Harzsch et al. 1998, Gerberding and Scholtz 2001, Harzsch 2001, Ungerer and Scholtz 2008), but by clusters of neurogenic cells, (Mittmann 2002, Stollewerk 2002, Stollewerk et al. 2003, Dove and Stollewerk 2003, Kadner and Stollewerk 2004, Stollewerk and Simpson 2005, Stollewerk and Chipman 2006). The structure of midline brain neuropils is similar in chelicerates and chilopods (but not diplopods) (Strausfeld 1998, Loesel et al. 2002). However, these traits may well be plesiomorphic, not shared derived properties, thus with no value to unite Myriapoda and Chelicerata in a convincing clade (Strausfeld 1998, Harzsch et al. 2005, Stollewerk and Chipman 2006, Stollewerk 2008). Following the Paradoxopoda hypothesis, the antenna, the mandible and the maxillae found in both Myriapoda and Tetraconata (= Crustacea + Hexapoda) have to be interpreted not as homologous organs, but due to evolutionary convergences.

Prpic and Tautz (2003) argued that the gnathobasic mandible could be convergent between myriapods and tetraconates in view that the maxilla also is gnathobasic in myriapods but is not in crustaceans and hexapods. However, this leaves aside the fact that in chelicerates, the corresponding appendages are neither gnathobasic nor telobasic mouthparts, they are legs actually. Hence, the myriapod mandible and maxilla are both derived appendages, and the tetraconate mandible and maxilla are derived as well. This does not exclude convergence, but does not allow comparison between the fate of mandibles and that of maxillae.

Another character of interest is provided by the leg's architecture. Although all arthropod legs are based on a roughly similar plan (see above), an additional podomere, the 'patella', is intercalated between femur and tibia in chelicerates. Pycnogonid legs also possess a patella, but differ from the other chelicerates' legs by additional segments at the proximal end (Snodgrass 1935, Chap. 5). In all other arthropods, a patella is absent from legs. Phylogenetic interpretation of this character is equivocal. If the arthropod primitive state does not include a patella, it can be viewed as a synapomorphy of the Chelicerata (including Pycnogonida). If the presence of a patella is plesiomorphic, its loss can be taken either as a synapomorphy of the Mandibulata or as a homoplasy.

We are thus faced to severe discrepancies between molecules and morphology. As there is but a single evolutionary history, one of the two evolutionary hypotheses, 'Mandibulata' vs. 'Paradoxopoda', must be wrong. In recent papers, Telford et al. (2008) and Rota-Stabelli and Telford (2008) showed that the molecular phylogeny of the diverse phyla and classes among the Panarthropoda strongly depends on the outgroup choice. When the Priapulida, a member of the Cycloneuralia with slow evolutionary rate, is taken as outgroup, then molecular phylogenies support the Mandibulata, reconciling molecular with morphological evidence.

Hexapods are Terrestrial Crustaceans

The idea that the Crustacea could be paraphyletic can be traced back as early as 1904 (Lankester 1904). E. Ray Lankester was the first to classify horseshoe crabs (*Limulus*) not within the Crustacea, as previously thought, but within the Chelicerata. In his view, all mandibulates were issued from a crustacean ancestor. More recently, Briggs and Fortey (1989), on the basis of a cladistic analysis of Cambrian and extant arthropods, pushed paraphyly in as much as to group all arthropod classes, including the Trilobita and the Chelicerata, as issued from a crustacean-like ancestor. However, this view received little agreement among zoologists, until the paraphyly of the Crustacea was resurrected in its present avatar: inclusion of the Hexapoda only within the Crustacea. Reviews and debates can be found in Fortey and Thomas (1997) and Deuve (2001).

The so-called 'Atelocerata' or 'Tracheata' assemblage (Myriapoda + Hexapoda) was supposed to be supported by morphological traits, such as tracheae and Malphighian tubules, absent from the Crustacea. However, these characters are thought to be convergences due to terrestrial life. Indeed, similar structures are also found in other panarthropod groups. Loss of the second antenna could be homoplastic, as often for character loss. The last character of contention was the mandible, which would be 'gnathobasic' in crustaceans, biting with the base of the mandibles, whereas myriapods and insects would have whole-limb mandibles that bite with their tips (so-called 'telognathic') (Fryer 1996 p. 2–7). However, recent works, using an antibody against the Distal-less protein known to be involved in appendage development (Panganiban et al. 1997), on a variety of myriapod, crustacean and hexapod species (Popadic et al. 1996, 1998, Scholtz et al. 1998, Prpic 2004), showed that mandibles of all mandibulates are of the gnathobasic type, thus supporting their homology (Scholtz 2001). This led to the demise of the 'Atelocerata' (Telford and Thomas 1995; but see in contrast Bitsch and Bitsch 2004).

At the same time, molecular phylogenies began to give support to a closer relation between Crustacea and Hexapoda than between the latter and Myriapoda (Turbeville et al. 1991, Boore et al. 1995, Friedrich and Tautz 1995). These first studies were based on a small amount of data and a very small sample of arthropod species. The close relation between hexapods and crustaceans has later been vindicated by numerous studies on a variety of taxa and molecules: ribosomal RNAs (Mallatt et al. 2004); various nuclear coding genes (Shultz and Regier 2000, Regier and Shultz 2001, Cook et al. 2001, Giribet et al. 2001, Pisani 2004, Baurain et al. 2007); mitochondrial DNA (Boore et al. 1998, Garcia-Machado et al., 1996, 1999, Wilson et al. 2000, Hwang et al. 2001, Lavrov et al. 2004, Cook et al. 2005, Hassanin 2006, Lim and Hwang 2006).

A number of morphological/developmental characters are shared by hexapods and crustaceans at the exclusion of chelicerates and myriapods. Most of them are related to the nervous system. In both Hexapoda and Crustacea, neurogenesis proceeds with the primary formation of neuroblasts, which are stem cells giving rise to neurons, glia and sensory cells. Moreover, based on their origin and position, some individual neuroblasts (Ungerer and Scholtz 2008) and a number of pioneer neurons are clearly homologous in both groups (Whitington 1995, 1996). They give rise to homologous nerve tracts (Harzsch 2004). At least some ganglia and neuropils composing the brain are also homologous (in addition to papers published in Fortey and Thomas (1997) and Deuve (2001), see Osorio and Bacon 1994, Osorio et al. 1995, Strausfeld 1998). In this respect, the visual system (eyes and optic lobes) is of particular interest.

Looking at Arthropod Eyes: Homology of Arthropod Compound Eyes?

Opticians distinguish 'conventional' cameras with a single aperture to 'compound' ones, with multiple apertures. The facetted compound eye of arthropods seems such an extraordinary organ as it deserves particular attention. Compound eyes are found in all four sub-phyla of the Arthropoda. However, they are found in a single group of chelicerates, the Xiphosura (horse-shoe crabs, e.g. *Limulus*), and in a single group of myriapods, the Scutigeromorpha (a family of chilopods, e.g. the centipede *Scutigera*). On the opposite, leaving aside blind animals, almost all crustaceans and hexapods have compound eyes.

Reciprocally, not all arthropod eyes are compound eyes. In addition to compound facetted eyes, simple eyes are present in arthropods. 'Simple' does not mean an eye with a single photoreceptor unit. Our own eyes are simple eyes; still our retina contains millions of photoreceptor cells. The simple ventral eye of the horseshoe crab *Limulus* contains about 300 photoreceptor cells (Battelle 2006). The 'simple' frontal eyes of wolf spiders possess quite a sophisticated optic apparatus (Williams and McIntyre 1980). Besides simple and compound eyes, I distinguish a third type that I call 'multiple eyes'. They are formed by groups of ocelli (small eyes). They are diversely located, at the lateral sides of the head as the stemmata of holometabolous insect larvae, or at the dorsal side as the median eye of the crustacean nauplius larva. These three different types of eyes may be found in the same species, either simultaneously or at different stages of its life cycle.

It is far beyond the scope of the present chapter to review in detail the diversity of arthropod eyes. In the following sections, I will focus on a single issue, related to the present topic: Are compound eyes homologous throughout the Arthropoda?

The classical view posits that the compound eye is a synapomorphy of the Arthropoda (e.g. Nielsen 2001). By itself the diversity of arthropod eyes and their dispatching between the various taxonomic groups question this hypothesis: homology implies that the common ancestor of all arthropods possessed a compound eye, then, loss would have happened a number of times during the evolution of the Arthropoda. Conversely, non-homology implies that compound eyes would have been designed *de novo* several times.

The basic unit of compound eyes is the ommatidium, a complex structure comprising several photoreceptor neurons (retinula cells) and other components such as non-photoreceptor pigment cells, lenses and accessory cells. Although by etymology 'ommatidium' means 'small eye', a compound eye is not the mere juxtaposition of small simple eyes. The intricate complexity of neuronal pathways of in the optic lobes as studied in *Drosophila* (Meinertzhagen and Hanson 1993) shows that vision is the result of integration of inputs from several ommatidia, if not from the whole eye. In other words, an ommatidium isolated from a compound eye cannot 'see' anything by itself. To discuss homology between the various arthropod compound eyes, it is thus necessary to examine the relations between the different elements of the ommatidium *and* the neuronal connections between retina and brain.

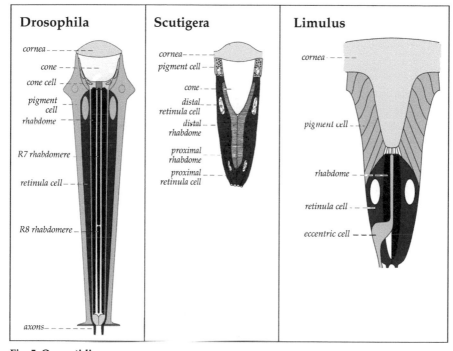

Fig. 5. Ommatidia.

A. The tetraconate ommatidium: ommatidium of the *Drosophila* eye; simplified from Wolff and Ready (1993); **B.** Ommatidium of the *Scutigera* eye; adapted from Paulus (2000); **C.** Ommatidium of the *Limulus* eye, adapted from Nilsson and Kelber (2007). (not to scale).

The structure of the ommatidium in compound eyes

Ommatidia from crustacean and hexapod compound eyes are strikingly similar (Melzer et al. 1997, Dohle 2001). Each ommatidium contains eight retinula cells (neuronal photoreceptor cells), in which the accumulation of opsin forms a rhabdome along the proximal to distal axis. At the distal part of the ommatidium, under the cornea, a crystalline cone forms a lens, produced by four cone cells. This led Dohle to propose the name 'Tetraconata' for the clade [Crustacea + Hexapoda] (Dohle 2001, Richter 2002), also called 'Pancrustacea'. This pattern of four cone cells and eight retinula cells is fairly constant in hexapods and malacostracans, but some minor variations can be found in non-malacostracan crustaceans (Hallberg and Elofsson 1983, Dohle 2001, Richter 2002, Oakley and Cunningham 2002, reviewed in Bitsch and Bitsch 2005, discussion in Nilsson and Kelber 2007). Paulus (2000, p. 190) concluded: "*the ommatidia of the Crustacea and Insecta are clearly homologous*". Structural similarities of the crustacean and hexapod eyes result from closely similar developmental processes (reviewed in Harzsch and Hafner 2006). The parallel between crustacean and insect eye formation has been extended to the developmental genetic level by the similar role of the *wingless* gene in eye growth in *Drosophila* and in the malacostracan crustacean *Mysidium* (Duman-Scheel et al. 2002).

The structure of the compound eye of scutigeromorph myriapods has been reviewed and revisited in detail by Müller et al (2003). The number and composition of cell types varies within the eye (central to marginal). Like the crustacean and hexapod ommatidia, *Scutigera*'s ommatidium comprises a crystalline cone made by four (sometimes five) cone cells. On this respect it deserves the name of 'tetraconate'. It differs from the pancrustacean ommatidium by a two-layered rhabdome: a distal open rhabdome made by 9 to 12 retinulae and a proximal closed rhabdome made by four retinula cells, instead of the single-layered rhabdome made by 8 retinula cells in Pancrustacea. Inter-ommmatidial cells are present. Some of which are pigment cells that isolate ommatidia from each other as in crustacean/hexapod eyes, while others are glandular cells, specific to scutigeromorph eyes. Summarising, the ommatidium of the scutigeromorph compound eye, although sharing some similarities, presents peculiar features that largely outcome the range of variations seen among ommatidia of crustaceans and hexapods.

The structure of the xiphosuran ommatidium is quite different (Fahrenbach 1968, 1969, reviewed in Paulus 2000, Bitsch and Bitsch 2005, Battelle 2006). First, there is no typical crystalline cone. Indeed, there is no differentiation between the cornea and the 'cone'. The cone of the *Limulus'* ommatidium is but an expansion of the cuticular cornea (Fahrenbach 1968). The number of retinula cells is variable, from 4 to as many as 20 (average 10–13) per ommatidium (Fahrenbach 1969). These retinula cells form a rhabdome in the centre of which a neuron, so-called 'eccentric cell' is located (sometimes two of them when the number of rhabdomeric cells is high). This eccentric cell is never seen in other arthropod compound eyes. The arrangement of the ommatidia is variable, not forming a regular hexagonal pattern on the whole surface of the eye (Fahrenbach 1969, Harzsch et al. 2006). These structural aspects result from differences in development (Harzsch et al. 2006, Harzsch and Hafner 2006).

Summarising, the ommatidium of the compound eye of the Xiphosura is so different from those of the other arthropods that, on both structural and developmental criteria, it is likely evolutionary convergent. This is in agreement with both the 'Paradoxopoda' and the 'Mandibulata' phylogenetic hypotheses. The great similarity of the ommatidium between Crustacea and Hexapoda brings support to the 'Tetraconata/Pancrustacea' hypothesis. The scutigeromorph ommatidium presents some similarities with the pan-crustacean one. This leaves the position of the Myriapoda equivocal: if the scutigeromorph compound eye and ommatidium are thought to be part of the myriapod ground plan, this would favour the

'Mandibulata' (with loss of the compound eye in the other myriapod groups), whilst if on the opposite they are taken as scutigeromorph apomorphies within the Myriapoda, these characters will be of no value in the 'Paradoxopoda' vs. 'Mandibulata' issue.

Optic Lobes in Crustaceans and Hexapods

Hanström (1947, p. 42) underlined the similarity between the pattern of neuropils and nerve nets in hexapod and malacostracan optic lobes and in more recent times Osorio drew attention to its importance for phylogenetic relationships (Osorio and Bacon 1994, Osorio et al. 1995, Nilsson and Osorio 1997). Both groups of arthropods possess (at least) three neuropils that directly or indirectly receive neuronal inputs from the retina, i.e. going from the distal side (retina) to the proximal one (protocerebrum proper): lamina, medulla and lobula. Nerve tracks between these neuropils form two chiasmata, one between the lamina and the medulla and a second between the medulla and the lobula. Such an organization is never found in any other arthropod, including non-malacostracan crustaceans, so-called 'entomostracans'. Indeed, the latter have only two neuropils with no neuronal crossing in between.

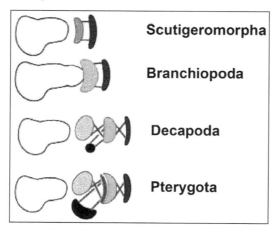

Fig. 6. Optic lobes: neuropils and projections from the retina.
Scutigeromorpha: a family of chilopods (centipedes), see text; Branchiopoda: a class of non-malacostracan crustaceans, e.g. *Artemia, Triops, Daphnia*; Decapoda: the most numerous class of malacostracan crustaceans, e.g. lobsters, crabs, crayfish; Pterygota: winged insects. In Scutigeromorphs and in non-malacostracan crustaceans, the lamina, the first optic neuropil, is followed by an unique second neuropil, with no chiasma in between. In malacostracans and in hexapods, the lamina is followed through a chiasma by the medulla (divided in medulla interna and medulla externa in insects), which in turn is followed by the lobula through a second chiasma and by a fourth neuropil, the deep lobula (malacostracans) or lobula plate (insects). Simplified from Sinakevitch et al. (2003).

Sinakevitch et al. (2003) re-examined in detail the morphology of the optic lobes in a variety of crustaceans and insects. They confirmed that the architecture of the optic lobes of branchiopod crustaceans and of scutigeromorphs greatly differs from that of malacostracans and hexapods. The main outcome of this study is that, in addition to the 'classical' lamina, medulla and lobula, a fourth neuropil is retrieved in these latter two groups, called the lobula plate in dipteran insects and lobula satellite, sublobula or deep lobula in the other species. Despite some diversity, a common and likely homologous plan can be derived.

Strikingly similar morphogenetic movements during development of the optic lobes in the malacostracan shrimp *Penaeus* (Elofsson 1969) and in the fly *Drosophila* (Meinertzhagen and Hanson 1993) result in the formation of chiasmata. The process of optic lobe development is different in branchiopods (Elofsson and Dahl 1970, Harzsch and Walossek 2001). Although convergence cannot be formally excluded (Bitsch and Bitsch 2005), their development and the relations between their various elements strongly argue in favour of homology of optic lobes between malacostracans and hexapods; for a more complete survey see Harzsch (2002).

On the molecular phylogenetic side, a closer relationship between the Hexapoda and the Malacostraca is found only with mitochondrial DNA (Garcia-Machado et al. 1996, 1999, Wilson et al. 2000, Lim and Hwang 2006), not with phylogenies based on other molecules.

Are Hexapoda Polyphyletic? The Collembolan Controversy

In 2003, Nardi et al. claimed that the 'Hexapoda' were a polyphyletic assemblage. Indeed, in their phylogenetic analysis of mitochondrial DNA, the Collembola (springtails) stemmed from the base of the Tetraconata clade, leaving the Crustacea and the rest of the hexapods as sister-groups (Fig. 7A). The collembolans, with proturans and diplurans, belonging to two other orders of wingless hexapods, have their mouth pieces located inside a cephalic cavity, contrary to all other hexapods, in which the mouth pieces can be seen outside the head capsule, thus forming the Ectognatha (= Insecta *sensu stricto*) clade. However, the relative positions of the three endognathous orders at the base of the hexapod phylogenetic tree are still unresolved or controversial.

Nardi et al.'s analysis soon raised critiques about biases, mainly because of heterogeneity in base composition of the mtDNA sequences used (Delsuc et al. 2003). Later, Cook et al. (2005) phylogenetic analyses of mtDNA sequences also suggested a polyphyletic 'Hexapoda' but with a pattern different from that of Nardi et al.: Collembola branched this time not as a sister-group of all other tetraconates, but as a sister group to the

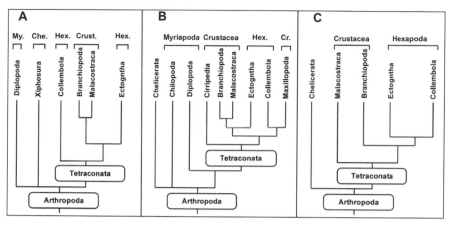

Fig. 7. The position of the Collembola within the Arthropoda.
A. Polyphyletic hypothesis 1, based on mtDNA (simplified from Nardi et al. 2003, Fig. 2);
B. Polyphyletic hypothesis 2, based on mtDNA (simplified from Cook et al. 2005, Fig. 2);
C. Monophyletic hypothesis, based on ribosomal proteins (simplified from Timmermans et al. 2008, Fig. 2).

so-called 'maxillopod' crustaceans, while the 'true' Insecta grouped with branchiopod and malacostracan crustaceans (Fig. 7B). Data from mtDNA alone cannot resolve the relationships between Crustacea, Collembola and Insecta (Cameron et al. 2004), even when the compositional bias of the mtDNA is taken into account (Holland et al. 2005, Hassanin et al. 2005, Hassanin 2006). In contrast, ribosomal RNA-based phylogenies (Kjer 2004, Mallatt et al. 2004, Luan et al. 2005, Mallatt and Giribet 2006, Misof et al. 2007) and nuclear protein-coding genes (Regier et al. 2005, Timmermans et al. 2008) retrieve Hexapoda, including Collembola, monophyletic with high support (Fig. 7C).

Development and Body Plan of the Collembolans

In fact, the idea that collembolans do not belong to the Hexapoda was not new (see Kristensen 1995 and Bitsch et al. 2004 for references and discussion, with morphological data supporting hexapod monophyly). Whatsoever, the development and the body plan of the Collembola are worth examining. They are unique in presenting a total cleavage (Jura 1972), while the early development of all other hexapod species begins with superficial (partial) cleavage, the first divisions of the zygote forming a syncytial embryo. However, superficial cleavage is found in several crustacean species (Anderson 1973, p. 278), making the type of cleavage a homoplastic character (see discussion in Scholtz 1997). Collembolan embryos possess appendage buds not only on the thorax, but also on all abdominal segments. However, this again is not specific to collembolans, the same being found in other hexapod orders, including within the Ectognatha, such as the Archeognatha (Jura 1972).

Regarding their body plan, collembolans possess three pairs of legs, thus deserving the name of 'hexapods'. This is not as self-evident as it may seem: indeed, although the crustacean classes present a wide panel of tagmata, segments with and/or without trunk appendages in varying numbers, there is no crustacean with three pairs of legs. More importantly, collembolans are unique among the Hexapoda in having at most six abdominal segments, whereas the basic hexapod number is 11 (Brusca and Brusca 2003). As adults, they bear a curious appendage, the collophore, on the first abdominal segment, another, the retinaculum, on the third abdominal segment, and on the fourth a special appendage, the furcula, used for jumping, hence their name of 'springtails'.

Summary and Conclusions

In the present chapter, I have addressed the following issues:

1. What is the position of the Arthropoda in the animal kingdom?
2. Are Arthropoda monophyletic?
3. Which, between Tardigrada and Onychophora, is the sister group of the Euarthropoda?
4. Are Pycnogonida the sister-group of the Euarthropoda?
5. What are the relations between the four arthropod subphyla, i.e. 'Paradoxopoda' *vs.* 'Mandibulata'?
6. Are Myriapoda monophyletic?
7. Are hexapods terrestrial crustaceans?
8. Are Hexapoda monophyletic?

In my opinion, some of these questions have received an answer thanks to advances during the recent decades in phylogenetic analyses, in morphology due to new and improved techniques, and in comparative studies of development, including in their genetic aspects ('evo-devo'). I favour those answers where there is congruence between molecules and morphology, 'evo-devo' being integration of both molecular and morphological studies. So, I think there is enough consistent data favouring the following answers:

1. Arthropoda is a member of the Ecdysozoa, probably issued from a cycloneuralian stem.
2. Arthropoda is a monophylum, with little doubt.
4. Pycnogonida is a member of the Chelicerata, not the sister-group of all other 'Euarthropoda'.
5. Mandibulata is a monophyletic group, Myriochelata/Paradoxopoda is likely a methodological artefact in molecular analyses.

7. Hexapoda is issued from a crustacean stem, thus making the 'Crustacea' *sensu stricto* a paraphyletic assemblage.
8. Hexapoda is a monophylum, claims for polyphyly are likely due to reconstruction artefacts in molecular phylogenies.

Some questions will hopefully receive a clear answer within the next few years:

3. Which, Onychophora or Tardigrada, is the sister group of Euarthropoda?

   At present, this issue is still open. Answering requires morphological investigations on more species, molecular analyses with more species and molecules using as out-groups members of the Cycloneuralia.

4. Are Pycnogonida the sister-group of all other chelicerates or are they nested within the Arachnida?

   The precise position of the Pycnogonida needs to be re-examined on a molecular ground using various molecules, not only mitochondrial, more sea-spider species, more other chelicerate species and appropriate out-groups. Morphological characters have also to be evaluated more precisely with more data using modern methods. Whatsoever, there is little doubt that Pycnogonida do belong to the Chelicerata, and thus the term 'Cheliceriformes' is obsolete.

6. Are the Myriapoda monophyletic? In my opinion, the question is still open. More studies are needed, with morphological, molecular and evo-devo approaches, on more 'myriapod' species, combined with parallel studies on closer and multiple outgroups.

7. Are Malacostraca the closest relatives to the Hexapoda?

   Morphology and development, most of the nervous and ocular systems, provide convincing data for such a relation. However, it is congruent only with mitochondrial DNA-based molecular phylogenies, which is unsatisfactory. It should be said that molecular phylogenies have been up to now desperately unable to resolve with confidence, because of lack of resolution and/or inconsistency between the various analyses/authors, the relationships not only between hexapods and crustaceans, but also between the various crustacean groups. Endeavour in this field is awaited.

Acknowledgements

Many thanks to Gerhard Scholtz for his comments and advice on a previous draft of this manuscript and to Louis Deharveng and Cyrille D'Haese for reading and help on the collembolan section.

References

Abu-Shaar, M. and R.S. Mann. 1998. Generation of multiple antagonistic domains along the proximodistal axis during *Drosophila* leg development. Development 125: 3821–3830.

Abzhanov, A. and T.C. Kaufman. 2000. Homologs of *Drosophila* appendage genes in the patterning of arthropod limbs. Dev. Biol. 227: 673–689.

Adoutte, A. and G. Balavoine, N. Lartillot, O. Lespinet, B. Prud'homme, and R. de Rosa. 2000. The new animal phylogeny: reliability and implications. Proc. Natl. Acad. Sci. USA 97: 4453–4456.

Aguinaldo, A.M. and J.M. Turbeville, L.S. Linford, M.C. Rivera, J.R. Garey, R.A. Raff, and J.A. Lake. 1997. Evidence for a clade of nematodes, arthropods and other moulting animals. Nature 387: 489–493.

Amundson, R. Richard Owen and animal form. pp. XV-LI. *In:* R. Amundson. [ed.] 2007. R. Owen. On the nature of limbs. Chicago University Press, Chicago.

Anderson, D.T. 1973. Embryology and Phylogeny in Annelids and Arthropods. Pergamon Press, Oxford.

Averof, M. and S.M. Cohen. 1997. Evolutionary origin of insect wings from ancestral gills. Nature 385: 627–630.

Ballard, J.W.O. and G.J. Olsen, D.P. Faith, W.A. Odgers, D.M. Rowell, and P.W. Atkinson. 1992. Evidence from 12S ribosomal RNA sequences that Onychophorans are modified arthropods. Science 258: 1345–1348.

Battelle, B.A. 2006. The eyes of *Limulus polyphemus* (Xiphosura, Chelicerata) and their afferent and efferent projections. Arthropod. Struct. Dev. 35: 261–274.

Baurain, D. and H. Brinkmann, and H. Philippe. 2007. Lack of resolution in the animal phylogeny: closely spaced cladogeneses or undetected systematic errors? Mol. Biol. Evol. 24: 6–9.

Bitsch, C. and J. Bitsch. 2004. Phylogenetic relationships of basal hexapods among the mandibulate arthropods: a cladisitic analysis based on comparative morphological characters. Zool. Script. 33: 511–550.

Bitsch, C. and J. Bitsch. Evolution of eye structure and arthropod phylogeny. pp. 185–214. *In:* S. Koenemann and R.A. Jenner. [eds.] 2005. Crustacea and Arthropod Relationships. CRC Press, Boca Raton, USA.

Boore, J.L. and T.M. Collins, D. Stanton, L.L. Daehler, and W.M. Brown. 1995. Deducing the pattern of arthropod phylogeny from mitochondrial DNA rearrangements. Nature 376: 163–165.

Boore, J.L. and D.V. Lavrov, and W.M. Brown. 1998. Gene translocation links insects and crustaceans. Nature 392: 667–668.

Briggs, D.E.G. and R.A. Fortey. 1989. The early radiation and relationships of the major arthropod groups. Science 246: 241–243.

Brusca, R.C. and G.J. Brusca. 2003. Invertebrates. Sinauer, Sunderland MA, USA.

Bullock, T.H. and G.A. Horridge. 1965. Structure and Function in the Nervous System of Invertebrates. Freeman & Co., San Francisco, USA.

Casares, F. and R.S. Mann. 1998. Control of antennal versus leg development in *Drosophila*. Nature 392: 723–726.

Casares, F. and R.S. Mann. 2001. The ground state of the ventral appendage in *Drosophila*. Science 293: 1477–1480.

Cook, C.E. and M.L. Smith, M.J. Telford, A. Bastianello, and M. Akam. 2001. Hox genes and the phylogeny of the arthropods. Curr. Biol. 11: 759–763.

Cook, C.E. and Q. Yue, and M. Akam. 2005. Mitochondrial genomes suggest that hexapods and crustaceans are mutually paraphyletic. Proc. Biol. Sci. 272: 1295–1304.

Costello, D.P. and C. Henley. 1976. Spiralian development: A perspective. Am. Zool. 16: 277–291.

Cuvier, G. 1817. Le règne animal, distribué d'après son organisation, pour servir de base à l'histoire naturelle et d'introduction à l'anatomie comparée. Deterville, Paris, France.

Damen, W.G. and M. Hausdorf, E.A. Seyfarth, and D. Tautz. 1998. A conserved mode of head segmentation in arthropods revealed by the expression pattern of Hox genes in a spider. Proc. Natl. Acad. Sci. USA 95: 10665–10670.

Damen, W.G. and T. Saridaki, and M. Averof. 2002. Diverse adaptations of an ancestral gill: a common evolutionary origin for wings, breathing organs, and spinnerets. Curr. Biol. 12: 1711–1716.

Darwin, C. 1872. The Origin of Species. Murray, London, UK.

de Beer, G.R. Embryology and evolution. pp. 57–78. *In:* G.R. de Beer. [ed.] 1938. Evolution. Essays on Aspects of Evolutionary Biology. Clarendon Press, Oxford, UK.

de Rosa, R. and J. Grenier, T. Andreeva, C. Cook, A. Adoutte, M. Akam, S. Carroll, and G. Balavoine. 1999. Hox genes in brachiopods and priapulids and protostome evolution. Nature 399: 772–776.

Deuve, T. 2001. The origin of the Hexapoda. Ann. Soc. Entomol. Fr. vol 37. Paris, France.

Dewell, R.A. and D.R. Nelson, and W.C. Dewell. Tardigrada. pp. 143–183. *In:* F.W. Harrison [ed.] 1993. Microscopic Anatomy of Invertebrates. Wiley-Liss, New York.

Dohle, W. Myriapod-insect relationships as opposed to an insect-crustacean sister group relationship. pp. 305–315. *In:* R.A. Fortey and R.H. Thomas. [eds.] 1997. Arthropod Relationships. Chapman & Hall, London.

Dohle, W. 2001. Are the insects terrestrial crustaceans? A discussion of some new facts and arguments and the proposal of the proper name 'Tetraconata' for the monophyletic unit Crustacea + Hexapoda. Ann. Soc. Entomol. Fr. 37: 85–103.

Dove, H. and A. Stollewerk. 2003. Comparative analysis of neurogenesis in the myriapod *Glomeris marginata* (Diplopoda) suggests more similarities to chelicerates than to insects. Development 130: 2161–2171.

Duman-Scheel, M. and N. Pirkl, and N.H. Patel. 2002. Analysis of the expression pattern of *Mysidium columbiae wingless* provides evidence for conserved mesodermal and retinal patterning processes among insects and crustaceans. Dev. Genes Evol. 212: 114–123.

Dunlop, J.A. and C.P. Arango. 2005. Pycnogonid affinities: a review. J. Zool. Syst. Evol. Research 43: 8–21.

Dunn, C.W. and A. Hejnol, D.Q Matus, K. Pang, W.E. Browne, S.A. Smith, E. Seaver, G.W. Rouse, M. Obst, G.D. Edgecombe, et al. 2008. Broad phylogenomic sampling improves resolution of the animal tree of life. Nature 452: 745–749.

Edgecombe, G.D. and S. Richter, and G.D.F. Wilson. 2003. The mandibular gnathal edges: Homologous structures throughout Mandibulata? African Invert. 44: 115–135.

Edgecombe, G.D. 2004. Morphological data, extant Myriapoda and the myriapod stem group. Contrib. Zool. 73 (3): 2.

Elofsson, R. 1969. The development of the compound eyes of *Penaeus duorarum* (Crustacea: Decapoda) with remarks on the nervous system. Z. Zellforsch. Mikrosk. Anat. 97: 323–350.

Elofsson, R. and E. Dahl. 1970. The optic neuropiles and chiasmata of Crustacea. Z. Zellforsch. Mikrosk. Anat. 107: 343–360.

Elofsson, R. 2006. The frontal eyes of crustaceans. Arthropod. Struct. Dev. 35: 275–291.

Eriksson, B.J. and G.E. Budd. 2000. Onychophoran cephalic nerves and their bearing on our understanding of head segmentation and stem-group evolution of Arthropoda. Arthropod. Struct. Dev. 29: 197–209.

Fahrenbach, W.H. 1968. The morphology of the eyes of *Limulus*. I. Cornea and epidermis of the compound eye. Z. Zellforsch. 87: 278–291.

Fahrenbach, W.H. 1969. The morphology of the eyes of *Limulus*. II. Ommatidia of the compound eye. Z. Zellforsch. 93: 451–483.

Fortey, R.A. and R.H. Thomas. 1997. Arthropod Relationships. The Systematics Association. vol. 55. Chapman & Hall, London, UK.

Friedrich, M. and D. Tautz. 1995. Ribosomal DNA phylogeny of the major extant arthropod classes and the evolution of myriapods. Nature 376: 165–167.

Friedrich, M. 2006. Continuity versus split and reconstitution: exploring the molecular developmental corollaries of insect eye primordium evolution. Dev. Biol. 299: 310–329.

Fryer, G. 1996. Reflections on arthropod evolution. Biol. J. Linnean Soc. 58: 1–55.

Fryer, G. A defence of arthropod polyphyly. pp. 23–33. *In:* R.A. Fortey and R.H. Thomas. [eds.] 1997. Arthropod Relationships. Chapman & Hall, London, UK.

Garcia-Machado, E. and N. Dennebouy, M.O. Suarez, J.C. Mounolou, and M. Monnerot. 1996. Partial sequence of the shrimp *Penaeus notialis* mitochondrial genome. C. R. Acad. Sci. Paris 319: 473–486.

Garcia-Machado, E. and M. Pempera, N. Dennebouy, M. Oliva-Suarez, J.C. Mounolou, and M. Monnerot. 1999. Mitochondrial genes collectively suggest the paraphyly of Crustacea with respect to Insecta. J. Mol. Evol. 49: 142–149.

Garey, J.R. and M. Krotec, D.R. Nelson, and J. Brooks. 1996. Molecular analysis support a tardigrade-arthropod association. Invert. Biol. 115: 79–88.

Garey, J.R. 2001. Ecdysozoa: The relationship between Cycloneuralia and Panarthropoda. Zool. Anz. 240: 321–330.

Geoffroy Saint-Hilaire, E. 1818. Philosophie anatomique, 1er vol. Méquignon-Marvis, Paris, France.

Gerberding, M. 1997. Germ band formation and early neurogenesis of *Leptodora kindti* (Cladocera): first evidence for neuroblasts in the entomostracan crustaceans. Inv. Reprod. Dev. 32: 63–73.

Gerberding, M. and G. Scholtz. 2001. Neurons and glia in the midline of the higher crustacean *Orchestia cavimana* are generated via an invariant cell lineage that comprises a median neuroblast and glial progenitors. Dev. Biol. 235: 397–409.

Giribet, G. and S. Carranza, J. Baguna, M. Riutort, and C. Ribera. 1996. First molecular evidence for the existence of a Tardigrada + Arthropoda clade. Mol. Biol. Evol. 13: 76–84.

Giribet, G. and G.D. Edgecombe, and W.C. Wheeler. 2001. Arthropod phylogeny based on eight molecular loci and morphology. Nature 413: 157–161.

Hallberg, E. and R. Elofsson. 1983. The larval compound eye of barnacles. J. Crust. Biol. 3: 17–24.

Hanström, B. 1947. The brain, the sense organs and the incretory organs of the head in the Crustacea Malacostraca. Lunds Univ. Arsskrift. N.F. 43: 1–45.

Harzsch, S. and J. Miller, J. Benton, R.R. Dawirs, and B. Beltz. 1998. Neurogenesis in the thoracic neuromeres of two crustaceans with different types of metamorphic development. J. Exp. Biol. 201: 2465–2479.

Harzsch, S. 2001. Neurogenesis in the crustacean ventral nerve cord: homology of neuronal stem cells in Malacostraca and Branchiopoda? Evol. Dev. 3: 154–169.

Harzsch, S. and D. Walossek. 2001. Neurogenesis in the developing visual system of the branchiopod crustacean *Triops longicaudatus* (LeConte, 1846): corresponding patterns of compound-eye formation in Crustacea and Insecta? Dev. Genes Evol. 211: 37–43.

Harzsch, S. 2002. The phylogenetic significance of crustacean optic neurophils and chiasmata: a re-examination. J. Comp. Neurol. 453: 10–21.

Harzsch, S. 2004. Phylogenetic comparison of serotonin-immunoreactive neurons in representatives of the Chilopoda, Diplopoda, and Chelicerata: implications for arthropod relationships. J. Morphol. 259: 198–213.

Harzsch, S. and C.H. Muller, and H. Wolf. 2005. From variable to constant cell numbers: cellular characteristics of the arthropod nervous system argue against a sister-group relationship of Chelicerata and "Myriapoda" but favour the Mandibulata concept. Dev. Genes Evol. 215: 53–68.

Harzsch, S. and G. Hafner. 2006. Evolution of eye development in arthropods: Phylogenetic aspects. Arthropod. Struct. Dev. 35: 319–340.

Harzsch, S. and K. Vilpoux, D.C. Blackburn, D. Platchetzki, N.L. Brown, R. Melzer, K.E. Kempler, and B.A. Battelle. 2006. Evolution of arthropod visual systems: development of the eyes and central visual pathways in the horseshoe crab *Limulus polyphemus* Linnaeus, 1758 (Chelicerata, Xiphosura). Dev. Dyn. 235: 2641–2655.

Hassanin, A. and N. Leger, and J. Deutsch. 2005. Evidence for multiple reversals of asymmetric mutational constraints during the evolution of the mitochondrial genome of metazoa, and consequences for phylogenetic inferences. Syst. Biol. 54: 277–298.

Hassanin, A. 2006. Phylogeny of Arthropoda inferred from mitochondrial sequences: strategies for limiting the misleading effects of multiple changes in pattern and rates of substitution. Mol. Phylogenet. Evol. 38: 100–116.

Hejnol, A. and G. Scholtz. 2004. Clonal analysis of *Distal-less* and *engrailed* expression patterns during early morphogenesis of uniramous and biramous crustacean limbs. Dev. Genes Evol. 214: 473–485.

Hwang, U.W. and M. Friedrich, D. Tautz, C.J. Park, and W. Kim. 2001. Mitochondrial protein phylogeny joins myriapods with chelicerates. Nature 413: 154–157.

Jager, M. and J. Murienne, C. Clabaut, J. Deutsch, H. Le Guyader, and M. Manuel. 2006. Homology of arthropod anterior appendages revealed by Hox gene expression in a sea spider. Nature 441: 506–508.

Jenner, R.A. and G. Scholtz. Playing another round of metazoan phylogenetics: historical epistemology, sensitivity analysis, and the position of Arthropoda within the Metazoa on the basis of morphology. pp. 355–385. In: S. Koenemann and R.A. Jenner. [eds.] 2005. Crustacea and Arthropod Relationships. CRC Press, Boca Raton, USA.

Jockusch, E.L. and C. Nulsen, S.J. Newfeld, and L.M. Nagy. 2000. Leg development in flies versus grasshoppers: differences in *dpp* expression do not lead to differences in the expression of downstream components of the leg patterning pathway. Development 127: 1617–1626.

Jockusch, E.L. and T.A. Williams, and L.M. Nagy. 2004. The evolution of patterning of serially homologous appendages in insects. Dev. Genes Evol. 214: 324–338.

Kadner, D. and A. Stollewerk. 2004. Neurogenesis in the chilopod *Lithobius forficatus* suggests more similarities to chelicerates than to insects. Dev. Genes Evol. 214: 367–379.

Kraus, O. 2001. "Myriapoda" and the ancestry of the Hexapoda. Ann. Soc. Entomol. Fr. 37: 105–127.

Kusche, K. and T. Burmester. 2001. Diplopod hemocyanin sequence and the phylogenetic position of the Myriapoda. Mol. Biol. Evol. 18: 1566–1573.

Lankester, E.R. 1870. On the use of the term homology in modern zoology and the distinction between homogenetic and homoplastic agreements. Ann. Mag. Nat. Hist. Ser. 4. 6: 34–43.

Lankester, E.R. 1904. The structure and classification of the Arthropoda. Quart. J. Microscop. Sci. 47: 523–576.

Lavrov, D.V. and W.M. Brown, and J.L. Boore. 2004. Phylogenetic position of the Pentastomida and (pan)crustacean relationships. Proc. Biol. Sci. 271: 537–544.

Lewis, E.B. 1978. A gene complex controlling segmentation in *Drosophila*. Nature 276: 565–570.

Lim, J.T. and U.W. Hwang. 2006. The complete mitochondrial genome of *Pollicipes mitella* (Crustacea, Maxillopoda, Cirripedia): non-monophylies of maxillopoda and crustacea. Mol. Cells 22: 314–322.

Loesel, R. and D.R. Nassel, and N.J. Strausfeld. 2002. Common design in a unique midline neurophil in the brains of arthropods. Arthropod. Struct. Dev. 31: 77–91.

Mallatt, J.M. and J.R. Garey, and J.W. Shultz. 2004. Ecdysozoan phylogeny and Bayesian inference: first use of nearly complete 28S and 18S rRNA gene sequences to classify the arthropods and their kin. Mol. Phylogenet. Evol. 31: 178–191.

Manuel, M. and M. Jager, J. Murienne, C. Clabaut, and H. Le Guyader. 2006. Hox genes in sea spiders (Pycnogonida) and the homology of arthropod head segments. Dev. Genes Evol. 216: 481–491.

Maxmen, A. and W.E. Browne, M.Q. Martindale, and G. Giribet. 2005. Neuroanatomy of sea spiders implies an appendicular origin of the protocerebral segment. Nature 437: 1144–1148.

Meinertzhagen, I.A. and T.E. Hanson. The development of the optic lobe. pp. 1363–1491. *In:* M. Bate and A. Martinez Arias. [eds.] 1993. The Development of *Drosophila melanogaster*. CHCL Press, Cold Spring Harbor NY, USA.

Melzer, R.R. and R. Diersch, D. Nicastro, and U. Smola. 1997. Compound eye evolution: Highly conserved retinula and cone cell patterns indicate a common origin of the insect and crustacean ommatidium. Naturwissenschaften 84: 542–544.

Mittmann, B. 2002. Early neurogenesis of the horseshoe crab *Limulus polyphemus* and its implication for arthropod relationships. Biol. Bull. 203: 221–222.

Mittmann, B. and G. Scholtz. 2003. Development of the nervous system in the "head" of *Limulus polyphemus* (Chelicerata: Xiphosura): morphological evidence for a correspondence between the segments of the chelicerae and of the (first) antennae of Mandibulata. Dev. Genes Evol. 213: 9–17.

Morgan, T.H. 1928. The Theory of the Gene. Yale University Press, New Haven.

Müller, C.H. and J. Rosenberg, S. Richter, and V.B. Meyer-Rochow. 2003. The compound eye of *Scutigera coleoptrata* (Linnaeus, 1758) (Chilopoda: Notostigmophora): an ultrastructural reinvestigation that adds support to the Mandibulata concept. Zoomorphology 122: 191–209.

Negrisolo, E. and A. Minelli, and G. Valle. 2004. The mitochondrial genome of the house centipede *Scutigera* and the monophyly versus paraphyly of myriapods. Mol. Biol. Evol. 21: 770–780.

Nielsen, C. 2001. Animal Evolution. Interrelationships of the living Phyla. Oxford University Press, Oxford, UK.

Nielsen, C. 2003. Proposing a solution to the Articulata-Ecdysozoa controversy. Zool. Script. 32: 475–482.

Nilsson, D.-E. and D. Osorio. Homology and parallelism in arthropod sensory processing. pp. 333–347. *In:* R.A. Fortey and R.H. Thomas [eds.] 1997. Arthropod Relationships. Chapman & Hall, London.

Nilsson, D.E. and A. Kelber. 2007. A functional analysis of compound eye evolution. Arthropod. Struct. Dev. 36: 373–385.

Nulsen, C. and L.M. Nagy. 1999. The role of *wingless* in the development of multibranched crustacean limbs. Dev. Genes Evol. 209: 340–348.

Nüsslein-Volhard, C. and E. Wieschaus. 1980. Mutations affecting segment number and polarity in *Drosophila*. Nature 287: 795–801.

Oakley, T.H. and C.W. Cunningham. 2002. Molecular phylogenetic evidence for the independent evolutionary origin of an arthropod compound eye. Proc. Natl. Acad. Sci. USA 99: 1426–1430.

Osorio, D. and J.P. Bacon. 1994. A good eye for arthropod evolution. BioEssays 16: 419–424.

Osorio, D. and M. Averof, and J.P. Bacon. 1995. Arthropod evolution: great brains, beautiful bodies. Trends Ecol. Evol. 10: 449–454.

Owen, R. 1849. On the Nature of Limbs. London John van Voorst, London, reprinted 2007, The Chicago University Press, Chicago..

Panganiban, G. and A. Sebring, L. Nagy, and S. Carroll. 1995. The development of crustacean limbs and the evolution of Arthropods. Science 270: 1363–1366.

Panganiban, G. and S.M. Irvine, C. Lowe, H. Roehl, L.S. Corley, B. Sherbon, J.K. Grenier, J.F. Fallon, J. Kimble, M. Walker, et al. 1997. The origin and evolution of animal appendages. Proc. Natl. Acad. Sci. USA 94: 5162–5166.

Park, S.J. and Y.S. Lee, and U.W. Hwang. 2007. The complete mitochondrial genome of the sea spider *Achelia bituberculata* (Pycnogonida, Ammotheidae): arthropod ground pattern of gene arrangement. BMC Genomics 8: 343.

Paulus, H.F. 2000. Phylogeny of the Myriapoda—Crustacea—Insecta: a new attempt using photoreceptor structure. J. Zool. Syst. Evol. Res. 38: 189–208.

Peterson, K.J. and D.J. Eernisse. 2001. Animal phylogeny and the ancestry of bilaterians: inferences from morphology and 18S rDNA gene sequences. Evol. Dev. 3: 170–205.

Philippe, H. and N. Lartillot, and H. Brinkmann. 2005. Multigene analyses of bilaterian animals corroborate the monophyly of Ecdysozoa, Lophotrochozoa, and Protostomia. Mol. Biol. Evol. 22: 1246–1253.

Pisani, D. 2004. Identifying and removing fast-evolving sites using compatibility analysis: an example from the Arthropoda. Syst. Biol. 53: 978–989.

Pisani, D. and L.L. Poling, M. Lyons-Weiler, and S.B. Hedges. 2004. The colonization of land by animals: molecular phylogeny and divergence times among arthropods. BMC Biol. 2: 1.

Popadic, A. 2005. Global evolution of *nubbin* expression patterns in arthropods: emerging view. Evol. Dev. 7: 359–361.

Popadic, A. and G. Panganiban, D. Rusch, W.A. Shear, and T.C. Kaufman. 1998. Molecular evidence for the gnathobasic derivation of arthropod mandibles and for the appendicular origin of the labrum and other structures. Dev. Genes Evol. 208: 142–150.

Popadic, A. and D. Rusch, M. Peterson, B.T. Rogers, and T.C. Kaufman. 1996. Origin of the arthropod mandible. Nature 380: 395.

Prpic, N.M. and R. Janssen, B. Wigand, M. Klingler, and W.G. Damen. 2003. Gene expression in spider appendages reveals reversal of *exd/hth* spatial specificity, altered leg gap gene dynamics, and suggests divergent distal morphogen signaling. Dev Biol. 264: 119–140.

Prpic, N.M. and D. Tautz. 2003. The expression of the proximodistal axis patterning genes *Distal-less* and *dachshund* in the appendages of *Glomeris marginata* (Myriapoda: Diplopoda) suggests a special role of these genes in patterning the head appendages. Dev. Biol. 260: 97–112.

Prpic, N.M. 2004. Homologs of *wingless* and *decapentaplegic* display a complex and dynamic expression profile during appendage development in the millipede *Glomeris marginata* (Myriapoda: Diplopoda). Front Zool. 1: 6.

Prpic, N.M. and W.G. Damen. 2004. Expression patterns of leg genes in the mouthparts of the spider *Cupiennius salei* (Chelicerata: Arachnida). Dev. Genes Evol. 214: 296–302.

Prpic, N.M. and W.G.M. Damen. Arthropod appendages: a prime example for the eovlution of morphological diversity and innovation. pp. 381–398. *In:* A. Minelli and G. Fusco. [eds.] 2008. Evolving Pathways. Cambrige Univ. Press, Cambridge, UK.

Prpic, N.M. and M.J. Telford. 2008. Expression of *homothorax* and *extradenticle* mRNA in the legs of the crustacean *Parhyale hawaiensis*: evidence for a reversal of gene expression regulation in the pancrustacean lineage. Dev. Genes Evol. 218: 333–339.

Regier, J.C. and J.W. Shultz. 2001. Elongation factor-2: a useful gene for arthropod phylogenetics. Mol. Phylogenet. Evol. 20: 136–148.

Regier, J.C. and H.M. Wilson, and J.W. Shultz. 2005. Phylogenetic analysis of Myriapoda using three nuclear protein-coding genes. Mol. Phylogenet. Evol. 34: 147–158.

Richter, S. 2002. The Tetraconata concept: hexapod-crustacean relationships and the phylogeny of Crustacea. Org. Divers. Evol. 2: 217–237.

Roeding, F. and S. Hagner-Holler, H. Ruhberg, I. Ebersberger, A. von Haeseler, M. Kube, R. Reinhardt, and T. Burmester. 2007. EST sequencing of Onychophora and phylogenomic analysis of Metazoa. Mol. Phylogenet. Evol. 45: 942–951.

Ronco, M. and T. Uda, T. Mito, A. Minelli, S. Noji, and M. Klingler. 2008. Antenna and all gnathal appendages are similarly transformed by *homothorax* knock-down in the cricket *Gryllus bimaculatus*. Dev. Biol. 313: 80–92.

Rota-Stabelli, O. and M.J. Telford. 2008. A multi-criterion approach for the selection of optimal outgroups in phylogeny: Recovering some support for Mandibulata over Myriochelata using mitogenomics. Mol. Phylogenet. Evol. 48: 103–111.

Ryu, S.H. and J.M. Lee, K.H. Jang, E.H. Choi, S.J. Park, C.Y. Chang, W. Kim, and U.W. Hwang. 2007. Partial mitochondrial gene arrangements support a close relationship between Tardigrada and Arthropoda. Mol. Cells 24: 351–7.

Schmidt-Rhaesa, A. 2001. Tardigrades: Are they really miniaturized dwarfs? Zool. Anz. 240: 549–555.

Schmidt-Rhaesa, A. and T. Bartolomeus, C. Lemburg, U. Ehlers, and J.R. Garey. 1998. The position of the Arthropoda in the phylogenetic system. J. Morphol. 238: 263–285.

Scholtz, G. Cleavage, germ-band formation and head segmentation: the ground pattern of Euarthropoda. pp. 317–332. *In:* R.A. Fortey and R.H. Thomas. [eds.] 1997. Arthropod Relationships. Chapman & Hall, London.

Scholtz, G. and B. Mittmann, and M. Gerberding. 1998. The pattern of *Distal-less* expression in the mouthparts of crustaceans, myriapods and insects: new evidence for a gnathobasic mandible and the common origin of Mandibulata. Int. J. Dev. Biol. 42: 801–810.

Scholtz, G. 2001. Evolution of developmental patterns in arthropods—the analysis of gene expression and its bearing on morphology and phylogenetics. Zoology 103: 99–111.

Scholtz, G. 2002. The Articulata hypothesis—or what is a segment? Org. Divers. Evol. 2: 197–215.

Scholtz, G. Is the taxon Articulata obsolete? Arguments in favour of a close relationship between annelids and arthropods. pp. 489–501. *In:* A. Legakis, S. Sfenthorakis, R. Polymeni and M. Thessalou-Legaki. [eds.] 2003. The New Panorama of Animal Evolution. Pensoft, Sofia-Moscow.

Scholtz, G. 2005. Homology and ontogeny: Pattern and process in comparative developmental biology. Theor. Biosci. 124: 121–143.

Scholtz, G. and G.D. Edgecombe. 2006. The evolution of arthropod heads: reconciling morphological, developmental and palaeontological evidence. Dev. Genes Evol. 216: 395–415.

Sewell, W. and T. Williams, J. Cooley, M. Terry, R. Ho, and L. Nagy. 2008. Evidence for a novel role for *dachshund* in patterning the proximal arthropod leg. Dev. Genes Evol. 218: 293–305.

Shear, W.A. 1992. End of the 'Uniramia' taxon. Nature 359: 477–478.

Shultz, J.W. and J.C. Regier. 2000. Phylogenetic analysis of arthropods using two nuclear protein-encoding genes supports a crustacean + hexapod clade. Proc. Biol. Sci. 267: 1011–1019.

Sinakevitch, I. and J.K. Douglass, G. Scholtz, R. Loesel, and N.J. Strausfeld. 2003. Conserved and convergent organization in the optic lobes of insects and isopods, with reference to other crustacean taxa. J. Comp. Neurol. 467: 150–172.

Snodgrass, R.E. 1935. Principles of Insect Morphology. McGraw-Hill, New York, USA.

Stollewerk, A. 2002. Recruitment of cell groups through Delta/Notch signalling during spider neurogenesis. Development 129: 5339–5348.

Stollewerk, A. and D. Tautz, and M. Weller. 2003. Neurogenesis in the spider: new insights from comparative analysis of morphological processes and gene expression patterns. Arthropod. Struct. Dev. 32: 5–16.

Stollewerk, A. Evolution of neurogenesis in arthropods. pp. 359–380. *In:* A. Minelli and G. Fusco [eds.] 2008. Evolving pathways. Cambrige Univ. Press, Cambridge, UK.

Stollewerk, A. and A.D. Chipman. 2006. Neurogenesis in myriapods and chelicerates and its importance for understanding arthropod relationships. Integr. Comp. Biol. 46: 195–206.

Stollewerk, A. and P. Simpson. 2005. Evolution of early development of the nervous system: a comparison between arthropods. Bioessays 27: 874–883.

Strausfeld, N.J. 1998. Crustacean-insect relationships: the use of brain characters to derive phylogeny amongst segmented invertebrates. Brain Behav. Evol. 52: 186–206.

Strausfeld, N.J. 2005. The evolution of crustacean and insect optic lobes and the origins of chiasmata. Arthropod. Struct. Dev. 34: 235–256.

Telford, M.J. and R.H. Thomas. 1995. Systematics: Demise of the Atelocerata? Nature 376: 123–124.

Telford, M.J. and R.H. Thomas. 1998. Expression of homeobox genes shows chelicerate arthropods retain their deutocerebral segment. Proc. Natl. Acad. Sci. USA 95: 10671–10675.

Telford, M.J. and S.J. Bourlat, A. Economou, D. Papillon, and O. Rota-Stabelli. 2008. The evolution of the Ecdysozoa. Philos. Trans. R. Soc. Lond. B Biol. Sci. 11: 1529–1537.

Timmermans, M.J.T.N. and D. Roelofs, J. Marien, and N.M. van Straalen. 2008. Revealing pancrustacean relationships: Phylogenetic analysis of ribosomal protein genes places Collembola (springtails) in a monophyletic Hexapoda and reinforces the discrepancy between mitochondrial and nuclear DNA markers. BMC Evol. Biol. 8: 83.

Turbeville, J.M. and D.M. Pfeifer, K.G. Field, and R.A. Raff. 1991. The phylognetic status of Arthropods, as inferred from 18S rRNA sequences. Mol. Biol. Evol. 8: 669–686.

Ungerer, P. and G. Scholtz. 2008. Filling the gap between identified neuroblasts and neurons in crustaceans adds new support for Tetraconata. Proc. Biol. Sci. 275: 369–376.

Wegerhoff, R. and O. Breidbach. Comparative aspects of the chelicerate nervous system. pp. 159–179. *In:* O. Breidbach and W. Kutsch. [eds.] 1995. The Nervous System of Invertebrates: An Evolutionary and Comparative Approach. Birkhauser, Basel.

Whitington, P.M. Conservation *versus* change in early axogenesis in arthropod embryos: A comparison between myriapods, crustaceans and insects. pp. 181–219. *In:* O. Breidbach and W. Kutsch [eds.] 1995. The Nervous System of Invertebrates: An Evolutionary and Comparative Approach. Birkhauser, Basel.

Whitington, P.M. 1996. Evolution of neural development in arthropods. Sem. Cell Dev. Biol. 7: 605–614.

Williams, D.S. and P. McIntyre. 1980. The principal eyes of a jumping spider have a telephoto component. Nature 288: 578–580.

Williams, T.A. 1998. *Distal-less* expression in crustaceans and the patterning of branched limbs. Dev. Genes Evol. 207: 427–434.

Williams, T.A. 1999. Morphogenesis and homology in arthropods limbs. Amer. Zool. 39: 664–675.

Williams, T.A. and C. Nulsen, and L.M. Nagy. 2002. A complex role for *Distal-less* in crustacean appendage development. Dev. Biol. 241: 302–312.

Williams, T.A. 2008. Early *Distal-less* expression in a developing crustacean limb bud becomes restricted to setal-forming cells. Evol. Dev. 10: 114–120.

Wilson, K. and V. Cahill, E. Ballment, and J. Benzie. 2000. The complete sequence of the mitochondrial genome of the crustacean *Penaeus monodon*: are malacostracan crustaceans more closely related to insects than to branchiopods? Mol. Biol. Evol. 17: 863–874.

Zantke, J. and K. Wolff, and G. Scholtz. 2008. Three-dimensional reconstruction of the central nervous system of *Macrobiotus hulefandi* (Eutardigrada, Parchela): implications for the phylogenetic position of Tardigrada. Zoomorphology 127: 21–36.

Zrzavy, J. and V. Hypsa, and M. Vlaskova. Arthropod phylogeny: taxonomic congruence, total evidence and conditional combination approaches to morphological and molecular data sets. pp. 97–107. *In:* R.A. Fortey and R.H. Thomas [eds.] 1997. Arthropod Relationships. Chapman & Hall, London, UK.

Section 2

The Earliest Animals: From Genes to Transitions

Chapter 5

The Pre-Nervous System and Beyond—Poriferan Milestones in the Early Evolution of the Metazoan Nervous System*

Michael Nickel

Introduction

Metazoan nervous systems are structurally and functionally diverse and highly complex. Even the simplest nervous systems, the nerve nets of Cnidaria and Ctenophora, showcase a structural complexity on the cellular and sub-cellular level which implies a long evolutionary history. The nervous system in a strict sense, as defined by most authors (for overviews see Bullock and Horridge 1969, Hanström 1928, Schmidt-Rhaesa 2007), is an autapomorphy of the Eumetazoa. One of the definitions is provided by Schmidt-Rhaesa (2007): "Nervous systems conduct information in a directed way through the body. This is done by electrical and/or chemical signals and by cells specialized for these functions". The classic neurone doctrine by Ramon y Cayal (1937), however, is strictly morphology-based and does not consider the nature of the signal. Bullock and Horridge (1969) summarized Ramon y Cayals's doctrine by stating that "all nervous systems consist in essence (whatever other, nonnervous elements may be

Institut für Spezielle Zoologie und Evolutionsbiologie Friedrich-Schiller-Universität Jena, Erbertstr. 1, D-07743 Jena Germany.
E-mail: nickel@porifera.net

*This chapter is dedicated to Stephen Hillenburg, who greatly directed the interest of a whole future scientist generation towards sponges and the marine life.

present) of distinct cells called neurons, which are specialized for nervous functions and which produce prolongations and branches". As we will see, this very formal definition resulted in an academic dispute if there is a nervous system in sponges. This question was negated based on solid arguments (Jones 1962), however it tends to reappear in literature until today (e.g. Meech and Mackie 2007, Perovic et al. 1999).

The phylogenetic context of the early evolution of the nervous system is difficult to consider, since sponges (Porifera) and Placozoa do not display distinct nervous system structures, namely defined nerve cells and synapses. For this reason, the evolution of the central nervous systems (CNS) of the Bilateria acquires the most attention (Bullock and Horridge 1969, Hanström 1928, Schmidt-Rhaesa 2007). However the evolutionary tendency for centralization is already intrinsic within the Cnidaria (Holland 2003). Nonetheless, 'simple' nerve nets and the properties of their neurons are generally the focus of literature on early nervous system evolution (Mackie 1970, Meech and Mackie 2007, Pantin 1952, 1956, Parker 1910, 1919, Passano 1963). This comprehension of 'early nervous system evolution' is synonymous with 'early evolution of electrical excitability'.

If we want to understand the basics, we have to consider how the first metazoa, the LCAM (last common ancestor of metazoa), sensed external stimuli, how it facilitated intercellular signaling and integration. Therefore, we will have to reflect on the evolution of pre-nervous integration systems. Even though sponges lack a nervous system, poriferan milestones of nervous system evolution will provide decent information in helping to reconstruct the LCAM's integrative capabilities and possible pre-nervous system intermediates.

The reason, why sponges are important in this context is simple: they display movements without muscles and signal integration without nerves (Nickel 2007). Their coordinated movements are extremely slow: contractions often follow rhythmic patterns and are easier to recognize by the human eye than the extremely slow locomotion of the body (Nickel 2004, 2006). The first reports of these behaviors, dating back to Aristotle, gave rise to one of the longest scientific debates in the history of science, the question if sponges are animal or plants or something in-between. In addition, this behavior and the underlying biological mechanisms and principles are the key to our main questions on how integration systems evolved from the LCAM to the Cnidaria.

Here, I aim to provide an overview of the current knowledge on sponge contraction, locomotion and integration, which will provide first evidence for the early co-evolution of contraction and integration in Metazoa. From this background, I will deduce a model for a pre-nervous system of the demosponges and I will develop a hypothesis on the evolutionary steps from the pre-nervous to the nervous system. The chapter will start with an overview on the long history of research on contraction and integration. This historical view will facilitate the understanding of present concepts, new hypotheses and the needs for future research. Even though I will correct some historical miss-interpretations (e.g. Jones 1962, concerning Aristotle's views on sponges), readers who are well aware about the historical context might want to continue directly with the paragraphs on current concepts and new conclusions below.

The History of Research on Contractile and Integrative Systems in Sponges

Δοκεῖ δὲ καὶ ὁ σπόγγος ἔχειν τινὰ αἴσθησιν[1]

Aristotle 350 B.C. (from Barthélemy-Saint Hilaire 1883)

Contraction: A Hint from the Classical Antiquity

The fact that sponges were for long periods regarded as plants (Linné 1767) or "Pflanzenthiere" (Esper 1794–1805) points out the difficulties of categorizing sponges. Generations of biologists seem to have had difficulties in how to interpret sponges on the scale from plant to animals (for an early review see Johnston 1842). One must not however take this as a sign of the superiority of present day science or scientific advancement over the centuries. It is likely more likely of a sign for varying perceptive concepts in biological science over the centuries. In fact, going back to the first written reports on sponges, we find that sponges were recognized as animals very early on. As early as 350 B.C., in his "Natural History of Animals", Aristotle (384–322 B.C.) addresses sponges as animals (Aristotle 1498). Unfortunately, this historically important fact is not generally recognized by zoologists (e.g. Meech 2008 argues that Aristotle regarded sponges as vegetables, which is only valid for parts of Aristotle's manuscripts). Aristotle mentions sponges in his first chapter of book one, characterizing them as sedentary marine animals, and attributing sensibility to them. This sensibility is expressed by the fact that, when not done carefully, they are hard to remove from the substrate. From chapter 16 of his book five (Fig. 1), which almost exclusively deals with sponges, it is evident that he is referring to sponge contraction. Aristotle reiterates that sponges are sensitive and he relates contractile behavior to sponge fishing and environmental conditions. He states that they are able to sense when they are about to be collected by sponge fisherman, and how they

---

[1] "And, by the way, the sponge appears to be endowed with a certain sensibility"; Aristotle, 1910. The History of Animals. Transl. by D'Arcy Wentworth Thompson. Oxford: Clarendon Press. [Electronic reprint]. The University of Adelaide Library. http://etext.library.adelaide.edu.au/a/aristotle/history/.

88 *Key Transitions in Animal Evolution*

LIBER QVINTVS zi

De urticis & fpongiax genere:quoûe modo gignant. CAP. XVI.

Odus idem gignêdi iis etiam eſt quæ nulla teſta integuntur:ut urtichs: & ſpongiis: qui teſta
incluſis:per rimas enim & cauernas:& fauces ſaxorum oriuntur. Genus duplex urticæ eſt:
aliæ enim ſinuoſis adhærêt quæ nûquâ á ſaxis abſoluuntur:aliæ plana & littora amât:quæ
ſuis abiunctæ ſedibus uagantur:Patellæ quoog abſolui ſolent: aliog tranſire: naſci etiam in
eubilibus ſpongiarum:nouimus:beſtiolam aranei ſimilem: nomen pinnotera mutuſtrem: degit hæc
intra ſpongiarû cubilæ:& aperiendo:claudendog:piſciculos capit: aperit antequâ ingrediâtur claudit
& contrahit cum ingreſſi ſunt.Genus ſpongiarum triplex ſtatuitur:nam aliæ raræ:aliæ ſpiſſæ: aliæ qs
nominât achilleas.Tenuiſſimum genus id tertium & ſpiſſiſſimum & ualidiſſimû eſt:quod galeis ocre
iſog inſeritur:eog minus ciere ſtrepitum poſſe rotatum eſt:genus hoc inuêtu per quâ rarum eſt.Quæ
autem in genere illo ſpiſſo præduræ ſunt atog aſperæ:nomine hirci nuncupatur: quæ quidem omnes
aut ad ſaxa naſcuntûr:aut iuxta litus lutog aluntur:cuius rei argumentû eſt og captæ limo gſertæ oês
cernuntur:quod centre indicat:cæteris quoog adhærentibus cibum per ipſum adnexum hauriri. Imbe-
cilliores propterea ſunt ſpiſſæ q rare:quia minus alto radicis hæſu inituntur.Senſum etiâ ſpongiis eſſe
aiunt:argumento qd'ad e uulloris acceſſum côtrahuntur.Ita ut euelli difficile ſit:quod idem ét faciunt
quotiens flatus tempeſtaſog urget:ne ſua de ſede pellantur.Sed ſunt qui de hoc dubitent:ut qui toro-
nam incolunt.Narrant tamen proculdubio beſtiolas quaſdam:uelut tineas lumbricoſue & eiuſmodi
alias conſiſtere intra ſpongias:atog ali:quas & euulſis ſpongiis piſciculi ſaxatiles deuorent: qui uel radi-
ces abſumunt totas:quæ inhærentes ſaxis remanſerint.Si euenerit forte ut ſpongia abrumpatur: reſi-
duo item renaſcitur & completur.Magnitudine âmpliſſima ſolutæ illæ & raræ ſpongiæ augentur plu
rimæ:quæ circa lyciam ſunt.Sed molliſſimæ quæ ſpiſſæ:nam achilleæ toroſiores iis côſtant: omnino
quæ altis trâquilliſog inſunt gurgitibus molliſſimæ ſunt:flatus.n.ac tempeſtates ſpongias quoogaut cæ
tera altilia reddunt duriores:& incrementum ipediunt:quamobrem ſpongiæ helleſpôti ſpiſſæ: ac du-
ræ ſunt:& omnino quas mare ultra maleam promontorium citraog fert:differunt iter ſe mollitie duri
tateog:nec calorem immodicum ſpôgiæ patiuntur:ſit enim eo ut more pullulantium putreſcant: quo
circa optime iuxta oras,comperiuntur:ſpongiæ ſi gurgite alto demerſæ ſunt: commode enim tempe-
râtur:propter altitudinem gurgitis.Color illotis uiuiſog nigricâs eſt:adhæret nec parte:nec toto.Inter-
ſunt enim fiſtulæ quædam inanes: ſed pluribus paſſim particulis hæſitât:& quaſi membrana ex-
têta ſubeſſe radicibus earum uidetur:ſuperne autem cæteri meatus côncreti propemodû latent.At ue-
ro quaterni:aut quini patent per quos paſci exiſtimantur.Genus item aliud eſt: quod nomê aphlyſias
inde accepit:quia nunquam lauetur quaſi illotariam uoces:habet hocampliores illos meatus: ſed reli-
quo toto corpore denſum:é nihilo tamê denſius:q quæ ante enumerauimus genera:quanq lêtius: &
ex toto pulmoni ſimilius cernitur.Senſum id genus habere diuog uiuere iter oês præcipue côuenit: eo
gnoſci præ cæteris ſpongiis eo facile poteſt documento:og cum ille albicêt limo ſubſidête id ſemper ni-
grorem oſtêdit.Quod ad ſpongias & teſtatorum generationem pertinet res ita ſe habet.

Fig. 1. Facsimile of Aristotle's History of the animals in the Latin translation, printed in 1498, page from book 5, chapter 16 which exclusively deals with sponges. The highlighted passage reads: "*It is said that the sponge is sensitive; and as a proof of this statement they say that if the sponge is made aware of an attempt being made to pluck it from its place of attachment it draws itself together, and it becomes a difficult task to detach it. It makes a similar contractile movement in windy and boisterous weather, obviously with the object of tightening its hold. Some persons express doubts as to the truth of this assertion; as, for instance, the people of Torone.*" (Aristotle, 1910; translation by D'Arcy Wentworth Thompson). For almost 2200 years, between its genesis 350 B.C. to around 1850, 'Aristotle's History' served as an integral source of zoological knowledge and was regularly cited (compare Johnston, 1842). Modified scan from Aristotle (1498).

react to this with contraction. This results in increasing efforts for the fishermen to remove them from the substrate. The same behavior can also be found during turbulent seas, when adhesion is even more enhanced. Interestingly, he adds that this behavior is questioned by the fishermen of Toroni; who presumably, did not share this opinion (Fig. 1).

These short passages by Aristotle are interesting in multiple ways, for science history as well as for zoology in general. First, he regards sponges as animals, a classification which was later refused until the mid 18th century, when Ellis (1755), Pallas (1766) as well as Ellis and Solander (1786) reestablished this view. Second, he describes sponge contractile behavior, which doesn't reappears in scientific literature before Pallas (1787). In addition, Aristotle interprets this behavior as a clear sign of integration of information in the sponge. The dissent of the Toroni fishermen he mentions may be seen as a first sign of a much later scientific debate on the sponge nervous system (see below).

Similar reports on sponges can be found in texts of various authors of the classical antique. For example, Claudius Aelianus (approx. 175–235 A.C.) mentions very similar aspects in his book "De Natura Animalium" (book 8, chapter 16). The context suggests that his is a review of Aristotle's text, rather than an independent report (Aelianus 1832). More than a century earlier, Plinius the older (23–79 A.C.) reported on sponges' contractile behavior as well as their ability to regenerate from their "roots" (Plinus 1855).

Sponge body contractions: Perception in modern biology

The earliest modern report on sponge contraction seems to be the one of Pallas (1787). Most likely inspired by the three antique documents mentioned above, he investigated sponge behavior and observed oscular contraction. Also the slightly earlier work of Ellis and Solander (1786) mentions Aristotle's description of sponge contraction, although they did not add their own observations to the scientific record. Their work demonstrates par excellence that the knowledge of the antique natural history influenced natural sciences at least until the mid 18th century.

It took some more decades, before zoologists started to examine the details of sponge contractions themselves on a broader basis. As we learn from Johnston (1842), sponge contractility played a certain role in the academic discussion on the animal or plant nature of sponges until the late 18th century. However, he himself stated in his nicely detailed review: "Sponges are unmoving and unirritable: hence they ever remain rooted to the place of their germination, and are incapable either of contracting or dilating themselves, or even of moving any fibre or portion of their mass" (Johnston 1842). Only some years later, Bowerbank (1856) presented his extensive long-time observations of oscules opening and closing as well as surface contractions. Lieberkühn (1859) reported the contractility of the body surface (exopinacoderm) of *Spongia tupha* PALLAS 1766. A few years later, Schmidt was the first to postulate a cellular mechanism for contraction. He suggested contractile, muscle like fibers, which he saw as the "erste wohl constatirte Beispiel des Vorkommens (willkürlich) contractiler Fasern im Körper der Spongien" [first well documented example of the existence of (voluntarily) contractile fibers in the body of the sponges]. In relation to *Tethya* sp. he spoke of the presence of muscle fibers within the cortex (Schmidt 1866).

One of the first graphical representations of a contracted sponge was published by the end of the 19th century on the zoological wall charts edited by Rudolf Leuckart (see Redi et al. 2002). Drawn and described by Marshall (1885), chart 47 compared a contracted and expanded specimen of *Tethya* sp. in cross sections. His remarkable drawings show the contraction of the cortex as well as parts of the aquiferous system (Fig. 2). The first detailed drawing of the contraction of a calcareous sponge was presented by Minchin (1900). He demonstrated the contraction of the pinacoderm as well as the resulting displacement of spicules.

Muscle fiber cells, or rather spindle shaped myocytes were postulated by Schmidt (1866), Sollas (1888) and other authors (for an overview see Jones 1962). Whereas Minchin (1900) and Wilson (1910) observed pinacoderm contraction. Consequently, two competing hypotheses on the

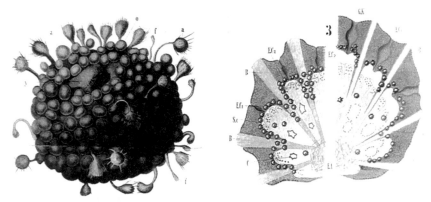

Fig. 3. *Tethya sp.* von Corfu, im Längsschnitt (wahrscheinlich T. lyncurium jun.) ohne Mund und Magenraum, natürliche Grösse 10 mm. Rechts in ausgedehntem, links in zusammengezogenem Zustande; Rindenschicht aus elastischen Fasern, die in verschiedener Richtung verlaufen und nach aussen zu runden Zellen werden (Fig. 5). Zu äusserst in der Rindenschicht liegen sehr kleine Kieselsternchen, zu innerst grössere. Es wird die Schichte von den Einströmungskanälen erster Ordnung (EO¹.) durchsetzt, welche in expandirtem Zustande in Folge der Contraction senkrecht an sie herantretender Fasern (Muskelzellen ?) flaschenförmig erweitert sind und in die förmig erweitert sind und in die gleichfalls erweiterten subcorticalen Räume (Spaltungsräume des Mesoderms) überleiten, von denen aus erst die Einströmungscanäle (EO².) zweiter Ordnung in die innere Schwammsubstanz eindringen, welche von dem mit kleinen Kieselsternchen austapezierten Kanalsystem (C.) durchzogen ist. Die Rindenschicht und Innenmasse verbinden sich mittelst radiärer Bündel einaxiger Nadeln (B.), welchen die erstere nach innen folgt, so dass sie in letztere eindringt. Bei der Contraction schieben sich die Einaxer etwas zusammen, sodass die Bündel stärker werden als bei der Ausdehnung; die Kanäle der inneren Schwammmasse verengen inneren Schwammmasse verengen sich, und ebenso die Einströmungskanäle erster Ordnung unter Contractionserscheinungen von Fasern, die theils der Länge nach, theils auch in Ringform in ihre Wandungen eingelagert sind. Durch die kraftvolle Zusammenziehung der Rinde fallen auch die subcorticalen Hohlräume zusammen, und das in dem ganzen Canalsystem der Tethya enthaltene Wasser wird mit nicht unbedeutender Gewalt herausgetrieben, sodass es in Gestalt kleiner Fontainen aus den Einströmungsöffnungen hervorquillt. Die Mitte des basalen Theils enthält einen Haufen ungeordnet gelagerter Einaxer (E.A.), aus dem die radiären Bündel hervortreten. Original.

Fig. 2. Early published drawings by Marshalls (1885) showing for the first time morphological changes during contraction in *Tethya* sp. Collage of two drawings present on Leuckart's zoological wall chart 47. Below, the detailed explanatory note on wall chart figure 3 (contraction, upper right) originally printed in a leaflet accompanying the wall chart is presented (combined and modified from Redi 2002).

contractile effectors emerged: (1) mesohyle-mediated contraction based on myocytes (preferable terminus after Boury-Esnault and Rützler 1997: actinocytes); (2) pinacoderm-mediated contraction based on pinacocytes. Most of the later authors preferred the first hypothesis.

Among the more recent works, Bagby (1966) first pointed out the similarities between the spindle-shaped myocytes and smooth muscle cells, partly because of the staining characteristics using Mallory's trichrome. In addition, he ultrastructurally demonstrated the presence of actin filament bundles (Bagby 1966). Several years later, however, he found similar filament structures in pinacocytes (Bagby 1970), which was confirmed in subsequent studies (Matsuno et al. 1988). Also in the 1960's, Pavans de Cecatty became one of the main supporters of the myocyte hypothesis, connecting it to the discussion on the existence of a nervous system in sponges (see below: *The Quest for the Elementary Nervous System*). Mainly based on histological and ultrastructural evidence, e.g. the demonstration of cell-cell contacts and synapse-like structures, he defined myocytes as evolutionary prototypes of neuromuscular cells (Pavans de Ceccatty 1960, 1971, 1974a, 1979). Myocytes were found to be present in a number of sponges and accepted as contractile effectors by most authors (Lentz 1966, 1968, Prosser et al. 1962, Vacelet 1966). Contemporaneously, Jones (1962) pointed out missing functional proof-of-concepts regarding myocytes contractility. He referred to a number of publications which demonstrated contractility of pinacocytes directly or indirectly (Bidder 1937, Jones 1957, Parker 1910, Wilson 1910). Finally, after some years of discussion, Pavans de Cecatty integrated contractile pinacocytes into his neuromuscular model and demonstrated the presence of actin filaments in pinacocytes (Pavans de Ceccatty 1981, 1986). Nevertheless, he did not change his view about myocytes as the main contractile effectors in sponges.

Simpson (1984) compiled 'The Cell Biology of Sponges' and provided a unique secondary literature source. His book has had the highest possible impact on the sponge research community and is still a major work of reference. However, it seems that during the two decades following publication of his book, the quantity of publications on sponge cell biology decreased, apart from some exceptions (e.g. Weissenfels 1989). This might be a coincidence or a result of the surpassing richness in detail provided by Simpson. His work also might have provoked the assumption that most about sponge biology is already known. Interestingly, only few works on sponge contraction were published from 1984 to 2004 (Matsuno et al. 1988, Sarà et al. 2001, Weissenfels 1990). However, after 20 years of relative silence, a new interest in the topic has arisen (Elliott and Leys 2007, Meech 2008, Nickel 2004, Nickel et al. 2006a, Nickel et al. 2006b, Nickel et al. 2005; for details see below: *Current Concepts on Sponge Contraction, Locomotion and Integration*).

The Late discovery of sponge locomotion

The second behavioral aspect in sponges is locomotion. This was hidden from zoologists for a long time and is even today hard to find in zoology textbooks, although some notable exceptions should be mentioned (Ruppert et al. 2004, Van Soest 1996). Locomotion, however, is not the most obvious behavior, since poriferans are sessile animals. Almost all species are firmly attached to the substrate, with only a few species drifting passively. Nevertheless, many, if not all, sponge species display movements on the substrate. These movements are too slow to be observed by the human eye without the aid of the one or the other time-lapse techniques. The most simple techniques are drawing outlines (e.g. Arndt 1941, Bond and Harris 1988), but today digital time-lapse imaging is used (Nickel 2006). However, understanding sponge locomotion, even more than understanding sponge contraction, requires the adoption of a time philosophy and a time scale which differs significantly from all other animals.

The first published report on sponge locomotion seems to date back to Carter (1848). He recognized the slow displacement of adult freshwater sponges on the walls of water tanks. Lieberkühn (1863) followed with his observation of spongillid fragments moving apart from each other. Noll (1881) reported on the movement of a specimen of *Reniera* sp. growing on an oyster shell in his aquarium. Similar aquarium based observations were also reported by Arndt (1941), followed by Burton (1948), who included field observations. Most authors point out that shape changes are part or consequence of the locomotion sequences, suggesting the influence of differential growth (Arndt 1941). In contrast, very prominent movements which are not correlated with body shape changes were reported by Fishelson (1981). Individuals of *Tethya seychelensis* (WRIGHT 1841) and *T. aurantium* (PALLAS 1766) sized 20–30 mm moved 50–80 mm within 24 hours. Fishelson postulated that traction forces are generated by myocytes within the filament-like body extensions. He concludes that the extensions which are able to anchor to the substrate mediate physical forces between the sponge and its physical environment.

Kilian (1967) observed negative phototaxis for juvenile freshwater sponges (*Ephydatia* spp. and *Spongilla* spp.) after hatching from gemmules. In contrast, *Heteromeyena* sp. displayed positive phototaxis. The strength of phototactic behavior is positively correlated with concentration of associated green algae within the sponges. These movements might be explained by amoeboid movements of the basal pinacoderm, shifting the whole sponge (Brøndsted 1942). Several studies investigated the movements of sponge masses after hatching from gemmules or of sponge reaggregates after artificial dissociation (for an overview see Jones 1962,

Wintermann 1951). Most of these studies confirmed amoeboid movements. However, one has to be aware that in all animals morphogenetic processes are associated with amoeboid cell movements and body shape changes. This is not necessarily equivalent to the movements observed in adult sponges shown above.

As mentioned before, during the late 1980s interest in sponge cell biology seemed to declined (see: *Sponge Body Contractions: Perception in Modern Biology*), while the interest in sponge locomotion seemed to hold the same level. The reason might be found in the wider availability of analog and digital time-lapse imaging methods, which are used by recent researchers (Bond 1992, Bond and Harris 1988, Nickel 2006, Nickel and Brümmer 2002, Nickel and Brümmer 2004, Nickel et al. 2002, Pronzato 2004).

The Quest for the Elementary Nervous System

The question of an integrative system in sponges is closely related to the phenomena of contraction and locomotion. Consequently, questions and assumptions of a possible nervous system in sponges came up early. As outlined above, antique authors regarded sponges as 'sensitive'. However, these questions did not only arise in sponges, but also in other animal groups. Based on the growing knowledge of the cnidarian nervous systems at the beginning of the 20th century, the first models on the evolution of nervous systems and 'the elementary nervous system' were published. Starting with Parker (1910, 1919), a century of general publications on the 'primitive nervous system' followed, indicated by titles combined from words like 'sponges', 'integration', 'conduction' and 'coordination': Pantin (1952, 1956), Jones (1962), Passano (1963), Lentz (1968), Pavans de Cecatty (1974a, b, 1979, 1989), Mackie (1970, 1979, 1990), and most recently Leys and Meech (2006).

Based on experiments using cnidarians and poriferans, Parker tried to reconstruct the 'elementary nervous system': the evolutionary initial nervous system which structurally and functionally represents all elements of a nervous system in their most simple state. As a prerequisite, 'the nervous system' and its elements need to be fundamentally defined. This happened to be much easier in Parker's lifetime than today, as Mackie (1990) points out. The enormously grown and growing biological knowledge makes it difficult to compile a list of characters defining a nerve cell. This refers especially to the knowledge on the identity of genetic information in every cell of a single individual as well as differentiation mechanisms via regulated gene expression (Anderson 1990). This is caused by the fact that many so-called 'nerve cell typical' features and molecules play a role in other biological contexts. Prominent examples are voltage-gated ion channels or transmitter substances (Mackie 1990).

According to Parker (1919) the first nervous system evolved from biological necessity to control 'independent effectors', namely contractile cells. Consequently, muscle cells are the evolutionary nucleus, around which the nervous system evolved. From his own studies on *Stylotella* sp. he concluded that sponges possess independent contractile effectors (pinacocytes and porocytes). In addition, he demonstrated the transmission of a mechanical stimulus, which after several minutes caused the contraction of an oscule several millimeters away. He spoke about 'neuroid' signal transduction, which 'creeps' through the sponge tissue, and he interpreted this observation as a form of transduction within the effector tissue itself. Therefore, he clearly distinguished it from his concept of signal transduction through the neuronal reflex arc, which connects sensor cells or pacemaker cells and 'dependent effectors' (e.g. epithel-muscle cells in Cnidaria) via nerve cells. This concept will have to be discussed below in the context of new molecular and physiological results.

McNair (1923) reported similar results on the mechanical stimulation of contraction and the transmission through sponge tissue in *Ephydatia fluviatilis* (LINNAEUS 1759). The contractile wave traveled with a speed of 35 $\mu m \cdot sec^{-1}$ (downwards the oscule) and 17 $\mu m \cdot sec^{-1}$ (upwards the oscule). Similar speeds of contractile waves ranging from 8 to 30 $\mu m \cdot sec^{-1}$ were reported for *E. fluviatilis*, *Spongia* sp. and *Hippospongia* sp. by other authors (De Vos and Van De Vyver 1981, Pavans de Ceccatty 1971, 1974a). Jones (1962) tried to explain the directional differences in speed of the contractile waves observed by McNair (1923) by attributing upwards movements to the endopinacoderm and downwards movement to the exopinacoderm which are structurally different. He justified his view by referring to a report by Ankel (1948), who demonstrated structural differences between the two pinacoderm layers: exopinacocytes are oriented lengthwise in relation to the oscule, while endopinacocytes are oriented crosswise. Consequently, Jones argued, the signal crosses more cell borders inside the endopinacoderm. However, he did not explain the possible mechanisms limiting a signal to travel in one direction within a certain cell layer. In this context, his suggestion seems to be not very likely.

In their reviews, Pantin (1952, 1956) and Passano (1963) modified the hypothesis of Parker, mainly based on their work in cnidarians. Consequently, their hypotheses on the evolution of nervous systems focused on electrical conduction, like Jones (1962) did. Nevertheless, all of them admitted to an evolutionary precursor system for 'neuroid' conduction. Macke (1970) argued that electrical synapses (gap junctions) likely evolved before chemical synapses and are therefore candidates for a neuroid pre-nervous system. This view was supported by the report of electrical coupling of *Haliclona* sp. cells by Loewenstein (1967). Both works resulted in an intensified search for gap junctions in sponges. Admittedly, two independent groups found ultrastructural evidence for gap junctions (Gaino and Sarà 1976, Green and Bergquist 1979). However, this could not be unequivocally confirmed by the same and other authors (Garrone and Lethias 1990, Garrone et al. 1980, Green and Bergquist 1982, Lethias et al. 1983). No further work focusing on gap junctions in sponges has been published after 1990.

Until today, no direct electrical conduction has been shown in demosponges and calcareous sponges. This might be an effect of the small cell sizes (Leys and Meech 2006, Mackie 1990). However, action potentials pass on through the trabecular tissue of hexactinellid sponges and influence ciliar beating of choanocytes (Leys and Mackie 1997, Leys and Mackie 1999, Leys et al. 1999, Mackie et al. 1983). Nevertheless, due to the syncytical nature of the trabecular system in hexactinellids (Leys et al. 2007), no nervous system like structures (e.g. synapses) are necessary to distribute the action potential. Since ciliates produce action potentials, as well, it is evident that voltage-gated ion channels, namely voltage-gated Ca^{2+}-channels (Linder and Schultz 2002, Schultz et al. 1986), predate the evolution of animal multicellularity.

An alternative scenario for the early evolution of the nervous system emphasizes the role of secretion (Grundfest 1959, 1965, Horridge 1968). Lentz (1966, 1968) demonstrated histochemically acetylcholine esterase activity as well as the presence of adrenalin and serotonin in the calcareous sponge *Sycon* sp. Activity was mainly located to actinocytes, and granulated as well as multipolar cells. In demosponges, Acetylcholine esterase activity was also demonstrated: in *Hippospongia* sp. (Thiney 1972) as well as in *Tethya wilhelma* SARÀ, SARÀ, NICKEL & BRÜMMER 2001 (Nickel 2001). In physiological experiments, acetylcholine influenced the rhythm as well as the intensity of oscule contractions of *Spongia officinalis* LINNAEUS 1759 (Pavans de Ceccatty 1971). Evidence presented by Emson (1966) on the effect of the acetylcholine receptor antagonist nicotine upon the choanocytes of *Cliona celata* GRANT 1826 is questionable, since non-physiological concentrations were used, which might have causeed cytotoxic effects. However, in the same work Emson (1966) showed the influence of adrenaline on the pumping behavior of *C. celata*, while Pavans de Cecatty (1971) found an increased contractile behavior in *S. officinalis*. Weyrer et al. (Weyrer et al. 1999b) established serotonin in the larvae of *Tedania ignis* (DUCHASSAING & MICHELOTTI 1864). These scattered demonstrations of neuroactive substance effects in sponges was completed by the cloning of a metabotrophic GABA/glutamate receptor in *Suberites domuncula* (OLIVI 1792) and the demonstration of GABA effecting intracellular Ca^{2+} concentrations (Perovic et al. 1999).

Quite in parallel to the research situation on cell biology (see above: *Sponge Body Contractions*), the research interest on integration in sponges seems to have declined between the 1980s and the dawn of the 21st century. However, at least several groups have kept up discussions on the topic (Mackie 1990, Pavans de Ceccatty 1989). In addition, during that period, Leys and co-workers demonstrated intracellular electrical potential changes and action potentials in the trabecular tissue of hexactinellid sponges (Leys and Mackie 1997, Leys and Mackie 1999, Leys et al. 1999, Mackie et al. 1983). These results provide clear evidence for the presence of voltage-gated ion channels in sponges, at least the Hexactinellida, as shown above.

Taking into account the complete historic sequence of discussions about the evolution of the first nervous systems it becomes evident that they lead to a number of misunderstandings. This seems to be quite obvious, as shown by the more or less open verbal attacks amongst exponents of the various hypotheses between the 1960's and the 1980's. Despite of the polarization, it has to be emphasized that the work of Jones (1962) is still the most comprehensive work of the publications. Some of his comments and conclusions, however, are now outdated. Notably, Mackie and Pavans de Ceccatty attempt a constructive debate of the topic, especially in their reviews (Mackie 1970, 1979, 1990, Pavans de Ceccatty 1974a, b, 1979, 1989). Their intention was not to manifest a nervous system in sponges. This has sometimes been misinterpreted by other authors. Mackie, whose own work mainly reflected aspects of the nervous systems of cnidarians, emphasized repeatedly that the understanding of integration in sponges is the key to the understanding of the evolution of the first nervous systems (Mackie 1979, 1990), which seems to contrast the opinion of Jones. According to Mackie, the pharmacological and histological evidence supports that the same mechanisms favoring the evolution of nervous systems in other metazoans acted in sponges. In this sense, sponges represent a prime example on the mechanistic possibility of typical metazoan features like the integrated mechanical response upon external stimuli without nerves or muscles (Mackie 1990). Pavans de Ceccatty supported similar views. His comprising neuromuscular model for sponges includes hard-wired inter-cellular signaling in combination with permanently morphologically re-modulating signaling pathways (Pavans de Ceccatty 1974a). This model is still up-to-date and is neither proven in all details nor refuted. The recent review (Leys et al. 2007) and reiteration of some new evidence (Meech 2008) has demonstrated the actuality and the interest on the topic. Recent physiological and molecular evidence and further ongoing work will be needed to provide more detailed evidence to validate or disprove the historic views.

Current Concepts on Sponge Contraction, Locomotion and Integration

Sponge contraction and locomotion

Phylogenetic context

In which sponge groups do we find locomotion and contraction? In a phylogenetic context, contraction and locomotion are unevenly distributed among the recognized sponge clades.

Within all sponge groups, contraction seems to be distributed wider than locomotion. Contraction is frequently found in Demospongiae (e.g. Arndt 1941, Marshall 1885, Nickel 2004, Schmidt 1866, for an example see Fig. 3a), Homoscleromorpha (unpublished data on *Oscarella lobularis* (Schmidt 1862), Fig. 3b) and Calcispongia (Minchin 1900, and see 3c). No contraction has been reported for Hexactinellida to date, which is not surprising, considering their rigid skeletons and their relatively low amount of cellular mass and extracellular matrix in relation to the silica skeleton (Leys et al. 2007). However, the same can be said about a number of Demospongiae clades. Those with rigid, spicule rich skeletons, like Petrosiidae Van Soest 1980 or Suberitidae Schmidt 1870 do not display significant external body contractions. Nevertheless, most of the species, if not all in these groups, will show oscular contractions (unpublished observations during SCUBA diving). In an x-ray microtomography based comparison of a glutamate treated specimen of *Suberites domuncula* and a control specimen, the treated specimen displayed internal contraction of

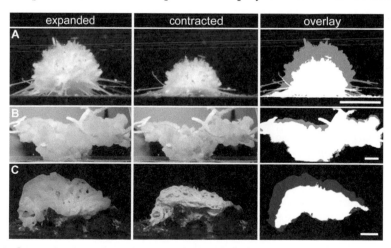

Fig. 3. Contraction in sponges. Line up of expanded (first column) and contracted states (second column) of sponge individuals and an overlay of both states (third column). Scale bars represent 5 mm. **A.** Demospongiae: *Tethya wilhelma* (see Nickel 2004). **B.** Homoscleromorpha: *Oscarella lobularis* (Nickel 2010). **C.** Calcispongia: *Clathrina clathrus* (Nickel 2010).

the aquiferous system (unpublished data). This situation is very similar to internal contractions in sponge species without such massive, rigid skeletons, which display external body contractions, e.g. *Ephydatia mülleri* (Lieberkühn 1855) (Elliott and Leys 2007). Whether Hexactinellida are able to perform such internal contractions or not has not been proven yet. However, it seems to be unlikely. If Hexactinellida do not contract, the question arises, whether (1) contraction is only a character of all Metazoa other than Hexactinellida or if (2) contraction in Metazoa is a primary character and Hexactinellida lost it secondarily. The later is more likely, since cellular contractility is also present in unicellular eukaryotes.

Movement of adult sponges on the substrate, in comparison, has only been demonstrated in demosponges (e.g. Bond 1992, Bond and Harris 1988, Fishelson 1981, Lieberkühn 1863, Nickel 2006, Pronzato 1999). It is likely that this will also be found in Homoscleromorpha, when investigated carefully under controlled conditions or *in situ*. Locomotion of adult or juvenile Hexactinellida or Silicispongia has not been reported to date. In contrast to contraction, which is quite widely distributed among sponges, locomotion in sponges seems to be an adaptation to unpredictable or unstable environments. This applies to many juvenile sponges, which have limited control over their environment chosen by phototactical or even chemotactical guidance of the larva prior to metamorphosis (e.g. compare Lieberkühn 1863 for movements of juvenile freshwater sponges). It also applies to moving species like *Tethya* spp. (Fishelson 1981, Sarà et al. 2001), presumably living in shallow coral reef communities or lagoons, where current, sediment and nutrient situation might fluctuate quickly on the microhabitat scale. Similar arguments can be addressed for light dependent species with associated photosynthetic microorganisms, like *Chondrilla nucula* Schmidt 1862 (Pronzato 2004), for which the light regime might change during the season, depending on growth competition with surrounding macroalgae.

Biomechanics of contraction

As shown above (see *Sponge body contractions: Perception in modern biology*), the question on the biomechanical mode of operation of contraction in sponges was frequently addressed and discussed contradictorily over the decades. However, most authors only focused on half of the biomechanical process, the contraction itself by the agonist system. Nevertheless, a further mechanism expanding the sponge during the second part of a contraction cycle has to be considered: the contraction antagonist.

For the contraction agonist, three mechanisms are possible and discussed: (1) mesohyle contraction by actinocytes (e.g. Elliott 2004, Pavans de Ceccatty 1960); (2) epithelial contraction by endopinacocytes (e.g. Bagby 1970, Leys 2007, Leys and Meech 2006, Nickel et al. 2006c);

or 3. a combination of both (e.g. Pavans de Ceccatty 1981, 1986). Present experimental data and theoretical biomechanical implications suggests that a third mode is most realistic: (1) contractile filaments are present in pinacocytes (Bagby 1970, Leys and Meech 2006, Matsuno et al. 1988, Pavans de Ceccatty 1981, 1986); (2) the density of actinocytes in the mesohyle of highly contractile sponges is not high enough to explain alone the strong contractility, leading to volume reductions of up to 70% (Nickel 2004); (3) the contraction primarily leads to a reduction of the volume of the canal system, while the mesohyle does not or only minimally contracts (Nickel et al. 2006c, Nickel 2010). All this evidence suggests that pinacocytes are indeed the major contractile agonists.

What about the antagonist system? If we regard the pinacocytes as the agonist, then we have to consider the following: a re-expansion of the pinacocytes after contraction would not necessarily produce the forces needed to reverse the narrowing of the aquiferous system elements. However, if one of the antagonist candidates is the system of choanocytes building up a water pressure inside the canals, then a second putative antagonist system is the mesohyle itself. Recent x-ray microtomography research demonstrated that during sponge contraction in *T. wilhelma* some regions of the mesohyle expanded rather than contracted (Nickel et al. 2006c, Nickel 2010). As a result, the network of extracellular matrix filaments is expanded, which was quantitatively confirmed by scanning electron microscopy (unpublished data). This phenomenon of elastic energy storage applies especially to very compact and collagen rich mesohyle regions of low cellular density. For mesohyle regions of lower collagen content but higher cellular density, the cells themselves may play active roles as antagonists. Expansion of the canals and consequently the whole sponge body might be supported by actinocytes within the mesohyle. During body contraction the actinocytes might be expanded or even stretched together with the mesohyle. During body expansion actinocytes might actively contract. However this is a speculative model supported by indirect evidence only. At present, *in vivo* microimaging of contracting actinocytes is missing. The only direct observation of contracting actinocytes demonstrated that they are involved in the active displacement of spicules during the formation of body filaments in *T. wilhelma* (Nickel and Brümmer 2004).

In fact, the kinetics of a complete contraction-expansion cycle in *T. wilhelma* (Nickel 2004) and *Ephydatia fluviatilis* (Weissenfels 1989, 1990) show a bi-phasic tendency for the expansion. This might be explained by a double-antagonist system: during the first phase of expansion, initiated by the reversion of the pinacocyte contraction, elastic energy stored in the mesohyle would be released quite quickly, leading to a first expansion phase. The second, slower phase of expansion, leading to the restoration of the full sponge body volume, would be caused by the pressure inside the aquiferous system produced by the choanocytes.

Biomechanics of locomotion

The biomechanics of sponge locomotion isn't as documented or understood like contraction. During locomotion over the substrate sponges do not alter the area of attachment to a significant degree. The first movement model was suggested by Fishelson (1981). His model is based on the assumption of muscle like forces created by actinocytes within body filaments extending from the globular sponge body to the surrounding substrate. This hypothesis was falsified by other authors who could not confirm any contractile tension within the filaments (Bond 1992, Bond and Harris 1988, Nickel 2006, Nickel and Brümmer 2004). It can be assumed that the movements of all sponges follow the same mechanisms, regardless of the presence of body filaments or not.

Bond and Harris (1988) were the first to suggest an amoeboid-like movement. The model is similar to the movement of fibroblast cells, which constantly reorganizes the cytoskeleton inside the cell during movement (Boal 2002). The model suggests that the movement of the sponge body is based on a constant reorganization of cells within the body (Bond 1992, Bond and Harris 1988). This mechanism is not to be confused with the growth of biomass by proliferation. Such a mechanism might play a role in body shape changes of some sponge species over very long time periods as observed in natural sponge communities (e.g. Pansini and Pronzato 1990). However, even in these cases, constant morphogenesis by cell movement plays an important role.

It was also suggested that sponge movement is dependent on body contraction. Such peristaltic modes of movement can be frequently found among invertebrates (Trueman 1975). However, this is not the case in sponges, where contraction and locomotion on the substrate are clearly independent (Bond and Harris 1988, Nickel 2006).

Open questions on contraction and locomotion

At present we can only exclude some hypothesis on the cellular mechanisms involved in sponge contraction and locomotion. It is clear, that no kind of muscular system is involved in either. We have seen that actinocytes, which have been morphologically compared to smooth muscle cells (e.g. Bagby 1966, Fishelson 1981), do not play a major role in contraction and locomotion.

The ongoing genome project on *Amphimedon queenslandica* HOOPER & VAN SOEST 2006 (aka *Reniera* sp. JGI-2005 WGS, searchable and downloadable at http://trace.ensembl.org/ and http://www.ncbi.nlm.nih.gov/Traces) will help to unravel cellular properties. This has been demonstrated in evolutionary developmental biology of sponges (Adamska et al. 2007a, Adamska et al. 2007b, Larroux et al. 2006). Using an *in silico* approach as described by Gauthier and Degnan (2008), preliminary local blast searches of the *A. queenslandica* WGS trace files using conserved sequences of smooth muscle cell typical proteins, like titin and desmin did not recover similar gene sequences (unpublished data).

Multidimensional life cell imaging methods (e.g. Hejnol and Schnabel 2006, Mori 1994) will be needed to assist in detailed analysis of the cellular contraction and movement dynamics during body contraction and locomotion. Intermediate filaments as well as receptor proteins and extracellular matrix elements involved in contraction in sponges are hardly ever investigated. Immunohistochemical methods using specific antibodies against these targets, eventually based on subsequent genomic analysis, will have to be forced (for an example see e.g. Wiens et al. 2001).

Integrative systems in sponges

Physiology of paracrine integration in sponges

As mentioned above, paracrine signalling is present in protists and is therefore evolutionary ancient. It is important to note that the release of any chemical substance by any cell is at first a result of the cell's present biochemical status and thus the cell's statement about it. Such a statement is *per se* not directed. It enters the level of a 'signal' only, if a 'receiver' is present, e.g. a receptor system on another cell. This principle holds true for uni- and multicellular organisms.

Obviously sponges react to environmental factors in a variety of ways, like all animals. All of these reactions involve intra- or intercellular signaling pathways, including second messenger systems, e.g. the temperature signaling cascade (Zocchi et al. 2003, Zocchi et al. 2001). However, such physiological pathways are not directly connected to coordination in the sense of nervous or nervous-like systems. Yet, such systems have been linked to discussions on the evolution of the nervous system (Jones 1962, Leys and Meech 2006). For the sake of clarity I will avoid mixing these topics.

At present, the physiology of sponge signal integration is mostly defined by pharmacological studies, on which I will be focusing on here. Earlier studies were reviewed by Jones (1962) as well as Leys and Meech (2006). However, in an extensive study by Ellwanger and Nickel (2006), quantitative data was added, which was not included in any previous reviews.

Among transmitter substances, Acetylcholine (ACh) is well investigated. In fact, the first pharmacological studies on receptors are related to the muscarinic and nicotinic Ach receptor types (Dale 1914). While the nicotinic receptors are ligand-gated ion channels, the muscarinic types are metabotrophic systems. They play important roles in the CNS of all phyla (Walker et al. 1996, Walker and Holden-Dye 1991). For sponges, specific acetylcholine esterase (AChE) activity has been found in *Sycon* sp. (Lentz 1966), *Hippospongia communis* (LAMARCK 1814) (Thiney 1972) and also in *T. wilhelma* (Nickel 2001). However, no AChE activity could be demonstrated in earlier studies, though several species were investigated: *Spongilla lacustris* (LINNAEUS 1759) (Mitropolitanskaya 1941), *Scypha* sp. (Bullock and Nachmansohn 1942), *Leuconia asperta* (SCHMIDT 1862) and *Siphonochalina crassa* TOPSENT 1925 (Bacq 1947). On the other hand, in physiological studies by Pavans de Ceccatty (Pavans de Ceccatty 1971), ACh increased the rhythm and intensity of contraction in *H. communis*. AChE activity is also in *T. wilhelma* cells (Nickel 2001). In contrast to these results, physiological tests did not reveal a significant correlation between ACh and contraction. ACh did not induce contraction in *T. wilhelma*, but it did alter contraction frequency (Ellwanger and Nickel 2006). Eventually, it was shown that ACh is involved in the regulation of body extension formation and retraction (Nickel 2001). In the same study, nicotine did not affect contraction and rhythm in any way (Ellwanger and Nickel 2006). Even though the results on ACh are contradictory, it is likely that specific ACh esterases and receptors are present in at least Demospongiae and Calcispongia.

Caffeine is a non-specific antagonist of adenosine-receptors in the mammalian brain. Its effect on the contractile behavior of *T. wilhelma* was also tested (Ellwanger and Nickel 2006). Caffeine induced contractions, increased the endogenous rhythm and attenuated the amplitude of endogenous contractions. Since the onset of caffeine-induced contraction seems to be decelerated, it must not act directly upon the contractile cells. From the disturbed endogenous rhythm and contraction amplitude, it seems likely that caffeine interferes with the regulation of endogenous contractions (Ellwanger and Nickel 2006). A putative adenosine receptor may be part of the pacemaker mechanism of endogenous contractions. In mammals, adenosine receptors are ubiquitous and important throughout all kind of tissues and cell types. They have also been shown to cause chronotropic effects in the heart (review in Klotz 2000). Adenosine receptors are all metabotrophic and link the ATP-driven energy metabolism to the signaling pathways of the cells. The adenosine signaling pathway strongly interacts with other messenger substances (Fredholm et al. 2000). Walker and co-authors (1996) concluded that such a central signaling mechanism evolved early and plays an important role in all invertebrates.

Amino acid transmitters have also proved to be effective in sponges. In *T. wilhelma*, glycine induced effects are similar to caffeine: it stimulated contraction, fastened the endogenous rhythm, and attenuated the amplitude (Ellwanger and Nickel 2006). Together with GABA and glutamate, which are also effective in sponges (Ellwanger et al. 2006, Perovic et al. 1999, Ramoino et al. 2007), glycine is one of the more important amino acid transmitter in the vertebrate brain (Oberdisse et al. 1997). It acts as a transmitter in a wide range of invertebrates, e.g. in *Hydra vulgaris* PALLAS 1766 as an inhibitory transmitter (Pierobon et al. 2001). The results of Ellwanger and Nickel (2006) indicate that glycine is involved in the regulation of endogenous rhythm, not upon the contractile cells directly. Similar results can be found for GABA- and glutamate-induced contractions in *T. wilhelma* (Ellwanger et al. 2006). From these results, together with the characterization of a GABA/glutamate-like receptor gene in *Geodia cydonium* (Perovic et al. 1999), the observation of a slower current under GABA exposure for *Cliona celata* (Emson 1966) and the immunohistochemical demonstration of GABA$_B$ receptors in *Chondrilla nucula* (Ramoino et al. 2007), it is evident that GABAergic signalling pathways exist in sponges.

The catecholamine adrenaline is another substance, which was tested on sponges independently. A slight reduction of pumping rates was demonstrated in *Cliona celata* (Emson 1966). Lentz specifically stained adrenaline-containing bipolar and multipolar cell types in *Sycon* sp., using histochemical methods (Lentz 1966). In *T. wilhelma*, the strict limitation of effects to the elongation of endogenous contraction cycle duration in *T. wilhelma* suggests that adrenalin is directly involved or at least interfering with the regulation of the endogenous contraction rhythm (Ellwanger and Nickel 2006). Adrenaline is known to upregulate genes of the molecular clock in mammals and is therefore involved in the feedback loop regulation of circadian rhythm (Shibata 2004). In this context, it is interesting that both, Reiswig and Nickel, independently reported on diurnal rhythms in *Cryptotethya* sp. and *T. wilhelma* (Nickel 2004, Reiswig 1971). In conclusion, adrenaline might directly be involved in the regulation of biological rhythms in sponges.

The biogenic amine serotonine was demostrated to be present in sponges using histochemical and imunohistochemical methods in *Sycon* sp. and *Tedania ignis* (Lentz 1966, Weyrer et al. 1999a). However, Emson found no effect upon *Cliona celata* (Emson 1966). In contrast, serotonine did immediately induce contraction in *T. wilhelma* (Ellwanger and Nickel 2006). The effect is limited: the endogenous contraction rhythm is not affected. However, the widespread findings of serotonine in all phyla of the metazoa, acting via ionotrophic and metabotrophic receptor types, points to an early evolution of this central signalling system (Walker et al. 1996, Walker and Holden-Dye 1991).

Nitric oxide (NO) is a gas which plays a central role in the regulation of a variety of physiological processes in vertebrates and invertebrates (Colasanti and Venturini 1998, Jacklet 1997). It interacts with a soluble guanylate cyclase (sGC) and is released by nitric oxide synthase (NOS) during conversion of l-arginine to l-citruline, mainly regulated by intracellular Ca^{2+}. NO diffuses freely within the animal body and its range is limited by its short half life. The molecule is supposed to be an ancient auto- and paracrine messenger (Walker et al. 1996). NOS was shown to be present in the sponge *Axinella polypoides* SCHMIDT 1862, and it was involved in the temperature signalling cascade (Giovine et al. 2001). Ellwanger and Nickel demonstrated that NO induces contraction in *T. wilhelma*, but also modulates the endogenous contraction rhythm and amplitude. Due to its fast diffusion, the short half life and the limited range, NO is an optimal candidate for a contraction-inducing messenger, which acts in an auto- and paracrine manner. It is likely involved in feedback loops in the cellular pacemaker machinery of endogenous rhythm (Ellwanger and Nickel 2006).

Contractions in sponges were also induced by cyclic AMP (Ellwanger and Nickel 2006). This is clearly an extracellular signaling event, which is most likely modulated by a specific cAMP receptor, similar to *Dictyostelium discoideum* RAPER 1935 (Lusche et al. 2005). Extracellular cAMP plays several regulative roles in Porifera: it is involved in the regulation of production and development of gemmules in freshwater sponges (Simpson and Rodan 1976), and it directly affects the locomotion of dissociated cells of *Clathrina cerebrum* (HAECKEL 1870) (Gaino and Magnino 1996). It may therefore link cellular mobility and body locomotion with contraction in sponges. All three phenomena are important aspects in the life of *Tethya* (Bond 1992, Bond and Harris 1988, Fishelson 1981, Nickel 2001, 2004, 2006).

Summarizing this still rather patchy image of paracrine integration in sponges, it is safe to conclude that a variety of paracrine pathways are present in sponges, even though the behavioral repertoire of sponges is relatively low. However, transmitter substances are known to play important roles in unicellular eukaryotes, which tells us that these paracrine pathways are evolutionary ancient (Walker et al. 1996, Walker and Holden-Dye 1991). Most likely, they first acted as autocrine or intracrine messenger in unicellular eukaryotes (for the principal compare e.g. Ramoino et al. 2003, Ramoino et al. 2004, Ramoino et al. 2005). Consequently, it is not surprising to find such mechanisms in action in sponges. However, a lot of integrative research (i.e. physiological, molecular and morphological studies) will be necessary to unravel the details of these signaling networks.

Physiology of electrical integration in sponges

Apart from the extensive work on electrical integration in Hexactinellida, reports on this type of integration in other poriferans are rare. The reports of action potentials and impulse conduction in glass sponges are well documented and reviewed elsewhere (Leys and Mackie 1997, Leys and Mackie 1999, Leys et al. 1999, Leys et al. 2007, Leys and Meech 2006, Mackie et al. 1983). For this reason, I will not go into further detail on this subject. Although the demonstration of action potentials in hexactinellids is spectacular *per se*, it does not provide the key for understanding the early evolution of the nervous system. Clearly it demonstrates that voltage-gated ion channels are present in Hexactinellids and most likely also in all other sponges. However, because the action potential travels through a syncytical tissue, signal transduction is not related to synaptic signal transition. This is displayed by the nervous system typical sequence of a double transition from an intracellular electric signal to an intercellular chemical signal and back to an intracellular electrical system again. On the other hand, action potentials and specific ion channels are known from protists, plants and bacteria (e.g. Mackie 1970, Schmidt-Rhaesa 2007, Schultz et al. 1986), so they are evolutionary ancient, too. For this reason, it is highly likely, that specific mechano-gated, voltage-gated and ligand-gated ion channels are present in all sponges.

No intracellular electrical recordings have been recorded so far from sponges except Hexactinellida. This is most likely due to problems related to cell size and the presence of mineral spicules in the tissue. However, experiments altering ion concentrations brought up evidence for the involvement of ion channels. Investigations upon the influence of extracellular ion concentrations pointed to the dependence of oscule contractions upon mono- and bivalent cations with interchangability for Na^+, K^+ and L^+, respectively Ca^{2+}, Mg^{2+} and Sr^{2+} in *Microciona* sp. (Prosser 1967, Prosser et al. 1962). In contrast, Na+ was not substitutable by K+ and Mg^{2+} or by Ca^{2+} in *S. officinalis* (Pavans de Ceccatty 1971). The direct application of electrical currents resulted in oscular contractions in the demosponge *Ephydatia fluviatilis* (McNair 1923).

Changes in cell potential are therefore likely to play a role in signal integration in all sponges. For this reason, electrical synapses (gap junctions) might also be involved in signal transduction, at least temporarily. Macke (1970) suggested that electrical synapses (gap junctions) likely predated chemical synapses in evolution and may have been the foundation of a neuroid pre-nervous system. This view was supported by Loewenstein (1967) who demonstrated the electrical coupling of *Haliclona* sp. cells. Both works resulted in the intensified search for gap junctions in sponges.

Although some ultrastructural evidence was presented for gap junctions in sponges (Gaino and Sarà 1976, Green and Bergquist 1979) it was not confirmed by the same and other researchers (Garrone and Lethias 1990, Garrone et al. 1980, Green and Bergquist 1982, Lethias et al. 1983). The presence of gap junctions in sponges remains an open question.

Morphology of integration systems in sponges

The morphology of a putative integrative system in sponges is still a matter of speculation. As described above (see *The quest for the elementary nervous system*), the observation of contraction led to a number of histological studies after 1880 aimed at finding the contractile cells and nerve cells in sponges. Early evidence was presented for bipolar and multipolar cells interpreted as sensory and ganglion cells (Lendenfeld 1889), fusiform cells seen as nerve cells ('aesthocytes', Sollas 1888), flask-like cells regarded as sensory cells (Dendy 1891) and so on. It can be safely stated that at least Demospongiae and Calcispongia posses a number of cell types which morphologically resemble cell types of the nervous system in other metazoan groups (for a compendium see Boury-Esnault and Rützler 1997, De Vos et al. 1991, Harrison and De Vos 1991). In most cases morphology was the only evidence for attributing functions of an integrating system (for a detailed review see Jones 1962, Lentz 1968).

Several histochemical methods which are used to specifically stain nerve cells (e.g. silver, Masson's trichrome, etc.) were found to stain a variety of sponge cell types in Demospongiae (Pavans de Ceccatty 1955, 1959, 1960) and Calcispongia (Pavans de Ceccatty 1955, Tuzet et al. 1952). Consequently they were regarded as nerve-like cell by several authors, even if the morphology did not match any known cell types of the nervous system of other animal groups (Jones 1962).

Some more specific evidence was presented by Lentz (1966, 1968), who used neurochemical methods to specifically demonstrate neurosecretory substances or enzymes in sponge cells. His results on acetylcholine esterase activity and the presence of serotonin in sponges was confirmed by others (Nickel 2001, Weyrer et al. 1999b). Although these results demonstrate that biochemical modules found in nervous systems are present in sponges, it does not necessarily mean that it is correct to regard these cells as nerve cells. More specific studies, combining immunohistochemistry and neurochemistry, are needed to reveal a more detailed image of how these modules are morphologically distributed among sponge cells.

Molecular biology of integration in sponges

A number of recent studies have focused on characterizing sponge genes, othologous or paralogous to nervous system specific genes of other metazoa. One of the earliest examples is the molecular and functional characterization of a metabotrophic glutamate/GABA receptor type in *Geodia cydonium* (Perovic et al. 1999). The authors concluded that "these findings suggest that the earliest evolutionary metazoan phylum, the Porifera, possesses a sophisticated intercellular communication and signaling system, as seen in the neuronal network of higher Metazoa". The concept of demonstrating typical molecular modules of the nervous system in sponges tempts to deduce a general architecture of an evolving nervous system in the 'Urmetazoa' or the LCAM. However, this is a path of trial and tribulation. As long as the detailed function of such sponge genes and their corresponding proteins within the cells and tissues are not characterized under a proper consideration of the morphological, ultrastructural and physiological context, the conclusions can only be preliminary. The same basic conditions apply for genome or large scale EST sequencing projects.

An EST project on the homoscleromorph *Oscarella carmella* MURICY & PEARSE 2004 provided support for the presence of components of vesiclerelated signaling pathways (Jacobs et al. 2007, Nichols et al. 2006). These included synaptogamin, SNARE-complex components similar to syntaxin, neurocalcin, as well as a slit-gene, typically involved in axon guidance (Jacobs et al. 2007). The authors of these reports stated that "these recent observations in sponges suggest the high activity of equipment involved in vesicle transport, and the presence of some synaptic and developmental signaling components typically associated with bilaterian neural systems. Given that sponges lack formal synapses, it is worth noting that non-synaptic communication between cells via calcium waves can occur through a variety of mechanisms." (Jacobs et al. 2007). As the authors correctly add, these genes also represent other functions in eukaryotic organisms. Final conclusions will have to wait until expression studies and pharmaceutical tests provide further supporting evidence.

By analyzing the genomes of demosponge *Amphimedon queenslandica* and the cnidarian *Nematostella vectensis* STEPHENSON 1935, Sakarya et al. (2007) demonstrated the presence of 36 members of gene families involved in the post-synaptic scaffold in so called 'lower metazoans'. Amongst them are voltage gated ion channels, ionotropic and metabotropic glutamate receptors. Obviously aware of the problem stated above, the authors cautiously stated that "Highly conserved protein interaction motifs and co-expression in sponges of multiple proteins whose homologs interact in eumetazoan synapses indicate that a complex protein scaffold was present at the origin of animals, perhaps predating nervous systems. A relatively small number of crucial innovations to this pre-existing structure may represent the founding changes that led to a post-synaptic element". The first assumption that the presence of the molecular modules of the scaffold predates the nervous system will have to be kept in mind. In fact, as we will see, this is a prerequisite for the evolution of nervous system *sensu stricto*. On the other hand, whether or not the presence of these proteins is equivalent to the presence of a highly organized structure like the post-synaptic scaffold is subject to discussion. I will return to this point in my consideration on the early evolution of a synapse-less pre-nervous integration system during phylogenesis of the Metazoa. Sakarya et al greatly demonstrated the value of genomic studies for reconstructing the biochemical repertoire of a theoretical LCAM *in silico*. This will inspire and provide a working hypothesis for a number of functional and morphological studies to further characterize the obvious integrative capacity of pre-nervous systems in those metazoan groups which first branched off during evolution.

Co-Evolution of Contraction and Integration: Pre-Nervous Milestones in Sponges

The lack of early evolutionary scenarios

The evolution of the nervous system has been discussed extensively in the last century by numerous authors from Parker (1910) to Meech and Mackie (2007). Nevertheless, there is still a significant lack of theories on the early evolution of integrative systems in Metazoa. Most of the publications focus on the evolution of excitability in Metazoa. As a result, the neuron concept (i.e. Barbara 2006, Cajal 1937) has played a major role in hypotheses on the nervous system evolution. This observation is valid for all major reviews and textbooks on neurobiology (e.g. Bullock and Horridge 1969, Grundfest 1959, 1965, Hanström 1928, Jones 1962, Leys 2007, Leys and Meech 2006, Mackie 1979, 1990, Meech and Mackie 2007, Pantin 1952, 1956, Parker 1910, 1919, Passano 1963, Schmidt-Rhaesa 2007). Consequently, most authors focus on the structural evolution of complexity in the nervous systems of bilaterian animals, trying to reconstruct "the nervous system of Urbilateria" (Ghysen 2003). The reason for this situation is probably best understood by referring to Bullock's and Horridge's (1969) statement in their attempt to present 'defining features of a nervous system': "A nervous system may be obvious and easily identified, as in higher groups, at least in its central parts; or it may be extremely difficult either to recognize or to deny, as in the multicellular sponges and the highly coordinated ciliate protozoans (heterotrichs)." By referring to the concept of 'dependant effectors' of Parker (1910, 1919), Pantin (1952, 1956) and Passano (1963), they continue to argue that "The relevance of the definition for sponges, and other metazoans in which a nervous system is to be demonstrated, is that it provides anatomical and physiological criteria to be satisfied by any true nervous system. Thus presumptive nerve cells must be shown to connect receptor sites with effectors or to connect with each other. Cells that send prolongations to end in open tissue or spaces or on spicules are unlikely to be nervous, unless other argument clearly exist." (Bullock and Horridge 1969).

Even though it is not clearly stated by most authors, they all assume the presence of chemical synapses in combination with electrical conductions over long distances. This should be a major defining character of any 'true nervous system' (i.e. Bullock and Horridge 1969, Grundfest 1959, Hanström 1928, Schmidt-Rhaesa 2007). This is influenced by the 'neurone doctrine' of Ramon y Cajal and Heinrich Wilhelm von Waldeyer-Hartz (for historic overviews see i.e. Barbara 2006, Smith 1996). A historic overview on basic concepts of nervous system evolution is given by Hanström (1928). Basically, all of the theories center on the differentiation of 'receptor' and 'effector' elements in the first nervous systems. Oscar and Richard Hertwig (1878) assumed that muscle cells, sensory cells and ganglion cells evolved from indifferent epithelial cells under constant interactions. Parker (1919) suggested that independent primitive receptor-effector cells received stimuli and react by appropriate activity, i.e. contraction. Parker assumed that in the course of metazoan evolution, receptors and effectors were separated in space and structure, but remained connected through axons. Although the term 'synapse' and the connected concept was given earlier (Sherrington 1897), it did not play a major role in the Parker's and subsequent theories.

But which evolutionary sequence resulted in this morphologically and physiologically highly specialized system? Grundfest (1959) suggested that neurosecretion—and not contraction—represents the ancient effector activity of Parker's primitive receptor-effector cells. He continues that "Each cell [of receptor and effector type] retained its activity both as a receptor and effector, though maintaining these functions in a modified and often specialized form." (Grundfest 1959). Lentz referred to Grundfest's conclusions when he reported on transmitter activity in sponges (Lentz 1966, 1968). However, from his results he concluded that synapses must be present in sponges and would have to be confirmed ultrastructurally or morphologically. At present, there is no ultrastructural evidence for synapses in sponges (Jones 1962, Leys et al. 2007, Mackie 1990, Meech and Mackie 2007, Simpson 1984). Due to genomic research (see above: *Molecular Biology of Integration in Sponges*) and the finding of a high number of genes related to the post-synaptic scaffold (Sakarya et al. 2007) and other synapse-related genes (Jacobs et al. 2007), the idea prevails that sponges might possess synapse-like cellular junctions: "The core potential for evolving synapses in sponges may extend to other types of junctions." (Sakarya et al. 2007). Statements like this reveal quite a traditional view on the evolution of the nervous system, which seems not to plumb the depths of possible scenarios. In this case, it is implied that synapses evolved from other type of cell junctions. Such a scenario is unlikely, given the ultra- and nanostructural and thus genetic differences between synapses and cell junctions like gap junctions, tight junctions, or desmosomes. In the following paragraphs, I will present a hypothesis on the early evolution of a (pre-)nervous integration system combining the views of Grundfest (1959), with those on compartmentalization of signaling in neurons (Mattson and Bruce-Keller 1999), genomic evidence (Jacobs et al. 2007, Sakarya et al. 2007), histological and morphological evidence (as reviewed by Jones 1962, Pavans de Ceccatty 1979, 1989, Simpson 1984) and physiological evidence (for details see Ellwanger et al. 2006, Ellwanger and Nickel 2006, Lentz 1966, 1968, Leys and Meech 2006 and other authors referenced in here).

The Paracrine-to-electrochemical-dominance transition hypothesis

Axioms

The following conclusions from the literature discussed above will serve as axioms for the hypothesis on the early evolution of the nervous system in Metazoa presented in the following two subchapters:

1. *Unicellular eukaryotes represent independent effectors or receptor-effector cells* sensu *Parker (1919)*. Explanation: ciliates are capable of producing action potentials by way of voltage-gated ion channels as an answer to external stimuli (Schultz et al. 1986).

2. *Unicellular eukaryotes use auto- and paracrine signaling pathways for intercellular communication*. Explanation: The release of signalling substances as statements of the biochemical status of a cell is pleisiomorphic. Ciliates and other protists are able to release several signaling substances and posses receptor systems for specific transmitter substances (Ramoino et al. 2003, Ramoino et al. 2004, Ramoino et al. 2005, Walker et al. 1996, Walker and Holden-Dye 1991).

3. *The sponge body constitutes a network of dependent effectors or receptor-effector cells* sensu *Grundfest (1959)*. Explanation: Sponge cells are capable of producing and releasing neurotransmitters (Lentz 1966, 1968, Nickel 2001, Thiney 1972, Weyrer et al. 1999b). They possess the gene and protein repertoire for neurotransmitter receptor systems (Jacobs et al. 2007, Perovic et al. 1999, Ramoino et al. 2007, Sakarya et al. 2007). They specifically react upon neurotransmitters and agonist/antagonists of receptor systems involved in the nervous system (Ellwanger et al. 2006, Ellwanger and Nickel 2006, Emson 1966, Perovic et al. 1999). This, however, does not imply the presence

of synapses or direct cell contacts. The network is mainly connected by chemical messengers (Ellwanger and Nickel 2006).

4. *Contraction in sponges is mainly mediated by epithelial cells.* Explanation: Pinacocytes are the main contractile effectors in sponges (e.g. Bagby 1970, Leys 2007, Leys and Meech 2006, Matsuno et al. 1988, Nickel et al. 2006c, Nickel 2010), eventually assisted by mesenchymal actinocytes (e.g. Pavans de Ceccatty 1981, 1986). No specialized muscles are present.

5. *Electrochemical potential changes coupled to intra and extracellular signaling pathways are a common character of all Metazoa.* Explanation: Hexactinellids display action potentials (Leys 2007, Leys and Mackie 1997, Leys and Mackie 1999, Leys et al. 1999, Mackie et al. 1983), Demospongiae display potential changes (Carpaneto et al. 2003, Loewenstein 1967). Similar mechanisms are present in ciliates (Schultz et al. 1986).

6. *Trophic functions of neurotransmitter-specific receptor systems predate integrative functions in metazoan (pre-)nervous systems.* Explanation: Neurotransmitters like GABA and glutamate are evolutionary ancient (Walker et al. 1996, Walker and Holden-Dye 1991). In Metazoa they are also involved in axon guidance (Ruediger and Bolz 2007) and trophic functions in the nervous system ontogeny (Nguyen et al. 2001), which are most likely evolutionary ancient functions, too.

7. *Synapses* sensu stricto *are an autapomorphy of the Eumetazoa.* Explanation: Synapse structures are present in Cnidaria and Ctenophora as well as in all Bilateria (Bullock and Horridge 1969, Hanström 1928, Schmidt-Rhaesa 2007).

The hypothetical paracrine pre-nervous system in the first metazoa (LCAM)

Unicellular eukaryotes like ciliates communicate via secretory pathways, coupled to metabotrophic intracellular signaling pathways and voltage-gated ion channels (Fig. 4A). Consequently, it is likely that the intercellular communication within the first Metazoa was based on the same mechanisms. Thus, a pre-nervous integration system consisting of receptor-effector cells *sensu* Grundfest (1959) must have been present from the beginning of multicellularity in the Metazoa. It can be regarded as a key feature of the *Last Common Ancestor of Metazoa* (LCAM). Its function must have closely resembled the auto- and paracrine signalling mechanisms used by unicellular eukaryotes, except that chemical messengers are released into the extracellular matrix and putative body cavities instead into the environment (Fig. 4B). Such extracellular but intracorporeal chemical signals dominate the whole integration system on the spatial

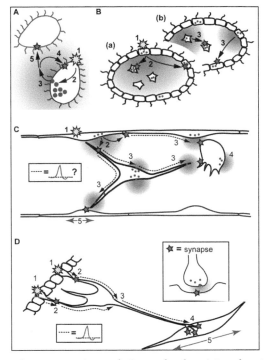

Fig. 4. Explanatory scheme presenting evolutionary levels as integral parts of the 'paracrine-to-electrochemical-dominance transition' hypothesis. For details refer to the text. Diffusion gradients **A.** Level of auto- and paracrine signalling in unicellular eukaryotes. (1) External stimulus followed by (2) a intracellular signal leading to (3) the release of a chemical messenger mediating (4) autocrine and (5) paracrine reception. **B.** Hypothetical level of the LCAM: (1) a sensed extracorporeal stimulus results in (2) the release of a paracrine messenger into the extracellular matrix (blue gradient). This diffusing signal reaches receptive cells which act as receptor-effectors *sensu* Grundfest including (3) secretion of secondary chemical signals (integration) or the same chemical substance (positive feedback). These secondary signal gradients (orange gradients) differ from the primary signal gradient and eventually reach other cells. **C.** Hypothetical intermediate level, eventually present in Demospongiae: again (1) an external stimulus is received by epithelial cells, followed by (2) the release of a chemical messenger (blue gradient). This signal is received by neighboring epithelial cells and mesenchymal multipolar cells. It locally triggers the opening of ligand-gated ion channel in the receptor cell, which (3) results in a potential change within the whole cell or a spreading local potential change over the cell body (action potential). The incurrent of specific ions (e.g. Ca^{2+}) and/or secondary messenger systems locally trigger the secondary release of extracellular messengers (orange gradients). During the proposed evolutionary transition, the secretory complexes are compartmentalized within the cells. The secondary signal is received by other cells and triggers secondary effects like (4) secretion, cell locomotion and (5) contraction of epithelial cells. **D.** Level of a simple nervous system representing the simple reflex-arc sensu Parker, from (1) the incoming external signal at epithelial receptor cells over (2) synapses to a nerve cell, generating excitatory or inhibitory postsynaptic potentials. This neuron (3) might integrate signals from separate sensory cells and generate an outgoing signal at (4) a synapse towards a muscle cell, representing a highly differentiated contractile effector cell (5).

Color image of this figure appears in the color plate section at the end of the book.

and temporal scale: signal molecules travel the distances between the cells by way of diffusion which is relatively slow. Consequently, chemical gradients of messenger molecules are present throughout the body.

Morphology strongly influences the spatial structure of the signal's diffusion gradients (e.g. cellular density, local extracellular matrix composition and density, body cavities, etc.). In this way the 3-dimensional diffusion gradients are shaped secondarily. Depending on differentiation, cells react upon incoming chemical signals by either the release of the same substances (positive feedback) or the release of other signal substances (chemical integration) including specific antagonists (negative feedback). More complex intracellular signaling pathways like molecular biological clocks (Ishida et al. 1999) lead to delayed release of substances (temporal integration). Cells dominated by such pathways can be regarded as the first pacemakers of the LCAM's pre-nervous system.

Electrochemical changes of the cell potential mediated by Ca^{2+}- and presumably other ion channels are likely to have played important roles in the control of intracellular biochemical pathways (Leys and Meech 2006). If syncytical tissue was present in the LCAM to a larger degree—for which we have no evidence—electrical conduction might have played an early role in intrasyncytical communication over longer distances, like in Hexactinellida (Leys et al. 2007). However, this seems to be a derived character of the later.

As stated by Grundfest (1959), contraction of certain specialized cells was the second effector mechanism in evolutionary early coordination systems besides the general ability of neurosecretion manifested in every cell type. In all probability, contraction was not the exclusive function of these cells. Instead, most likely the first contractile cells were the epithelial cells of the LCAM. The cellular contractile apparatus did not resemble the highly ordered, fast acting contractile intermediate filament system of muscle cells, but were less specialized like contractile endothelial cells of vertebrates or pinacocytes of sponges.

Recent results on morphology and physiology of integration systems in demosponges suggest that such a non-directional paracrine pre-nervous system is the main integration system (Ellwanger and Nickel 2006). Whether or not some groups of Demospongiae *sensu stricto*, Homoscleromorpha or Calcispongia took further evolutionary steps as described below remains an open question to date.

The paracrine-to-electrochemical-dominance transition

At some phases during the evolution of the nervous system *sensu stricto*, (Bullock and Horridge 1969, Hanström 1928) selective pressures must have favored (1.) directed communication and (2.) faster communication (in comparison to diffusion of molecules). This lead to further cellular specialization and compartmentalization as well as a transition of the temporal and spatial dominance from paracrine signal transduction to electrochemical signal transduction.

Contrary to Parker (1919), and following Hertwig and Hertwig (1878), I assume a parallel evolution of highly specialized contractile effectors (muscle cells), sensory cells and ganglion cells from non-specialized epithelial cells. A variety of selective pressures must have favored compartmentalization of the body and faster contractile responses to external stimuli. Namely, the spatial concentration of contractile cells into defined cellular layers and eventually contractile cellular sheets must have privileged a co-evolution of faster contraction mechanisms and directed intercellular communication. It is likely that the evolution of these cellular differentiations took place parallel to the evolution of the mesoderm. However, this will have to be proven, since the evolutionary origin of the mesoderm is presently under discussion (Martindale et al. 2004, Seipel and Schmid 2006, Spring et al. 2002).

The trophic and morphogenetic effects of several neurosecretory molecules predated the evolution of their transmitter function. They modulated growth and elongation of cells, frequently guided by gradients of neurosecretions. Like all other cells, these elongated cells were capable of reception and neurosecretion by itself. If faster signaling is favored by evolutionary pressures, slight accidental structural accumulation of membrane bound and cytoplasmic signaling pathway proteins would have enhanced the speed of intracellular signal transmission. Such intracellular accumulation might have been influenced by the mentioned neurosecretory cell guidance effects. However, slight unequal distributions of secretion and receptor complexes would lead to an alteration of the neurosecretion gradient fields around these cells (Fig 4C). The elongation of cells and the concentration of the neurosecretory machinery towards the cellular ends in conjunction with electric potential changes within the cells would have resulted in several benefits: (1) cost-benefit for neurosecretion would increase, since it is localized; (2) the temporal dominance of messenger diffusion would be balanced by a higher spatial share of the faster intracellular electric conduction; (3) autocrine effects (external feedback loops) would be minimized. A decrease of distance between the site of neurosecretion and putative receptor sites in neighboring cells would result in further effects: (1) it would increase signal specificity; (2) it would further increase cost benefit; due to a decreasing dilution over diffusion distance, it would be possible to obtain the same or even higher levels at the receptor sites, by releasing lower messenger concentrations; (3) autocrine effects would be completely eliminated, enabling strictly directed signal transduction between single cells.

Gradually such an evolutionary scenario would favor further compartmentalization within the cell and thus slowly result in the formation of pre-synaptic structures and finally synapses (Fig. 4D). During this evolutionary course, the dominance of paracrine signal transduction by means of time and distance is broken. In other words, the temporal and the spatial share of the paracrine subsystem of an evolutionary early (pre-) nervous system is minimized. Instead, the faster intracellular electrical signal transduction covers larger distances between the cells thus gaining temporal and spatial dominance within the metazoan integration system. This scenario is a prerequisite for the evolution of spatial focused two-way-transmission as found in the nerve nets of cnidarians and ctenophorans. This is also true for the polarized one-way-transmission which dominates centralized nervous systems (Anderson 1985, Bullock and Horridge 1969, Meech 2008).

Phylogenetic Considerations and Open Questions

This leads me to the obvious question: can we provide phylogenetic evidence for this scenario? Are we able to identify groups among recent Metazoa which can be regarded as representatives of intermediate stages in the paracrine-to-electrochemical-dominance transition? The fact that sponges possess synaptic protein complexes but do not display functional synapses qualifies them as candidates. Clearly, all higher Taxa —Hexactinellida, Demospongiae *sensu stricto*, Homoscleromorpha and Calcispongia—display pre-nervous integrative systems. However, we still know less about these systems in comparison to other integrative systems, like the nerve-nets of cnidarians. We will need to include expression studies as well as immunohistochemical cLSM and TEM studies to demonstrate the distribution of synaptic proteins among sponge cell types, and especially on the sub-cellular level.

The phylogenetic relations among all of the basal branching metazoan groups will also be essential. The phylogenetic position of Placozoa will be of special interest (e.g. Collins et al. 2005, Schierwater and DeSalle 2007), as well as the scenario of paraphyly among sponges (e.g. Borchiellini et al. 2001). Neither groups display nervous systems (Fig. 5a). At present, the literature on integrative systems in sponges hardly distinguishes any results between Hexactinellida, Demospongiae, Homoscleromorpha and Calcispongia. This will have to be sorted out. The paraphyly scenario that has each of these four clades branching out separately from the metazoan stem lineage, as well as the possibility of a sponge-like LCAM implies far-reaching possibilities. Each of the four taxa might in fact represent different graded stages of the early evolution of the pre-nervous system. Studies involving members of all taxa are necessary to complete this picture. On the other hand, recent phylogenomic anlyses supported the classic view of monophyletic Porifera (Philipp et al. 2009, Pick et al. 2010). In this case it will be interesting to investigate which of the poriferan pre-nervous system modules are dapomorphies and which arr plesiomorphies inherited from the LCAM (Fig. 5b).

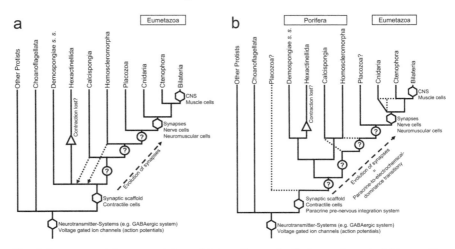

Fig. 5. Summary of the co-evolution of integration and effector systems in Metazoa. The phylogenetic tree follows (a) Erpenbeck and Wörheide (2007), Borchiellini et al. (2001) and others and (b) Philippe et al. (2009) and Pick et al. (2010). The evolution of the ultrastructure of synapses can presently not be resolved, but the synapse gene repertoire must be pleisiomorphic to all Metazoa. In this context, the relative phylogenetic position between sponges and the Placozoa is not crucial and resolving this problem will most likely not add any information to the question of synapse evolution. In contrast, the phylogenetic position of the Ctenophora provides essential information, since they display true synapses. For further details refer to the text.

Recently, Dunn et al. (Dunn et al. 2008) suggested that Ctenophora might represent the most basal branching metazoan clade. The support for this hypothesis is relatively weak at the moment. However, it is worth thinking about it in the context of nervous system evolution. Since ctenophorans posses synapses and nerve cells are arranged in a nerve nets (Bullock and Horridge 1969, Hernandez-Nicaise 1973, Schmidt-Rhaesa 2007) the result would be that it either evolved independently or that the LCAM possessed a synaptic nervous system. In this case, the synapse- and nerve-less condition is the sponge clades and the Placozoa would be secondary. This scenario seems to be less parsimonious than others, which consider sponges and Placozoa to represent the basal branches of the metazoan tree (Fig. 5).

Conclusions and Outlook

The history of perception of sponges as animals is full of controversial discussions and miss-interpretations. Interestingly, contraction and sensibility, or in other words, the questions of reception, integration and effector systems, have all played a major role in this controversy. As soon as the fine structure of nerve cells was described, the controversy became one on the presence of nerve cells in sponges. The statement of Jones (1962) that there is no nervous system in sponges subsided the discussions. The question of the presence of absence of a nervous system in sponges still arises regularly. Taking into account the behavioral, physiological, morphological, histological and genomic data on sponges reviewed above, my personal impression is that the question itself is the wrong one: "Is there a nervous system in sponges?"—This can clearly be denied based on present data. But what is the informative value of this answer?—Almost none. Understanding the evolution of biological systems does not work simply on the basis of present vs. absent statements. If the nervous system evolved, and it did, we will be able to find traces of the early evolution. Consequently, the question(s) will have to be different and probably more specific: Which nervous-system typical modules or pre-cursors of such modules can be found in those animal groups which do not possess a nervous system in the strictest sense? We are talking about all levels from genomics to the tissue or animal level. These questions have guided me when considering a pre-nervous integration system in sponges (Ellwanger and Nickel 2006 and this publication). Though this pre-nervous system is at present not yet well enough characterized in detail, it is possible to outline it on an integrative data basis. Clearly, recent genomic research (Jacobs et al. 2007, Nichols et al. 2006, Sakarya et al. 2007) has greatly stimulated the genesis of the 'paracrine-to-electrochemical-dominance transition' hypothesis of the early evolution of the nervous system from the pre-nervous system.

The open questions regarding the hypothesis presented here or other hypotheses on the early evolution of the nervous system will not be solved by pure genomic or other purely molecular oriented approaches (Jacobs et al. 2007, Sakarya et al. 2007), nor by pure physiological approaches (e.g. Elliott and Leys 2007, Ellwanger et al. 2006, Ellwanger and Nickel 2006). Ideally, future research will have to combine both approaches and add a third component: morphology. *In situ* hybridization as well as life cell imaging techniques in combination with physiological and (cell/tissue) behavioral studies will play major roles in an integrated approach. To give an example: The presence of certain genes related to the nervous system (Sakarya et al. 2007) or even the expression of some receptors (Perovic et al. 1999) in sponges do not necessarily exhibit the same functions like those in the nervous system of other metazoans. Regardless of whether or not the target genes are orthologs or paralogs of typical nervous system-related genes of other metazoans, these genes might represent different functionalities. Many elements of the nervous system are known from unicellular eukaryotes, e.g. the GABA-receptors in *Paramecium caudatum* MÜLLER 1773 (Ramoino et al. 2003, Ramoino et al. 2004, Ramoino et al. 2005). Consequently, we can expect *per se* to find similar genes in sponges. Thus it will be most important to support detailed genetic, functional and morphological characterizations of the contraction effectors, pacemakers of endogenous rhythms, and the signaling pathways in sponges. For this, phylogenetic considerations have to be taken into account. Ideally, such studies would include model species ranging from demosponges (e.g. *T. wilhelma*, *E. fluviatilis*, *S. lacustris* and others) to homoscleromorphs (e.g. *Oscarella* spp.) and calcareous sponges (*Clathrina* spp., *Sycon* spp. and others). Only in this way, can we expect to gather sequential evolutionary information on the early metazoan (pre-)nervous signalling system and the associated (pre-)muscular contraction systems.

Finally, it is interesting to point out that the pre-nervous system in sponges is strongly associated with the sponge epithelia, the pinacoderm. It was previously suggested that the early evolution of central nervous system brought about an "era of skin brains" (Holland 2003). By referring to the epithelia of *Hydra* Meinhardt (2002) stated that "an old body became a young brain". Following this perspective, we might regard the pinacoderm of sponges as an evolutionary 'very young skin brain'.

Acknowledgements

I thank Sally P. Leys (Edmonton, Canada), Dennis V. Lavrov (Ames, USA), Calhoun Bond (Greensboro, USA), Roberto Pronzato (Genova, Italy), Werner E.G. Müller (Mainz, Germany) as well as George O. Mackie and Henry Reiswig (both Victoria, Canada) for inspiring this work and for discussions and comments on the topic. I am grateful to Martin S. Fischer (Jena, Germany), Hans-Dieter Görtz and Franz Brümmer (both Stuttgart, Germany) for supporting my work and providing infrastructure. I thank Bernd Schierwater (Hannover, Germany) and Rob de Salle (New York, USA) for organising the SICB-symposium, Angie Faust (Hannover, Germany) for assistance during the editorial process of this chapter, and Katherine Zocholl (Ulm, Germany) for English language editing.

References

Adamska, M. and S.M. Degnan, K.M. Green, M. Adamski, A. Craigie, C. Larroux, and B.M. Degnan. 2007a. Wnt and TGF-ß expression in the sponge *Amphimedon queenslandica* and the origin of metazoan embryonic patterning. PLoS ONE 2: e1031.

Adamska, M. and D.Q. Matus, M. Adamski, K. Green, D.S. Rokhsar, M.Q. Martindale, and B.M. Degnan. 2007b. The evolutionary origin of hedgehog proteins. Curr. Biol. 17: R836–R837.

Aelianus, C. De natura animalium libri septemdecim. F. Jacobs. [ed.] 1832. Frommann, Jena.

Anderson, P.A. 1985. Physiology of a bidirectional, excitatory, chemical synapse. J. Neurophysiol. 53: 821–835.

Anderson, P.A.V. [ed.]. 1990. Evolution of the First Nervous Systems. Plenum, New York.

Ankel, W. 1948. Über Fragen der Formbildung und der Zell-Determination bei Süßwasserschwämmen. Verh. Dtsch. Zool. Ges. 1948: 58–66.

Aristotle. De natura animalium libri novem; de partibus animalium libri quatuor; de generatione animalium libri quinqui. In: T. Gaza [ed.] 1498. Domini Octaviani Scoti, 89 pp.

Aristotle. 1910. The History of Animals. Transl. by D'Arcy Wentworth Thompson. Oxford: Clarendon Press. [Electronic reprint]. The University of Adelaide Library. http://etext. library.adelaide.edu.au/a/aristotle/history/

Arndt, W. 1941. Lebendbeobachtungen an Kiesel- und Hornschwämmen des Berliner Aquariums. Zool. Garten (Leipzig) 13: 140–166.

Bacq, Z.M. 1947. L'acétylcholine et l'adrénaline chez lez Invertébrés. Biol. Rev. 22: 73–91.

Bagby, R.M. 1966. The fine structure of myocytes in the sponges *Microciona prolifera* (Ellis and Sollander) and *Tedania ignis* (Duchassaing and Michelotti). J. Morphol. 118: 167–182.

Bagby, R.M. 1970. The fine structure of pinacocytes in the marine sponge *Microciona prolifera*. Z. Zellforsch. Mikroskop. Anat. 105: 579–594.

Barbara, J.-G. 2006. The physiological construction of the neurone concept (1891–1952). Comptes Rendus Biologies 329: 437–449.

Barthélemy-Saint Hilaire, J. 1883. Histoire des Animaux D'Aristote [Electronic Ressource]. Libraire Hachette. http://remacle.org/bloodwolf/philosophes/Aristote/tableanimaux. htm.

Bidder, G.P. 1937. The perfection of sponges. Proc. Linn. Soc. London 149th session: 119–147.

Boal, D. 2002. Mechanics of the Cell. Cambridge University Press, Cambridge.

Bond, C. and A.K. Harris. 1988. Locomotion of sponges and its physical mechanism. J. Exp. Zool. 246: 271–284.

Bond, C. 1992. Continuous cell movements rearrange anatomical structures in intact sponges. J. Exp. Zool. 263: 284–302.

Borchiellini, C. and M. Manuel, E. Alivon, N. Boury-Esnault, J. Vacelet, and P.Y. Le. 2001. Sponge paraphyly and the origin of Metazoa. J. Evol. Biol. 14: 171–179.

Borchiellini, C. and C. Chombard, M. Manuel, E. Alivon, J. Vacelet, and N. Boury-Esnault. 2004. Molecular phylogeny of Demospongiae: implications for classification and scenarios of character evolution. Mol. Phylogenet. Evol. 32: 823–837.

Boury-Esnault, N. and K. Rützler. 1997. Thesaurus of sponge morphology. Smithsonian Contributions to Zoology 596: 1–55.

Boury-Esnault, N. 2006. Systematics and evolution of Demospongiae. Can. J. Zool. 84: 205–224.

Bowerbank, J.S. 1856. On the vital powers of the Spongiadae. Rep. Brit. Assoc. London 26: 438–451.

Brøndsted, H.V. 1942. Formbildungsprozesse bei einem sehr primitiven Metazoon, dem Süsswasserschwamm *Spongilla lacustris* (L.). Protoplasma 37: 244–257.

Bullock, T.H. and G. Horridge. 1969. Structure and Function of the Nervous Systems of Invertebrates. W.H. Freeman, San Fransico.

Bullock, T.H. and D. Nachmansohn. 1942. Coline esterase in primitive nervous systems. J. Cell. Comp. Physiol. 20: 239–242.

Burton, M. 1948. Observations on littoral sponges, including the supposed swarming of larvae, movement and coalescence in mature individuals, longevity and death. Proc. Zool. Soc. London 118: 893–915.

Carpaneto, A. and R. Magrassi, E. Zocchi, C. Cerrano, and C. Usai. 2003. Patch-clamp recordings in isolated sponge cells (*Axinella polypoides*). J. Biochem. Biophys. Methods 55: 179–189.

Carter, H.J. 1848. Notes on the species, structure, and animality of the freshwater sponges in the tanks of Bombay (genus *Spongilla*). Annals and Magazine of Natural History 2nd. ser. 1: 303–311.

Colasanti, M. and G. Venturini. 1998. Nitric oxide in invertebrates. Mol. Neurobiol. 17: 157–174.

Collins, A.G. and P. Cartwright, C.S. McFadden, and B. Schierwater. 2005. Phylogenetic context and basal metazoan model systems. Integr. Comp. Biol. 45: 585–594.

Dale, H.H. 1914. The action of certain esters and ethers of choline and their relation to muscarine. J. Pharmacol. Exp. Ther. 6: 147–190.

De Vos, L. and G. Van De Vyver. 1981. Étude de la contraction spontanée chez 'éponge d'eau douce *Ephydatia fluviatilis* cultivée in vitro. Ann. Soc. Roy. Zool. Bel. 111: 21–32.

De Vos, L. and K. Rützler, N. Boury-Esnault, C. Donaday, and J. Vacelet. 1991. Atlas of Sponge Morphology. Smithsonian Institution Press, Washington.

Dendy, A. 1891. Memoirs: Studies on the Comparative Anatomy of Sponges: III.—On the Anatomy of *Grantia labyrinthica*, Carter, and the so-called Family Teichonida. Q. J. Microsc. Sci. 32: 1–39.

Dunn, C.W. and A. Hejnol, D.Q. Matus, K. Pang, W.E. Browne, S.A. Smith, E. Seaver, G.W. Rouse, M. Obst, G.D. Edgecombe, M.V. Sorensen, S.H.D. Haddock, A. Schmidt-Rhaesa, A. Okusu, R.M. Kristensen, W.C. Wheeler, M.Q. Martindale, and G. Giribet. 2008. Broad phylogenomic sampling improves resolution of the animal tree of life. Nature 452: 745–749.

Elliott, G. 2004. The contractile apparatus of the sneezing freshwater sponge *Ephydatia muelleri*. Newsletter of the American Microscopical Society, Fall issue: 5–7.

Elliott, G.R.D. and S.P. Leys. 2007. Coordinated contractions effectively expel water from the aquiferous system of a freshwater sponge. J. Exp. Biol. 210: 3736–3748.

Ellis, J. 1755. An essay towards a natural history of the corallines and other marine products of the like kind, commonly found on the coasts of Great Britain and Ireland, London.

Ellis, J. and D. Solander. 1786. The Natural History of Many Curious and Uncommon Zoophytes, Collected from Various Parts of the Globe. Systematically Arranged and Described by the Late Daniel Solander. Benjamin White & Son, London , 206 pp and 63 Pl.

Ellwanger, K. and A. Eich, and M. Nickel. 2006. GABA and glutamate specifically induce contractions in the sponge *Tethya wilhelma*. J. Comp. Physiol. A 193: 1–11.

Ellwanger, K. and M. Nickel. 2006. Neuroactive substances specifically modulate rhythmic body contractions in the nerveless metazoon *Tethya wilhelma* (Demospongiae, Porifera). Frontiers in Zoology 3: 7.

Emson, R.H. 1966. The reactions of the sponge *Cliona celata* to applied stimuli. Comp. Biochem. Physiol. 18: 805–827.

Erpenbeck, D. and G. Wörheide. 2007. On the molecular phylogeny of sponges (Porifera). Zootaxa 1668: 107–126.

Esper, E.J. 1794–1805. Die Pflanzenthiere: in Abbildungen nach der Natur mit Farben erleuchtet nebst Beschreibungen. Raspe, Nürnberg.

Fishelson, L. 1981. Observations on the moving colonies of the genus *Tethya* (Demospongia, Porifera): 1. Behavior and cytology. Zoomorphology 98: 89–100.

Fredholm, B.B. and G. Arslan, L. Halldner, B.r. Kull, G. Schulte, and W. Wasserman. 2000. Structure and function of adenosine receptors and their genes. Naunyn Schmiedebergs Arch. Pharmacol. 362: 364–374.

Gaino, E. and M. Sarà. 1976. Ultrastructural observation on the cell junctions in aggregates of *Reniera fulva* (Porifera, Demospongiae). Boll. Mus. Ist. Biol. Univ. Genova 44: 5–10.

Gaino, E. and G. Magnino. 1996. Effects of exogenous cAMP on the morphology and behavior of dissociated cells of the sponge *Clathrina cerebrum* (Porifera, Calcarea). Eur. J. Cell Biol. 70: 92–96.

Garrone, R. and C. Lethias, Freeze fracture study of sponge cells. pp. 121–128. *In:* K. Rützler [ed.] 1990, New Perspectives in Sponge Biology. Smithsonian Institution Press, Washington.

Garrone, R. and C. Lethias, and J. Escaig. 1980. Freeze-fracture study of sponge cell membranes and extracellular matrix. Preliminary results. Biol. Cell. 38: 71–74.

Gauthier, M. and B. Degnan. 2008. The transcription factor NF-κB in the demosponge *Amphimedon queenslandica* : insights on the evolutionary origin of the Rel homology domain. Dev. Genes and Evol. 218: 23–32.

Ghysen, A. 2003. The origin and evolution of the nervous system. Int. J. Dev. Biol. 47: 555–62.

Giovine, M. and M. Pozzolini, A. Favre, G. Bavestrello, C. Cerrano, F. Ottaviani, L. Chiarantini, A. Cerasi, M. Cangiotti, E. Zocchi, S. Scarfi, M. Sara, and U. Benatti. 2001. Heat stress-activated, calcium-dependent nitric oxide synthase in sponges. Nitric Oxide 5: 427–431.

Green, C.R. and P.R. Bergquist. Cell membrane specializations in the Porifera. pp. 153–158. *In:* C. Lévi and N. Boury-Esnault [eds.] 1979. Biologie des Spongiaires. Editions du C.N.R.S., Paris.

Green, C. and P.R. Bergquist. 1982. Phylogenetic relationships within the Invertebrata in relation to the structure of septate junctions and the development of occluding junctional types. J. Cell Sci. 53: 270–305.

Grundfest, H. Evolution of conduction in the nervous system. pp. 43–86. *In:* A. Bass. [ed.] 1959. Evolution of Nervous Control from Primitive Organisms to Man. American Association for the Advancement of Science, Washington.

Grundfest, H. Evolution of electrophysiological varieties among sensory receptor systems. pp. 107–138. *In:* J. Pringle. [ed.] 1965. Essays on Physiological Evolution. The Macmillan Company, New York.

Hanström, B. 1928. Vergleichende Anatomie des Nervensystems der wirbellosen Tiere unter Berücksichtigung seiner Funktion. Springer, Berlin.

Harrison, F.W. and L. De Vos. Porifera. pp. 29–90. *In:* F.W. Harrison and J.A. Westfall. [eds.] 1991. Microscopic Anatomy of Invertebrates. Wiley-Liss, New York.

Hejnol, A. and R. Schnabel. 2006. What a couple of dimensions can do for you: Comparative developmental studies using 4D microscopy—examples from tardigrade development. Integr. Comp. Biol. 46: 151–161.

Hernandez-Nicaise, M.-L. 1973. The nervous system of ctenophores III. Ultrastructure of synapses. J. Neurocytol. 2: 249–263.

Hertwig, O. and R. Hertwig. 1878. Das Nervensystem und die Sinnesorgane der Medusen. Vogel, Leipzig.

Holland, N.D. 2003. Early central nervous system evolution: an era of skin brains? Nat. Rev. Neurosci. 4: 617–627.

Horridge, G. 1968. Interneurons: Their Origin, Action, Specificity, Growth, and Plasticity. WH Freeman, London.

Ishida, N. and M. Kaneko, and R. Allada. 1999. Biological clocks. Proceedings of the Natl. Acad. of Sci. 96: 8819–8820.

Jacklet, J.W. 1997. Nitric oxide signalling in invertebrates. Invert. Neurosci. 3: 1–14.

Jacobs, D.K. and N. Nakanishi, D. Yuan, A. Camara, S.A. Nichols, and V. Hartenstein. 2007. Evolution of sensory structures in basal metazoa. Integr. Comp. Biol. 47: 712–723.

Johnston, G. 1842. A History of British Sponges and Lithophytes. W.H. Lizard, Edinburgh.

Jones, W.C. 1957. The contractility and healing behaviour of pieces of *Leucosolenia complicata*. Q. J. Microsc. Sci. 98: 203–217.

Jones, C.W. 1962. Is there a nervous system in sponges? Biol. Rev. 37: 1–50.

Kilian, E.F. 1967. Ortsveränderungen von Süßwasserschwämmen unter dem Einfluß von Licht. Verh. Dtsch. Zool. Ges. 31 (Suppl.): 395–401.

Klotz, K.-N. 2000. Adenosine receptors and their ligands. Naunyn Schmiedebergs Arch. Pharmacol. 362: 382–391.

Larroux, C. and B. Fahey, D. Liubicich, V.F. Hinman, M. Gauthier, M. Gongora, K. Green, G. Worheide, S.P. Leys, and B.M. Degnan. 2006. Developmental expression of transcription factor genes in a demosponge: insights into the origin of metazoan multicellularity. Evol. & Devel. 8: 150–173.

Lendenfeld, R.V. 1889. A Monograph of the Horny Sponges. Trübner and Co, Ludgate Hill, EC, London.

Lentz, T.L. 1966. Histochemical localization of neurohumors in a sponge. J. Exp. Zool. 162: 171–180.

Lentz, T.L. 1968. Primitive Nervous Systems. Yale University Press, New Haven.

Lethias, C. and R. Garrone, and M. Mazzarana. 1983. Fine structure of sponge cell membranes: Comparative study with freeze-fracture and conventional thin section methods. Tissue & Cell 15: 523–536.

Leys, S.P. and G.O. Mackie. 1997. Electrical recording from a glass sponge. Nature 387: 29–30.

Leys, S.P. and G.O. Mackie. 1999. Propagated electrical impulses in a sponge. Memoirs of the Queensland Museum 4430: 342.

Leys, S.P. and G.O. Mackie, and R.W. Meech. 1999. Impulse conduction in a sponge. J. Exp. Biol. 202: 1139–1150.

Leys, S.P. and R.W. Meech. 2006. Physiology of coordination in sponges. Can. J. Zool. 84: 288–306.

Leys, S.P. Sponge coordination, tissues, and the evolution of gastrulation. In: M. Custódio, G. Lôbo-Hajdu, E. Hajdu and G. Muricy. [eds.] 2007. Porifera Reserach: Biodiversity, Innovation and Sustainability. Série Livros 28. Museu Nacional, Rio de Janeiro.

Leys, S.P. and G.O. Mackie, H.M. Reiswig, and W.S. David. 2007. The Biology of Glass Sponges. Advances in Marine Biology, Volume 52. Academic Press London, Oxford, Boston, New York and San Diego.

Lieberkühn, N. 1859. Neue Beiträge zur Anatomie der Spongien. Arch. Anat. Physiol. 353–358, 515–529.

Lieberkühn, N. 1863. Über Bewegungserscheinungen bei den Schwämmen. Archiv für Anatomie, Physiologie und wissenschaftliche Medicin, 30: 717–730.

Linder, J.U. and J.E. Schultz. 2002. Guanylyl cyclases in unicellular organisms. Mol. Cell. Biochem. 230: 149–158.

Linné, C. von. 1767. Systema Naturae Editio duodecima, reformata. Insecta, Vermes. Holmiae, Amsterdam.

Loewenstein, W.R. 1967. On the genesis of cellular communication. Dev. Biol. 15: 503–520.

Lusche, D. and K. Bezares-Roder, K. Happle, and C. Schlatterer. 2005. cAMP controls cytosolic Ca^{2+} levels in Dictyostelium discoideum. BMC Cell Biology 6: 12.

Mackie, G.O. 1970. Neuroid conduction and the evolution of conducting tissue. Q. Rev. Biol. 45: 319–332.

Mackie, G.O. Is there a conduction system in sponges? pp. 145–152. In: C. Lévi and N. Boury-Esnault. [eds.] 1979. Biologie des Spongiaires. Editions du C.N.R.S., Paris.

Mackie, G.O. and I.D. Lawn, and M.P. De Ceccatty. 1983. Studies on hexactinellid sponges: 2. Excitability, conduction and coordination of responses in Rhabdocalyptus dawsoni. Philos. Trans. R. Soc. Lond. B 301: 401–418.

Mackie, G.O. 1990. The elementary nervous system revisited. Am. Zoologist 30: 907–920.

Marshall, W. Coelenterata, Porifera, Tetractinellidae; Tafel XLVII. In: R. Leuckart. [ed.] 1885. Zoologische Wandttafeln der wirbellosen Thiere. Th. Fischer, Kassel.

Martindale, M.Q. and K. Pang, and J.R. Finnerty. 2004. Investigating the origins of triploblasty: `mesodermal' gene expression in a diploblastic animal, the sea anemone Nematostella vectensis (phylum, Cnidaria; class, Anthozoa). Development 131: 2463–2474.

Matsuno, A. and H. Ishida, M. Kuroda, and Y. Masuda. 1988. Ultrastructures of contractile bundles in epithelial cells of the sponge. Zoolog. Sci. 5: 1212.

Mattson, M.P. and A.J. Bruce-Keller. 1999. Compartmentalization of signaling in neurons: Evolution and deployment. J. Neurosci. Res. 58: 2–9.

McNair, G.T. 1923. Motor reactions of the fresh-water sponge *Ephydatia fluviatilis*. Biol. Bull. 44: 153–166.

Meech, R.W. and G.O. Mackie, Evolution of excitability in lower metazoans. *In:* G. North and J. Greenspann. [eds.] 2007. Invertebrate Neurobiology. Cold Spring Harbor Monograph Series 49. Cold Spring Harbor Laboratory Press, New York.

Meech, R.W. 2008. Non-neural reflexes: sponges and the origins of behaviour. Curr. Biol. 18: R70–R72.

Meinhardt, H. 2002. The radial-symmetric hydra and the evolution of the bilateral body plan: an old body became a young brain. BioEssays 24: 185–191.

Minchin, E. Sponges. pp. 1–178. *In:* E.R. Lancester. [ed.] 1900. A Treatise on Zoology. Part II. The Porifera and Coelentera. Adam & Charles Black, London.

Mitropolitanskaya, R. 1941. On the presence of acetylcholin and cholinesterases in the Protozoa, Spongia and Coelenterata. C. R. Acad. Sci. USSR 31: 717–718.

Mori, M. 1994. Electron microscopic and new microscopic studies of hepatocyte cytoskeleton: Physiological and pathological relevance. J. Electron. Microsc. (Tokyo) 43: 347–355.

Nguyen, L. and J.-M. Rigo, V. Rocher, S. Belachew, B. Malgrange, B. Rogister, P. Leprince, and G. Moonen. 2001. Neurotransmitters as early signals for central nervous system development. Cell Tissue Res. 305: 187–202.

Nichols, S.A. and W. Dirks, J.S. Pearse, and N. King. 2006. Early evolution of animal cell signaling and adhesion genes. Proc. Natl. Acad. Sci. USA 103: 12451–12456.

Nickel, M. 2001. Cell biology and biotechnology of marine invertebrates. Sponges (Porifera) as model organisms. Dissertation Thesis, Universität Stuttgart, Stuttgart.

Nickel, M. and M. Vitello, and F. Brümmer. 2002. Dynamics of cellular movements in the locomotion of the sponge *Tethya wilhelma*. Integr. Comp. Biol. 42: 1285.

Nickel, M. and F. Brümmer. 2002. Patterns of movement and their integration in sessile, nerveless metazoa illustrated by *Tethya wilhelma* (Porifera, Demospongiae). Zoology 105: 69.

Nickel, M. 2004. Kinetics and rhythm of body contractions in the sponge *Tethya wilhelma* (Porifera : Demospongiae). J. Exp. Biol. 207: 4515–4524.

Nickel, M. and F. Brümmer. 2004. Body extension types of *Tethya wilhelma*: cellular organisation and their function in movement. Boll. Mus. Ist. Biol. Univ. Genova 68: 483–489.

Nickel, M. and J. Hammel, T. Donath, and F. Beckmann, Quantitative morphometrics and contraction analysis ot the marine sponge *Tethya wilhelma*—using synchrotron radiation based x-ray microtomography and in vivo x-ray imaging. pp. 1065–1066, Print/CD-ROM / Online: http://hasyweb.desy.de/science/annual_reports/2005_report/part1/contrib/47/13996.pdf. *In:* U. Krell, J. Schneider and M. von Zimmermann [eds.] 2005, HASYLAB Annual Report 2005. Part I. DESY, Hamburg.

Nickel, M. 2006. Like a 'rolling stone': quantitative analysis of the body movement and skeletal dynamics of the sponge *Tethya wilhelma*. J. Exp. Biol. 209: 2839–2846.

Nickel, M. and E. Bullinger, and F. Beckmann. 2006a. Functional morphology of *Tethya* species (Porifera): 2. Three-dimensional morphometrics on spicules and skeleton superstructures of *T. minuta*. Zoomorphol. 125: 225–239.

Nickel, M. and T. Donath, M. Schweikert, and F. Beckmann. 2006b. Functional morphology of *Tethya* species (Porifera): 1. quantitative 3D-analysis of *T. wilhelma* by synchrotron radiation based x-ray microtomography. Zoomorphol. 209–223.

Nickel, M. and J. Hammel, T. Donath, and F. Beckmann. Elastic energy load into spicule reinforced extracellular matrix of the sponge *Tethya wilhelma?*—Evidence from synchrotron radiation based x-ray microtomography pp. 1279–1280, Print / CD-ROM / Online: http://hasyweb.desy.de/science/annual_reports/2006_report/part1/contrib/47/17965.pdf. *In:* U. Krell, J. Schneider and M. von Zimmermann. [eds.] 2006c, HASYLAB Annual Report 2006. Part I. DESY, Hamburg.

Nickel, M. 2007. Movements without muscles, information processing without nerves. JMBA GME 6: 8–9.

Nickel, M. 2010. Evolutionary emergence of synaptic nervous systems: what can we learn from the non-synaptic, nerveless Porifera?, Invertebrate Biology, 129: 1–16.

Noll, F.C. 1881. Mein Seewasser-Zimmeraquarium. Zool. Garten (Frankfurt) 22: 168–177.

Oberdisse, E. and E. Hackenthal, and K. Kuschinsky. 1997. Pharmakologie und Toxikologie. Springer-Verlag, Heidelberg.

Pallas, P. 1766. Elenchus Zoophytorum sistens generum adumbrationes generaliores et specierum cognitarum succinctas descriptiones cum selectis auctorum synonymis. P. van Cleef, The Hague.

Pallas, P. 1787. Charakteristik der Thierpflanzen. Raspesche Buchhandlung, Nürnberg.

Pansini, M. and R. Pronzato. 1990. Observations on the dynamics of a Mediterranean sponge community. K. Ruetzler. [ed.] New Perspectives in Sponge Biology; Third International Conference on the Biology of Sponges, Woods Hole, Massachusetts, USA, November 17–23, 1985. ix+533 pp. 404–415.

Pantin, C.F.A. 1952. The elementory nervous system. Proc. R. Soc. Lond. B. Biol. Sci. 140: 147–168.

Pantin, C.F.A. 1956. The origin of the nervous system. Pubblicazione della Stazione Zoologica di Napoli 28: 171–181.

Parker, G.H. 1910. The reactions of sponges, with a consideration of the origin of the nervous system. J. Exp. Zool. 8: 765–805.

Parker, G.H. 1919. The Elementary Nervous System. Lippincot, Philadelphia.

Passano, L. 1963. Primitive nervous systems. Proc. Natl. Acad. Sci. USA 50: 306–313.

Pavans de Ceccatty, M. 1955. Le systeme nerveux des éponges calcaires et siliceuses. 288.

Pavans de Ceccatty, M. 1959. Les structures cellulaires de type nerveux chez *Hippospongia communis* LMK. Ann. Sci. Nat. Zool. Biol. Anim. 12: 105–112.

Pavans de Ceccatty, M. 1960. Les structures cellulaires de type nerveux et de type musculaire de l'éponge siliceuse *Tethya lyncurium* Lmck. C. R. Acad. Sci. Paris 251: 1818–1819.

Pavans de Ceccatty, M. 1971. Effects of drugs and ions on a primitive system of spontaneous contractions in a sponge (*Euspongia officinalis*). Experientia 27: 57–59.

Pavans de Ceccatty, M. 1974a. Coordination in sponges the foundations of integration. Am. Zoologist 14: 895–903.

Pavans de Ceccatty, M. 1974b. The origin of the integrative systems a change in view derived from research on coelenterates and sponges. Perspect. Biol. Med. 17: 379–390.

Pavans de Ceccatty, M. Cell correlations and integrations in sponges. pp. 123–135. *In:* C. Lévi and N. Boury-Esnault. [eds.] 1979. Biologie des Spongiaires. Editions du CNRS, Paris.

Pavans de Ceccatty, M. 1981. Demonstration of actin filaments in sponge cells. Cell Biol. Int. 5: 945–952.

Pavans de Ceccatty, M. 1986. Cytoskeletal organization and tissue patterns of epithelia in the sponge *Ephydatia mülleri*. J. Morphol. 189: 45–66.

Pavans de Ceccatty, M. 1989. Les éponges, à l'aube des communications cellulaires. Pour la Science 142: 64–72.

Perovic, S. and A. Krasko, I. Prokic, I.M. Mueller, and W.E.G. Mueller. 1999. Origin of neuronal-like receptors in Metazoa: Cloning of a metabotropic glutamate/GABA-like receptor from the marine sponge *Geodia cydonium*. Cell Tissue Res. 296: 395–404.

Philippe, H., R. Derelle, P. Lopez, K. Pick, C. Borchiellini, N. Boury-Esnault, J. Vacelet, E. Renard, E. Houiliston, E. Quéinnec, C. Da Silva, P. Wincker, H. Le Guyader, S. Leys, D.J. Jackson, F. Schreiber, D. Erpenbeck, B. Morgenstern, G. Wörheide, and M. Manuel. 2009. Phylogenomics Revives Traditional Views on Deep Animal Relationships. Curr. Biol. 19: 1–7.

Pick, K.S., H. Philippe, F. Schreiber, D. Erpenbeck, D.J. Jackson, P. Wrede, M. Wiens, A. Alie, B. Morgenstern, M. Manuel, and G. Worheide. 2010. Improved phylogenomic taxon sampling noticeably affects non-bilaterian relationships. Mol. Biol. Evol.: msq089.

Pierobon, P. and R. Minei, P. Porcu, C. Sogliano, A. Tino, G. Marino, G. Biggio, and A. Concas. 2001. Putative glycine receptors in Hydra: a biochemical and behavioural study. Eur. J. Neurosci. 14: 1659–1666.

Plinus. The natural history. *In:* J. Bostock and H.T. Riley [eds.] 1855. Taylor and Francis, London.

Pronzato, R. 1999. Sponge-fishing, disease and farming in the Mediterranean Sea. Aquat. Conserv. 9: 485–493.

Pronzato, R. 2004. A climber sponge. Boll. Mus. Ist. Biol. Univ. Genova 68: 549–552.

Prosser, C.L. and T. Nagai, and R.A. Nystrom. 1962. Oscular contraction in sponges. Comp. Biochem. Physiol. 6: 69–74.

Prosser, C.L. 1967. Ionic analyses and effects of ions on contractions of sponge tissues. Z. Vgl. Physiol. 54: 109–120.

Ramoino, P. and P. Fronte, F. Beltrame, A. Diasporo, M. Fato, L. Raiteri, S. Stigliani, and C. Usai. 2003. Swimming behaviour regulation by $GABA_B$ receptors in *Paramecium*. Exp. Cell Res. 291: 398–405.

Ramoino, P. and L. Gallus, S. Paluzzi, L. Raiteri, A. Diaspro, M. Fato, G. Bonanno, G. Tagliafierro, C. Ferretti, and R. Manconi. 2007. The GABAergic-like system in the marine demosponge *Chondrilla nucula*. Microsc. Res. Tech. 70: 944–951.

Ramoino, P. and S. Scaglione, A. Diasporo, F. Beltrame, M. Fato, and C. Usai. 2004. $GABA_A$ receptor subunits identified in *Paramecium* by immunofluorescence confocal microscopy. FEMS Microbiol. Lett. 238: 449–453.

Ramoino, P. and C. Usai, F. Beltrame, M. Fato, L. Gallus, G. Tagliafierro, R. Magrassi, and A. Diaspro. 2005. $GABA_B$ receptor intracellular trafficking after internalization in *Paramecium*. Microsc. Res. Tech. 68: 290–295.

Ramon y Cayal, S. 1937. Recollections of my life. Memoirs of the American Philosophical Society 8: 1–638.

Redi, C.A. and S. Garagna, M. Zuccotti, E. Capanna, and H. Zacharias. 2002. Visual Zoology. The Pavia collection of Leuckart's zoological wall charts (1877). Ibis, Como.

Reiswig, H.M. 1971. In-situ pumping activities of tropical Demospongiae. Mar. Biol. 9: 38–50.

Ruediger, T. and J. Bolz, Neurotransmitters and the development of neuronal circuits. *In:* D. Bagnard [ed.] 2007. Axon Growth and Guidance. Springer Science+Business Media, LLC, Landes Bioscience, New York.

Ruppert, E. and R. Fox, and R. Barnes. 2004. Invertebrate Zoology (7. Auflage). Thomson Publishing, Brooks/Cole.

Sakarya, O. and K.A. Armstrong, M. Adamska, M. Adamski, I.F. Wang, B. Tidor, B.M. Degnan, T.H. Oakley, and K.S. Kosik. 2007. A post-synaptic scaffold at the origin of the animal kingdom. PLoS ONE 2: e506.

Sarà, M. and A. Sarà, M. Nickel, and F. Brümmer. 2001. Three new species of *Tethya* (Porifera: Demospongiae) from German aquaria. Stuttgarter Beitr. Naturk. Ser. A 631: 1–15.

Schierwater, B. and R. DeSalle. 2007. Can we ever identify the Urmetazoan? Integr.Comp. Biol. 47: 670–676.

Schmidt-Rhaesa, A. 2007. The Evolution of Organ Systems. Oxford University Press, Oxford.

Schmidt, O. 1866. Zweites Supplement der Spongien des Adriatischen Meeres enthaltend die Vergleichung der Adriatischen und Britischen Spongiengattungen. Verlag von Wilhelm Engelmann, Leipzig.

Schultz, J.E. and T. Pohl, and S. Klumpp, 1986. Voltage-gated Ca^{2+} entry into *Paramecium* linked to intraciliary increase in cyclic GMP. Nature 322: 271–273.

Seipel, K. and V. Schmid. 2006. Mesodermal anatomies in cnidarian polyps and medusae. Int. J. Dev. Biol. 50: 589–99.

Sherrington, C.S. 1897. Croonian Lecture: The Mammalian Spinal Cord as an Organ of Reflex Action. Proceedings of the Royal Society of London 61: 220–221.

Shibata, S. 2004. Neural regulation of the hepatic Circadian rhythm. Anat. Rec. Part A 280A: 901–909.

Simpson, T.L. and G.A. Rodan. 1976. Role of cyclic AMP in the release from dormancy of freshwater sponge gemmules. Dev. Biol. 49: 544–547.

Simpson, T.L. 1984. The Cell Biology of Sponges. Springer-Verlag, New York.

Smith, C. 1996. Sherrington's legacy: evolution of the synapse concept, 1890s–1990s. J. Hist. Neurosci. 5: 43–55.

Sollas, W.J. 1888. Report on the Tetractinellida collected by H.M.S. Challenger during the years 1873–1876. Report on the Scientific Results of the voyage of H.M.S. Challenger 25: 1–458.

Spring, J. and N. Yanze, C. Jösch, A.M. Middel, B. Winninger, and V. Schmid. 2002. Conservation of Brachyury, Mef2, and Snail in the Myogenic Lineage of Jellyfish: A Connection to the Mesoderm of Bilateria. Dev. Biol. 244: 372–384.

Thiney, Y. 1972. Morphologie et cytochimie ultrastructurale de l'oscule d'Hippospongia communis LMK et de sa régénération. PhD thesis. Université Claude Bernard (Lyon I): 1–63.

Trueman, E.R. 1975. The Locomotion of Soft-bodied Animals. Edward Arnold Limited, London.

Tuzet, O. and R. Loubatiäres, and M. Pavans de Ceccatty. 1952. Les cellulues nerveuses de l'éponge *Sycon raphanus* O.S. C. R. Acad. Sci. Paris 234: 1394–1396.

Vacelet, J. 1966. Les cellules contractiles de l'éponge cornée Verongia cavernicola Vacelet. C. R. Acad. Sci. Paris 263: 1330–1332.

Van Soest, R.W.M. Porifera, Schwämme. *In:* W. Westeheide and R. Rieger. [eds.] 1996. Spezielle Zoologie, Teil 1: Einzeller und Wirbellose. Fischer, Stuttgart.

Walker, R.J. and L. Holden-Dye. 1991. Evolutionary aspects of transmitter molecules, their receptors and channels. Parasitology 102: S7–S29.

Walker, R.J. and H.L. Brooks, and L. Holden-Dye. 1996. Evolution and overview of classical transmitter molecules and their receptors. Parasitology 113: S3–S33.

Weissenfels, N. 1989. Biologie und Mikroskopische Anatomie der Süßwasserschwämme (Spongillidae). Gustav Fischer Verlag, Stuttgart.

Weissenfels, N. 1990. Condensation rhythm of fresh-water sponges (Spongillidae, Porifera). Eur. J. Cell Biol. 53: 373–383.

Weyrer, S. and K. Rutzler, and R. Rieger. 1999a. Serotonin in Porifera? Evidence from developing *Tedania ignis*, the Caribbean fire sponge (Demospongiae). Memoirs of the Queensland Museum 44: 659–665.

Weyrer, S. and K. Rützler, and R. Rieger. 1999b. Serotonin in Porifera? Evidence from developing *Tedania ignis*, the Caribbean fire sponge (Demospongiae). Memoirs of the Queensland Museum 44: 659–665.

Wiens, M. and B. Diehl-Seifert, and W.E.G. Müller. 2001. Sponge Bcl-2 homologous protein (BHP2-GC) confers distinct stress resistance to human HEK-293 cells. Cell Death and Differentiation 8: 887–898.

Wilson, H.V. 1910. A study of some epitheloid membranes in monaxonid sponges. J. Exp. Zool. 9: 536–571.

Wintermann, G. 1951. Entwicklungsphysiologische Untersuchungen an Süßwasserschwämmen. Zool. Jahr. 71: 428–486.

Zocchi, E. and A. Carpaneto, C. Cerrano, G. Bavestrello, M. Giovine, S. Bruzzone, L. Guida, L. Franco, and C. Usai. 2001. The temperature-signaling cascade in sponges involves a heat-gated cation channel, abscisic acid, and cyclic ADP-ribose. Proc. Natl. Acad. Sci. USA 98: 14859–14864.

Zocchi, E. and G. Basile, C. Cerrano, G. Bavestrello, M. Giovine, S. Bruzzone, L. Guida, A. Carpaneto, R. Magrassi, and C. Usai. 2003. ABA- and cADPR-mediated effects on respiration and filtration downstream of the temperature-signaling cascade in sponges. J. Cell Sci. 116: 629–636.

Chapter 6

A Key Innovation in Animal Evolution, the Emergence of Neurogenesis: Cellular and Molecular Cues from Cnidarian Nervous Systems

Brigitte Galliot

Summary

The emergence of neurogenesis led to the acquisition of an efficient neuromuscular transmission in eumetazoans, as shown by cnidarians that use evolutionarily-conserved neurophysiological principles to crumple, feed, swim. However, the cnidarian neuroanatomies are quite diverse and reconstructing the urcnidarian nervous system is not an easy task. Three types of characters shared by anthozoans and medusozoans appear plesiomorphic: (1) three cell types forming the cnidarian nervous system, neurosensory cells, ganglionic neurons and nematocytes (cnidocytes) that combine mechano-chemosensation and venom secretion; (2) a chemical conduction through nerve nets and nerve rings, those being considered as annular central nervous systems; (3) a larval apical sensory organ that initiates metamorphosis. Other characters receive a disputed origin: (1) the neural stem cell(s), that are multipotent interstitial stem cell in hydrozoans, but not identified in other classes; (2) the electrical conduction

Department of Zoology and Animal Biology, Faculty of Science, University of Geneva, Sciences III, 30 quai Ernest Ansermet, CH-1211 Geneva 4, Switzerland.
E-mail: brigitte.galliot@unige.ch

through neurons and epithelial cells so far detected only in hydrozoans; (3) the embryonic origin of the nervous system; (4) the medusa sensory organs, ocelli or lens-eyes for light, statocysts for pressure, which are lacking in anthozoans. Interestingly numerous gene families that regulate neurogenesis in bilaterian are expressed during cnidarian neurogenesis, e.g. cnidarian eyes express *Pax, Six* and *opsin*, supporting a common origin for vision. However data establishing a clear picture of the cnidarian neurogenic circuitry are currently missing. In fact many "neurogenic" gene families likely arose and evolved in the absence of neurogenesis, as exemplified by Porifera that express them but lack synaptic transmission. Therefore some eumetazoan-specific families, missing in Porifera as *ParaHox/Hox*-like and *Otx*-like genes, might have contributed to the emergence of neurogenesis.

The Key Position of Cnidaria to Trace Back the First-Evolved Nervous Systems

Despite numerous controversies, recent phylogenomic approaches have convincingly shown that cnidarians form with ctenophores (combjellies) a unique clade named coelenterates; as bilaterians coelenterates differentiate nerves and muscles and therefore group in eumetazoans (Philippe et al. 2009). In contrast, poriferans (sponges) and placozoans would occupy more basal positions in metazoans (Schierwater et al. 2009). Poriferans likely differentiate a proto-neuronal system: they are capable of chemical conduction (Leys et al. 1999), they differentiate sensory-like cells and express regulatory genes that are neurogenic in bilaterians (Gazave et al. 2008, Richards et al. 2008). However they do not display any cell types exhibiting synaptic conduction and usually feed by passive filtration. Therefore coelenterates provide appropriate model systems to trace back the first-evolved nervous systems that rely on synaptic conduction (Anderson and Spencer 1989).

The coelenterate phylum is supposed to have diverged about 650 million years ago, preceding the Cambrian explosion, the period when ancestors to most extant bilaterian phyla arose from a common hypothetical ancestor named Urbilateria (De Robertis 2008). As a consequence Coelenterata form a sister clade to the bilaterians. Cnidarians are most often marine animals that are made up of two cell layers, the ectoderm and the endoderm, separated by an extracellular matrix named mesoglea (Bouillon 1994a). However this "diploblastic" criterion is disputed as numerous cnidarian species actually differentiate "mesodermal" derivatives as striated muscle at one or the other stage of their life cycle (Seipel and Schmid 2006). Cnidarian species cluster in two distinct groups (Bridge et al. 1995, Collins et al. 2006): anthozoans that live exclusively as polyps (corals, sea anemones) and medusozoans that display a complex life cycle with a pelagic parental medusa stage, a larval planula stage and a benthic polyp stage (Fig. 1). Among those, the cubozoans (box jellyfish) predominantly live as medusae, whereas scyphozoans and hydrozoan species usually follow a life cycle where they alternate between these two forms. The scyphozoan polyp produces through strobilation (transverse fission) a young flat ephyra that subsequently shape into medusa (Franc 1994) whereas the hydrozoan polyp produces through budding a young medusa (Galliot and Schmid 2002). The freshwater *Hydra* belongs to this latter group, but *Hydra* lost the medusa stage and the embryos produced by polyps develop directly, lacking the planula stage and the metamorphosis process. Cnidarian polyps are basically a tube with a single opening circled by a ring of tentacles, which has a mouth-anus function, whereas jellyfish display a more complex anatomy with the mouth-anus opening located at the extremity of the manubrium under the bell (Figs. 2Q, 2R, see also Fig. 2C in Galliot et al. 2009).

In the mid XIXe century, with limited imaging tools, Louis Agassiz identified for the first time the cnidarian nervous system on two live

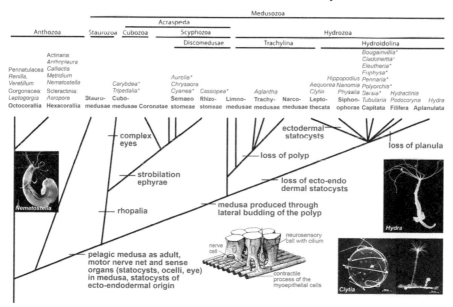

Fig. 1. Phylogeny of cnidarians and key steps in the evolution of their nervous systems.
This tree reproduces the phylogeny established by Collins et al. (2006). The main changes affecting development and/or differentiation of the nervous system are indicated. Species used for investigating the anatomy, the physiology or the differentiation of the nervous system are listed on the top. In hydrozoans, *Hydractinia* and *Hydra* have lost the medusa stage. Asterisks indicate species that differentiate photo-sensory organs (eye-spots that detect light as ocelli or lens eyes). The scheme showing the basic cellular organization supporting neuromuscular transmission in cnidarians is taken from Mackie and Passano (1968).

hydrozoan jellyfish, *Sarsia* and *Bougainvillia* (Agassiz 1850, Mackie 2004a). Twenty five years later the pulsated swimming behavior of jellyfish was investigated experimentally by George Romanes, who proved that pacemakers were actually quite different in scyphozoans and hydrozoans, restricted to sense organs (the rhopalia) in the former, more diffuse and extending along the bell ring in the latter (Romanes 1876, Romanes 1877). Surprised by what he named a dichotomy, his work undoubtedly highlighted the striking potential of studying "primitive" nervous systems. Since then the history of the emergence of neurogenesis in the animal kingdom led to the elaboration of successive scenarios (Passano 1963). The application of cellular and electrophysiological methods to coelenterates definitely proved that basically the same neurophysiological principles are valid from cnidarians to bilaterians (Horridge 1954, Passano and McCullough 1964, Anderson and Spencer 1989, Westfall 1996). And indeed Andy Spencer concluded from the electrophysiological behavior of cnidarians neurons that "*many of the basic synaptic mechanisms and properties that we associate with more "advanced" nervous systems, can be demonstrated in the Cnidaria. With some danger of oversimplifying, one could say that it was in this phylum that most of the important properties of synapses evolved, and that since that time, most evolutionary change in higher nervous systems has been with respect to the complexity of connections*" (Spencer 1989).

In this chapter, we will survey the following questions: What are the common anatomical and physiological characters shared by the cnidarian nervous systems? How, when and where neurogenesis and nematogenesis take place in cnidarians? What is currently known about the genetic circuitry that regulates these two processes at any stage of the cnidarian life cycle? What evolutionary scenario can account for the key transition represented by the emergence of neurogenesis, presumably in the last common eumetazoan? Beside the evolutionary history of neurophysiology and neurogenesis, which biological questions can be investigated in cnidarian model systems?

The Cellular Components of Cnidarian Nervous Systems

The nematocytes (or cnidocytes) are phylum-specific mechanoreceptor cells

Cnidarian nervous systems are made of mechanoreceptor cells named nematocytes or cnidocytes (giving their name to the phylum), and neurons. The nematocytes are highly specialized mechano-receptor cells, which play a key role in the capture of preys and defense (Figs. 2A-2E) —reviewed in (Bouillon 1994b, Tardent 1995). These stinging cells are abundant, representing 35% of the cells in *Hydra* (David 1973). Mature nematocytes function as receptor-effector cells, receptor thanks to their

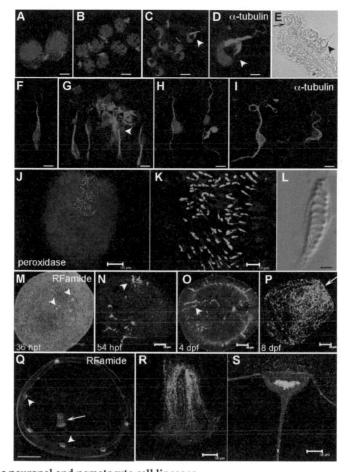

Fig. 2. The neuronal and nematocyte cell lineages.
(A-E) The *Hydra* nematocyte lineage: interstitial stem cells (A) synchronously divide, providing nematoblasts (B) that differentiate a typical capsule (nematocyst) as observed in mature nematocytes (C,D). Arrowheads: discharged nematocyst. (E) Bright-field view of a tentacle with nematocytes either undischarged (arrow) or discharged (arrowhead) embedded in large epithelial battery cells. **(F-I)** *Hydra* neurons detected after maceration with alpha-tubulin antibody (red) and DAPI (blue): sensory (F, G), bipolar (H) or multipolar (I) also named ganglionic cells. **(J-L)** *Nematostella* spirocytes detected in larva thanks to their peroxidase activity (green). Spirocytes are anthozoan mechanosensory cells involved in adhesion to prey and non-prey (Kass-Simon and Scappaticci 2002). **(M-P)** Neurogenesis in the developing *Nematostella*: Early neurons expressing the neuropeptide RFamide (arrowheads) appear in the endodermal layer (M), then migrate to the ectoderm (N) where they form a net (O). In the newly metamorphosed polyp (P), the density of RF-amide neurons is higher in the oral region (arrow). **(Q-S)** *Clytia* medusa showing a high density of RF-amide neurons (red) in the manubrium (Q arrow, R) and in the tentacle bulbs (Q arrowheads, S). Blue: DAPI staining ; green: endogenous GFP. Scale bars: 2 μm (A,B,D,L), 5 μm (E-I, C), 10 μm (K), 50 μm (J, M-P, R-S), 500 μm (Q).

Color image of this figure appears in the color plate section at the end of the book.

cnidocil that can be stimulated chemically or mecanically (by preys), effector thanks to their nematocyst (or cnidocyst), a thick-wall capsule that respond in nanoseconds by discharging its toxic content (Fig. 2C-2E) (Ozbek et al. 2009). This venom that is released as large droplets into the prey by an everting tubule, immobilises the prey, which, by releasing the peptide glutathione, induces the feeding response, i.e. tentacle bending and mouth opening (Loomis 1955, Lenhoff et al. 1982, Shimizu 2002). How the information sensed by the cnidocil apparatus is transduced to target the discharge function is not clear. The nematocyst discharge seems to operate in the absence of neuronal control indicating that nematocytes can behave autonomously (Aerne et al. 1991). However nematocytes are under neuronal control as evidenced by the presence of two-cell as well as three-cell synaptic pathways in the tentacle epidermis (Holtmann and Thurm 2001, Westfall et al. 2002). This neuronal control is actually rather playing an inhibitory function to reduce the spontaneous firing activity of nematocytes. Mechanosensory cells display variable morphologies and functions; in anthozoans, spirocytes are mechanosensory cells (Fig. 2J-L) involved in adhesion to prey and non-prey (Kass-Simon and Scappaticci 2002). Finally mechanoreception in cnidarians is likely not restricted to nematocytes, some medusa differentiate clusters of hair cells at the base of tentacles and on the velum, quite similar to the vertebrate hair cells of the inner ear (Arkett et al. 1988). These hair cells are possibly involved in the feeding response and the locomotion behavior.

Several neuronal cell types in cnidarians: neurosensory, sensory-motor and ganglionic

In *Hydra* neurons that can be bipolar, tripolar or multipolar, are of three distinct types, neurosensory, ganglionic and neurosecretory (Lentz and Barrnett 1965, Davis 1974). Sensory cells characterized by their cilium that reach the surface, are located within the ectodermal layer (Fig. 2F,G). By contrast, ganglion neurons (bi, tri or multipolar), which are the most common type of neuronal cells in this animal, are spread in both cell layers, along the mesoglea and function as interneurons (Fig. 2H,I). However the sensory cells might also be considered as multifunctional cells, which indeed receive external inputs but also produce secretory granules and possibly function as motorneuron and interneuron (Westfall and Kinnamon 1978). Similarly in sea anemones, sensory neurons also associate with smooth muscle fibers, suggesting that they behave as sensory and motoneurons (Grimmelikhuijzen et al. 1989). Medusozoans share a common structure, the motor nerve net, initially identified by Passano (who named it the giant fiber nerve net), which conducts the excitation through the subumbrellar muscles of the medusae (Satterlie 2002). In the jellyfish *Cyanea*, the motor nerve net establish bidirectional non-polarized synapses with their target cells, namely myoepithelial cells and nematocytes, thus functioning as sensory-motoneurons (Anderson 1985).

A common progenitor cell for nematocytes and neurons?

In *Hydra*, neurons and nematocytes derive from the same multipotent stem cell, the interstitial stem cell, which also provides progenitors for two other cell lineages, the gland cells and the gametes (Bode 1996). However contrary to Lentz's statement, interstitial stem cells in *Hydra* cannot differentiate into epithelial cells. This restricted multipotent property of intersitial stem cells was nicely demonstrated in "nerve-free" or "epithelial" polyps obtained after chemical (Campbell 1976, Yaross and Bode 1978) or genetic (Sugiyama and Fujisawa 1978, Terada et al. 1988) elimination of the interstitial stem cells. As a result the interstitial derivatives, namely nematocytes and neurons, progressively disappear, the animals become completely epithelial as evidenced by the loss of autonomous feeding behavior and the need for force-feeding to maintain them alive. Upon transplantation of interstitial stem cells in such animals a complete nervous system differentiate *de novo* (Minobe et al. 1995). Neuronal progenitors and nematoblasts, which express a common set of regulatory genes (Fig. 3), might actually share a common bipotent progenitor that expresses *Gsx/cnox-2* (Miljkovic-Licina et al. 2007). All together these data suggest that nematocytes and neurons can be considered as "sister cell types that evolved from a common precursor by cell type diversification" (Arendt 2003).

But here again the same rules do not seem to apply to all cnidarians: First in some hydrozoan species (*Pennaria, Physalia*), sensory cells but neither ganglionic neurons nor nematocytes can differentiate in the absence of interstitial stem cells, suggesting that these sensory cells arise from epithelial cells (Martin and Thomas 1981, Thomas 1987). Whether the experimental conditions that eliminate the intersitial cells (colchicine treatment, surgical resection of the entoderm) enhance the plasticity of the epithelial cells is not known. Therefore a separate origin for sensory cells and ganglionic neurons should definitively be tested in wild-type conditions. Second, interstitial stem cells were only characterized in hydrozoans so far and it was proposed that epithelial cells would provide the progenitors for the nematocyte and neuronal cell lineages in anthozoans (Marlow et al. 2009), as well as in cubozoans and scyphozoans (Nakanishi et al. 2008). However cell lineage analyses were mostly performed in hydrozoan species, namely *Hydra*, therefore similar studies are definitively needed to clarify what is shared between cnidarian species during nematogenesis and neurogenesis.

The complex and highly variable anatomic organization of cnidarian nervous systems

Classically the cnidarian nervous system is described as a "diffuse nerve net" (Pantin 1952), which can indeed be visualised by the neuron-specific RFamide immunostaining as in developing *Nematostella* (Fig. 2M-P). However in adults and larvae, the distribution of neurons is neither random nor uniform. First, significant variations of the general neuron density can easily be noted along the anatomy: in medusae neurons are denser in tentacle bulbs and tentacles, in the nerve rings along the bell margin and in the manubrium as in *Clytia* medusa where the RF-amide neurons are denser in the manubrium and the tentacle bulbs (Fig. 2Q-S), in polyps neurons are denser at the extremities than in the body column (at least six fold higher in the head region in *Hydra*). Second, immunostaining help identify distinct subsets of neurons with specific spatial distribution (Grimmelikhuijzen et al. 1989, Koizumi et al. 1990), suggesting a much higher complexity than anticipated. Third, nerve rings that correspond to compression of the nerve-net architecture were identified at the base of tentacles in some *Hydra* species (Koizumi et al. 1992), around the mouth opening in *Nematostella* (Marlow et al. 2009). Nerve rings were also characterized along the bell margin of some but not all medusae (Table 1): In hydromedusae, there are two nerve rings (inner and outer), in cubozoans a single one and none in scyphozoans (Satterlie 2002). These nerve rings that are connected to the sense organs allow a fast conduction and coordinate the swimming behaviors. More generally in all classes nerve rings are considered as an annular form of central nervous system (Passano and McCullough 1965, Koizumi 2007, Garm et al. 2007). These observations clearly indicate that cnidarian nervous systems are already quite sophisticated, organized in a more complex fashion than random nerve nets. However given the numerous variations in their anatomies, capturing the portrait of the ancestral cnidarian nervous system remains quite elusive. We will discuss that point once we will know more about behaviors, cell differentiation and gene regulatory pathways.

Complex Behaviors Rely on Chemical and Electrical Conduction in Cnidarians

Contraction bursts and elongation in polyps, crumpling behavior in hydromedusae

Abraham Trembley, who was the first to definitively establish the animality of *Hydra*, reported about its spontaneous contractile activity, having observed that polyps regularly contract to form a ball before extending again (Trembley, 1744). Passano applied electrophysiological methods to *Hydra* to

Table 1. Anatomical and physiological differences between the nervous systems of the four cnidarian classes (Satterlie 2002). Concerning electrical conduction and gap junctions, please refer to the text.

	Anthozoans	*Scyphozoans*	*Cubozoans*	*Hydrozoans*
Medusa anatomy	(no medusa stage)	no velum	velum	velum
Nerve nets (NN)	Diffuse NN	Motor NN Diffuse NN	Motor NN	Motor NN
Nerve rings (NR) at the bell margin	(no medusa stage)	none	a single NR	two NRs: inner (subumbrellar), outer (exumbrellar)
Swim pacemakers	(no swim)	rhopalia (8, 16 or >4x4)	rhopalia (4x, each with 6 eyes)	inner nerve ring
Neuronal electrical conduction	(?)	(?)	(?)	Electrical and dye coupling in neurons
Epithelial / neuroid conduction	inherent muscle contraction	(?)	(?)	(+) myoid conduction
Gap junctions (GJ)	no conventional GJ	no conventional GJ	no conventional GJ	neuro-epithelial GJ, neuro-neuronal GJ, epithelial GJ

show that indeed these contraction bursts and extension movements that represent the main activity of the polyps, are initiated by two interacting pacemakers, one named CB (Contraction Bursts) located in the nerve ring at the base of the hypostome and inducing contractions of the longitudinal myofibers, and a second one, named RP (Rhythmic Potential), endodermal, close to the basal disc that leads to extension by inducing contraction of the circular myofibers (Passano and McCullough 1964, Passano and McCullough 1965, McCullough 1965). The contraction bursts are enhanced by touch or water movement, repressed under starvation, transiently inhibited by blue light, whereas a third pacemaker, the tentacle pulse (TB) induces tentacle contraction. More recently pharmacological approaches showed that GABA, glutamate and glycine receptors modulate the activity of these pacemakers (Kass-Simon et al. 2003, Ruggieri et al. 2004). Sea anemones (anthozoans) also display spontaneous contraction bursts for which pacemakers and conduction systems with similarities to those characterized in *Hydra* were identified (McFarlane 1974a). Hydromedusae also display a spontaneous contractile activity named "crumpling" when they contract their tentacles, manubrium and umbrella margin. However this behavior, which is supposed to be protective, involves epithelial conduction through the radial muscles but no pacemaker activity (Mackie and Passano 1968, Mackie 1975, Spencer 1979).

The swimming behavior of medusae

Over the past 50 years, neurobiologists characterized in several cnidarian species the complex wiring that support the swimming behavior, they characterized the pacemaker activity that initiates this behavior but also the motor nerve net (MNN) that transmits the excitation, the various conduction pathways that enhance or inhibit this transmission, and finally the neuro-muscular transmission (Satterlie 2002). As anticipated from Romanes' work, these studies confirmed the anatomical and physiological variations that can be observed between hydromedusae and cubomedusae /scyphomedusae (see Table 1). Moreover some species evolved unique behaviors, such as the hydrozoan jellyfish *Aglantha* that can swim either fast or slow, fast to escape and slow similarly to the other medusae (Mackie 2004b). This behavioral plasticity relies on a complex neural circuitry with giant neurons involved in fast swimming and 14 distinct but interacting conduction systems.

The feeding response

Beside swimming, cnidarians also actively catch their food, a behavior that requires venom discharge from the nematocytes to immobilize the preys (Tardent 1995, Ozbek et al. 2009) and coordinated movements of the tentacles to bring the food at the mouth opening to be ingested (Westfall and Kinnamon 1984, Mackie et al. 2003). Similarly in medusae, one or several tentacles bend towards the manubrium to transfer the prey to the digestive track and this feeding behavior usually slow downs or inhibits swimming. In the sea anemone *Calliactis* three conduction systems interact to regulate the feeding behavior (McFarlane 1975). Once the preys are ingested, the animals exhibit satiety (Grosvenor et al. 1996) and the digestion of nutrients requires neurally controlled movements of the digestive tract (Shimizu et al. 2004).

The light response

Finally the light response is another important aspect of cnidarian behaviors: light can regulate the rhythmic contractions of *Hydra* polyps (Passano and McCullough 1962) and induce phototaxic behavior as demonstrated by *Hydractinia* planulae or *Tripedalia* medusae that are attracted by light (Plickert and Schneider 2004). Indeed electrophysiological analyses confirmed that visual inputs regulate swimming of hydromedusae and cubomedusae (Anderson and Mackie 1977, Garm and Bielecki 2008). In fact many medusozoan species differentiate light-sensing organs at the adult medusa stage, organs that are connected to nerve rings. These organs are either simple ocelli (multicellular organ composed of several photosensitive cells) as in *Aurelia*, or camera lens-eyes as in *Cladonema* and *Tripedalia* (Martin 2002) (Fig. 1). Moreover cubozoans and scyphozoans differentiate multifunctional sense organs named rhopalia where structures sensing gravity and light group together with the swimming pacemaker (Nakanishi et al. 2009). In cubozoans, the rhopalia are the most complex, containing each a statolith, sensory epithelia, two lensed eyes and up to four pigment-cup ocelli; they regulate obstacle avoidance and mating, and were proposed to be part of the central nervous system (Garm et al. 2007). However light regulation can also take place in the absence of light-sensing organs as in *Hydra* and *Hydractinia* where neurosecretory cells possibly regulate myoepithelial activity through RFamide neuropeptides (Plickert and Schneider 2004). Alternatively light can also directly act on photosensitive neurons (Mackie 1975, Anderson and Mackie 1977).

Chemical conduction relies on classical neurotransmitters and neuropeptides

The electrophysiological work first clearly demonstrated the role of chemical synapses in conduction systems as the motor nerve net in hydromedusae (Anderson 1985), Then it was completed by pharmacological and cellular studies that searched for the chemical support of this conduction (see the detailed review by (Kass-Simon and Pierobon 2007). In short, these studies showed that ion channels (calcium, potassium and sodium) function similarly in jellyfish and vertebrates (Grigoriev et al. 1999, Jeziorski et al. 1999, Spafford et al. 1999), that neurotransmitters like glycine (Pierobon et al. 2001), nitric oxide (Colasanti et al. 1997), endocannabinoid (De Petrocellis et al. 1999), glutamate (Bellis et al. 1991) and acetylcholine (Kass-Simon and Pierobon 2007, Denker et al. 2008a) likely play a physiological role in hydrozoan behavior. Indeed NO affects the swimming behavior of *Aglantha digitale* (Moroz et al. 2004) whereas pharmacologically-induced modulations of the glutamate, glycine and GABA receptors affect the pacemaker activity in *Hydra* (Bellis et al. 1991, Pierobon et al. 1995, Kass-Simon et al. 2003, Pierobon et al. 2004, Ruggieri et al. 2004). Similarly in anthozoans aminergic-like receptors were pharmacologically characterized and a G-protein coupled receptor was identified (Bouchard et al. 2003). One striking aspect of cnidarian nervous systems is that they are strongly peptidergic, i.e. neurosecretory cells release peptides as signaling molecules (Grimmelikhuijzen et al. 2002, Koizumi 2002, Fujisawa 2008), which is providing an additional level of complexity. At the neurophysiological level, neuropeptides indeed affect the neuromuscular transmission (McFarlane et al. 1991, McFarlane et al. 1993, Takahashi et al. 1997, Yum et al. 1998, Takahashi et al. 2003, Hayakawa et al. 2007), the visual and photic response (Plickert and Schneider 2004, Parkefelt and Ekstrom 2009) and possibly the feeding response (Pernet et al. 2004).

Conventional gap junctions and electrical conduction were identified only in hydrozoans

Beside chemical conduction, electrical conduction takes place in cnidarian nervous systems as demonstrated by the electrical and dye coupling observed in photosensitive neurons of *Polyorchis* that allows the photoregulation of the swimming behavior (Anderson and Mackie 1977, Spencer and Satterlie 1980). This electrical conduction presumably relies on gap junctions and indeed neuro-epitheliomuscular and neuro-neuronal gap junctions were observed in *Hydra*; the same neurons using actually both chemical and electrical conduction (Westfall et al. 1980).

However gap junctions are also involved in electrical conduction between non-neuronal cells, allowing epithelial or neuroid conduction. This type of conduction was proposed to be at work in hydrozoans as *Hydra* and siphonophores (Passano and McCullough 1965, Mackie 1965). In hydromedusae (*Sarsia, Euphysa*) and *Hydra*, gap junctions between epithelial cells were identified (Mackie and Passano 1968, Hand and Gobel 1972) and myoid conduction could be recorded in *Nanomia* as well as in *Euphysa* (Spencer 1971, Josephson and Schwab 1979). In *Hydra*, dissociated myoepithelial cells exhibit electrical activity (Kass-Simon and Diesl 1977) and nerve-free animals are excitable (Campbell et al. 1976), suggesting that this neuroid conduction indeed relies on epithelial gap junctions although their presence is disputed (de Laat et al. 1980). Nevertheless molecular data seem to support this scenario, gap junctions are built on connexins in vertebrates, innexins in invertebrates; innexins are indeed expressed in *Hydra* and the fusion protein Inx1-GFP was found along membranes of epithelial cells (Alexopoulos et al. 2004). Therefore it is conceivable that epithelial conduction makes use of similar cellular and molecular basis since the eumetazoan ancestor.

However again hydrozoans seem to play solo: Similarly to *Hydra* "inherent muscle contractions" occur in the absence of any nerve-net activity in the sea anemone *Calliactis* (McFarlane 1974b), but in contrast to hydrozoans, evidences for conventional gap junctions and innexin genes are missing in anthozoans (Mackie et al. 1984, Magie and Martindale 2008). Conventional gap junctions are also missing in scyphozoans and cubozoans. However genes encoding non-conventional gap junction proteins were cloned from two distinct anthozoan species (*Renilla* and *Nematostella*) (Germain and Anctil 1996, Magie and Martindale 2008); hence non-conventional gap junctions might account for neuroid conduction in anthozoans. At the evolutionary level, neuroid conduction, which is able to rapidly integrate external information, is present in plants, protists, sponges, coelenterates and a variety of bilaterian phyla (molluscs, tunicates, amphibians); therefore neuroid conduction was proposed to have evolved earlier and independently of synaptic conduction (Horridge 1968, Mackie 1970, Leys et al. 1999). If true this would mean that neuroid conduction was maintained in hydrozoans but lost or significantly diverged in other cnidarian classes. Alternatively electrical conduction would have arisen multiple times independently.

Where and When Neurogenesis and Nematogenesis Take Place in Cnidarians?

Neurogenesis and nematogenesis during embryonic and larval development

Data about neurogenesis and nematogenesis during development are available in a number of hydrozoan species as *Clytia (Phialidium)* (Thomas 1987), *Pennaria (Halocordyle)* (Martin and Thomas 1981, Kolberg and Martin 1988), *Hydractinia* (Plickert 1989), *Hydra* (Brumwell and Martin 1992), *Podocoryne* (Groger and Schmid 2001), but also anthozoan species as *Anthopleura* (Chia and Koss 1979), *Nematostella* (Marlow et al. 2009), and scyphozoan species as *Aurelia* (Yuan et al. 2008, Nakanishi et al. 2008). These analyses clearly established that anthozoan, hydrozoan and scyphozoan larvae differentiate a nervous system. In these species neuronal markers as RFamide start to be expressed quite early, at the late gastrula stage (Figs. 2M-2O). In hydrozoans, some endodermal cells, named interstitial stem cells give rise to nematoblasts and neuroblasts, which rapidly migrate towards the ectodermal layer (Martin et al. 1997, Groger and Schmid 2001) (see Fig. 2A in Galliot et al. 2009). In *Hydractinia*, this differentiation pathway is regulated by the canonical Wnt pathway (Teo et al. 2006).

In *Podocoryne* planula, tyrosin-tubulin neurons develop progressively, forming repetitive units from anterior to posterior, reminiscent of the formation of the central nervous system in bilaterians. A similar asymmetric development was noted in *Aurelia* (scyphozoan) (Nakanishi et al. 2008) and *Acropora* (anthozoan) where the multipolar and sensory nerve cells expressing the *Gsx* ortholog *cnox-2Am* are restricted to the mid-body region (Hayward et al. 2001), while the sensory nerve cells expressing RF-amide, *Pax-C* or *Emx* are denser at the aboral pole but rare or absent from the oral pole (Miller et al. 2000, de Jong et al. 2006). However in non hydrozoan classes, evidences for interstitial stem cells are missing and the neurons are supposed to differentiate from epithelial cells; moreover in *Aurelia* (scyphozoan) and *Nematostella* (anthozoan) the sensory nervous system is believed to differentiate from the ectodermal layer, a situation that is again reminiscent to that observed in vertebrates (Nakanishi et al. 2008, Marlow et al. 2009). Therefore an asymmetry in the distribution of neuronal populations along the anterior to posterior axis appears to be the rule in the cnidarian larvae and further analyses in representative cnidarian and ctenophore species should tell us whether the ectodermal origin of sensory cells can be considered as plesiomorphic.

The apical tuft, a larval chemosensory organ shared by anthozoans and medusozoans

Most cnidarian species undergo metamorphosis during their development, i.e. the swimming larvae transform into sessile polyps, a complex process that invoves cell death, cell proliferation and morphogenesis. Upon meta-morphosis, the anterior pole of the larva will become the aboral region of the polyp (also named foot) while the posterior pole of the larva will provide the oral region (also named head). The nervous system plays an essential role in this transition: anthozoan and scyphozoan larvae differentiate at the anterior/aboral larval pole a transient neuronal structure, named the apical tuft or the apical sensory organ (Yuan et al. 2008, Marlow et al. 2009). In hydrozoans, sensory neurons also densely pack at the aboral pole. This sensory anterior structure, which differentiates under the control of the FGF pathway (Matus et al. 2007, Rentzsch et al. 2008), can sense environmental cues to promotes the settlement of the swimming larva before it undergoes metamorphosis into polyp (Pang et al. 2004). Several molecular components involved in this response were identified in the hydrozoan *Hydractinia* (Walther et al. 1996, Frank et al. 2001). External clues such as lipids of bacterial source actually trigger a signaling cascade that leads to the release of LWamide neuropeptides (Leitz et al. 1994, Leitz and Lay 1995, Schmich et al. 1998, Plickert et al. 2003). These neuropeptides synchronize the cellular response that leads to metamorphosis. Serotonin was also proposed to be part of this response in *Phialidium* (McCauley 1997). In addition RFamide neuropeptides can inhibit the process (Katsukura et al. 2003). During metamorphosis of hydrozoan and scyphozoan larvae, large parts of the larval nervous system degenerate and a new wave of neuronal differentiation is observed with complex migration patterns (Kroiher et al. 1990, Martin 2000, Nakanishi et al. 2008). A similar process also probably occurs in metamorphosing anthozoans (de Jong et al. 2006), suggesting that developmental neurogenesis follow similar rules in cnidarians.

Adult neurogenesis and nematogenesis in polyps

The cnidarian polyps continuously differentiate a nerve net that exhibits an oral-aboral polarity, with nerve rings on the oral/apical side, pharyngeal and oral in *Nematostella*, apical in *Hydra* but no sensory organs as recognized in medusae (Galliot et al. 2009, Marlow et al. 2009). In some *Hydra* species the apical nerve ring was not characterized anatomically but Passano considered that even in such species a "functional" apical nerve ring is actually active, corresponding to a less-compressed network of ganglionic

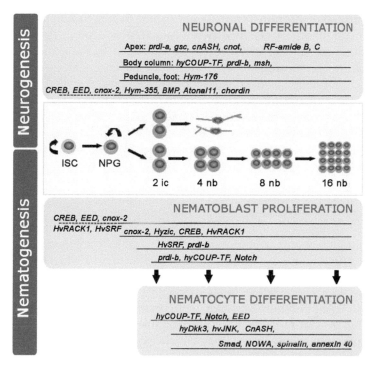

Fig. 3. Neurogenesis and nematogenesis in *Hydra*.
In hydrozoans interstitial stem cells (ISC) provide progenitors for neurons and nematocytes, but also for gland cells and gametes (not shown). ISC express orthologs to *RACK1* (Hornberger and Hassel 1997), *SRF* (Hoffmann and Kroiher 2001), EED (Genikhovich et al. 2006), CREB (Chera et al. 2007). Neuronal progenitors (NPG) are fast-cycling cells that differentiate into neurons and nematocytes; they express the ParaHox gene *Gsx/cnox-2* (Miljkovic-Licina et al. 2007). Hym-176 is a myoactive neuropeptide but the Hym-355 and RF-amide neuropeptides regulate neuronal differentiation positively and negatively respectively. The homeobox genes *msh*, *prdl-b* (Miljkovic-Licina et al. 2004) and the nuclear receptor *COUP-TF* (Gauchat et al. 2004) are restricted to neurogenesis in the body column whereas *prdl-a* (Gauchat et al. 1998), *Gsc* (Broun et al. 1999), *cnASH* (Hayakawa et al. 2004), *cnot* (Galliot et al. 2009) and *Gsx/cnox-2* (Miljkovic-Licina et al. 2007) are expressed in distinct neuronal subsets in the apical region. Regulation of neurogenesis by epithelial or gland cells (Guder et al. 2006, Fujisawa 2008) are not indicated here.
During nematogenesis, *Gsx/cnox-2* (Miljkovic-Licina et al. 2007), *Hyzic* (Lindgens et al. 2004) and *prdl-b* (Gauchat et al. 2004) are successively required for keeping the clustered nematoblasts synchronously dividing (horizontal arrows). In contrast, *hyCOUP-TF* that acts as a transcriptional repressor (Gauchat et al. 2004), *HvJNK* (Philipp et al. 2005) and the Notch pathway (Kasbauer et al. 2007, Khalturin et al. 2007) likely promote arrest of proliferation and entry into differentiation, a switch that can take place after 2, 3, 4 or 5 runs of cell division (vertical arrows). The nematocytes that differentiate the typical venom-filled capsule (nematocyst) express Annexin40 (Schlaepfer et al. 1992), the Achaete scute homolog *CnASH* (Grens et al. 1995, Lindgens et al. 2004), *Smad* (Hobmayer et al. 2001) and *dickkopf-3* (Fedders et al. 2004). A number of phylum/species-specific genes are expressed in nematocytes (Hwang et al. 2007), among them spinalin (Koch et al. 1998) and NOWA (Engel et al. 2002) are structural proteins of the nematocyst.

cells (personal communication). Over the past decades adult neurogenesis and nematogenesis were mostly investigated in *Hydra*; the interstitial cells, which are multipotent stem cells restricted to the ectoderm of the body column, continuously providing neurosensory cells, neurosecretory cells, ganglionic cells, mechanoreceptor cells (nematocytes), gland cells and gametes when the animals follow the sexual cycle (Bode 1996). These interstitial stem cells that can now be traced in transgenic *Hydra* (Khalturin et al. 2007), divide faster than the epithelial stem cells, once a day. In non-hydrozoan species, interstitial stem cells were not characterized so far and neurons are supposed to differentiate directly from the epithelial cells.

In *Hydra*, the nematocyte and neuronal differentiation pathways follow distinct regulations (Fig. 4): interstitial cells committed to the nematocyte lineage undergo up to five synchronous cell cycle divisions, forming clusters of syncitial nematoblasts. Once they stop proliferating, the nematoblasts start differentiating their nematocyst vacuole, which can be of four distinct types (Holstein and Emschermann 1995). Differentiated nematocytes then migrate to their definitive location, namely the tentacles, according to a process that relies on contact guidance from surrounding tentacles (Campbell and Marcum 1980). In the tentacles, nematocytes are embedded within large epithelial cells named battery cells, each battery cell containing several nematocytes, themselves connected to sensory neurons by synapses. After discharge of their capsule, nematocytes are eliminated and replaced by new ones.

In contrast, the differentiation of neuronal cell appears simpler: neuronal progenitors are located along the body column, more numerous in the upper and lower (peduncle) regions but absent from the tentacles or the basal disc. Indeed transplantation of interstitial stem cells in nerve free *Hydra* have shown that neurogenesis but not nematogenesis is strongly influenced by the position of the graft along the body column, i.e. enhanced at the lower and upper positions of the body column where nerve cell density is higher (Yaross and Bode 1978). Interestingly this position-dependent regulation of neurogenesis seems to be largely under the control of epithelial cells (Koizumi et al. 1990). Neuronal progenitors get arrested in G2 until a signal let them divide and terminally differentiate as a sensory or ganglionic cell (Schaller et al. 1989, Bode 1996). Mature neurons receive signals from the head and foot regions to migrate, explaining the higher neuronal densities recorded at the extremities. After mid-gastric bisection, nematocytes and neurons disappear from the head regenerating tip and a wave of *de novo* neurogenesis occurs in the presumptive head region on the second day (Fig. D-G), preceding the emergence of the regenerated head (Miljkovic-Licina et al. 2007). A similar wave of *de novo* neurogenesis is also observed in the presumptive head region during budding, the asexual form of reproduction in *Hydra*. Therefore *Hydra* provides a model system where adult neurogenesis can be investigated in homeostatic and two distinct developmental contexts.

Differentiation and regeneration of light-sensing organs in medusae

In cnidarians, sensory organs that detect light and pressure differentiate at the time the medusae develop as ephyrae in scyphozoans or as buds in hydrozoans. The sensory organs that can detect light exhibit a variable complexity in their anatomy (Martin 2002): they can be clustered photoreceptor cells named ocelli or more complex lens-eyes. A distinct sensory organ named statocyst (or lithocyst) can also measure pressure. In scyphozoans and cubozoans these two types of sensory organs are grouped together within a structure named rhopalia that also contains the swim pacemaker. In *Aurelia* (syphozoan) the gravity-sensing organ, the swim pacemaker and the ocelli differentiate following a strict temporal order (Nakanishi et al. 2009). Whether the differentiation of cubozoan rhopalia follows a similar temporal order is currently unknown. In all medusae analyzed so far the photoreceptor cells are ciliated as in vertebrate visual photoreceptor cells and not rhabdomeric as predominantly observed in invertebrate ones. Moreover cnidarians express c-opsins, suggesting that these two types of photoreceptors were already present in the Cnidaria Bilateria ancestor (Suga et al. 2008). Also the cubozoan and hydrozoan eyes express *Pax* and *Six* genes (Kozmik et al. 2003, Stierwald et al. 2004), suggesting a common regulation of vision in eumetazoans. Nevertheless there is one strange case, that of *Tripedalia* (cubozoan), whose larva exhibits pigmented rhabdomeric photoreceptor cells, while the adult medusa eye uses ciliated ones (Nordstrom et al. 2003, Kozmik et al. 2008). As cnidarian larvae do not differentiate eyes, the *Tripedalia* larva might thus represent an ancestral rhabdomeric light-sensing organ that was lost in most cnidarian species. Finally lens-eyes can regenerate as in *Cladonema* (Stierwald et al. 2004), which thus offers an experimental model system to investigate the specification of eyes in developmental and regenerative context in cnidarians.

Adult neurogenesis and nematogenesis in medusae

In the mature medusa, the manubrium and the tentacle bulbs are the sites of intense production of neurons and nematocytes as observed in the hydrozoan jellyfish (Figs. 2Q-S). In contrast to *Hydra* polyps where all stages of nematogenesis overlap along the body column, the differentiation stages in *Clytia* follow a proximo-distal gradient along the tentacle bulbs (Denker et al. 2008b). Moreover the tentacle bulb isolated from the medusa has the capacity to survive for several days in culture, opening the possibility for manipulations and functional studies.

A Tentative View of Neurogenesis in Early Eumetazoan Evolution

Can we trace back in cnidarians an ancestral neurogenic circuitry?

In the absence of genetically tractable model systems in cnidarians, the characterization of a neurogenic circuitry shared by cnidarians first, by coelenterates second and by coelenterates and bilaterians third, will rely on five criteria:

1) The orthologous character of the cnidarian and bilaterian gene families;
2) for each gene family, the stage and cell type specific regulation in the nervous systems of anthozoan and medusozoan species;
3) the biochemical characterization of the functional domains of a given gene product;
4) the characterization of the regulatory elements and trans-acting factors involved in cell-specific expression;
5) the functional proof of the neurogenic function through loss-of-function (possibly completed by gain-of-function assays) in anthozoan and medusozoan species.

So far studies that would fulfill all these criteria are missing but tools are now available in several cnidarian species to investigate each of these criteria.

Gene cloning in cnidarians was initially targeted to evolutionarily-conserved developmental genes, then gene sequences were obtained through EST projects as in *Hydra* (Hwang et al. 2007), *Clytia* (Jager et al. 2006), *Hydractinia* (Soza-Ried et al. 2009), *Acropora* and *Nematostella* (Technau et al. 2005). Given the importance of peptides as neurotransmitters in cnidarians, a peptide sequencing project was launched, which identified significant regulators of neurophysiology, neurogenesis and morphogenesis (Takahashi et al. 1997, Fujisawa 2008). In parallel genome sequencing projects were launched in *Nematostella* (Putnam et al. 2007) and *Hydra* (http://hydrazome.metazome. net). Analyses from high throughput sequencing have provided two major conclusions, first that the complexity of a large number of gene families was already established in the Cnidaria Bilateria ancestor (Technau et al. 2005, Putnam et al. 2007); second that this ancestral diversity was secondarily reduced in some bilaterian phyla as Ecdysozoa (Miller et al. 2005). The gene families that regulate neurogenesis in bilaterians, the origin of most of them predates Porifera divergence (Fig. 5) (Sakarya et al. 2007, Simionato et al. 2007, Larroux et al. 2008, Richards et al. 2008, Sullivan et al. 2008) and their diversity was indeed already established in cnidarians, although it was still increased after coelenterate divergence (Galliot et al. 2009).

Regarding the hydrozoan adult nervous system, four types of cellular expression patterns were recorded (Fig. 3): *(1) restricted to the neuronal cell*

lineage, (2) restricted to the nematocyte lineage, (3) co-expressed in these two
cell lineages, (4) expressed in the nervous system but not restricted to it (Seipel
et al. 2004, Miljkovic-Licina et al. 2004, Gauchat et al. 2004, Galliot et al.
2006, Chera et al. 2007, Denker et al. 2008b, Galliot et al. 2009). It should be
noted that expression in the nematocyte lineage can often mask neuronal
expression, often more difficult to detect. Therefore genes from the second
class might actually belong to the third one. In summary the candidate

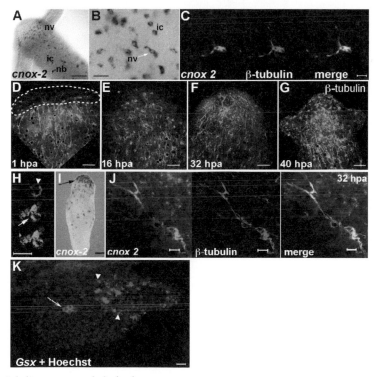

Fig. 4. Adult neurogenesis in hydrozoans.
(A-C) Neurons (arrow) located at the base of the head region between the tentacle roots
express the ParaHox *Gsx* orthologue, *cnox-2*. Along the body column, *cnox-2*+ cells are
interstitial cells (ic) and proliferating nematoblasts (nb). **(D-G)** *De novo* neurogenesis during
Hydra head regeneration. Following bisection the tip of the head-regenerating half is
depleted of neurons (D, outline), progressively repopulated with neuronal progenitors **(U)**
and mature neurons (V) detected with anti beta-tubulin (green). The apical nervous system
at 40 hpa is still less dense than in adult polyps (W). hpa : hours post-amputation. **(H-J)**
During head regeneration, *cnox-2* is expressed in progenitors and differentiating neurons
in the presumptive head region (arrow) shown here at 32 hpa. I) Apical BrdU labeled cells
(red) expressing *cnox-2* (green). J) Apical neurons co-expressing *cnox-2* transcripts (red) and
beta-tubulin (green). **(K)** In the *Clytia* medusa, the proliferative (arrow) and differentiating
(arrowheads) zones of the tentacle bulbs express the ParaHox gene *Gsx*, detected here with
FastRed (red). Scale bars: 200 μm (A), 10 μm (B, I, J), 20 μm (K), 50 μm (D-G).

Color image of this figure appears in the color plate section at the end of the book.

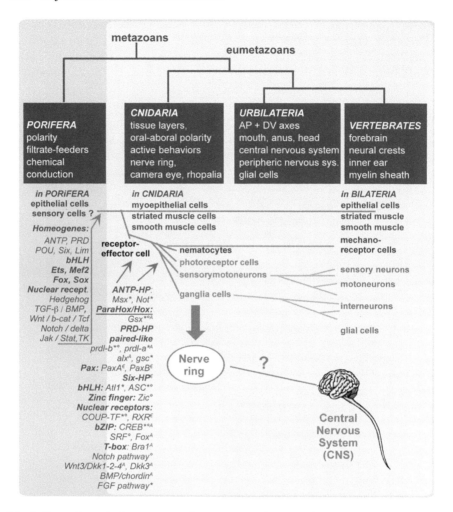

Fig. 5. **The early evolutionary steps of neurogenesis from Porifera to Bilateria.**
Sponges (porifera) that differentiate sensory-like cells but no nervous system feed through passive filtration. Emergence of cells with synaptic conduction likely preceded the divergence of Cnidaria. The various active behaviors of cnidarians rely on complex conduction systems using nerve nets, nerve rings and sensory organs as well as on mechanoreceptor cells (nematocytes) that are densely packed in the tentacles. Cnidarian neurons are already diversified in sensorymotoneurons, ganglionic cells and photoreceptor cells that in medusae can cluster to form ocelli or more complex camera-type eyes. Ganglionic cells that function as interneurons can compress together to form nerve rings, recognized as an annular form of cephalisation. However this scheme does not reflect the wide anatomical and physiological diversity among cnidarian nervous systems (see text). The cnidarian genes expressed during neurogenesis (*), nematogenesis (°), eye differentiation (ϵ) and/or apical patterning (A) are listed (brown). The presence in sponges (red) and placozoans (Schierwater et al. 2009) of gene families that regulate neurogenesis in bilaterians, indicates that their conservation across evolution was constrained even in the absence of neurogenesis.

regulators of neurogenesis in cnidarians are bHLH (*Achaete-Scute, atonal-like*), ANTP-class (*Gsx, emx, msx, not*), PRD-class (*aristaless*-like, *gsc, rx, repo, PaxA/C, PaxB*) and SIN-class (*Six1/2, Six3/6, Six4/5*) transcription factors whereas nematogenesis would specifically require *Zic, Dickkopf3, JNK, Smad, CnASH/Ash1, Notch* (see Fig. 3).

To better characterize the cellular status of the cells expressing a given gene, protein colocalization and cellular analyses can be coupled to in situ hybridization. Except one candidate in *Polyorchis* (Lin et al. 2001), no pan-neuronal markers that would cross-hybridize between cnidarian species was reported yet but several antibodies provide useful tools to detect large subsets of neurons (Fig. 2; Fig. 4). Moreover cell proliferation markers as *in vivo* BrdU-labeling or anti-phoshoH3, which detect S-phase cells and mitotic cells respectively, can also be combined to expression analyses (Fig. 4H). Such approaches, well established in *Hydra* (Koizumi 2002, Lindgens et al. 2004, Gauchat et al. 2004, Miljkovic-Licina et al. 2007) and *Clytia* (Denker et al. 2008b), tell us whether a given gene is rather expressed in proliferating progenitors, differentiating cells and/or terminally differentiated cells. Moreover studying cellular behaviors live is now possible in *Hydra* thanks to the transient expression of reporter constructs (Bottger et al. 2002, Miljkovic et al. 2002, Muller-Taubenberger et al. 2006) or the obtention of stable transgenic lines (Wittlieb et al. 2006, Khalturin et al. 2007).

Biochemical analyses as gel retardation are useful *in vitro* approaches as they can identify among tissue extracts the protein complexes that are activated at specific stages. Such methods also confirmed the expected function of predicted protein domains as the specific DNA-binding activity of the CREB, bHLH, paired-like, Pax, RXR, COUP-TF transcription factors that are expressed in cnidarian nervous systems (Galliot et al. 1995, Grens et al. 1995, Gauchat et al. 1998, Kostrouch et al. 1998, Miller et al. 2000, Sun et al. 2001, Plaza et al. 2003, Kozmik et al. 2003, Gauchat et al. 2004, Kaloulis et al. 2004). The transactivation potential of some of these could also be tested in transfected mammalian cells (Kozmik et al. 2003, Gauchat et al. 2004) or in bilaterian model systems (see Table 2).

A limited number of autologous functional studies have been performed so far, they provided results (Table 2) regarding the following aspects: (1) the RGamide and PWamide peptides play opposite roles on neuronal differentiation in *Hydra*; (2) the LWamide and RFamide neuropeptides play opposite roles on metamorphosis in *Hydractinia*; (3) the canonical Wnt3 pathway regulates the stock of interstitial stem cells in *Hydractinia*; (4) the inhibition of the Wnt3 pathway might specify a neurogenic region in *Hydra* (Guder et al. 2006); (5) the FGF pathway controls aboral specification in the *Nematostella* planula; (6) the *Gsx/cnox-2* homeobox gene controls the proliferation of bipotent neuronal progenitors

Table 2. Cnidarian genes regulating neurogenesis in cnidarians and/or in bilaterians as deduced from functional analyses.

Gene name	Cnidarian species	Function in cnidarians	References
LWamide (Hym-54)	Hydra vulgaris, Hydractinia	neuropeptide required for metamorphosis	(Takahashi et al. 1997, Plickert et al. 2003)
RFamide	Hydractinia	neuropeptide inhibiting metamorphosis	(Katsukura et al. 2003)
RGamide (Hym-355)	Hydra vulgaris	neuropeptide enhancing neurogenesis	(Takahashi et al. 2000)
PWamide (Hym-33H)	Hydra vulgaris	epitheliopeptide inhibiting neurogenesis	(Takahashi et al. 1997, Takahashi et al. 2009)
Notch pathway	Hydra vulgaris	Required for the post-mitotic differentiation of nematocytes	(Kasbauer et al. 2007, Khalturin et al. 2007)
Frizzled, Wnt3	Hydractinia	Wnt3 overactivation by paullones induces nerve cell and nematocyte differentiation	(Teo et al. 2006)
Wnt3, Dickkopf1/2/4	Hydra vulgaris	Dickkopf1/2/4 antagonize Wnt3 in the body column, inducing a neurogenic zone	(Guder et al. 2006)
FGF (NvFGFa1, NvFGFa2), FGFR (NvFGFRa)	Nematostella vectensis	NvFGFa1, NvFGFRa support apical sensory organ formation in Nv planula; NvFGFa2 inhibits its ectopic formation	(Rentzsch et al. 2008)
Zic	Hydra vulgaris	Promotes proliferation of nematoblasts and prevents their differentiation	(Lindgens et al. 2004)
ParaHox Gsx (cnox-2)	Hydra vulgaris	Promotes proliferation of progenitors for apical neurons and nematoblasts in intact and regenerating Hydra; upstream to Zic	(Miljkovic-Licina et al. 2007)
		Overexpression in bilaterians	
Achaete-scute (CnASH)	Hydra vulgaris	Proneural activity in Drosophila: induction of ectopic sensory organs, partial rescue of the achaete and scute double mutant	(Grens et al. 1995)
Brachyury (HyBra2)	Hydra vulgaris	Neural inducing activity in Xenopus	(Bielen et al. 2007)
Pax B (Pax2/5/8 - Pax6 like)	Tripedalia cystophora	Proneural activity in Drosophila: induction of small ectopic eyes, partial rescue of spa(pol), a Pax2 eye mutant.	(Kozmik et al. 2003)

in *Hydra*; (7) the Zic gene promotes proliferation of nematoblasts and the Notch pathway supports nematocyte differentiation in *Hydra*.

Which of the candidate genes mentioned above belong to evolutionarily-conserved genetic circuitries? It is certainly too early to discuss any picture but some candidates are promising as the Pax/Six/opsin cascade in eye specification. Similarly in developing mammals the *Gsx* orthologs specify brain neuronal progenitors (Toresson and Campbell 2001, Yun et al. 2003), whereas in *Drosophila* and mammals *Gsx/Ind* promotes dorsal ventral patterning along the neural tube through negative epistatic relationships with the *NK2/Vnd* and *Msx/msh* homeobox genes (Mieko Mizutani and Bier 2008). As all these genes are expressed in cnidarian species, including the *Hydra msx* in neurons, further studies testing the regulatory interactions between *Gsx/cnox-2*, *msx* and *NK-2* genes in cnidarian nervous systems should tell us whether this block of genes was already functional in the Cnidaria Bilateria ancestor.

How to draw the Urcnidarian nervous system?

Given the variations observed in the neuroanatomies and neurophysiologies of cnidarians, drawing the ancestral cnidarian nervous system is not an easy task. The common features that are shared by anthozoans and medusozoans include four main characters, (1) *the differentiation of three main cell types*, the ectodermal sensory cells that are often multifunctional, the basal ganglionic cells and the nematocytes; (2) the functional organization of these cell types in *conduction systems that use nerve nets and nerve rings*; (3) the conduction through *chemical synapses* that display strikingly evolutionarily-conserved properties with bilaterian synapses; (4) the development of *an anterior sensory organ* in the swimming larva required for settlement and metamorphosis.

Now a number of characters that are not shared between anthozoans and medusozoans receive a controversial origin. Two scenarios can be discussed for each of them, either they arose by convergent evolution, i.e. were acquired independently in various phyla, or they are considered as homologous and were submitted to divergent evolution, meaning in Cnidaria, some loss in one or the other cnidarian group. In this review, we have identified at least four of these convergent/divergent characters:

1) *the stem cell(s) of the neuronal and interstitial cell lineages*, identified as interstitial in hydrozoans, and possibly epithelial in other cnidarian classes. Also the question of a unique stem cell for all cells of the various hydrozoan nervous systems remains open as a possible epithelial origin of sensory cells was documented. For both questions further comparative cellular and molecular studies will help clarify these issues;

2) *the embryonic origin of the nervous system* is also disputed: endodermal in hydrozoans and possibly ectodermal in anthozoans and scyphozoans, at least for some cell lineages. If confirmed, this would indicate that the hydrozoan situation is derived and that an ancestral ectodermal origin might actually correspond to the prevalent bilaterian situation;

3) *the electrical conduction through neurons and epithelial cells* appears to be restricted to hydrozoans. Given that gap junctions support this electrical conduction and that innexins that form conventional gap junctions are widely conserved between invertebrates including hydrozoans, convergent evolution seems unlikely. Consequently, conventional gap junctions would have been lost in anthozoans and hydrozoans would better represent the ancestral situation for this feature;

4) *the presence of light-sensing organs and vision in medusozoans*, but not in anthozoans. Light sensing is widely spread in non-metazoan species but the clustering of photoreceptor cells to form sensory organs was a major innovation in animal evolution. The question of a unique origin for all animal eyes or a repeatedly convergent evolution is a long-standing one (Nilsson and Arendt 2008). However several key components of the genetic circuitry driving eye specification and phototransduction in bilaterians (*Pax, Six, c-opsins*) are already available and properly regulated in cnidarians, strongly supporting the hypothesis of a unique origin in the Cnidaria Bilateria ancestor. Nevertheless the repetitive recruitment of orthologous genes to perform the same task was proposed (Kozmik et al. 2008), suggesting then some higher hierarchical order in the accessibility to developmental processes.

Perspectives: the paradigmatic value of the cnidarian nervous system

Cnidarians nervous systems provide unique experimental paradigms not only to trace back the evolutionary history of their differentiation and their physiology, but also to decipher some striking properties that might have major biomedical impact. In fact cellular studies in *Hydra* polyps have demonstrated an unusual plasticity of the nervous system as neurons constantly change their phenotype while migrating towards the extremities (Bode et al. 1986, Bode 1992). In animals totally depleted of their neuronal progenitors after exposure to antineoplastic drugs, differentiated neurons of the body column change their phenotype in de novo regenerated heads. More strikingly, ganglionic neurons of the body column can transdifferentiate into apical neurosensory cells after regeneration, as evidenced by the de novo differentiation of a cilium

(Koizumi et al. 1988). Similarly the striated muscle cells of the jellyfish *Podocoryne* can be induced to transdifferentiate to neurons and smooth muscle cells (Schmid and Reber-Muller 1995). Whether this cellular plasticity is restricted to hydrozoans remains to be investigated.

Beside transdifferentiation, cnidarians permanently renew their nervous system in adulthood and can even regenerate it after injury, including complete eyes. The biology of stem cells at the various stages of the life cycle is one aspect of this potential that requires careful comparative investigations. Whether there are some common molecular basis in cnidarians for this unusual regenerative potential is at the moment not clear, but the fact that cnidarian genomes contain most of the signaling pathways and regulatory genes active in bilaterians, together with the recent development of potent functional tools are promising conditions to help uncover some ancient principles about developmental and adult neurogenesis.

Acknowledgements

The author is very grateful to L.M. Passano for having kindly introduced her to cnidarian neurophysiology. The work in my laboratory is supported by the Swiss National Fonds, the Canton of Geneva, the Claraz Donation and the Geneva Academic Society.

References

Aerne, B.L. and R.P. Stidwill, P. Tardent. 1991. Nematocyte discharge in Hydra does not require the presence of nerve cells. J. Exp. Zool. 258: 137–141.
Agassiz, L. 1850. Contributions to the natural history of the acalephae of North America, Part 1: On the naked-eyed medusae of the shores of Massachusetts, in their perfect state of development. Memoirs of the American Academy of Arts and Sciences 4: 221–316.
Alexopoulos, H. and A. Bottger, S. Fischer, A. Levin, A. Wolf, T. Fujisawa, S. Hayakawa, T. Gojobori, J.A. Davies, CN. David, and J.P. Bacon. 2004. Evolution of gap junctions: the missing link? Curr. Biol. 14: R879–880.
Anderson, P.A. and G.O. Mackie. 1977. Electrically coupled, photosensitive neurons control swimming in a jellyfish. Science 197: 186–188.
Anderson, P.A. 1985. Physiology of a bidirectional, excitatory, chemical synapse. J. Neurophysiol. 53: 821–835.
Anderson, P.A. and A.N. Spencer. 1989. The importance of cnidarian synapses for neurobiology. J Neurobiol 20: 435–457.
Arendt, D. 2003. Evolution of eyes and photoreceptor cell types. Int. J. Dev. Biol. 47: 563–571.
Arkett, S.A. and G.O. Mackie, R.W. Meech. 1988. Hair cell mechanoreception in the jellyfish *Aglantha digitale*. J. Exp. Biol. 135: 329–342.
Bellis, S.L. and W. Grosvenor, G. Kass-Simon, D.E. Rhoads. 1991. Chemoreception in *Hydra vulgaris* (attenuata): initial characterization of two distinct binding sites for L-glutamic acid. Biochim. Biophys. Acta 1061: 89–94.
Bielen, H. and S. Oberleitner, S. Marcellini, L. Gee, P. Lemaire, H.R. Bode, R. Rupp, and U. Technau. 2007. Divergent functions of two ancient Hydra Brachyury paralogues suggest specific roles for their C-terminal domains in tissue fate induction. Development 134: 4187–4197.

Bode, H. and J. Dunne, S. Heimfeld, L. Huang, L. Javois, O. Koizumi, J. Westerfield, and M. Yaross. 1986. Transdifferentiation occurs continuously in adult hydra. Curr. Top. Dev. Biol. 20: 257–280.

Bode, H.R. 1992. Continuous conversion of neuron phenotype in hydra. Trends Genet. 8: 279–284.

Bode, H.R. 1996. The interstitial cell lineage of hydra: a stem cell system that arose early in evolution. J. Cell. Sci. 109: 1155–1164.

Bottger, A. and O. Alexandrova, M. Cikala, M. Schade, M. Herold, and C.N. David. 2002. GFP expression in Hydra: lessons from the particle gun. Dev. Genes. Evol. 212: 302–305.

Bouchard, C. and P. Ribeiro, F. Dube, and M. Anctil. 2003. A new G protein-coupled receptor from a primitive metazoan shows homology with vertebrate aminergic receptors and displays constitutive activity in mammalian cells. J. Neurochem. 86: 1149–1161.

Bouillon, J. Embranchement des cnidaires (Cnidaria). P.P. 1–28. *In:* P.P. Grassé. [ed.] 1994a. Traité de Zoologie. Cnidaires, Ctenaires, Masson Paris.

Bouillon, J. Cellules urticantes. pp. 65–107. *In:* P.P. Grassé. [ed.] 1994b. Traité de Zoologie. Anatomie, Systématique, Biologie. CNIDAIRES, Hydrozoaires, Scyphozoaires, Cubozoaires, CTENAIRES, Masson, Paris.

Bouillon, J. Embranchement des cnidaires (Cnidaria). pp. 1–28. *In:* P.P. Grassé. [ed.] 1994b. Traité de Zoologie. Cnidaires, Cténaires, Masson, Paris.

Bridge, D. and C.W. Cunningham, R. DeSalle, and L.W. Buss. 1995. Class-level relationships in the phylum Cnidaria: molecular and morphological evidence. Mol. Biol. Evol. 12: 679–689.

Broun, M. and S. Sokol, and H.R. Bode.1999. Cngsc, a homologue of goosecoid, participates in the patterning of the head, and is expressed in the organizer region of Hydra. Development 126: 5245–5254.

Brumwell, G.B. and V.J. Martin. 1992. Immunocytochemically defined populations of neurons progressively increase in size through embryogenesis of *Hydra vulgaris*. Biol. Bull. 203: 70–79.

Campbell, R.D. 1976. Elimination by Hydra interstitial and nerve cells by means of colchicine. J. Cell. Sci. 21: 1–13.

Campbell, R.D. and R.K. Josephson, W.E. Schwab, and N.B. Rushforth. 1976. Excitability of nerve-free hydra. Nature 262: 388–390.

Campbell, R.D. and B.A. Marcum. 1980. Nematocyte migration in Hydra: evidence for contact guidance in vivo. J. Cell. Sci. 41: 33–51.

Chapman, J.A. and E.F. Kirkness, O. Simakov, S.E. Hampson, T. Mitros, T. Weinmaier, T. Rattei, P.G. Balasubramanian, J. Borman, D. Busam, K. Disbennett, C. Pfannkoch, N. Sumin, G.G. Sutton, L.D. Viswanathan, B. Walenz, D.M. Goodstein, U. Hellsten, T. Kawashima, S.E. Prochnik, N.H. Putnam, S. Shu, B. Blumberg, C.E. Dana, L. Gee, D.F. Kibler, L. Law, D. Lindgens, D.E. Martinez, J. Peng, P.A. Wigge, B. Bertulat, C. Guder, Y. Nakamura, S. Ozbek, H. Watanabe, K. Khalturin, G. Hemmrich, A. Franke, R. Augustin, S. Fraune, E. Hayakawa, S. Hayakawa, M. Hirose, J.S. Hwang, K. Ikeo, C. Nishimiya-Fujisawa, A. Ogura, T. Takahashi, P.R. Steinmetz, X. Zhang, R. Aufschnaiter, M.K. Eder, A.K. Gorny, W. Salvenmoser, A.M. Heimberg, B.M. Wheeler, K.J. Peterson, A. Bottger, P. Tischler, A. Wolf, T. Gojobori, K.A. Remington, R.L. Strausberg, J.C. Venter, U. Technau, B. Hobmayer, T.C. Bosch, T.W. Holstein, T. Fujisawa, H.R. Bode, C.N. David, D.S. Rokhsar, and R.E. Steele. 2010. The dynamic genome of Hydra. Nature 464(7288): 592–596.

Chera, S. and K. Kaloulis, and B. Galliot. 2007. The cAMP response element binding protein (CREB) as an integrative HUB selector in metazoans: clues from the hydra model system. Biosystems 87: 191–203.

Chia, F.S. and R. Koss. 1979. Fine-structural studies of the nervous-system and the apical organ in the planula larva of the sea-anemone *Anthopleura elegantissima*. J Morph 160: 275–298.

Colasanti, M. and G. Venturini, A. Merante, G. Musci, and G.M. Lauro. 1997. Nitric oxide involvement in *Hydra vulgaris* very primitive olfactory-like system. J. Neurosci. 17: 493–499.

Collins, A.G. and P. Schuchert, A.C. Marques, T. Jankowski, M. Medina, and B. Schierwater. 2006. Medusozoan phylogeny and character evolution clarified by new large and small subunit rDNA data and an assessment of the utility of phylogenetic mixture models. Syst. Biol. 55: 97–115.

David, C.N. 1973. A quantitative method for maceration of hydra tissue. Wilhelm Roux' Archives of Developmental Biology 171: 259–268.

Davis, L.E. 1974. Ultrastructural studies of development of nerves in hydra. Am. Zool. 14: 551–573.

de Jong, D.M. and N.R. Hislop, D.C. Hayward, J.S. Reece-Hoyes, P.C. Pontynen. E.E. Ball, and D.J. Miller. 2006. Components of both major axial patterning systems of the Bilateria are differentially expressed along the primary axis of a 'radiate' animal, the anthozoan cnidarian *Acropora millepora*. Dev. Biol. 298: 632–643.

de Laat, S.W. and L.G. Tertoolen, and C.J. Grimmelikhuijzen. 1980. No junctional communication between epithelial cells in hydra. Nature 288: 711–713.

De Petrocellis, L. and D. Melck, T. Bisogno, A. Milone, and V. Di Marzo. 1999. Finding of the endocannabinoid signalling system in Hydra, a very primitive organism: possible role in the feeding response. Neuroscience 92: 377–387.

De Robertis, E.M. 2008. Evo-devo: variations on ancestral themes. Cell 132: 185–195.

Denker, E. and A. Chatonnet, and N. Rabet. 2008a. Acetylcholinesterase activity in *Clytia hemisphaerica* (Cnidaria). Chem. Biol. Interact. 175: 125–128.

Denker, E. and M. Manuel, L. Leclere, H. Le Guyader, and N. Rabet. 2008b. Ordered progression of nematogenesis from stem cells through differentiation stages in the tentacle bulb of *Clytia hemisphaerica* (Hydrozoa, Cnidaria). Dev. Biol. 315: 99–113.

Engel, U. and S. Oezbek, R. Engel, B. Petri, F. Lottspeich, and T.W. Holstein. 2002. Nowa, a novel protein with minicollagen Cys-rich domains, is involved in nematocyst formation in Hydra. J. Cell. Sci. 115: 3923–3934.

Fedders, H. and R. Augustin, and T.C. Bosch. 2004. A Dickkopf- 3-related gene is expressed in differentiating nematocytes in the basal metazoan Hydra. Dev. Genes. Evol. 214: 72–80.

Franc, A. La strobilation. pp. 635–638. *In:* P.P. Grassé. [ed.] 1994. Traité de Zoologie. Cnidaires, Cténaires. Masson, Paris.

Frank, U. and T. Leitz, and W.A. Muller. 2001. The hydroid Hydractinia: a versatile, informative cnidarian representative. BioEssays 23: 963–971.

Fujisawa, T. 2008. Hydra peptide project 1993–2007. Dev. Growth Differ. 50 Suppl 1: S257–68.

Galliot, B. and M. Welschof, O. Schuckert, S. Hoffmeister, and H.C. Schaller. 1995. The cAMP response element binding protein is involved in hydra regeneration. Development 121: 1205–1216.

Galliot, B. and V. Schmid. 2002. Cnidarians as a model system for understanding evolution and regeneration. Int. J. Dev. Biol. 46: 39–48.

Galliot, B. and M. Miljkovic-Licina, R. de Rosa, and S. Chera. 2006. Hydra, a niche for cell and developmental plasticity. Semin. Cell Dev. Biol. 17: 492–502.

Galliot, B. and M. Quiquand, L. Ghila, R. de Rosa, M. Miljkovic-Licina, and S. Chera. 2009. Origins of neurogenesis, a cnidarian view. Dev. Biol. 332: 2–24.

Garm, A. and Y. Poussart, L. Parkefelt, P. Ekstrom, and D.E. Nilsson. 2007. The ring nerve of the box jellyfish *Tripedalia cystophora*. Cell Tissue Res. 329: 147–157.

Garm, A. and J. Bielecki. 2008. Swim pacemakers in box jellyfish are modulated by the visual input. J. Comp. Physiol. A Neuroethol. Sens. Neural. Behav. Physiol. 194: 641–651.

Gauchat, D. and S. Kreger, T. Holstein, and B. Galliot. 1998. prdl-a, a gene marker for hydra apical differentiation related to triploblastic paired-like head-specific genes. Development 125: 1637–1645.

Gauchat, D. and H. Escriva, M. Miljkovic-Licina, S. Chera, M.C. Langlois, A. Begue, V. Laudet, and B. Galliot. 2004. The orphan COUP-TF nuclear receptors are markers for neurogenesis from cnidarians to vertebrates. Dev. Biol. 275: 104–123.

Gazave, E. and P. Lapebie, E. Renard, C. Bezac, N. Boury-Esnault, J. Vacelet, T. Perez, M. Manuel, and C. Borchiellini. 2008. NK homeobox genes with choanocyte-specific expression in homoscleromorph sponges. Dev. Genes Evol. 218: 479–489.

Genikhovich, G. and U. Kurn, G. Hemmrich, and T.C. Bosch. 2006. Discovery of genes expressed in Hydra embryogenesis. Dev. Biol. 289: 466–481.

Germain, G. and M. Anctil. 1996. Evidence for intercellular coupling and connexin-protein in the luminescent endoderm of Koellikeri (Cnidaria, Anthozoa). Biol. Bull. 191: 353–366.

Grens, A. and E. Mason, J.L. Marsh, and H.R. Bode. 1995. Evolutionary conservation of a cell fate specification gene: the Hydra achaete-scute homolog has proneural activity in *Drosophila*. Development 121: 4027–4035.

Grigoriev, N.G. and J.D. Spafford, and A.N. Spencer. 1999. Modulation of jellyfish potassium channels by external potassium ions. J. Neurophysiol. 82: 1728–1739.

Grimmelikhuijzen, C.J. and D. Graff, O. Koizumi, J.A. Westfall, and I.D. McFarlane. 1989. Neurons and their peptide transmitters in Coelenterates. pp. 95–109. *In:* Serie A: Life Sciences Vol. 188, ed. NA Series, Plenum Press, New York.

Grimmelikhuijzen, C.J. and M. Williamson, and G.N. Hansen. 2002. Neuropeptides in cnidarians. Can. J Zool. 80: 1690–1702.

Groger, H. and V. Schmid. 2001. Larval development in Cnidaria: a connection to Bilateria? Genesis 29: 110–114.

Grosvenor, W. and D.E. Rhoads, and G. Kass-Simon. 1996. Chemoreceptive control of feeding processes in hydra. Chem. Senses 21: 313–321.

Guder, C. and S. Pinho, T.G. Nacak, H.A. Schmidt, B. Hobmayer, C. Niehrs, and T.W. Holstein. 2006. An ancient Wnt-Dickkopf antagonism in Hydra. Development 133: 901–911.

Hand, A.R. and S. Gobel. 1972. The structural organization of the septate and gap junctions of Hydra. J. Cell. Biol. 52: 397–408.

Hayakawa, E. and C. Fujisawa, and T. Fujisawa. 2004. Involvement of Hydra achaete-scute gene CnASH in the differentiation pathway of sensory neurons in the tentacles. Dev. Genes Evol. 214: 486–492.

Hayakawa, E. and T. Takahashi, C. Nishimiya-Fujisawa, and T. Fujisawa. 2007. A novel neuropeptide (FRamide) family identified by a peptidomic approach in *Hydra magnipapillata*. Febs. J. 274: 5438–5448.

Hayward, D.C. and J. Catmull, J.S. Reece-Hoyes, H. Berghammer, H. Dodd, S.J. Hann, D.J. Miller, and E.E. Ball. 2001. Gene structure and larval expression of cnox-2Am from the coral *Acropora millepora*. Dev. Genes Evol. 211: 10–19.

Hobmayer, B. and F. Rentzsch, and T.W. Holstein. 2001. Identification and expression of HySmad1, a member of the R-Smad family of TGFbeta signal transducers, in the diploblastic metazoan Hydra. Dev. Genes Evol. 211: 597–602.

Hoffmann, U. and M. Kroiher. 2001. A possible role for the cnidarian homologue of serum response factor in decision making by undifferentiated cells. Dev. Biol. 236: 304–315.

Holstein, T. and P. Emschermann. Zytologie. pp 5–15. *In:* J. Schwoerbel and P. Zwick [eds.] 1995. Cnidaria: Hydrozoa, Kamptozoa, Gustav Fisher Verlag, Stuttgart.

Holtmann, M. and U. Thurm. 2001. Mono- and oligo-vesicular synapses and their connectivity in a Cnidarian sensory epithelium (*Coryne tubulosa*). J. Comp. Neurol. 432: 537–549.

Hornberger, M.R. and M. Hassel. 1997. Expression of HvRACK1, a member of the RACK1 subfamily of regulatory WD40 proteins in *Hydra vulgaris*, is coordinated between epithelial and interstitial cells in a position-dependent manner. Dev. Genes Evol. 206: 435–446.

Horridge, G.A. 1954. The nerves and muscles of medusae. I. Conduction in the nervous system of *Aurellia aurita* Lamarck. J. Exp. Biol. 31: 594–600.

Horridge, G.A. The origins of the nervous system. pp. 1–31. *In:* G.H. Bourne [ed.] 1968.The Structure and Function of Nervous Tissue, Academic Press Inc, New York.

Hwang, J.S. and H. Ohyanagi, S. Hayakawa, N. Osato, C. Nishimiya-Fujisawa, K. Ikeo, C.N. David, T. Fujisawa, and T. Gojobori. 2007. The evolutionary emergence of cell type-specific genes inferred from the gene expression analysis of Hydra. Proc. Natl. Acad. Sci. USA 104: 14735–14740.

Jager, M. and E. Queinnec, E. Houliston, and M. Manuel. 2006. Expansion of the SOX gene family predated the emergence of the Bilateria. Mol. Phylogenet. Evol. 39: 468–477.

Jeziorski, M.C. and R.M. Greenberg, and P.A. Anderson. 1999. Cloning and expression of a jellyfish calcium channel beta subunit reveal functional conservation of the alpha1-beta interaction. Recept Chann. 6: 375–386.

Josephson, R.K. and W.E. Schwab. 1979. Electrical properties of an excitable epithelium. J. Gen. Physiol. 74: 213–236.

Kaloulis, K. and S. Chera, M. Hassel, D. Gauchat, and B. Galliot. 2004. Reactivation of developmental programs: The cAMP-response element-binding protein pathway is involved in hydra head regeneration. Proc. Natl. Acad. Sci. USA 101: 2363–2368.

Kasbauer, T. and P. Towb, O. Alexandrova, C.N. David, E. Dall'armi, A. Staudigl, B. Stiening, and A. Bottger. 2007. The Notch signaling pathway in the cnidarian Hydra. Dev. Biol. 303: 376–390.

Kass-Simon, G. and V.K. Diesl. 1977. Spontaneous and evoked potentials from dissociated epithelial cells of Hydra. Nature 265: 75–77.

Kass-Simon, G. and A.A. Scappaticci. 2002. The behavioral and developmental physiology of nematocysts. Can. J. Zool. 80: 1772–1794.

Kass-Simon, G. and P. Pannaccione, and P. Pierobon. 2003. GABA and glutamate receptors are involved in modulating pacemaker activity in hydra. Comp. Biochem. Physiol. A Mol. Integr. Physiol. 136: 329–342.

Kass-Simon, G. and P. Pierobon. 2007. Cnidarian chemical neurotransmission, an updated overview. Comp. Biochem. Physiol. A Mol. Integr. Physiol. 146: 9–25.

Katsukura, Y. and C.N. David, C.J. Grimmelikhuijzen, and T. Sugiyama. 2003. Inhibition of metamorphosis by RFamide neuropeptides in planula larvae of *Hydractinia echinata*. Dev. Genes Evol. 213: 579–586.

Khalturin, K. and F. Anton-Erxleben, S. Milde, C. Plotz, J. Wittlieb, G. Hemmrich, and T.C. Bosch. 2007. Transgenic stem cells in Hydra reveal an early evolutionary origin for key elements controlling self-renewal and differentiation. Dev. Biol. 309: 32–44.

Koch, A.W. and T.W. Holstein, C. Mala, E. Kurz, J. Engel, and C.N. David. 1998. Spinalin, a new glycine- and histidine-rich protein in spines of Hydra nematocysts. J. Cell. Sci. 111: 1545–1554.

Koizumi, O. and S. Heimfeld, and H.R. Bode. 1988. Plasticity in the nervous system of adult hydra. II. Conversion of ganglion cells of the body column into epidermal sensory cells of the hypostome. Dev. Biol. 129: 358–371.

Koizumi, O. and H. Mizumoto, T. Sugiyama, and H.R. Bode. 1990. Nerve net formation in the primitive nervous system of Hydra—an overview. Neurosci. Res. Suppl. 13: S165–170.

Koizumi, O. and M. Itazawa, H. Mizumoto, S. Minobe, L.C. Javois, C.J. Grimmelikhuijzen, and H.R. Bode. 1992. Nerve ring of the hypostome in hydra. I. Its structure, development, and maintenance. J. Comp. Neurol. 326: 7–21.

Koizumi, O. 2002. Developmental neurobiology of hydra, a model animal of cnidarians. Can. J. Zool. 80: 1678–1689.

Koizumi, O. 2007. Nerve ring of the hypostome in hydra: is it an origin of the central nervous system of bilaterian animals? Brain Behav. Evol. 69: 151–159.

Kolberg, K.J. and V.J. Martin. 1988. Morphological, cytochemical and neuropharmacological evidence for the presence of catecholamines in hydrozoan planulae. Development 103: 249–258.

Kostrouch, Z. and M. Kostrouchova, W. Love, E. Jannini, J. Piatigorsky, and J.E. Rall. 1998. Retinoic acid X receptor in the diploblast, tripedalia cystophora. Proc. Natl. Acad. Sci. USA 95: 13442–13447.

Kozmik, Z. and M. Daube, E. Frei, B. Norman, L. Kos, L.J. Dishaw, M. Noll, and J. Piatigorsky. 2003. Role of pax genes in eye evolution: a Cnidarian PaxB gene uniting Pax2 and Pax6 functions. Dev. Cell. 5: 773–785.

Kozmik, Z. and J. Ruzickova, K. Jonasova, Y. Matsumoto, P. Vopalensky, I. Kozmikova, H. Strnad, S. Kawamura, J. Piatigorsky, V. Paces, and C. Vlcek. 2008. Assembly of the cnidarian camera-type eye from vertebrate-like components. Proc. Natl. Acad. Sci. USA 105: 8989–8993.

Kroiher, M. and G. Plickert, and W.A. Müller. 1990. Pattern of cell proliferation in embryogenesis and planula development of *Hydractinia echinata* predicts the postmetamorphic body pattern. Roux´s Arch. Dev. Biol. 199: 156–163.

Larroux, C. and G.N. Luke, P. Koopman, D.S. Rokhsar, S.M. Shimeld, and B.M. Degnan. 2008. Genesis and expansion of metazoan transcription factor gene classes. Mol. Biol .Evol. 25: 980–996.

Leitz, T. and K. Morand, and M. Mann. 1994. Metamorphosin A: a novel peptide controlling development of the lower metazoan *Hydractinia echinata* (Coelenterata, Hydrozoa). Dev. Biol. 163: 440–446.

Leitz, T. and M. Lay. 1995. Metamorphosin A is a neuropeptide. Roux' Arch. Dev. Biol. 204: 276–279.

Lenhoff, H.M. and W. Heagy, and J. Danner. Bioassay for, and characterization of, activators and inhibitors of the feeding response. *In:* H.M. Lenhoff [ed.] 1982. Hydra: Research Methods, Plenum Press, New York, pp. 443–451.

Lentz, T.L. and R.J. Barrnett. 1965. Fine structure of the nervous system of Hydra. Am. Zool. 5: 341–356.

Leys, S.P. and G.O. Mackie, and R.W. Meech. 1999. Impulse conduction in a sponge. J. Exp. Biol. 202 (Pt 9): 1139–1150.

Lin, Y.C. and W.J. Gallin, and A.N. Spencer. 2001. The anatomy of the nervous system of the hydrozoan jellyfish, *Polyorchis penicillatus*, as revealed by a monoclonal antibody. Invert. Neurosci. 4: 65–75.

Lindgens, D. and T.W. Holstein, and U. Technau. 2004. Hyzic, the Hydra homolog of the zic/odd-paired gene, is involved in the early specification of the sensory nematocytes. Development 131: 191–201.

Loomis, W.F. 1955. Glutathione control of the specific feeding reactions of hydra. Ann. N. Y. Acad. Sci. 62: 209–228.

Mackie, G.O. 1965. Conduction in the nerve-free epithelia of siphonophores. Am. Zool. 5: 439–453.

Mackie, G.O. and L.M. Passano. 1968. Epithelial conduction in hydromedusae. J. Gen. Physiol. 52: 600–621.

Mackie, G.O. 1970. Neuroid conduction and the evolution of conducting tissues. Q. Rev. Biol. 45: 319–332.

Mackie, G.O. 1975. Neurobiology of stomotoca. II. Pacemakers and conduction pathways. J. Neurobiol. 6: 357–378.

Mackie, G. and P.A. Anderson, and C.L. Singla. 1984. Apparent absence of gap junctions in two classes of Cnidaria. Biol. Bull. 167: 120–123.

Mackie, G.O. and R.M. Marx, and R.W. Meech. 2003. Central circuitry in the jellyfish *Aglantha digitale* IV. Pathways coordinating feeding behaviour. J. Exp. Biol. 206: 2487–2505.

Mackie, G. 2004a. The first description of nerves in a cnidarian: Louis Agassiz's account of 1850. Hydrobiologia 530/531: 27–32.

Mackie, G.O. 2004b. Central neural circuitry in the jellyfish *Aglantha*: a model 'simple nervous system'. Neurosignals 13: 5–19.

Magie, C.R. and M.Q. Martindale. 2008. Cell-cell adhesion in the cnidaria: insights into the evolution of tissue morphogenesis. Biol. Bull. 214: 218–232.

Marlow, H.Q. and M. Srivastava, D.Q. Matus, D. Rokhsar, and M.Q. Martindale. 2009. Anatomy and development of the nervous system of *Nematostella vectensis*, an anthozoan cnidarian. Dev. Neurobiol. 69: 235–254.

Martin, V.J. and M.B. Thomas. 1981. The origin of the nervous-system in *Pennaria tiarella*, as revealed by treatment with colchicine. Biol. Bull. 160: 303–310.

Martin, V.J. and C.L. Littlefield, W.E. Archer, and H.R. Bode. 1997. Embryogenesis in hydra. Biol. Bull. 192: 345–363.

Martin, V.J. 2000. Reorganization of the nervous system during metamorphosis of a hydrozoan planula. Inv. Biol. 119: 243–253.

Martin, V.J. 2002. Photoreceptors of cnidarians. Can. J. Zool. 80: 1703–1722.

Matus, D.Q. and G.H. Thomsen, and M.Q. Martindale. 2007. FGF signaling in gastrulation and neural development in *Nematostella vectensis*, an anthozoan cnidarian. Dev. Genes Evol. 217: 137–148.

McCauley, D.W. 1997. Serotonin plays an early role in the metamorphosis of the hydrozoan *Phialidium gregarium*. Dev. Biol. 190: 229–240.

McCullough, C.B. 1965. Pacemaker interaction in Hydra. Am. Zool. 5: 499–504.

McFarlane, I.D. 1974a. Control of the pacemaker system of the nerve-net in the sea anemone *Calliactis parasitica*. J. Exp. Biol. 61: 129–143.

McFarlane, I.D. 1974b. Excitatory and inhibitory control of inherent contractions in the sea anemone *Calliactis parasitica*. J. Exp. Biol. 60: 397–422.

McFarlane, I.D. 1975. Control of mouth opening and pharynx protrusion during feeding in the sea anemone *Calliactis parasitica*. J. Exp. Biol. 63: 615–626.

McFarlane, I.D. and P.A. Anderson, and C.J. Grimmelikhuijzen. 1991. Effects of three anthozoan neuropeptides, Antho-RWamide I, Antho-RWamide II and Antho-RFamide, on slow muscles from sea anemones. J. Exp. Biol. 156: 419–431.

McFarlane, I.D. and D. Hudman, H.P. Nothacker, and C.J. Grimmelikhuijzen. 1993. The expansion behaviour of sea anemones may be coordinated by two inhibitory neuropeptides, Antho-KAamide and Antho-RIamide. Proc. Biol. Sci. 253: 183–188.

Mieko Mizutani, C. and E. Bier. 2008. EvoD/Vo: the origins of BMP signalling in the neuroectoderm. Nat. Rev. Genet. 9: 663–677.

Miljkovic, M. and F. Mazet, and B. Galliot. 2002. Cnidarian and bilaterian promoters can direct GFP expression in transfected hydra. Dev. Biol. 246: 377–390.

Miljkovic-Licina, M. and D. Gauchat, and B. Galliot. 2004. Neuronal evolution: analysis of regulatory genes in a first-evolved nervous system, the hydra nervous system. Biosystems 76: 75–87.

Miljkovic-Licina, M. and S. Chera, L. Ghila, and B. Galliot. 2007. Head regeneration in wild-type hydra requires de novo neurogenesis. Development 134: 1191–1201.

Miller, D.J. and D.C. Hayward, J.S. Reece-Hoyes, I. Scholten, J. Catmull, W.J. Gehring, P. Callaerts, J.E. Larsen, and E.E. Ball. 2000. Pax gene diversity in the basal cnidarian Acropora millepora (Cnidaria, Anthozoa): implications for the evolution of the Pax gene family. Proc. Natl. Acad. Sci. USA 97: 4475–4480.

Miller, D.J. and E.E. Ball, and U. Technau. 2005. Cnidarians and ancestral genetic complexity in the animal kingdom. Trends Genet 21: 536–539.

Minobe, S. and O. Koizumi, and T. Sugiyama. 1995. Nerve cell differentiation in nerve-free tissue of epithelial hydra from precursor cells introduced by grafting. I. Tentacles and hypostome. Dev. Biol. 172: 170–181.

Moroz, L.L. and R.W. Meech, J.V. Sweedler, and G.O Mackie. 2004. Nitric oxide regulates swimming in the jellyfish *Aglantha digitale*. J. Comp. Neurol. 471: 26–36.

Muller-Taubenberger, A. and M.J. Vos, A. Bottger, M. Lasi, F.P. Lai, M. Fischer, and K. Rottner. 2006. Monomeric red fluorescent protein variants used for imaging studies in different species. Eur. J. Cell. Biol. 85: 1119–1129.

Nakanishi, N. and D. Yuan, D.K. Jacobs, and V. Hartenstein. 2008. Early development, pattern, and reorganization of the planula nervous system in *Aurelia* (Cnidaria, Scyphozoa). Dev. Genes Evol. 218: 511–524.

Nakanishi, N. and V. Hartenstein, and D.K. Jacobs. 2009. Development of the rhopalial nervous system in *Aurelia* sp.1 (Cnidaria, Scyphozoa). Dev. Genes Evol. 219: 301–317.

Nilsson, D.E. and D. Arendt. 2008. Eye evolution: the blurry beginning. Curr. Biol. 18: R1096–1098.

Nordstrom, K. and R. Wallen, J. Seymour, and D. Nilsson. 2003. A simple visual system without neurons in jellyfish larvae. Proc. Biol. Sci. 270: 2349–2354.

Ozbek, S. and P.G. Balasubramanian, and T.W. Holstein. 2009. Cnidocyst structure and the biomechanics of discharge. Toxicon 54: 1038–1045.

Pang, K. and D.Q. Matus, and M.Q. Martindale. 2004. The ancestral role of COE genes may have been in chemoreception: evidence from the development of the sea anemone, *Nematostella vectensis* (Phylum Cnidaria; Class Anthozoa). Dev. Genes Evol. 214: 134–138.

Pantin, C.F. 1952. The elementary nervous system. Proc. R. Soc. Lond. B Biol. Sci. 140: 147–168.

Parkefelt, L. and P. Ekstrom. 2009. Prominent system of RFamide immunoreactive neurons in the rhopalia of box jellyfish (Cnidaria: Cubozoa). J. Comp. Neurol. 516: 157–165.

Passano, L.M. and C.B. McCullough. 1962. The light response and the rhythmic potentials of Hydra. Proc. Natl. Acad. Sci. USA 48: 1376–1382.

Passano, L.M. 1963. Primitive nervous systems. Proc. Natl. Acad. Sci. USA 50: 306–313.

Passano, L.M. and C.B. McCullough. 1964. Co-ordinating systems and behaviour in Hydra. I. Pacemaker system of the periodic contractions. J. Exp. Biol. 41: 643–664.

Passano, L.M. and C.B. McCullough. 1965. Co-ordinating systems and behaviour in Hydra. II. The Rhythmic potential system. J. Exp. Biol. 42: 205–231.

Pernet, V. and M. Anctil, and C.J. Grimmelikhuijzen. 2004. Antho-RFamide-containing neurons in the primitive nervous system of the anthozoan *Renilla koellikeri*. J. Comp. Neurol. 472: 208–220.

Philipp, I. and T.W. Holstein, and B. Hobmayer. 2005. HvJNK, a Hydra member of the c-Jun NH(2)-terminal kinase gene family, is expressed during nematocyte differentiation. Gene. Expr. Patterns 5: 397–402.

Philippe, H. and R. Derelle, P. Lopez, K. Pick, C. Borchiellini, N. Boury-Esnault, J. Vacelet, E. Renard, E. Houliston, E. Queinnec, C. Da Silva, P. Wincker, H. Le Guyader, S. Leys, D.J. Jackson, F. Schreiber, D. Erpenbeck, B. Morgenstern, G. Worheide, and M. Manuel. 2009. Phylogenomics revives traditional views on deep animal relationships. Curr. Biol. 19: 706–712.

Pierobon, P. and A. Concas, G. Santoro, G. Marino, R. Minei, A. Pannaccione, M.C. Mostallino, and G. Biggio. 1995. Biochemical and functional identification of GABA receptors in *Hydra vulgaris*. Life Sci. 56: 1485–1497.

Pierobon, P. and R. Minei, P. Porcu, C. Sogliano, A. Tino,, G. Marino, G. Biggio and A. Concas. 2001. Putative glycine receptors in Hydra: a biochemical and behavioural study. Eur. J. Neurosci. 14: 1659–1666.

Pierobon, P. and C. Sogliano, R. Minei, A. Tino, P. Porcu, G. Marino, C. Tortiglione and A. Concas. 2004. Putative NMDA receptors in Hydra: a biochemical and functional study. Eur. J. Neurosci. 20: 2598–2604.

Plaza, S. and D.M. De Jong, W.J. Gehring, and D.J. Miller. 2003. DNA-binding characteristics of cnidarian Pax-C and Pax-B proteins in vivo and in vitro: no simple relationship with the Pax-6 and Pax-2/5/8 classes. J. Exp. Zool. B Mol. Dev. Evol. 299: 26–35.

Plickert, G. 1989. Proportion-altering factor (PAF) stimulates nerve cell formation in *Hydractinia echinata*. Cell Differ. Dev. 26: 19–27.

Plickert, G. and E. Schetter, N. Verhey-Van-Wijk, J. Schlossherr, M. Steinbuchel, and M. Gajewski. 2003. The role of alpha-amidated neuropeptides in hydroid development— LWamides and metamorphosis in *Hydractinia echinata*. Int. J. Dev. Biol. 47: 439–450.

Plickert, G. and B. Schneider. 2004. Neuropeptides and photic behavior in Cnidaria. Hydrobiologia 530/531: 49–57.

Putnam, N.H. and M. Srivastava, U. Hellsten, B. Dirks, J. Chapman, A. Salamov, A. Terry, H. Shapiro, E. Lindquist, V.V. Kapitonov, J. Jurka, G. Genikhovich, I.V. Grigoriev, S.M. Lucas, R.E. Steele, J.R. Finnerty, U. Technau, M.Q. Martindale and D.S. Rokhsar. 2007. 2007. Sea anemone genome reveals ancestral eumetazoan gene repertoire and genomic organization. Science 317: 86–94.

Rentzsch, F. and J.H. Fritzenwanker, C.B. Scholz, and U. Technau. 2008. FGF signalling controls formation of the apical sensory organ in the cnidarian *Nematostella vectensis*. Development 135: 1761–1769.

Richards, G.S. and E. Simionato, M. Perron, M. Adamska, M. Vervoort, and B.M. Degnan. 2008. Sponge genes provide new insight into the evolutionary origin of the neurogenic circuit. Curr. Biol. 18: 1156–1161.

Romanes, G.J. 1876 Preliminary observations on the locomotor system of medusae. Phil. Trans. Roy. Soc. Lond. [B] 166: 269–313.

Romanes, G.J. 1877. Further observations on the locomotor system of medusae. Phil. Trans. Roy. Soc. Lond. [B] 167: 659–752.

Ruggieri, R.D. and P. Pierobon, and G. Kass-Simon. 2004. Pacemaker activity in hydra is modulated by glycine receptor ligands. Comp. Biochem. Physiol. A Mol. Integr. Physiol. 138: 193–202.

Sakarya, O. and K.A. Armstrong, M. Adamska, M. Adamski, B. Tidor, B.M. Degnan, T.H. Oakley, and K.S. Kosik. 2007. A post-synaptic scaffold at the origin of the animal kingdom. PLoS ONE 2: e506.

Satterlie, R.A. 2002. Neuronal control of swimming in jellyfish: a comparative story. Can. J. Zool. 80: 1654–1669.

Schaller, H.C. and S.A. Hoffmeister, and S. Dubel. 1989. Role of the neuropeptide head activator for growth and development in hydra and mammals. Development 107: 99–107.

Schierwater, B. and M. Eitel, W. Jakob, H.J. Osigus, S.L. Dellaporta, S.O. Kolokotronis, and R. Desalle. 2009. Concatenated analysis sheds light on early metazoan evolution and fuels a modern "urmetazoon" hypothesis. PLoS Biol 7: e20.

Schlaepfer, D.D. and H.R. Bode, and H.T. Haigler. 1992. Distinct cellular expression pattern of annexins in *Hydra vulgaris*. J. Cell. Biol. 118: 911–928.

Schmich, J. and S. Trepel, and T. Leitz. 1998. The role of GLWamides in metamorphosis of *Hydractinia echinata*. Dev. Genes Evol. 208: 267–273.

Schmid, V. and S. Reber-Muller. 1995. Transdifferentiation of isolated striated muscle of jellyfish in vitro: the initiation process. Semin. Cell. Biol. 6: 109–116.

Seipel, K. and N. Yanze, and V. Schmid. 2004. The germ line and somatic stem cell gene *Cniwi* in the jellyfish *Podocoryne carnea*. Int. J. Dev. Biol. 48: 1–7.

Seipel, K. and V. Schmid. 2006. Mesodermal anatomies in cnidarian polyps and medusae. Int. J. Dev. Biol. 50: 589–599.

Shimizu, H. 2002. Feeding and wounding responses in Hydra suggest functional and structural polarization of the tentacle nervous system. Comp. Biochem. Physiol. A Mol. Integr. Physiol. 131: 669–674.

Shimizu, H. and O. Koizumi, and T. Fujisawa. 2004. Three digestive movements in Hydra regulated by the diffuse nerve net in the body column. J. Comp. Physiol. A Neuroethol. Sens. Neural. Behav. Physiol. 190: 623–630.

Simionato, E. and V. Ledent, G. Richards, M. Thomas-Chollier, P. Kerner, D. Coornaert, B.M. Degnan, and M. Vervoort. 2007. Origin and diversification of the basic helix-loop-helix gene family in metazoans: insights from comparative genomics. BMC Evol. Biol. 7: 33.

Soza-Ried, J. and A. Hotz-Wagenblatt, K.H. Glatting, C. Del Val, K. Fellenberg, H.R. Bode, U. Frank, J.D. Hoheisel, and M. Frohme. 2009. The transcriptome of the colonial marine hydroid *Hydractinia echinata*. FEBS. J. 277: 197–209.

Spafford, J.D. and A.N. Spencer, and W.J. Gallin. 1999. Genomic organization of a voltage-gated Na+ channel in a hydrozoan jellyfish: insights into the evolution of voltage-gated Na+ channel genes. Recept. Chann. 6: 493–506.

Spencer, A.N. 1971. Myoid conduction in the siphonophore *Nanomia bijuga*. Nature 233: 490–491.

Spencer, A.N. 1979. Neurobiology of polyorchis. II. Structure of effector systems. J. Neurobiol. 10: 95–117.

Spencer, A.N. and R.A. Satterlie. 1980. Electrical and dye coupling in an identified group of neurons in a coelenterate. J. Neurobiol. 11: 13–19.

Spencer, A.N. Chemical and electrical synaptic transmision in the Cnidaria. pp. 33–53. *In:* P.A.V. Anderson [ed.] 1989. Evolution of the First Nervous Systems, Plenum Press, New York.

Stierwald, M. and N. Yanze, R.P. Bamert, L. Kammermeier, and V. Schmid. 2004. The Sine oculis/Six class family of homeobox genes in jellyfish with and without eyes: development and eye regeneration. Dev. Biol. 274: 70–81.

Suga, H. and V. Schmid, and W.J. Gehring. 2008. Evolution and functional diversity of jellyfish opsins. Curr. Biol. 18: 51–55.

Sugiyama, T. and T. Fujisawa. 1978. Genetic analysis of developmental mechanisms in Hydra. II. Isolation and characterization of an interstitial cell-deficient strain. J. Cell. Sci. 29: 35–52.

Sullivan, J.C. and D. Sher, M. Eisenstein, K.. Reitzel, H. Marlow, D. Levanon, Y. Groner, J.R. Finnerty, and U. Gat. 2008. The evolutionary origin of the Runx/CBFbeta transcription factors--studies of the most basal metazoans. BMC Evol. Biol. 8: 228.

Sun, H. and D.P. Dickinson, J. Costello, and W.H. Li. 2001. Isolation of cladonema pax-b genes and studies of the dna-binding properties of cnidarian pax paired domains. Mol. Biol. Evol. 18: 1905–1918.

Takahashi, T. and Y. Muneoka, J. Lohmann, M.S. Lopez de Haro, G. Solleder, T.C. Bosch, C.N. David, H.R. Bode, O. Koizumi, H. Shimizu, M. Hatta, T. Fujisawa, and T. Sugiyama. 1997. Systematic isolation of peptide signal molecules regulating development in Hydra: LWamide and PW families. Proc. Natl. Acad. Sci. USA 94: 1241–1246.

Takahashi ,T. and O. Koizumi, Y. Ariura, A. Romanovitch, T.C. Bosch, Y. Kobayakawa, S. Mohri, H.R. Bode, S. Yum, M. Hatta, and T. Fujisawa. 2000. A novel neuropeptide, Hym-355, positively regulates neuron differentiation in Hydra. Development 127: 997–1005.

Takahashi, T. and Y. Kobayakawa, Y. Muneoka, Y. Fujisawa, S. Mohri, M. Hatta, H. Shimizu, T. Fujisawa, T. Sugiyama, M. Takahara, K. Yanagi, and O. Koizumi. 2003. Identification of a new member of the GLWamide peptide family: physiological activity and cellular localization in cnidarian polyps. Comp. Biochem. Physiol. B Biochem. Mol. Biol. 135: 309–324.

Takahashi, T. and O. Koizumi, E. Hayakawa, S. Minobe, R. Suetsugu, Y. Kobayakawa, T.C. Bosch, C.N. David and T. Fujisawa. 2009. Further characterization of the PW peptide family that inhibits neuron differentiation in Hydra. Dev. Genes Evol. 219: 119–129.

Tardent, P. 1995. The cnidarian cnidocyte, a high-tech cellular weaponry. BioEssays 17: 351–362.

Technau, U. and S. Rudd, P. Maxwell, P.M. Gordon, M. Saina, L.C. Grasso, D.C. Hayward, C.W. Sensen, R. Saint, T.W. Holstein, E.E. Ball and D.J. Miller. 2005. Maintenance of ancestral complexity and non-metazoan genes in two basal cnidarians. Trends Genet. 21: 633–639.

Teo, R. and F. Mohrlen, G. Plickert, W.A. Muller, and U. Frank. 2006. An evolutionary conserved role of Wnt signaling in stem cell fate decision. Dev. Biol. 289: 91–99.

Terada, H. and T. Sugiyama, and Y. Shigenaka. 1988. Genetic analysis of developmental mechanisms in Hydra. XVIII. Mechanism for elimination of the interstitial cell lineage in the mutant strain Sf-1. Dev. Biol. 126: 263–269.

Thomas, J.T. 1987. The embryonic origin of neurosensory cells and the role of nerve-cells in metamorphosis in *Phialidium gregarium* (Cnidaria, Hydrozoa). Int. J. Invertebr. Reprod. Dev. 11: 265–285.

Toresson, H. and K. Campbell. 2001. A role for Gsh1 in the developing striatum and olfactory bulb of Gsh2 mutant mice. Development 128: 4769–4780.

Trembley, A. 1744. Mémoires pour servir à l'histoire d'un genre de polypes d'eau douce, à bras en forme de cornes. Leiden.

Walther, M. and R. Ulrich, M. Kroiher, and S. Berking. 1996. Metamorphosis and pattern formation in *Hydractinia echinata*, a colonial hydroid. Int. J. Dev. Biol. 40: 313–322.

Weber, C. 1981. Structure, histochemistry, ontogenetic development and regeneration of the Ocellus of *Cladonema radiatum* Duj. (Cnidaria, Hydrozoa, Anthomedusae). J. Morphol. 167: 313–331.

Westfall, I.A. 1996. Ultrastructure of synapses in the first-evolved nervous systems. J. Neurocytol. 25: 735–746.

Westfall, J.A. and C.F. Elliott, and R.W. Carlin. 2002. Ultrastructural evidence for two-cell and three-cell neural pathways in the tentacle epidermis of the sea anemone *Aiptasia pallida*. J. Morphol. 251: 83–92.

Westfall, J.A. and J.C. Kinnamon. 1978. A second sensory—motor—interneuron with neurosecretory granules in Hydra. J. Neurocytol. 7: 365–379.

Westfall, J.A. and J.C. Kinnamon, and D.E. Sims. 1980. Neuro-epitheliomuscular cell and neuro-neuronal gap junctions in Hydra. J. Neurocytol. 9: 725–732.

Westfall, J.A. and J.C. Kinnamon. 1984. Perioral synaptic connections and their possible role in the feeding behavior of Hydra. Tissue Cell 16: 355–365.

Wittlieb, J. and K. Khalturin, J.U. Lohmann, F. Anton-Erxleben, and T.C. Bosch. 2006. Transgenic Hydra allow in vivo tracking of individual stem cells during morphogenesis. Proc. Natl. Acad. Sci. USA 103: 6208–6211.

Yaross, M.S. and H.R. Bode. 1978. Regulation of interstitial cell differentiation in *Hydra attenuata*. III. Effects of I-cell and nerve cell densities. J. Cell. Sci .34: 1–26.

Yuan, D. and N. Nakanishi, D.K. Jacobs, and V. Hartenstein. 2008. Embryonic development and metamorphosis of the scyphozoan *Aurelia*. Dev. Genes Evol. 218: 525–539.

Yum, S. and T. Takahashi, O. Koizumi, Y. Ariura, Y. Kobayakawa, S. Mohri and T. Fujisawa. 1998. A novel neuropeptide, Hym-176, induces contraction of the ectodermal muscle in Hydra. Biochem. Biophys. Res Commun. 248: 584–590.

Yun, K. and S. Garel, S. Fischman, and J.L. Rubenstein. 2003. Patterning of the lateral ganglionic eminence by the Gsh1 and Gsh2 homeobox genes regulates striatal and olfactory bulb histogenesis and the growth of axons through the basal ganglia. J. Comp. Neurol. 461: 151–165.

Chapter 7

From Cnidaria to "Higher Metazoa" in One Step

Ferdinando Boero[1] and *Stefano Piraino*[2]

Introduction

Metazoan evolution is traditionally seen as a long road where the main steps of the increasing addition of features leading from simple to complex organisms are called milestones. This is the way zoology is treated in textbooks (see Valentine 2004 for a review). The first chapters are dedicated to the simplest animals, and the diversity of animal phyla is ordered along a series of bauplans of gradually increasing complexity.

Protozoans usually introduce the story, since it is highly reasonable that the Metazoa originated from a colonial protozoan, probably a choanoflagellate, that evolved cellular differentiation. Cellular differentiation is the initial milestone leading, from a colony of individual cells, to an individual and multicellular organism. This organizational level has been achieved by sponges, most probably from choanoflagellates. It is difficult to "see" in sponges the premises of all the astonishing features that define what we call animals, as they lack, for instance, true nervous cells, the trademark of animals. Nevertheless, some sponges do have real tissues (see Wang and Lavrov 2007 and references therein), and tissue formation is probably another sponge milestone. Recent findings also showed that sponges possess at least some of the molecular equipments associated with development of sense organs (reviewed in Jacobs et al. 2007) and synaptic connections

---
[1]DiSTeBa, University of Salento, 73100 Lecce, Italy.
E-mail: boero@unisalento.it
[2]DiSTeBa, University of Salento, 73100 Lecce, Italy.
E-mail: stefano.piraino@unisalento.it

(Sakarya et al. 2007). Finally, studies on sponge nuclear and mitochondrial genomes (Larroux et al. 2006, Erpenbeck et al. 2007) confirmed that Porifera are true metazoans.

Peramorphosis vs. Paedomorphosis

It is paradoxical that ecology and embryology-developmental biology, the disciplines that contributed more to Darwinian thinking, did not participate much to the neo-Darwinian synthesis. Gould (1977) replaced developmental biology into an evolutionary framework and proposed heterochrony, and especially paedomorphosis, as a key evolutionary mechanism. From Gould's seminal book, many ideas stemmed, all proposing evolutionary processes that started from complex organisms that became simpler through neoteny or progenesis and that, then, gained new complexity from the re-elaboration of the simplified states.

However, to simplify something to reach then a higher complexity, requires a certain degree of complexity that, presumably, was reached by adding complexity to a simpler organization. Peramorphosis is just this: the addition of new features to old ones. If metazoan evolution is inferred by looking at vertebrates, or insects, or molluscs, or even nematodes, it is obvious that the story runs by the re-elaboration of features that were there already. Paedomorphosis alone, at this level of complexity, makes perfect sense. But if we concentrate on "simple" organisms, the situation becomes radically different: it is difficult to subtract when no additions have been made. Sponges added multicellularity and cell differentiation, but all the other metazoan features seem to be missing. The organization of adult sponges provides no hint about the origin of the other milestones of metazoan evolution. As suggested by Maldonado (2004), the key to identify sponge relationships with the rest of the Metazoa is their larva. Some sponge larvae, in fact, are very similar to cnidarian planulae, or to acelous turbellaria, or to placozoa. They might be the beginning of subsequent Metazoan evolution. They are clearly individuals, they have a one-way movement, with a front and a rear, they move and make choices (the substrate where to settle). If we subtract most of sponge development, what remains is a planula-like organism. From there, we might hope to see the rest of the path that leads to "higher" metazoa.

No matter what techniques are used to build up the tree of metazoan evolution, the base is invariably occupied by a group of phyla: sponges, placozoans, ctenophores, and cnidaria. Molecular data about "who came first" might be controversial, according to the chosen markers or software algorhytms. The reason for this might be linked to some mismatch between the pace of evolution, intended as introduction of novelties, and that of phylogeny, intended as time of divergence between two monophyletic clades.

Micro- vs. Macroevolution Through the Inverted Cone of Development

The identification of jumps in evolutionary patterns, or punctuated equilibria, derived from the analysis of the fossil record and led, for some time, to the dismissal of Darwinian gradualism. The view was that organisms remained more or less unvaried (the equilibrium) and then, all of a sudden (in geological terms), they became something else, with an evolutionary jump. These jumps "punctuate" the otherwise prevailing equilibrium. The introduction of the concept caused a healthy burst of scientific debate, linked to the coupling or uncoupling of micro- and macroevolution. For some evolutionists (especially neontologists), microevolution was the variation within species, and macroevolution was the formation of new species. For some other evolutionists (especially paleontologists), microevolution was the origin of species and macroevolution was the origin of higher taxa.

The introduction of the inverted cone metaphor by Arthur (1997) reconciled the role of developmental biology and of genetics in proposing a theoretical framework for macroevolution and for its coupling with microevolution, if any. The inverted cone (Fig. 1) implies that development can be seen as starting with a totipotent cell, the zygote, whose genome will be expressed in its whole potential.

Cleavage, and the consequent differentiation, reduce the potency of the new, differentiated cells that, in spite of still having the whole information, have the possibility of expressing just a part of it, exemplified by the number of other cell types that will derive from them. At the end

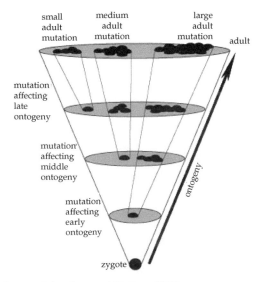

Fig. 1. The inverted cone of development (Arthur, 1977).

of development, completely differentiated cell types will lead just to a reiteration of the same cell type. A quantitatively identical genetic mutation affecting early ontogeny, then, will have a much greater impact on the final result of development than a mutation of the same size occurring at intermediate or final steps of ontogeny. With this conceptual tool at hand, it is much easier to envisage what might be the result of mutations at different ontogenetic stages.

Figure 2 shows how species 1 (S1) can give rise to species 2 (S2), the two remaining in the same genus (G1). This episode of "simple speciation" implies a mutation that becomes effective near the end of ontogeny. Species 1 and 2 share most of their ontogenetic patterns and differ for just some adult features, at the end of the inverted cone.

The founder of a monophyletic genus is a new species, but its evolution from the most proximate ancestor (referable to another genus, otherwise its descendant would not be the founder of the new one) requires a greater change than simple speciation. Species 3 (S3) does not derive from S2 due to the same evolutionary mechanism that led from S1 to S2. The mutation, even of the same size, acts earlier in ontogeny, and had a bearing on a greater part of development and of the resulting adult stage.

A new family, as depicted for the origin of species 4 (S4), requires a much earlier-in-ontogeny mutation, and even earlier are the mutations leading to a new order (from S4 to S5), a new class (from S5 to S6), or a new phylum (from S6 to S7). The first representative of a new phylum, e.g. S7 in fig. 2, is a new species that is also ascribed to a new class, a new order, a new family, a new genus. Phyla are not formed by a posteriori assemblages of the independently-formed lower taxa.

This aspect is particularly important, since genetic mutations of the same size (in terms of mutated nucleotides) might have completely different impacts on the body plan according to the ontogenetic stage in which they act. If we measure evolution according to the genetic distances obtained by

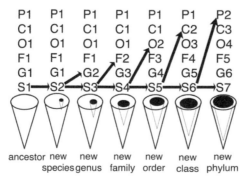

Fig. 2. Adaptation of the theory of the inverted cone of development into the formation of new taxa of different hierarchical order (redrawn from Boero et al. 2005).

comparing genomes, we obtain the overall genetic similarity among taxa, which reflects the divergence in time, measured as if evolution had a steady and gradual pace. But one thing is having diverged since some time, and another thing is to have acquired a sudden mutation, presumably irrelevant to the overall similarity of genomes, that leads to a new milestone. The gradual evolution of neutral genetic divergence might be uncoupled with the saltational evolution linked to the origin of higher taxa.

The Evolutionary Cone

It has often been said that higher taxa are the product of our way of thinking and that species are the only "real" evolutionary units. But if higher taxa are monophyletic, then speciation events, as seen in Fig. 2, do have different bearings on the descendants of the newly formed species, according to the developmental changes that generated them. One might say, re-evaluating Haeckel's recapitulation principle, that higher taxa can be identified according to the shared percentage of ontogenetic patterns. The more ontogenies differ, the higher is the "higher taxon" distinction among lineages. And higher taxa represent clades deriving from these speciations of higher rank than "simple speciations" leading to the origin of new species referable to the same genus of the ancestor.

A monophyletic phylum starts with one species (Fig. 3), representing the founder of the phylum. That species will represent also a new genus, a new family, a new order, and a new class. From this, all the subsequent radiation, if any, will take place. Leading to the distinction of new lower-level taxa, each one starting, however, with a single species. Depending on the magnitude of the genetic (and phenotypic) modifications, new higher-order taxa can evolve suddenly, without the gradual pattern shown in Fig. 3. Evolution can be both gradual and saltational, even though gradual evolution, leading to simple speciation, is much more probable than saltational one, leading to speciation originating higher-order taxa, since the viability of the result of large genetic modifications might be impaired by the change. The possibility that the sum of an enormous number of little variations might lead to the emergence of higher-order taxa is not reasonable in the light of the inverted cones of both development and evolution. Simple speciation events, in fact, do not radically alter the body plan of the descendents in respect to that of the ancestors, only leading to a simple series of variations on the same theme. Higher-order taxa, to evolve, need radical changes, occurring towards the tip of the inverted cone, far from the end of ontogeny.

Alternative Morphs—Combined Cones

The theory of alternative morphs (West-Eberhard 2003) was proposed to explain how evolutionary novelties can emerge in taxa with complex life

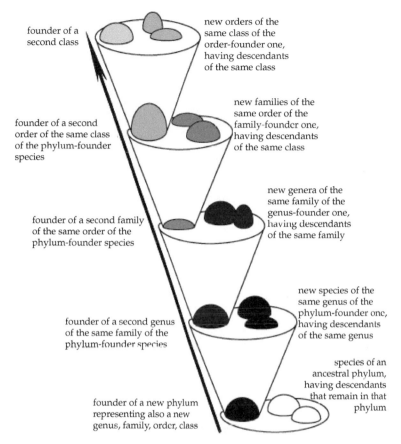

founder of a
second class

new orders of the
same class of the
order-founder one,
having descendants
of the same class

new families of the
same order of the
family-founder one,
having descendants
of the same class

founder of a second
order of the same class
of the phylum-founder
species

new genera of the
same family of the
genus-founder one,
having descendants
of the same family

founder of a second family
of the same order of the
phylum-founder species

new species of the
same genus of the
phylum-founder one,
having descendants
of the same genus

founder of a second genus
of the same family of the
phylum-founder species

species of an
ancestral phylum,
having descendants
that remain in that
phylum

founder of a new phylum
representing also a new
genus, family, order, class

Fig. 3. Radiation of biodiversity explained by the metaphore of the inverted cone. Genetic changes of different magnitude may take place at different developmental timings, explaining both saltational and gradual radiation of taxa.

cycles. The rationale of the theory is that the strongest form of selection is stabilizing selection, with a strong tendency to conserve extant genetic and morphological states. To circumvent stabilizing selection, allowing neutral evolution only, some morphs can be temporarily reduced or suppressed (by paedomorphosis) and they can then be re-expressed in a modified form. The same idea, at a genetic level, is that of gene duplication. Important genes cannot be changed, but they can evolve by duplication, so that one copy does the "original work" and the others are free to change, leading to novelties.

If paedomorphosis and peramorphosis can modulate change even in species with direct development, this is even simpler for species with complex life cycles, leading to different morphs. Translated into the inverted cone metaphor (Fig. 4), the alternative morphs suggest that mutations can affect a larval stage of the cycle (for instance the polyp stage in a medusozoan), or the adult (for instance the medusa stage of a medusozoan), or both, according to what will be the contribution of the cell altered by the mutation to the rest of development. Regulatory genes, furthermore, can modulate the expression of the various stages, a rather widespread phenomenon in the Cnidaria, and especially in the Hydrozoa, where medusa suppression occurs very easily (Bouillon et al. 2006).

The suppression of medusae does not imply the elimination of the genetic information specifying for them. Being unexpressed, this information is not subjected to natural selection and is free to change with no limits. If re-expressed, these modified genes can originate much different morphs from the original ones, as probably occurred for the medusae of *Obelia* (Boero et al. 1996, Govindarajan et al. 2006) (Fig. 5).

In spite of being the classical textbook example of the Hydrozoa, the medusae of *Obelia* have a radically different body plan from that of all other medusae: their umbrella is flat and not concave, they lack a velum, their tentacles have a chordal structure, identical to polyp tentacles. The hydroids of *Obelia*, instead, are undistinguishable from those of other members of the family, the Campanulariidae. Whereas the hydroids are highly conservative, the medusa is greatly innovative.

The probable reason for this is that the genes of the *Obelia* medusae have been re-activated after having been inactivated in an ancestral species with suppressed medusae. *Obelia* is probably one of the best examples of the origin of a new body plan through a mechanism that is perfectly

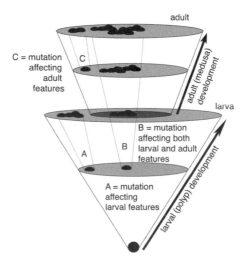

Fig. 4. The theory of alternative morphs in complex life cycles (e.g. hydrozoan life cycle) explained by the metaphore of the inverted cone. The outcome of genetic changes depends on the stage in the life cycle (larval: planula and polyp, or adult: medusa) where they take place.

explained by the theory of alternative morphs. The polyp morph, in fact, is almost indistinguishable from the polyps of other genera of the Campanulariidae, and it is for this reason that the genus is placed into this family. The medusa, however, differs in such a way from all other hydromedusae (first of all by not having a velum, the trademark of hydromedusae) that, if the polyp were unknown, it should be placed into a separate medusozoan class.

In the light of the inverted cone model (Fig. 6), the developmental cone of the hydroid remained unmodified. The medusa stage was suppressed and its genetic substrate mutated freely. When the medusa was re-expressed, its inverted cone was radically modified, resulting into a new body plan by re-elaboration of the existing genetic material.

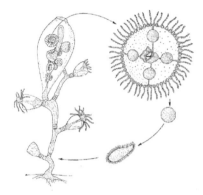

Fig. 5. The life cycle of *Obelia* sp., a classical model of hydrozoan life cycle.

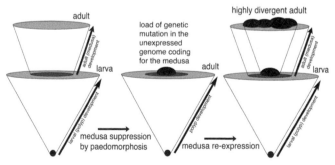

Fig. 6. High morphological divergence in the adult stage of *Obelia* sp. explained by alternation of medusa suppression, silent accumulation of genetic changes in the unexpressed part of the genome, and medusa re-expression.

The Cnidaria are the Basis of Metazoan Evolution

Boero et al. (1998, 2005, 2007) developed the idea that most of the features that are considered as the "milestones" of metazoan evolution evolved in the Cnidaria and that the so-called Cambrian explosion had the Cnidaria as its explosive, since representatives of the phylum are in the fossil record before all other present-day animals. The proposal stemmed from comparative embryology and anatomy and was later corroborated by molecular studies (see Seipel and Schmid 2005 and references therein, Matus et al. 2006), even if for some time it was not favorably accepted (e.g. Nielsen 2001). However, alternative molecular approaches can support different views, as a recent one (Dunn et al. 2008) viewing the Ctenophora as the stem group of the Metazoa, among the traditional "radial" groups.

The disparity of phylogenetic reconstructions calls for a double check of their validity, especially when molecular markers presume a more or less continuous change (i.e. gradual evolutions) that might not cope with sudden genetic changes leading to sharp evolutionary novelties by saltational events, as proposed in the first part of this chapter. Comparative morphology, under this respect, still has a lot to say. Just as molecular biology, though. The two must go together. Eye evolution, for instance, is with all probability a cnidarian affair, given the number of conserved critical genes to the regulatory developmental cascade (e.g. *Brain3*, *eyes absent*, *Six/sine oculis*) leading to eye formation throughout the metazoa. If the genetic control is the same, eyes are with all probability monophyletic, even though studies based on comparative morphology suggested analogy. A *PaxB* cubozoan gene, expressed in the rhopalia, retain combined features of Pax6- and Pax 2/5/8 types, respectively two master genes for eye and ear/gravity organs formation (Gehring 2005, O'Brien and Degnan 2003). These arguments are in favour of a shared ancestry of sense organs between cnidarians and bilaterians (Jacobs et al. 2007).

The milestones of metazoan evolution that have or might have their first appearance in the cnidaria are, according to Boero et al. (2007):

The eyes, supported by the conservation of the genetic control of eye development by PAX genes;

The statocysts, as equilibrium organs, first developed in jellyfish;

The bilateral symmetry, as documented by the finding of fundamental genetic controls for the establishment of a secondary main body axis (reviewed in Boero et al. 2007) and by some anatomical evidences, such as the dramatic polarization of nerve cells in the planula larva that, for the first time in the Metazoa, exhibits a central nervous system with an anterior, brain-like structure, and longitudinal neural projections towards the posterior pole (Piraino, unpublished data);

The mesoderm, whose homology with the mesoderm of "triploblasts" has been repeatedly demonstrated at anatomical, embryological, and molecular levels (see Seipel and Schmid 2005, 2006);

The coelom, as formed during the early ontogeny of the subumbrellar chamber of the hydromedusae, lined by a mesoderm-derived tissue, the striated muscle layer;

The chitinous exoskeleton, present in hydrozoan polyp stages;

The calcium-based skeleton, present in the Madreporaria and in some Hydrozoa;

Metamery and modularity is also a possible milestone, if the repetition of modular parts (polyps) along an axis (colony) will be demonstrated to be regulated by the same genetic controls regulating the formation of metameres or repeated modules in individual organisms.

If all or most of the metazoan evolution milestones cited above prove to be homologous, as the eyes and the mesoderm have been, it can be argued that most of the key features of metazoan groups are present in the Cnidaria and that the various metazoan phyla have simply developed some cnidarian feature while not having others. The chitinous exoskeleton, for instance, characterizes the Ecdysozoa, whereas a calcium-based skeleton characterizes the Deuterostomia.

The phylogenetic tree of the Metazoa is usually divided into three megabranches, representing groups of phyla that share the same basic features (milestones). They are united at base by being bilateral, as opposed to the stem groups, that are called Radiata. The bilateral planula-like adult or larva of these radial forms was not taken as a sign of bilaterality in these groups since the bilateral stages lacked a central nervous system. The recent discovery of a central nervous system in a Hydrozoan planula sets back the origin of both bilaterality and the central nervous system. The most ancient Metazoa with a bilateral symmetry and a central nervous system are the Cnidaria. The recent description of Cambrian jellyfish fossils almost identical to recent ones (Cartwright et al. 2007) suggests that jellyfish were present even before the Cambrian explosion.

A somehow bold statement: The era of the Bilateria was set 540 million years ago, by paedomorphosis, when a planula became sexually competent, stopping its development but retaining the genetic information of all the milestones already present in the Cnidaria. The development of a chitinous exoskeleton would have started the Ecdysozoa, whereas that of a calcium-based skeleton would have started the Deuterostomia, all with their coelom, and mesoderm, and sense organs, and metamery (already evolved by the Cnidaria).

Conclusion: Testing Hypotheses

The above hypothesis, consisting of a constellation of sub-hypotheses, stems from comparative morphology and can be tested by investigating the genetic control that leads to the formation of the milestones, forming the sub-hypotheses. The first tests already suggested homology for both eyes and mesoderm. If most of the other milestones will prove to be homologous, the Cnidaria will acquire a pivotal role in metazoan evolution, being not only the base of their tree, but also the basis of their evolutionary novelties. The mechanisms allowing for the radiation of the main metazoan lineages from the Cnidaria are the suppression by paedomorphosis of some acquired features, the retention of their genetic specification, their re-elaboration followed by their re-expression in a modified form. This mechanism is suggested by the inverted cones of development and evolution, linked to the paedomorphic and the alternative morphs models of evolution.

If these hypotheses will be validated, the evolution of the Metazoa will result as deriving from a single, major step: the evolution of Cnidaria. And zoology textbooks will require some re-writing.

Acknowledgements

This is dedicated to the memory of Volker Schmid (1939–2008). We worked together for more than ten years, searching for the secrets of the immortal jellyfish, but we were too slow in getting them. Volker started it all by stating, one day, that he was teaching his students that the subumbrellar cavity of hydromedusae is a coelom. When FB asked him why he hadn't published anything about that, he said: But it is so obvious! It was obvious for you, Volker. And also for us, one second after you simply said that sentence. Everything became clear in a single shot. But it was worth while stating it formally for the unbelieving fans of the "higher" taxa, and then you were so kind to demonstrate it also with some molecules. Thank you. The unconventional scientist refuses to die.

This is also a contribution to the European Network of Excellence in Marine Biodiversity and Ecosystem Functioning, and to SESAME, an Integrated Project of the EU. Even if that money is gone, this is also a product of the PEET project of the National Science Foundation of the USA. The Italian Ministry of University and Research (PRIN, FIRB projects) also supported experimental work on the brain of hydrozoan planula larvae.

Bernd Schierwater (together with Stephen Dellaporta and Rob DeSalle) was kind enough to invite us to write this chapter. He helped us in an earlier version of the development of these ideas, with deep insight on the Placozoa.

Jean Bouillon (1926–2009) even without knowing it, provided continuous inspiration and support, and will never be thanked enough Jean's work will stand for a very long time as a beacon for present and future Hydrozoan researchers.

We also thank Walter Gehring who read and commented on the early draft of this manuscript.

References

Arthur, W. 1997. The Origin of Animal Body Plans. Cambridge University Press, Cambridge, UK.

Boero, F. and J. Bouillon, and S. Piraino. 1996. Classification and phylogeny in the Hydroidomedusae (Hydrozoa, Cnidaria). Sci.Mar. 60: 7–16.

Boero, F. and C. Gravili, P. Pagliara, S. Piraino, J. Bouillon, and V. Schmid. 1998. The cnidarian premises of metazoan evolution: from triploblasty, to coelom formation, to metamery. Ital. J. Zool. 65: 5–9.

Boero, F. and J. Bouillon, and S. Piraino. 2005. The role of Cnidaria in evolution and ecology. Ital. J. Zool. 72: 65–71.

Boero, F. and B. Schierwater, and S. Piraino. 2007. Cnidarian milestones in metazoan evolution. Integr. Comp. Biol. 47: 693–700.

Bouillon, J. and C. Gravili, F. Pagès, J-M. Gili, and F. Boero. 2006. An Introduction to Hydrozoa. Mémoires du Musèum National d'Histoire Naturelle 194: 1–593.

Cartwright P. and S.L. Halgedahl, J.R. Hendricks, R.D. Jarrard, A.C. Marques, A.G. Collins, and B. Lieberman. 2007. Exceptionally Preserved Jellyfishes from the Middle Cambrian. PLoS ONE 2(10): e1121.

Dunn, C.W. and A. Hejnol, D.Q. Matus, K. Pang, W.E. Browne, S.A. Smith, E. Seaver, G.W. Rouse, M. Obst, G.D. Edgecombe, M.V. Sorensen, S.H. Haddock, A. Schmidt-Rgaesa, A. Okusu, R.M. Kristensen, W.C. Wheeler, M.Q. Martindale, and G. Giribet. 2008. Broad phylogenomic analysis improves the resolution of the animal tree of life. Nature 452: 745–749.

Erpenbeck, D. and O. Voigt, M. Adamski, M. Adamska, J.N. Hooper, G. Worheide, and B.M. Degnan. 2007. Mitochondrial diversity of early-branching metazoa is revealed by the complete mt genome of a haplosclerid demosponge. Mol. Biol. Evol. 24: 19–22.

Gehring, W. 2005. New perspectives on eye development and the evolution of eyes and photoreceptors. J. Heredity 96: 1–13.

Govindarajan, A. and F. Boero, and K. Halanych. 2006. Phylogenetic analysis with multiple markers indicates repeated loss of the adult medusa stage in Campanulariidae (Hydrozoa, Cnidaria). Mol. Phyl. Evol. 38: 820–834.

Jacobs, D.K. and N. Nakanishi, D. Yuan, A. Camara, S.A. Nichols, and V. Hartenstein. 2007. Evolution of sensory structures in basal metazoa. Integr. Comp. Biol. 47: 712–723.

Kusserow, A. and K. Pang, C. Sturm, M. Hrouda, J. Lentfer, H.A. Schmidt, U. Technau, A. von Haeseler, B. Hobmayer, M.Q. Martindale, and T.W. Holstein. 2005. Unexpected complexity of the Wnt gene family in a sea anemone. Nature 433: 156–160.

Larroux, C. and B. Fahey, D. Liubicich, V.F. Hinman, M. Gauthier, M. Gongora, K. Green, G. Wörheide, S.P. Leys, and B.M. Degnan. 2006. Developmental expression of transcription factor genes in a demosponge: insights into the origin of metazoan multicellularity. Evol. Dev. 8: 150–173.

Maldonado, M. 2004. Choanoflagellates, choanocytes, and animal multicellularity. Invert. Biol. 123: 1–22.

Matus, D.Q. and K. Pang, H. Marlow, C.W. Dunn, G.H. Thomsen, and M.Q. Martindale. 2006. Molecular evidence for deep evolutionary roots of bilaterality in animal development. Proc. Natl. Acad. Sci. USA 103: 11195–11200.

Nielsen, C. 2001. Animal Evolution, Interrelationships of the Living Phyla. Oxford Univ. Press, Oxford.

O'Brien, E.K. and B.M. Degnan. 2003. Expression of *Pax2–5–8* in the gastropod statocyst: insights into the antiquity of metazoan geosensory organs. Evol. Dev. 5: 572–578.

Sakarya, O. and K.A. Armstrong, M. Adamska, M. Adamski, I-F. Wang, B. Tidor, B.M. Degnan, T.H. Oakley, and K.S. Kosick. 2007. A post-synaptic scaffold at the origin of the animal kingdom. PLoS ONE 2: e506.

Seipel, K. and V. Schmid. 2005. Evolution of striated muscle: jellyfish and the origin of triploblasty. Dev. Biol. 282: 14–26.

Seipel, K. and V. Schmid. 2006. Mesodermal anatomies in cnidarian polyps and medusae. Int. J. Dev. Biol. 50: 589–599.

Wang, X. and D.V. Lavrov. 2007. Mitochondrial genome of the homoscleromorph *Oscarella carmela* (Porifera, Demospongiae) reveals unexpected complexity in the common ancestor of sponges and other animals. Mol. Biol. Evol. 24: 363–373.

West-Eberhard, M.J. 2003. Developmental Plasticity and Evolution. Oxford University Press, Oxford.

Chapter 8

Basal Metazoan Sensory Evolution

D.K. Jacobs,[1]*[†] D.A. Gold,* N. Nakanishi,* D. Yuan,* A. Camara,*
S.A. Nichols,[¥] and V. Hartenstein[†]

Introduction

Cnidaria have traditionally been viewed as the most basal animals with complex, multicellular structures dedicated to sensory perception. However, sponges also have a surprising range of the genes required for sensory and neural functions in Bilateria. We develop arguments explaining the shared aspects of developmental regulation across sense organs and between sense organs and other structures, focusing on explanations that involve divergent evolution from a common ancestral condition. In Bilateria, distinct sense-organ types share components of developmental-gene regulation. These regulators are also present in basal metazoans, suggesting evolution of multiple bilaterian organs from a smaller number of antecedent sensory structures in a metazoan ancestor. More specifically, we hypothesize that developmental genetic similarities between sense organs and appendages may reflect descent from closely associated structures, or a composite organ, in the common ancestor of Cnidaria and Bilateria, and we argue that such similarities between

*Department of Ecology and Evolutionary Biology, UCLA, 621 Young Drive South, Los Angeles CA 90095-1606.
[†]Department of Molecular, Cellular and Developmental Biology, UCLA, 621 Young Drive South, Los Angeles CA 90095-1606.
[¥]Department of Molecular and Cell Biology, 142 Life Sciences Addition, University of California, Berkeley, CA 94720.
[1]E-mail: djacobs@ucla.edu

bilaterian sense organs and kidneys may derive from a multifunctional aggregations of choanocyte–like cells in a metazoan ancestor. We hope the speculative arguments presented here will stimulate further discussion of these and related questions.

The word "animal" implies muscle-driven motility coordinated by neural integration of sensory stimuli, which is produced in specialized multicellular sensory structures. Consequently, a number of sets of questions spring to mind when considering evolution of Metazoan sensation: Where on the tree of animal life did the first sense organs evolve? Do sense organs share a common evolutionary origin with other structures or organs? What type of sense organ evolved first and how are different classes of sense organs related to one another? Are bilaterian sense organs related to the sensory features in the more basal radiate taxa? Does the placement of Scyphozoa and Hydrozoa together in a medusozoan group support a derived condition for cnidarian sense organs? How does evidence suggesting common origin of bilaterian and cnidarian sense organs relate to the presence of bilaterian-like dorso-ventral axial organization in Cnidaria?

Not all of the preceding questions can be definitively answered at this time. However, developmental gene-expression studies, genome sequencing, and expressed-sequence-tag studies are shedding light on some of these issues. Interestingly, the initial answers to these questions are not always consistent with *a priori* expectations. For example, one might expect that evolution of genes thought to be explicitly involved in the development of sense organs would coincide with the evolution of the radiates, as Cnidaria and Ctenophora are the most basally branching lineages with specialized sense "organs". This expectation is not met; regulatory genes involved in sense-organ development in "higher" Metazoa are present more basally in sponges, as are genes considered essential for synaptic function. Although not explicitly muscular or neural, sponges exhibit coordinated contraction as well as coordinated cessation of pumping. Thus, a view of sponges as more active is replacing an older perception that held sponges to be virtually "inanimate".

In this work we touch on the features that distinguish sense organs, independent of classic definitions. We then consider the questions listed above in the context of the basal branches of the metazoan tree, focusing on the cnidarians and sponges. In cnidarians we address the relationship between cnidarian and bilaterian sensory structures, as well as shared aspects of sense organs and appendages. In the sponges we discuss the possible evolutionary antecedents of sense organs. Lastly we consider how different reconstructions of the metazoan tree could effect these interpretations. The speculative hypotheses presented here emphasize differential persistence and modification of an ancestral condition, rather than invoking wholesale "cooptation" of genes, as an explanation for conflicting patterns of gene expression and morphology observed across the metazoan tree. In each instance considered, many other hypotheses could be advanced, and we encourage others to generate specific competing hypotheses.

What do Sense Organs Have in Common?

Cells generally have an ability to assay aspects of their surroundings. However, multicellular organisms have the challenge of differential exposure of cells to external and internal environments, as well as the opportunity to have cells with specialized sensory functions. Sensory structures that form part of the epidermis are found in all animal phyla from cnidarians onward. In cnidarians and some basal bilaterian groups (e.g. acoels, platyhelminths, nemertines), sensory structures consist of "naked" sensory neurons whose dendrite is formed by a modified cilium (Chia and Koss 1979). The cell bodies of sensory neurons are often sunken beneath the level of the epidermis, or can even reside within the central nervous system. From these "naked" sensory neurons one distinguishes sensilla and sensory organs. Sensilla constitute individual sensory neurons, or small arrays of sensory neurons, which are integrated with specialized non-neuronal cells that typically function in particular sensory modalities, such as light reception, mechanoreception (auditory/inertial/touch/stretch/vibration), and chemoreception (taste/smell). Finally, sense organs are large assemblies of sensory neurons and non-neuronal cells that form macroscopic structures. Highly developed sensory organs are widespread and exist for all sensory modalities in bilaterians. In some cases, such as the compound eyes and auditory organs of arthropods, arrays of contiguous sensilla are integrated into large sensory organs. In this view "sense organs" already exist in cnidarians in the form of eyes and statocysts, despite the lack of mesoderm often invoked as a required condition for organ systems. In many instances, sensory organs and sensilla coexist with naked sensory neurons in the same animal.

The sensory neurons of a sensory organ or sensillum usually bear cilia and/or microvillar structures on their apical surfaces, and these surfaces are often modified into complex membrane features (e.g. Fain 2003). Photo-reception and chemo-reception involve seven-pass transmembrane G protein-coupled receptors (GPCRs), and mechanical stimuli require membrane-bound ion channels (other sensory-cell types can detect ionic concentrations or electrical fields). Such sensory neurons then communicate by electrical potential, either through axons that are components of the sensory cells themselves (the typical invertebrate condition), or via synapses on the cell bodies to adjacent neural cells (a frequent vertebrate condition, as in the hair cells of the inner ear).

It is important to note that not all GPCRs are involved in photo- and chemo-sensory organs or "sensory" perception. Multiple independent classes of these receptors are involved in synaptic, hormonal, and developmental signaling internal to the organism (e.g. http://www.sdbonline.org/fly/aignfam/gpcr.htm), and the proliferation of multiple classes of GPCRs appears to be a critical distinctive feature of animals relative to other eukaryotes (http://drnelson.utmem.edu/MHEL.7TM. html, Alvarez 2008, Römpler et al. 2008). Thus, sense organs are distinct in the particular application of GPCRs to external chemical and photoreception.

Despite these underlying similarities uniting the different sensory organs, traditional morphological comparisons have suggested that many of these structures evolved independently in multiple classes of metazoans (e.g. Salvini-Plawen and Mayr 1977). This view has been changing, largely in part to shared aspects of developmental gene expression in sense organs across the bilaterian tree, and across classes of sensory structures in a single animal. We discuss aspects of this shared genetic regulation in Bilateria; we then explore how these bilaterian-based inferences play out when compared to the limited cnidarians and sponge information. Our primary objective is to treat the range of multicellular sensory structures rather than naked sensory cells or simple sensilli.

Common Aspects of Sense-Organ Developmental Gene Regulation in Bilaterians

A suite of interactive developmental regulatory genes are highly conserved in sensory organogenesis, both between the different sensory systems, and between diverse bilaterian clades. For example, the basic helix-loop-helix gene *atonal* and its multiple vertebrate homologues are expressed in, and function in, the development of virtually all sense organs in *Drosophila* and vertebrates. *Atonal* is required for the development of the placodally derived eye, ear, and nose in vertebrates (Baker and Bronner-Fraser 2001). In *Drosophila*, *atonal* defines sense organs that consist of closely stacked sensory units, such as chordotonal organs found in stretch receptors, auditory organs, or the ommatidia of the insect compound eye (e.g. Jarman and Ahmed 1998). Initiation of development in these organs has been traced to the 3′ cis region of *atonal*; only after *atonal* expression in the imaginal disks do other "master" genes, such as *eyes absent*, specify which particular sense organ will develop (Niwa et al. 2004) suggesting that multiple sensory systems evolved from a single undifferentiated, *atonal*-dependant structure, an idea that will be expanded upon later in this text.

In addition to *atonal*, a number of other genes, initially identified by the loss of eyes in *Drosophila* mutants, function in the regulatory cascades governing the development of multiple classes of sense organ. These include *eyes absent* and *dachshund,* as well as members of the *Six* gene-family—a distinctive group of homeodomain-containing genes that includes *sine oculis* and *optix.* In addition, genes such as Brain3 are required for specifying aspects of sensory-cell and sensory-nerve-cell differentiation in multiple classes of sense organs (auditory, olfactory and visual). Like *atonal,* these downstream regulators are conserved throughout the bilateria, and retain functions similar to *Drosophila.* For example, vertebrate placodes, the regions of ectoderm that give rise to sensory systems in the head, are defined and differentiated by members of the *Six* and *Eya* families, among other genes (Schlosser 2007). Mouse Brain3 mutants are deaf and blind, and lack balance due to the absence of hair cells in the semicircular canals (e.g. Pan et al. 2005). Over-expression studies illuminate some of the commonality and combinatorial function of these genes. Famously, expression of the vertebrate homologue of *eyeless* (PAX6) successfully rescues eyes in *eyeless* mutants of *Drosophila.* However, over-expression experiments (that deliver the gene product throughout the organism) convert chordotonal organs to eyes (e.g. Halder et al. 1995). This conversion illustrates the shared developmental genetic regulation present in multiple classes of sense organ, as well as the role that Pax genes, such as *eyeless,* play in determining a subset of sense organs that includes eyes (e.g. Schlosser 2006). Thus a substantial list, including upstream regulatory genes and downstream genes with sensory-cell-type specificity, are common features of a wide range of sensory organs (Schlosser 2006) provides a summary of shared regulatory-gene control across vertebrate sensory structures).

Sharing of Developmental Regulatory Genes Across Systems

The preceding section presented a picture of the regulation of sense-organ development across divergent bilaterians. However, additional complexities intrude on this seemingly rational hierarchical organization. Developmental genes often serve multiple functions, thus hypotheses regarding common ancestry of function with distantly related organisms are not necessarily straightforward. They require attention to other lines of evidence that may suggest which facets of expression are likely to reflect shared ancestry. Many of the genes involved in the development of sensory organs are also involved in the development of structures that are not, or might not, typically be considered sense organs. These cases can be divided into: (1) cases with obvious functional and developmental connections to sense organs, such as nerve and muscle development, and (2) those where a developmental connection to sensory structures is less apparent, such as kidney development. Common attributes of distinctly different organs are often dismissed as cooptation, but this is too easy; how such cooptation occurs is critical to understanding evolution. *We argue that whether considered cooptive or not, overlap of gene function likely reflects aspects of shared ancestry of some components of the system,* and that this common origin may be supported by examination of the basal lineages. We first touch on the "expected example" of nerve and muscle development, before considering three more challenging examples where unexpected structures share genetic overlap with sensory organs: the pituitary gland, the kidneys, and invertebrate limbs. We argue that these seemingly unrelated structures aren't as distinct as they appear to be, and that pituitary glands, kidneys, and invertebrate limbs share attributes, and potentially ancestry, with sensory organs.

The relationship between sensory system, nervous system and muscle development

Overlap of expression of sense-organ regulatory genes with muscles and nerves is perhaps to be expected, given the functional and synaptic connections between these systems. In addition, gene duplication appears to have generated multiple players with separate functions in sensory cells, nerves and muscles. There are many examples of this overlap in groups of genes that evolved basal to the radiation of bilaterians. The Six gene classes *sine oculis* and *optix* are primarily involved in the development of sense organs in *Drosophila*, while the class *myotonix* is primarily involved in muscle development. In the NK2 homeodomain genes, *tinman* and *bagpipe* are involved in the differentiation of cardiac and smooth muscles, while *vnd* functions in the development of the medial nervous system (discussed in Jacobs et al. 1998). In Vertebrates, separate copies of Brain3 appear to have distinct functions, seemingly coincident with the division of neural and sensory cell types in the vertebrate nervous system, relative to the single neurosensory cell that performs this combined function in most invertebrate sensory neurons. The above examples of gene family function and gene duplication suggest some of the typical and more prosaic ways in which genes involved in sense-organ developmental regulation appear to "coopt" new functions in their evolutionary history. However, this cooptation does not preclude the possibility that these sensory, neural, and muscular controls are related by a common ancestral cell-type. As mentioned above, invertebrate sensory cells also have neuronal processes, while vertebrates sensory cells and neurons are distinct. This idea will gain further support when we discuss more basally branching taxa, such as the cnidarians and sponges.

The relationship between sensory system and pituitary development

A number of sense-organ-specific genes such as the Six genes, as well as *eyes absent*, and *dachshund* homologues (e.g. Schlosser 2006) are expressed in the pituitary as well as in sensory structures, as is the POU gene PIT1, the most closely related POU gene to Brain3. This developmental-genetic overlap between sensory systems and the pituitary gland is surprising on its face, but proves consistent with the evolution of the adenohypophyseal component of the pituitary, from an external chemosensory to an internal endocrine organ in chordate lineage (e.g. Gorbman 1995, Jacobs and Gates 2003). Thus, the presence of the gene Pit1 in more basal taxa, including cnidarians and sponges (Jacobs and Gates 2003), is consistent with an evolutionarily antecedent to the vertebrate pituitary, perhaps involved in external reproductive communication. And like vertebrate sensory organs, the pituitary is placoidally derived. Other structures derived from cephalic placodes in vertebrates share aspects of regulation with formal sense organs (Schlosser 2006) and also likely have a common evolutionary origin with sensory structures.

The relationship between sensory system and kidney development

Both sense organs and kidneys express the same suite of regulators in development, and there are a number of diseases that effect the ear and kidney in particular leading to the biomedical term otic-renal complex (see Izzedine et al. 2004 for review). In this unexpected case—the commonality of kidneys and sense organs—we argue below that this could reflect cellular organization in sponges, in which groupings of choanocytes may serve multiple functions and subsequently evolved into the *sense organs and kidneys* in bilaterians. We advance this particular argument based on the initially surprising commonality of sensory regulation and disease in distinctly different organs. However, this does not limit the possibility that many other systems in higher Metazoa may also have common origins; given the small set of differentiated cell and tissue types found in sponges, this may be necessarily the case. For an additional example, the expression of *atonal* homologues associated with the neuroendocrine cells of the gut (e.g. Yang et al. 2001, Bjerknes and Cheng 2006) also suggests to us that they could logically be derivative from the choanocyte cell component.

The relationship between sensory system and limb development

Vertebrate limbs are novel derived feature of gnathostome vertebrates; consequently, the pharyngeal arches are the vertebrate structure most directly related to invertebrate appendages (e.g. Shubin et al. 1997). So in an evolutionary context, the inner ear, derived from the pharyngeal arches, is an appendage-derived sensory structure. Moreover, common developmental gene expression and motor proteins, such as prestin and myosinVIIa, function in both vertebrate "ears" and the hearing organs found in the joints of *Drosophila* appendages (e.g. the Johnston's organ; Todi et al. 2004, Kamakouchi et al. 2009) arguing for evolutionary continuity through a shared ancestral auditory/inertial or comparable mechano-sensory structure (Boekhoff-Falk 2005, Fritsch et al. 2006) borne in this "appendage" context.

Besides the appendage joint/hearing organ association found in *Drosophila*, many other examples of sensory-appendages exist in the fly. For example, fringe and associated regulators function along the equator (akin to a dorso-ventral compartment boundary) of the *Drosophila* eye, as well as in the evolutionarily secondary *Drosophila* wing, where they are responsible for defining the wing margin, which itself bears a row of sensory bristles. The eye imaginal disk in drosophila is a positional equivalent serial homologue of the walking appendages. There is a well-documented relationship between eye, antennae, and appendage formation in *Drosophila*, and ectopic expression of *Antennapedia* can convert *Drosophila* antennae into second thoracic legs (Schneuwly et al. 1987), as well as stop eye development through mutual inhibition with *eyeless* (Plaza, 2001). The fly antenna, an "olfactory appendage" with multiple odorant receptors (Vosshall et al. 1999, Laissue et al. 2008) as well as an auditory and gravity measuring structure(Todi et al. 2004, Kamakouchi et al. 2009), shares the same imaginal disk with the compound eye. Misexpression of one gene, *Dip3*, is sufficient to convert eyes into antennae (Duong et al. 2008). More generally, homology of anterior sensory structures with more posterior appendages in the segmental series are standard inferences across the arthropods, and can be further supported by the presence of chemosensory (Zacharuk 1980) and mechanosensory (McIver 1985, Keil 1997) sensilla on the posterior appendages.

Finally, this sensory-appendage overlap is not unique to the arthropods. The presence of eyes on the parapodia in some species of polychaete (e.g. Verger-Bocquet 1981, Purshce 2005) documents evolutionary conversion of limbs to sense-organ-bearing structures. They are evolutionary "phenocopies", producing phenotypes comparable to those engendered by *eyeless* over-expression that convert limb-borne chordotonal organs to eyes, as was discussed above. The presence of eyes on the terminal tube feet (appendages) near the ends of the "arms" (axial structures) of sea stars (e.g. Mooi et al. 2005, Jacobs et al. 2005) provides an instance in yet another bilaterian phylum where sense organs and are associated with "appendages".

Cnidaria: Homology of Medusan Sensory Structures with the Bilateria

Having considered these common aspects of bilaterians development, we now compare this information to the cnidarian condition. The outgroup status of Cnidaria should help polarize the evolutionary changes in Bilateria, testing whether the relationships discussed above are in fact ancestral or derived. Sense organs of the cnidarian medusa are highly developed and distributed across scyphozoan, hydrozoan and cubazoan lineages. The rhopalium, the sense organ bearing structure of Scyphozoa and Cubozoa (a modified group within the scyphozoans), is borne on the margin of the bell in the medusa, and contains the statocyst and eyes. The rhopalia of cubozoan medusae contain eyes with lenses, the most dramatic of cnidarian sense organs. These eyes presumably facilitate swimming in these very active medusae with extremely toxic nematocysts. Other cnidarian eyes are simpler. These eyes tend to be simple eyespots or pinhole camera eyes that lack true lenses (see Martin 2002, Piatagorsky and Kozmik 2004 for review). In the scyphozoan *Aurelia*, the statocyst is effectively a "rock on a stalk", with a dense array of mechanosensory cells that serve as a "touch plate" at the base of the stalk where it can contact the overlying epithelium of the rhopalium (e.g. Spangenberg, et al. 1996, Arai 1997).

Several studies document expression of regulatory genes in Cnidaria that typically function in the development of bilaterian sense organs. These studies document a common aspect of gene expression, albeit with significant variation. In the scyphozoan *Aurelia* a homologue of *sine oculis* is expressed in the rhopalia (Bebeneck et al. 2004), as is the case for Brain3 (Jacobs and Gates 2001) and *eyes absent* (Nakanishi et al. in prep). Six-class genes are also expressed in the development of the eyes in the hydrozoan *Cladonima* (Stierwald et al. 2004). These sorts of data, taken together, provide a substantial argument for a shared ancestry between bilaterian and Cnidarian sense organs generally. Shared ancestry of specialized classes of sensory organs, such as eyes, also appears likely. The evolution of the light-sensitive GPCR opsins is complex, as there are many functions and families of the related GPCR receptors. However, recent analyses support a sister taxon relationship between the ciliary opsins of Cnidaria and those of bilaterians (e.g. Suga et al. 2008) strongly suggesting a shared ancestry of this major mode of photo-sensation.

In Cubozoa a paired-class gene has been identified that is expressed in sense-organ development (Kosmik et al. 2003). Interestingly, this PaxB gene does not appear to be a simple homologue of eyeless/Pax6, as it contains an *eyeless*/Pax6 type homeodomain combined with a paired domain typical of PAX 2/5/8—a regulatory gene more closely associated with ear development that is also expressed in statocysts in mollusks (O'Brien and Degnan 2004). Statocysts are ear-like in their inertial function and are localized with the eye in the cnidarian rhopalium. Given that cubozoan statocyst expresses PAXB along with the eye, a PaxB-type gene appears to have undergone duplication and modification in the evolution of the bilaterian condition such that eyes and ears are differentially regulated by separate PAX6 and PAX 2/5/8genes. This evolution in the ancestry of eyeless/Pax6 contrasts with a number of other sense-organ regulatory genes such as *sine oculis* (Bebeneck et al. 2004), Brain3 (Jacobs and Gates 2001) and *eyes absent* (Nakanishi et al. *in prep*), all of which appear to be extremely similar in their functional domains to specific bilaterian homologues. Thus, *eyeless*/PAX6 may have evolved more recently into its role in the eye development than other regulatory genes that also function in other sense organs.

The Issue of Exclusivity of Sensory Structures to the Medusa Stage

In opposition to the above arguments for a shared ancestry of bilaterian-cnidarian sensory systems is the perception that cnidarian sense organs are exclusive to the medusa, and that the medusan phase is derived given the basal placement of the Anthozoa, which lack such a stage in their life cycle (e.g. Bridge et al. 1992, Collins et al. 2006). However, a variety of arguments limit the strength of support for completely *de novo* evolution of cnidarian sense organs. Neither the polyp nor the sense-organ containing medusa are present in outgroups, consequently the power of tree reconstruction to resolve the presence or absence of medusa or polyp is minimal (Jacobs and Gates 2003). This, combined with the frequency of loss of the medusa phase in hydrozoan lineages, limits confidence in the inferred absence of a medusa-like form in the common ancestor. In addition, features that may merit consideration as sense organs are present in planula and polyps. In particular, statocysts are found in some unusual hydrozoan polyps (Campbell 1972) and ocelli associate with the tentacle bases in some stauromedusan (Scyphozoa) polyps (Blumer 1995). Thus, the emphasis on the medusan phase of the life history may be unwarranted.

   Numerous opsins have been discovered in the anthozoan *Nematostella*, and from the hydrozoan *Hydra*, neither of which has "eyes" in the traditional sense (Suga et al. 2008). Suga et al. (2008) have shown the expression of comparable ciliary opsins in the eyes of the hydozoan *Cladonema* as well as in other potentially sensory structures such as tentacles and the manubrium (oral structure), a point developed further below.

Additional Cnidarian Sensory Systems

The cnidocytes of Cnidaria are innnervated (e.g. Anderson et al. 2004) and have triggers that respond to sensory stimuli. In some instances they synaptically connect with adjacent sensory cells (Westfall 2004). Thus, cnidocytes are, at once, a potential source of sensory stimulation and, presumably, modulate their firing in response to neuronal stimuli (e.g. Anderson et al. 2004). Having acknowledged this complexity, we set it aside and limit the discussion to the integration of more traditional sensory cells into what may be considered sense organs.

In the planula larvae of Cnidaria, FMRF-positive sensory cells are found in a belt running around the locomotory "forward" end (aboral after polyp formation) of the planula ectoderm (e.g. Martin 1992, 2002). The axons of these cells extend "forward" along the basement membrane of the ectoderm and are ramified, forming what appears to be a small neuropile at the aboral pole of the planula. This feature varies among taxa; in hydrozoans such as *Hydractinia,* the array of sensory cells appears closer to the aboral end of the elongate planula. There is also ontogenetic variation in which the sensory cells move closer to the aboral end prior to settlement (Nakanishi et al. 2008). Strictly speaking, the sensory neurons of the cnidarian planula correspond to the "naked" sensory neurons discussed previously; however, one might consider dense arrays of such chemoreceptive and/or mechanoreceptive neurons as "precursors" of sense organs. Expression data for *atonal* in hydrozoan planulae (Seipel et al. 2004) also suggest that this integrated array of sensory cells could merit "sense organ" status.

The hypostome and manubrium, oral structures of the polyp and medusa respectively, may also rise to the status of sense organs. In *Aurelia* ephyrae (early medusa), sensory cells are present in rows on the edges of both the ectoderm and endoderm of the manubrium. POU genes such as Brain3 (unpublished) are expressed in the manubriium of *Aurelia,* as is a homologue of *sine oculis* (Bebenek et al. 2004). Similar *sine oculis* expression in the manubrium is evident in the hydrozoan *Podycoryne,* but this may not be the case in *Cladonema* where a related Six gene *myotonix/ Six4,5* is expressed in the manubrium (Steirwald 2004). *Cladonema* does, however, exhibit manubrium-specific opsins that are distinct from those found in the eyes, gonads, or ubiquitously across the body, suggesting that the manubrium has a distinct photosensory role (Suga et al. 2008). In *Podycoryne,* limited expression of *atonal* is evident in the manubrium (Seipel et al. 2004), and PaxB is expressed in the manubrium and hypostome (Groger et al. 2000).

Tentacles as Appendages and Sense Organs

The shared developmental aspects between sense organs and appendages discussed in the bilaterians above, is also found in the Cnidaria. Cnidarian tentacles are variable; ectoderm and endoderm layers and a central lumen connected to the gastrovascular cavity are typical of anthozoan tentacles. In contrast, polyp tentacles of scyphozoans and some hydrozoans lack a lumen; these tentacles have a single row of large vacuolated endodermal cells as the core of a slender tentacle. A variety of tentacle morphologies are also present in medusae. We discuss whether tentacles are (1) sense organs, (2) sense organ bearing structures, and (3) whether tentacles and rhopalia (that bear sense organs in scyphozoans) are alternative developmental outcomes of an initially common developmental field or program.

Ultrastructural studies as well as markers such as FMRF that typically recognize sensory cells and neurons document arrays of sensory cells in tentacles that are substantially denser than those found in the body wall of the polyp or in the medusan bell. *Optix* homologues are also expressed in certain presumed sensory neurons or cnidocytes in tentacles of *Podocornyne* (Stierwald 2004). Sensory cells form concentrations at the base of the tentacle or, in some instances, at the tips of the tentacles (e.g. Holtman and Thurm 2001); these concentrations merit consideration as sense "organs".

Sense-organ-related genes are preferentially expressed near the bases of hydrozoan tentacles; *sine oculis* and PAXB are expressed here in *Podocoryne*, a hydrozoan medusa that lacks eyes (Steirwald et al. 2004, Groger et al. 2000). Sensory gene expression associated with tentacle bases is not exclusive to medusae. In the anemone *Nematostella*, PaxB homologues are expressed adjacent to the tentacles (Matus et al. 2007). In addition, the base of the tentacle is the locus of ocelli in some unusual polyps as discussed above (Blumer 1995). Thus, a developmental field specialized for the formation of sensory organs appears to be associated with the bases of cnidarian tentacles, but concentrations of sensory cells at the end of the tentacle also occur, as is the case in the ployp tentacles of the hydrozoan *Coryne* (e.g. Holtmann and Thurm 2001).

In *Hydra* an *aristaless* homologue is expressed at the base of tentacles (Smith et al. 2000), comparable to the proximal component of expression seen in arthropod limbs (Campbell et al. 1993). TGF beta expression always precedes tentacle formation in tentacle induction experiments (Reinhardt et al. 2004) and continues to be expressed at the tentacle base. Both *decapentplegic* and *aristaless* are involved in the localization and outgrowth of the appendages in flies (e.g. Campbell et al. 1993, Crickmore and Mann 2007). Thus there are also common aspects of bilaterian appendage and cnidarian tentacle development.

As noted above, in typical Scyphozoa, rhopalia alternate with tentacles in a comparable bell-margin position; in Hydrozoa, sense organs associate with the tentacle bases. Overall, there is support for a common appendage/sense-organ field in Cnidaria comparable to that evident in

Bilateria as discussed above. In those hydrozoans with a medusa stage, many have eyes associated with the tentacle base. The relative position of the eye and tentacle appears to be evolutionarily plastic; the necto-benthic *Polyorchis penicillatus* feeds on the bottom and its eyes are on the oral side, presumably aiding in prey identification on the bottom, whereas the nektonic *P. monteryensis* (e.g. Gladfelter 1972) has eyes on the aboral side of the tentacle, presumably aiding in identification of prey in the water column. Nevertheless, the hydrozoan eye appears to be closely associated with the base of the tentacle. In Scyphozoa there are typically eight rhopalia that alternate with eight tentacles around the bell margin. Cubozoa have four rhopalia that similarly alternate with tentacles. Although, there are exceptions to this alternating tentacle/rhopalia pattern (e.g. Russell 1970) they appear to be derived. Thus appendages in the form of tentacles and the sense organ bearing rhopalia occupy a similar position/field that appears to assume alternative fates in development. This is consistent with the arguments relating appendages and sense organs in Bilateria developed above, and relates to our discussion of tentacles considered as appendages as well as sense organs in cnidarians.

Sensory Attributes of Sponges

Sponges are thought to constitute the most basal branch, or branches, of the animal tree, and a progressivist view of evolution has long treated them as primitively simple (Jacobs and Gates 2003). Yet, there is increasing evidence that sponges are not as simple as often anticipated: (1) some sponge lineages exhibit coordinated motor response to sensory stimuli and others posses an electrical-conduction mechanism; (2) sponges have genes encoding proteins that function in a range of bilaterian developmental processes; and (3) sponges have many of the genes employed in the development of sense organs. The presence of genes known to function in eumetazoan sense-organ development in a group lacking formal sense organs presents interpretive challenges. Certain sets of larval cells or the grouping of choanocytes into functional arrays represent possible sponge structures potentially related to eumetazoan sense organs. We discuss these briefly and explore the possibility that multiple organs, including kidneys and sense organs, may share ancestry with ensembles of choanocytes.

Motor Coordination of Sponges

Sponges exhibit contractile behaviors (reviewed by Leys and Meech 2006, Elliot and Leys 2003). In the small, freshwater sponge *Ephydatia*, an inhalent expansion phase precedes a coordinated contraction that forces water out of the osculum. This contractile activity generates high-velocity flow in the finer channel systems that then propagate toward the osculum.

Effectively, this seems to be a "coughing" mechanism that eliminates unwanted material, chemicals or organisms from the vasculature. Sponges are known to have specialized contractile cells, termed myocytes, which have been compared to smooth-muscle cells; however, other eptithelial cell types (pinacocytes, actinocytes) contribute to contractile behavior (reviewed by Leys and Meech 2006).

Given that sponges lack formal synapses it is worth noting that non-synaptic communication between cells via calcium waves can occur through a variety of mechanisms. One such class of mechanism involves gap junctions or gap junction components, but these have yet to be documented in sponges and are presumed absent. Others involve the vesicular release of molecules such as ATP, which can operate through receptors associated with calcium channels or through specific classes of GPCRs (see North and Verkohtsky 2006 for review of purinergic communication). Such receptors are known to permit non-synaptic intercellular communications in nerves and non-neuron components such as between glial cells. In hexactinellids "action potentials" that appear to involve calcium propagate along the continuous membranes of the syncytium that constitutes the inner and outer surface of these sponges (Leys and Mackie 1997). This propagation of signals along the syncytium permits rapid coordinated choanocyte response to environmental stimuli in hexactinellids. In other classes of sponges, propagation of information appears to involve calcium dependent cell/cell communication (Leys and Meech 2006). Mechanisms of this sort, involving non-synaptic vesicular release of signaling molecules and a "calcium wave" propagation, seem broadly consistent with available information communication in cellular sponges reviewed in detail by Leys and Meech (2006).

Sensory Systems in Sponges

The ring-cells around the posterior pole (relative to direction of motion) of the parenchyma larva of the demosponge *Amphimedon* has been shown to be photosensitive and to respond to blue light (Leys et al. 2002, see Maldonado et al. 2003 for observations on other demosponge larvae). These cells effectively steer the sponge, using long cilia providing for a phototactic response. Sakarya et al. (2007) document that flask cells of larval sponges express proteins involved in postsynaptic organization in Bilateria, and speculate that these cells are sensory. These larval sensory attributes are of interest as larvae provide a likely evolutionary link with the radiate and bilaterian groups (e.g. Maldonado 2004).

Groups of choanocyte cells in adult sponges also bear some similarity to eumetazoan sensory structures as: (1) choanocytes are crudely similar in morphology to sensory cells, particularly mechanosensory cells; (2) the deployment of sponge choanocytes in chambers is similar to the array of sensory cells in sense organs; and (3) choanocytes are a likely source of stimuli that produce the contractions and electrical communications as noted above. Choanocytes of sponges and choanoflagellates present a cilium/flagellum surrounded by a microvillar ring on the apex of the cell, which bears at least superficial similarity to the typical organization of many sensory cells, such as those of the ear (e.g. Fritsch et al. 2006, Fain 2003). Clearly chemical signals in the water can induce contractile responses in demosponges (e.g. Ellwanger et al. 2007, Leys and Meech 2006). In addition it appears likely that mechano and chemosensory responses to particles would be necessary for the feeding function of the choanocyte and that communication between adjacent choanocytes in the choanosome structure would also be essential to feeding. Feeding behavior appears coordinated across sponges rather than just within choanosomes as different types of particles are preferred under different circumstance (e.g. Yahel et al. 2006, 2007).

Genetic Control of Sponge Neural Development

The molecular complexity of sponges exceeds that expected based on their presumed "primitive" nature. Nichols et al. (2006) reported a range of extracellular matrix proteins as well as components of the major intercellular signaling pathways operative in metazoan development from their EST study of the demosponge *Oscarella*. Larroux et al. (2006, 2008) reported a diverse array of homeodomains and other DNA-binding regulatory genes from the demosponge *Amphimedon queenslandica* (formerly *Reneira*). Thus, sponges possess a significant subset of the equipment used to differentiate cells and tissues in Bilateria and Cnidaria (see Ryan et al. 2006 for a recent survey of cnidarian homeodomans from the *Nematostella* genome and Simionato et al. [2007] for survey of bHLH regulators across Metazoa, including cnidarians and demosponge genomic data).

Recent work by Sakarya et al. (2007) documents the presence of "post-synaptic" proteins and argues that these proteins are organized into a post-synaptic density comparable to that found in eumetazoan synapses. This suggests surprising functionality given the absence of formal synapses in sponges. An EST study of the demosponge *Oscarella* provides additional support for the presence of molecular components that are required for vesicle related signaling function (Jacobs et al. 2007). These recent observations in sponges suggest a high activity of equipment involved for vesicle transport, and the presence of some synaptic and developmental signaling components typically associated with bilaterian neural systems.

Turning to sense-organ-associated regulators, *sine oculis* homologues are present in all classes of sponges (Bebeneck et al. 2004), as are homologues of Brain3 (Jacobs and Gates 2001, 2003 & unpublished). Similarly, relatives of *atonal* are present in demosponges (Simionato et al. 2007). Richards et al. (2008) demonstrated expression of Notch-Delta signaling and atonal-like basic helix loop helix neurogenic genes in the Demnosponge *A. queenslandica*. NK2 genes play important roles in bilaterian mesoderm and neural differentiation (Jacobs et al. 1998), and it has recently been argued that the presence of NK2 gene expression in Homoscleromorph choanocytes is consistent with an ancestral neural/ sensory function of this cell type (Gazave et al. 2008). Thus, sponges, and associations of choanocytes in particular, appear to have many components of the regulatory gene cascades associated with sense-organ development in Eumetazoa.

Choanocytes as the Ancestral Sensory Structure

As noted above, vertebrate sensory organs have a surprising amount in common with the kidney; both, for example, express Pax6, *eyes absent* and *sine oculis* in development, and numerous genetic defects affect both structures (e.g. Izzedine et al. 2004). Consideration of sense organs, and organs that eliminate nitrogenous waste, both as evolutionary derivatives or relatives of a choanocyte chambers, may help explain these commonalities. The fluid motion engendered by choanocyte chambers renders these structures the central agency in nitrogenous waste excretion, in addition to their other functions (e.g. Laugenbruch and Weissenfels 1987); vacuoles involved in the excretion of solids following phagocytic feeding presumably represent a separate aspect of waste disposal (e.g. Willenz and Van De Vwer 1986). In a number of bilaterian invertebrates, nitrogen-excreting protonephridia consist of specialized ciliated flame cells that generate the flow and pressure differential critical for initial filtration, much as sponge choancytes generate flow in feeding. These systems appear intermediate between choanocytes and metanephridia that rely on blood pressure for filtration (e.g. Bartolomaeus and Quast 2005). Thus we draw attention to the potential evolutionary continuity of function and structure between associations of choanocytes and protonephridia, and ultimately metanephridia. These are of interest in the context of the potential for explaining the common features of sense organs and kidneys (e.g. Izzedine et al. 2004). Such explanations are necessarily speculative, but will soon be subject to more detailed test with an increasing knowledge of gene expression and function in sponges. It should also be noted that this argument does not negate the possibility that a number of other structures such as the neurendocrine structure of the gut epithelium, as mentioned above, might also derive from or share ancestry with the choanosome.

Tree Topology

Tree topology is critical to evolutionary interpretation of the events surrounding the evolution of sensory systems in the basal Metazoa. Most continue to treat sponges as basal in the Metazoa (Srivastava et al. 2008, Ruiz-Trillo et al. 2008). Other works (Borchiellini et al. 2001, Medina et al. 2001, Sperling et al. 2007) suggest that Eumetazoa derive from a paraphyletic sponge group. These analyses tend to place the Eumatazoa as sister to the calcareous sponges. Sponge paraphyly implies that the ancestral eumetaozan was sponge-like, with choanocytes and other broadly distributed attributes of sponges, lending credence to arguments that choanosome development may have contributed to the evolution of sensory structures as argued above (Sperling et al. 2007). Additionally, unique demnosponge derived steranes constitute the earliest evidence of animals in the rock record, potentially supporting earlier evolution of sponges relative to other animal groups (Love et al. 2009). The placement of Cnidaria as sister to the Bilateria has also received recent support in other studies (e.g. Halanych 2004, Baguñà et al. 2008), as well as the relationships between the classes of cnidaria as discussed above.

However, one recent analysis placed Ctenophores basal on the animal tree (Dunn et al. 2008), while others have suggested that Placozoans are basal in a clade composed of Placozoa, Cnidaria, Ctenophora and the sponges, which itself is sister to the Bilatera (Schierwater et al. 2009, Signorovitch et al. 2007, Blackstone 2009). The enigmatic Placozoa are certainly of interest, as they may provide information on the nature of the stem of the metazoan tree and potentially permitting interpretation of Vendian (late Precambrian) fossils (e.g. Conway-Morris 2003). The large size of the placozoan mitochondrial genome is comparable to those found in protista, suggesting that Placozoa may be the most basal branch of the Metazoa. Conversely, Ruiz-Trillo et al. (2008) placed the Placozoans as sister to the bilaterians. Ribosomal genes place Placozoa in a variety of basal postions (e.g. Borchiellini et al. 2001, Hallanych 2004), but are largely consistent with the basal placement and/or paraphyly of sponges discussed above. There is evidence for PAX-like genes in the presumptively basal Placozoa (Hadrys et al. 2005), as well as basic helix–loop–helix family genes, POU-homeobox genes, and most of the processes necessary for neural formation and conduction (Schierwater et al. 2008). This is broadly consistent with the evolution of many major classes of metazoan regulatory proteins in the stem lineage, prior to the radiation of modern metazoan phyla (Derelle et al. 2007 provides a recent analysis of homeobox gene families in this context).

Whether sponges, placozoans, or ctenophores are the most basally branching members of the animal tree has limited effect on the arguments presented in this paper, as the available evidence is pointing to a basal animal node that is complexely endowed with the regulatory apparatus that is know to function in bilateria sense organ development.

Summary

We have argued that many aspects of sense organ evolution preceded the evolution of formal organs in the triploblastic Bilateria. Clearly Cnidaria have well-developed neural and sensory features, some of which may merit treatment as "organs". However even sponges appear to have precursory elements of sensory organization. In addition, sense-organs share attributes with endocrine structures, appendages and kidneys. We argue that these similarities are a product of derivation from common ancestral structures. In a more general sense, as one compares structures in divergent ancient lineages such as the basal lineages of the Metazoa, we feel that similarities that are the product of shared ancestry are likely to be manifest in surprising and subtle ways. Thus, neither inferences of similarity as indicative of strict homology nor dismissal of similarity as products of convergence or cooptation should meet with facile acceptance.

Acknowledgements

We thank Sally Leys for discussions, Chris Winchell and anonymous reviewers for their helpful critique, and NASA Astrobiology Institute for support.

References

Anderson, P.A.V. and L.F. Thompson, and C.G. Moneypenny. 2004. Evidence for a common pattern of peptidergic innervation of cnidocytes. Biol. Bull. 207: 141–146.

Alvarez, C. 2008. On the origins of arrestin and rhodopsin. BMC Evol. Biol. 8: 222.

Arai, M.N. 1997. A Functional Biology of Scyphozoa. Chapman and Hall, London and New York.

Baguñá, J. and P. Martínez, J. Paps, and M. Riutort. 2008. Back in time: a new systematic proposal for the Bilateria. Phil. Trans. R. Soc. B 363: 1481–1491.

Baker, C.V. and M. Bronner-Fraser. 2001. Vertebrate cranial placodes I. Embryonic induction. Dev. Biol. 232: 1–61.

Bartolomaeus, T. and B. Quast. 2005. Structure and development of nephridia in Annelida and related taxa. Hydrobiologia 535: 139–165.

Bebenek, I.G. and R.D. Gates, J. Morris, V. Hartenstein, and D.K. Jacobs. 2004. *Sine oculis* in basal Metazoa. Dev. Genes Evol. 214: 342–351.

Bjerknes, M. and H. Cheng. 2006. Neurogenin 3 and the enteroendocrine cell lineage small intestinal epithelium. Devel. Biol. 300: 722–735.

Blackstone, N.W. 2009. A New Look at Some Old Animals. PLoS Biol 7(1) e1000007.

Blumer, M.J.F. and L.V. Salvini-Plawen, R. Kikinger, and T. Buchinger. 1995. Ocelli in a Cnidaria polyp: the ultrastructure of the pigment spots in *Stylocoronella riedli* (Scyphozoa, Stauromedusae). Zoomorphology 115: 221–227.

Boekhoff-Falk, G. 2005. Hearing in *Drosophila*: Development of Johnston's organ and emerging parallels to vertebrates ear development. Dev. Dynam. 232: 550–558.

Borchiellini, C. and M. Manuel, E. Alivon, N. Boury-Esnault, J. Vacelet, and Y. Le Parco. 2001. Sponge paraphyly and the origin of Metazoa. J. Evol. Biol. 14: 171–179.

Bridge, D. and C.W. Cunningham, B. Schierwater, R. DeSalle, and L.W. Buss. 1992. Class-level relationships in the phylum Cnidaria: evidence from mitochondrial gene structure. Proc. Natl. Acad. Sci. USA 89: 8750–8753.

Campbell, G. and T. Weaver, and A. Tomlinson. 1993. Axis specification in the developing *Drosophila* appendage: the role of wingless, decapentalegic, and the homeobox gene aristaless. Cell. 74: 1113–1123.

Chia, F.S. and R. Koss. 1979. Fine structural studies of the nervous system and the apical organ in the planula larva of the sea anemone *Anthopleura elegantissima*. J. Morphol. 160: 275–297.

Collins, A.G. and P. Schuchert, A.C. Marques, T. Jankowski, M. Medina, and B. Schierwater. 2006. Medusozoan phylogeny and character evolution clarified by new large and small subunit rDNA data and an assessment of the utility of phylogenetic mixture models. Syst. Biol. 55: 97–115.

Conway-Morris, S. 2003. The Cambrian "explosion" of metazoans and molecular biology: would Darwin be satisfied? Int. J. Dev. Biol. 47: 505–515.

Crickmore, M.A. and R.S. Mann. 2007. Hox control of morphogen mobility and organ development through regulation of glypican expression. Development 134: 327–334.

Derelle, R. and P.L. Herve´ Le Guyader, and M. Manuel. 2007. Homeodomain proteins belong to the ancestral molecular toolkit of Eukaryotes. Evol. Dev. 9: 212–219.

Dunn, C.W. et al. 2008. Broad phylogenomic sampling improves resolution of the animal tree of life. Nature 452: 745–749.

Duong, H.A. et al. 2008. Transformation of eye to antenna by misexpression of a single gene. Mech. of Develop. 125: 130–141.

Ellwanger, K. and A. Eich, and M. Nickel. 2007. GABA and glutamate specifically induce contractions in the sponge *Tethya wilhelma*. J. Comp. Physiol. A 193: 1–11.

Fain, G.L. 2003. Sensory Transduction. 288pp. Sinauer Associates, Inc, Sunderland, MA.

Fritzsch, B. and S. Pauley, and K.W. Beisel. 2006. Cells, molecules and morphogenesis: the making of the vertebrate ear. Brain Res. 1091: 151–171.

Gladfelter, W.B. 1972. Structure and function of the locomotory system of *Polyorchis montereyensis* (Cnidaria, Hydrozoa). Helgoland Mar. Res. 23: 38–79.

Gorbman, A. 1995. Olfactory origins and evolution of the brain-pituitary endocrine system: facts and speculation. Gen. Comp. Endocrinol. 97: 171–178

Groger, H. and P. Callaerts, W.J. Gehring, and V. Schmid. 2000. Characterization and expression analysis of an ancestor-type Pax gene in the hydrozoan jellyfish *Podocoryne carnea*. Mech. Develop. 94: 157–169.

Hadrys, T. and R. DeSalle, S. Sagasser, N. Fischer, and B. Schierwater. 2005 The Trichoplax PaxB gene: A putative proto-PaxA/B/C gene predating the origin of nerve and sensory cells. Mol. Biol. Evol. 22: 1569–1578.

Halanych, K.M. 2004 The new view of animal phylogeny. Annu. Rev. Ecol. Syst. 35: 229–256.

Halder, G. and P. Callaerts, and W.J. Gehring. 1995. Induction of ectopic eyes by targeted expression of the *eyeless* gene in *Drosophila*. Science 267: 1788–1792.

Holtman, M. and U. Thurm. 2001. Variations of concentric hair cells in a cnidarian sensory epithelium. J. Comp. Neurol. 432: 550–563.

Izzedine, H. and F. Tankere, V. Launay-Vacher, and G. Deray. 2004. Ear and kidney syndromes: Molecular versus clinical approach. Kidney Int. 65: 369–385.

Jacobs, D.K. and S.E. Lee, M.N. Dawson, J.L. Staton and K.A Raskoff. The history of development through the evolution of molecules: Gene trees, hearts, eyes, and dorsoventral inversion, pp. 323–357. *In:* R. DeSalle and B. Schierwater. [eds.] 1998. Molecular Approaches to Ecology and Evolution. Birkhauser, Basel.

Jacobs, D.K. and R.D. Gates. 2001. Evolution of POU/homeodomains in basal Metazoa: Implications of the evolution of sensory systems and the pituitary. Dev. Biol. 235: 241.

Jacobs, D.K. and R.D. Gates. 2001. Is reproductive signaling antecedent to metazoan sensory and neural organization? Am. Zool. 41: 1482–1482.

Jacobs, D.K. and R.D. Gates. 2003. Developmental genes and the reconstruction of metazoan evolution—implications of evolutionary loss, limits on inference of ancestry and type 2 errors. Integr. Comp. Biol. 43: 11–18.

Jacobs, D.K. and N.C. Hughes, S.T. Fitz-Gibbon, and C.J. Winchell. 2005. Terminal addition, the Cambrian radiation and the Phanerozoic evolution of bilaterian form. Evol. Dev. 7: 498–514.

Jacobs, D.K. et al. 2007. Evolution of sensory structures in basal metazoa. Integr. Comp. Biol. 47: 712–723.

Jarman, A.P. and I. Ahmed. 1998. The specificity of proneural genes in determining *Drosophila* sense organ identity. Mech. Dev. 76: 117–125.

Kamikouchi, A. and H.K. Inagaki, T. Effertz, O. Hendrich, A. Fiala, M.C. Göpfert, and K. Ito. 2009. The neural basis of *Drosophila* gravity-sensing and hearing. Nature 458: 165–169.

Keil, T.A. 1997. Functional morphology of insect mechanoreceptors. Microsc. Res. Tech., 39(6): 506–531.

Kozmik, Z. and M. Daube, E. Frei, B. Norman, L. Kos, L.J. Dishaw, M. Noll, and J. Piatigorsky. 2003. Role of Pax genes in eye evolution: a cnidarian PaxB gene uniting Pax2 and Pax6 functions. Dev. Cell 5: 773–785.

Laissue, P.P. et al. 2008. The olfactory sensory map in *Drosophila*. Adv. Ex. Med. Biol. 628: 102.

Langenbruch, P.F. and N. Weissenfels. 1987. Canal systems and choanocyte chambers in freshwater sponges (Porifera, Spongillidae). Zoomorphology 107: 11–16.

Larroux, C. and B. Fahey, D. Liubicich, V.F. Hinman, M. Gauthier, M. Gongora, K. Green, G. Wörheide, S.P. Leys, and B.M. Degnan. 2006. Developmental expression of transcription factor genes in a demosponge: insights into the origin of metazoan multicellularity. Evol. Develop. 8: 150–173.

Larroux, C. and G.N. Luke, P. Koopman, D.S. Rokhsar, S.M. Shimeld, and B.M. Degnan et al. 2008. Genesis and expansion of metazoan transcription factor gene classes. Mol. Biol. Evol. 25: 980–996.

Leys, S.P. and G.O. Mackie. 1997. Electrical recording from a glass sponge. Nature 387: 29–30.

Leys, S.P. and T.W. Cronin, B.M. Degnan, and J.N. Marshall. 2002. Spectral sensitivity in a sponge larva. J. Comp. Physiol. A 188: 199–202.

Leys, S.P. and R.W. Meech. 2006. Physiology of coordination in sponges. Can. J. Zool. 84: 288–306.

Love, G.D. et al. 2009. Fossil steroids record the appearance of Demospongiae during the Cryogenian period. Nature 457: 718–721.

Maldonado, M. and M. Dunfort, D.A. McCarthy, and C.M. Young. 2003. The cellular basis of photobehavior in the tufted parenchymella larva of demosponges. Mar. Biol. 143: 427–441.

Maldonado, M. 2004. Choanoflagellates, choanocytes, and animal multicellularity. Invertebr. Biol. 123: 1–22.

Martin, V.J. 1992. Characterization of RFamide-positive subset of ganglionic cells in the hydrozoan planular nerve net. Cell Tissue Res. 269: 431–438.

Martin, V.J. 2002. Photoreceptors of cnidarians. Can. J. Zool. 80: 1703–1722.

Matus, D.Q. and K. Pang, H. Marlow, C.W. Dunn, G.H. Thomsen, and M.Q. Martindale. 2006. Molecular evidence for deep evolutionary roots of bilaterality in animal development. Proc. Natl. Acad. Sci. USA 103: 11195–11200.

Matus D.Q. and K. Pang, M. Daly, and M.Q. Martindale. 2007. Expression of Pax gene family members in the anthozoan cnidarian, *Nematostella vectensis*. Evol. Dev. 9: 25–38.

McIver, S.B. Mechanoreception. pp. 71–132. *In:* G.A. Kerkut and L.I. Gilbert. [eds.] 1985. Comprehensive Insect Physiology, Biochemistry, and Pharmacology. Pergamon, Oxford.

Medina, M. and A.G. Collins, J.D. Silberman, and M.L. Sogin. 2001. Evaluating hypotheses of basal animal phylogeny using complete sequences of large and small subunit rRNA. Proc. Natl. Acad. Sci. USA 98: 9707–9712.

Mooi, R. and B. David, and G.A. Wray. 2005. Arrays in rays: terminal addition in echinoderms and its correlation with gene expression. Evol. Dev. 7: 542–555.

Nakanishi, N. et al. 2008. Early development, pattern, and reorganization of the planula nervous system in Aurelia (Cnidaria, Scyphozoa). Develop. Genes and Evol. 218: 511.

Nichols, S.A. and W. Dirks, J.S. Pearse, and N. King. 2006. Early evolution of animal cell signaling and adhesion genes. PNAS 103: 12451–12456.

Niwa, N. et al. 2004. A conserved developmental program for sensory organ formation in *Drosophila melanogaster*. Nat. Genet. 36: 293–297.

North, R.A. and A. Verkhratsky. 2006. Purinergic transmission in the central nervous system. Pflugers Arch-Eur. J. Physiol. 452: 479–485.

O'Brien, E.K. and B.M. Degnan. 2003. Expression of Pax258 in the gastropod statocyst: insights into the antiquity of metazoan geosensory organs. Evol. Dev. 5: 572–578.

Pan, L. and Z.Y. Yang, L. Feng, and L. Gan. 2005. Functional equivalence of Brn3 POU-domain transcription factors in mouse retinal neurogenesis. Development 132: 703–712.

Piatigorsky, J. and Z. Kozmik. 2004. Cubozoan jellyfish: an Evo/Devo model for eyes and other sensory systems. Int. J. Dev. Biol. 48: 719–729.

Plaza, S. 2001. Molecular basis for the inhibition of *Drosophila* eye development by Antennapedia. The EMBO J. 20: 802–811.

Purschke, G. 2005. Sense organs in polychaetes (Annelida). Hydrobiologia 535/536: 53–78.

Reinhardt, B. et al. 2004. HyBMP5-8b, a BMP5-8 orthologue, acts during axial patterning and tentacle formation in hydra. Develop. Biol. 267: 43–59.

Richards, G.S. et al. 2008. Sponge genes provide new insight into the evolutionary origin of the neurogenic circuit. Curr. Biol. 18: 1156–1161.

Römpler, H. et al. 2007. G Protein-coupled time travel: evolutionary aspects of GPCR research. Mol. Interv. 7: 17–25.

Ruiz-Trillo, I. and A.J. Roger, G. Burger, M.W. Gray, and B.F. Lang. 2008. A phylogenomic investigation into the origin of Metazoa. Mol. Biol. Evol. 25: 664–672.

Russell, F.S. 1970. The Medusae of the British Isles. Vol. 2. Cambridge University Press, Cambridge.

Ryan, J.F. et al. 2006. The cnidarian-bilaterian ancestor possessed at least 56 homeoboxes: evidence from the starlet sea anemone, *Nematostella vectensis*. GenomeBiology.com, 7: R64.

Sakarya, O. and K.A. Armstrong, M. Adamska, M. Adamski, I-F. Wang, B. Tidor, B.M. Degnan, T.H. Oakley, and K.S. Kosick. 2007. A post-synaptic scaffold at the origin of the animal kingdom. PloS ONE 2: e506.

Salvini-Plawen, L.V. and E. Mayr. 1977. On the evolution of photoreceptors and eyes. Evol. Biol. 10: 207–263.

Schlosser, G. 2007. How old genes make a new head: redeployment of Six and Eya genes during the evolution of vertebrate cranial placodes. Integr. Comp. Biol. 47: 343–359.

Schlosser, G. 2006. Induction and specification of cranial placodes. Dev. Biol. 294: 303–351.

Schierwater, B. and D. de Jong, and R. DeSalle. 2009. Placozoa and the evolution of Metazoa and intrasomatic cell differentiation. Int. J. Biochem. Cell Biol. 41: 370–379.

Schierwater, B. and K. Kamm, M. Srivastava, D. Rokhsar, R.D. Rosengarten, and S.L. Dellaporta. 2008. The early ANTP gene repertoire: insights from the placozoan genome. PloS ONE, 3: e2457.

Schneuwly, S. 1987. Redesigning the body plan of *Drosophila* by ectopic expression of the homoeotic gene *Antennapedia*. Nature 325: 816–818.

Seipel, K. and N. Yanze, and V. Schmid. 2004. Developmental and evolutionary aspects of the basic helix-loop-helix transcription factors *Atonal-like 1* and *Achaete-scute homolog 2* in the jellyfish. Dev. Biol. 269: 331–345.

Signorovitch, A.Y. and L.W. Buss, and S.L. Dellaporta. 2007. Comparative Genomics of Large Mitochondriain Placozoans. PLOS Genetics 3: e13.

Simionato, E. et al. 2007. Origin and diversification of the basic helix-loop-helix gene family in metazoans: insights from comparative genomics. BMC Evol. Biol. 7: 33.

Shubin, N. and C. Tabin, and S. Carroll. 1997. Fossils, genes and the evolution of animal limbs. Nature 388: 639–648.

Smith, K.M. and L. Gee, and H.R. Bode. 2000. HyAlx, an aristaless-related gene, is involved in tentacle formation in *hydra*. Development 127: 4743–4752.

Spangenberg, D.B. and E. Coccaro, R. Schwarte, and B. Lowe. 1996. Touch-plate and statolith formation in graviceptors of ephyrae which developed while weightless in space. Scan. Micro. 10: 875–888.

Sperling, E.A. and D. Pisani, and K.J. Peterson. 2007. Poriferan paraphyly and its implications for Precambrian paleobiology. Geol. Soc. Lond. Spec. Publ. 286: 355–367.

Srivastava, M. et al. 2008. The Trichoplax genome and the nature of placozoans. Nature 454: 955–960.

Stierwald, M. and N. Yanze, R.P. Bamert, L. Kammermeier, and V. Schmid. 2004. The *Sine oculis/Six* class family of homeobox genes in jellyfish with and without eyes: development and eye regeneration. Dev. Biol. 274: 70–81.

Suga, H. et al. 2008. Evolution and functional diversity of jellyfish opsins. Curr. Biol. 18: 51–55.

Todi, S.V. and Y. Sharma, and D.F. Eberl. 2004. Anatomical and molecular design of the *Drosophila* antenna as a flagellar auditory organ. Microsc. Res. Techniq. 63: 388–399.

Verger-Bocquet, M. 1981. Etude comparative, au niveau infrastructural, entre l'ꞏ il de souche et les taches oculaires du stolon chez *Syllis spongicola* Grübe (Annélide Polychète). Archives de Zoologie Expérimentale et Génerale 122: 253–258.

Vosshall, L.B. and H. Amrein, P.S. Morozov, A. Rzhetsky, and R. Axel. 1999. A spatial map of olfactory receptor expression in the *Drosophila* antenna. Cell 96: 725–736.

Westfall, J.A. 2004. Neural pathways and innervation of cnidocytes in tentacles of sea anemones. Hydrobiologia 530/531: 117–121.

Willenz, P. and G. Van De Vwer. 1986. Ultrastructural evidence of extruding exocytosis of residual bodies in the freshwater sponge *Ephydatia*. J. Morphol. 190: 307–318.

Yahel, G. and D.I. Eerkes-Medrano, and S.P. Leys. 2006. Size independent selective filtration of ultraplankton by hexactinellid glass sponges. Aquat. Microb. Ecol. 45: 181–194.

Yahel, G. et al. 2007. *In situ* feeding and metabolism of glass sponges (Hexactinellida, Porifera) studied in a deep temperate fjord with a remotely operated submersible Limnol. Oceanogr. 52: 428.

Yang, Q. and A.N. Bermingham, M.J. Finegold, and H.Y. Zoghbi. 2001. Requirement of *Math1* for secretory cell lineage commitment in the mouse intestine. Science 294: 2155–2158.

Zacharuk, R.Y. 1980. Ultrastructure and function of insect Chemosensilla. Ann. Rev. Entomol. 25: 27–47.

Chapter 9

Cnidarian Gene Expression Patterns and the Origins of Bilaterality—Are Cnidarians Reading the same Game Plan as "Higher" Animals?

Eldon E. Ball,[1] Danielle M. de Jong,[1,2] Bernd Schierwater,[3]
Chuya Shinzato,[2] David C. Hayward[1] and David J. Miller[2]*

Introduction

The past few years have seen a dramatic increase in the available data on gene sequence and gene expression for cnidarians and other "lower" Metazoa, and a flurry of recent papers has drawn on these to address the origins of bilaterality. Cnidarian homologs of many genes that play key roles in the specification of both the A/P and D/V axes of bilaterians have been characterized, and their patterns of expression determined. Some of these expression patterns are consistent with the conservation of function between

[1]Evolution, Ecology and Genetics, Research School of Biology, Bldg 46, Australian National University, Canberra, ACT 0200, Australia.
E-mail: eldon.ball@anu.edu.au
[2]ARC Centre of Excellence in Coral Reef Biology and Comparative Genomics Centre, James Cook University, Townsville, Queensland 4811, Australia.
E-mail: david.miller@jcu.edu.au
[3]ITZ, Ecology and Evolution, Tierärztliche Hochschule Hannover, D-30559, Hannover, Germany.
E-mail: bernardo@trichoplax.com

Cnidaria and Bilateria, but others clearly differ. Moreover, in some cases very different interpretations have been made on the basis of the same, or similar, data. In part, these differences reflect the inevitable uncertainties associated with the depth of the divergence between cnidarians and bilaterians. In this paper we briefly summarize the cnidarian data on gene expression and organization relevant to axis formation, the varying interpretations of these data, and where they conflict. Our conclusion is that the oral-aboral axis probably does correspond to the anterior-posterior axis of bilaterians, but that its polarity remains uncertain, and that many of the same genes are involved in determining the directive axis of cnidarians and the dorsal-ventral axis of bilaterians, but with sufficient differences in expression that exact homologies are uncertain.

Relationships Between the Axes of Bilateria and Radiata

Although still frequently described as radially symmetrical and diploblastic, there is a long-standing debate over the axes and the body layers of cnidarians and how it/they relate to those of Bilateria. Do some modern cnidarians illustrate an evolutionary stage through which bilaterians passed on the way to becoming overtly bilateral (Boero et al. 1998, Seipel and Schmid 2005), or are they an evolutionary offshoot that evolved from an already bilateral common ancestor? The standard textbook view has been that the Cnidaria are separated from the Bilateria by two major dichotomies. Firstly, they are considered to be radial, while Bilateria are bilateral. Secondly, they are considered to be diploblastic (have two body layers) in contrast to the triploblastic Bilateria. Further, it is commonly assumed that the oral end of Cnidaria, where the mouth lies, corresponds to the anterior end of Bilateria. Although these are widely prevalent views, none of them has gone undebated.

The Fossil Record and the Anatomy of the Ancestral Cnidarian

Before evaluating the textbook generalizations outlined above, there are several matters that should be considered. Firstly, the nature of the common ancestor of Cnidaria and Bilateria is still a matter of active debate; the possibilities being a medusoid form (see Seipel and Schmidt 2005 for a summary of arguments), a polypoid form or a planuloid form. The fossil record does not provide any guidance as to the nature of the common ancestor since the earliest known fossil assemblages, from the Neoproterozoic Ediacaran deposits, "included a mixture of stem- and crown-group radial animals, stem-group bilaterian animals, 'failed experiments' in animal evolution, and perhaps representatives of other eukaryotic kingdoms" (Narbonne 2005). Secondly, there is a hypothesis, traced by Willmer (1990) from Haeckel (1874, 1875), through Hyman

(1940) to Salvini-Plawen (1978) which proposes that cnidarians arose from a settling biradial planula with a single pair of tentacles. This hypothesis and related scenarios (including the placula hypothesis; Bütschli 1884) sit fairly comfortably with the molecular evidence presented below. If the common ancestor of Cnidaria and Bilateria was indeed such an animal this would help explain the lack of candidate fossils, since preservation of such an organism would require highly specialized conditions.

All extant cnidarians are likely to have some derived characteristics, so any conclusions about ancestral states based solely on living cnidarians are open to question. Nevertheless, with that caveat, the Anthozoa, which for reasons summarized below, is now thought to best reflect the ancestral condition, sits least comfortably within the textbook radial- diploblastic scheme outlined above. In contrast, the Hydrozoa, on whose characteristics the scheme is mostly based, is now considered to be the most derived. Today the most commonly cited arguments supporting the proposition that the Anthozoa comprises the least derived of modern forms are molecular, and include mitochondrial genome structure, which is circular in Anthozoa and Bilateria, but linear in Medusozoa (the Classes Hydrozoa, Cubozoa and Scyphozoa), and phylogenetic analyses of sequence data from several classes of genes. Even in the premolecular era, however, many scholars took this point of view. For example, Pantin (1966) argued that both differentiation of nematocysts and, above all, the simplicity of the anthozoan nervous system, which in contrast to that of hydrozoans, really is a nerve net showing few signs of centralization except an increased density of neurons around the mouth, supported this point of view. Willmer (1990) has succinctly summarized the non-molecular evidence for the primitive nature of the Anthozoa as follows: simpler life cycles, lesser ability to cope with physiologically difficult environments, and less elaborate and diverse cnidoblasts. Hydrozoans, by contrast, have more complex musculature and nervous systems, as well as often having specialized sense organs, all of which point to a higher degree of derivation.

Not only do the lines of evidence cited above point to anthozoan morphology as best reflecting ancestral characteristics, but also this is in agreement with the fossil record. Conway-Morris' recent (2006) summary of Cambrian fossils lists numerous anthozoan groups as probable but fails to list any hydrozoans and states that there are no convincing Cambrian scyphozoans or cubozoans (but see Hagadorn 2002, Pickerill 1982). This presents a challenge to the commonly accepted idea of a radially symmetrical ancestral cnidarian since most extant anthozoans are either biradial or bilateral. Obvious examples of bilaterality can be found among corals, both ancient and modern (e.g. Fig. 1A, B) but most other anthozoans show similar tendencies, having an elongate mouth with specializations at one or both ends (Fig. 1C,D) and corresponding asymmetries in septal musculature.

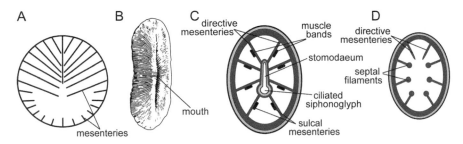

Fig. 1. Many members of the Class Anthozoa exhibit bilaterality. **A.** Mesentery pattern in an ancient rugose coral, a now extinct group which flourished during the Paleozoic (redrawn from Willmer 1990). **B.** Skeleton of a modern solitary fungiid coral (Anthozoa; Scleractinia) showing extreme elongation of the mouth. **C, D.** Cross sections of an *Alcyonium* (Anthozoa; Octocorallia) polyp cut at the level of the pharynx (C) and further proximally (D), showing several aspects of bilaterality (redrawn from Russell-Hunter 1979).

As might be expected, this bilaterality is the result of asymmetrical gene activity, as will be discussed below, following consideration of patterning of the oral/aboral axis and the possible role of Hox genes in this process.

Does Oral-Aboral Correspond to Anterior-Posterior and, If So, What is The Correspondence?

It has commonly been assumed that oral-aboral does correspond to anterior-posterior. Leaving aside the correctness of this assumption, if this is the case then how do the two axes match up (i.e. is oral anterior or posterior)? The case becomes particularly complicated when it is considered that some of the answers will depend on whether the planula or the polyp stage is considered and there is no clear answer as to whether the ancestral cnidarian was planuloid, polypoid or even medusoid. Considering the planula first, some of the arguments for the aboral end being anterior are that it is anterior as the animal swims, it frequently has a concentration of sensory neurons, and in planulae of *Podocoryne* (Class Hydrozoa) the nervous system develops sequentially from that end (Gröger and Schmid 2001), as is the case in bilaterians. The main counter argument appears to be that the blastopore becomes the oral pore, which in turn becomes the mouth after settlement, and the mouth is generally toward the anterior end in bilaterians. However, as Rieger et al. (2005) pointed out, the mouth is not actually AT the anterior pole. In fact, in acoel flatworms, which may be the most similar living forms to the ancestral bilaterian, the mouth is located ventrally on the posterior third of the animal. If one considers the polyp to be the ancestral life form then the argument that oral corresponds to anterior becomes stronger, since in many cnidarians there is a concentration of neurons associated with the mouth (Grimmelikhuizen and Westfall 1995).

One way of assessing axial correspondence is to look at the patterns of gene expression of cnidarian homologs of genes that are characteristically expressed either anteriorly or posteriorly in bilaterians. This apparently straightforward approach, however, has a potential flaw in that expression is sometimes found to switch ends at settlement. Thus, both *Podocoryne Cnox2-pc* and *Acropora Pax-Dam* switch from aboral to oral expression at the time of settlement (Fig. 2). In the absence of functional studies it is unclear whether this change is associated with a change in gene function.

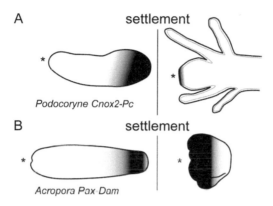

Fig. 2. A source of potential confusion in using the position at which a gene is expressed along the oral-aboral axis to homologize this to the A/P axis is that this position is sometimes reversed at the time of settlement. Two genes exhibiting this phenomenon are: **(A)** *Podocoryne Cnox2-Pc* (drawn from Masuda-Nakagawa, 2000), and **(B)** *Acropora Pax-Dam* (drawn from de Jong et al. 2006). Expression is shown in black.

What is a "Hox" gene?

The cnidarian "Hox" complement has been intensively studied, but the data have been interpreted in very different ways (e.g. compare Finnerty et al. 2004, Kamm et al. 2006, Chourrout et al. 2006, Ryan et al. 2007). A contributing factor has been the shifting definition of "Hox" which continues to plague the field today. The definition originally given to the word by a distinguished group of investigators was very broad: "any homeobox containing gene or genomic fragment may be given the designation Hox (Homeobox) providing that a significant fraction of the amino acids it encodes are identical to those of the homeobox in the *Drosophila Antennapedia* gene" (Martin et al. 1987). However, by the mid-1990s, the recognition of orthologous clusters of genes performing similar functions in flies and mice (Scott et al. 1992) led to a much more restrictive definition of the term "Hox gene". From the mid 1990s and into the first few years of this century, definitions of Hox encompassed not only sequence similarity with specific Hox classes from *Drosophila* and mammals, but also clustering and colinearity (Ruddle et al. 1994, Akam 1995); because colinearity was discovered in organisms with clusters of linked Hox genes, there was an underlying assumption that the latter was mandatory for the former. As a consequence of this paradigm, much of the cnidarian "Hox" literature of this period dealt with classification of Hox-like genes as anterior, posterior, intermediate, or ParaHox, and with the (fruitless) search for clusters.

In recent years, the discovery of secondarily fragmented, or "atomised" (Duboule 2007), clusters in a range of bilaterians has led to a re-evaluation of the significance of linkage and indirectly to the relaxation of criteria for membership of the "Hox" club. In *Oikopleura* (Seo et al. 2004) and other tunicates, the Hox cluster has undergone losses and is highly fragmented, and yet the remaining Hox genes are expressed in patterns that would be expected based on their fly and mammalian orthologs. Hox cluster integrity is therefore not mandatory for spatial colinearity, as it seems only to be essential for maintenance of temporal colinearity (e.g. Kmita and Duboule 2003)—in fact the Hox "cluster" of *Drosophila* is fragmented and some of its former members clearly no longer function as Hox genes.

In the context of the loosening of the definition of "Hox" by removal of the requirement for linkage, it is important to understand that "atomised" state of *Oikopleura* and other tunicates IS secondary—tunicates unquestionably had an ancestor in which the Hox genes were clustered (albeit in what Duboule has called a "disorganised" cluster). However, there is no evidence for such an ancestral condition in the case of the cnidarian Hox-like genes—in this case, the dispersed/"atomised" state is likely to be ancestral. Because it is not easy to capture this idea that not all "atomised" clusters are equal in a definition, the term "Hox" is now very imprecise. The definition employed in the annotation of the *Nematostella* genome (Putnam et al. 2007) more or less reflects a return to the original definition—an Antennapedia class homeobox gene that is more similar to Hox genes than to non-Hox genes (Miller and Ball 2008).

Axial Patterning Predating the Canonical Hox System?

Although our primary focus is on the Cnidaria, several recent whole genome sequencing projects have provided important insights into the origins of the bilaterian Hox cluster. Turning first to sponges, no Hox, ParaHox or extended Hox genes are present in the genome of *Amphimedon* (formerly *Reniera*) *queenslandica* (Larroux et al. 2006, 2007), and studies focussed on several other species imply that this is true of the phylum in general (e.g. Seimiya 1994, Manuel and Le Parco 2000). Yet the larva of *Amphimedon* is clearly axially organized (Degnan et al. 2005, Adamska et al. 2007) and bears a remarkable resemblance to the planula larva of *Acropora*. Indeed, Lévi (1963) has characterized sponge larvae as having axial, or even tetraradial, symmetry. Thus, in sponges it appears that there is a clear axial (oral-aboral) organization, which is likely controlled, at least in part, by Wnt and TGFβ (Adamska, et al. 2007) without the participation of Hox genes, since they are absent from sponges (Larroux et al. 2007).

In Placozoa, which are morphologically polarized relative to the substratum but otherwise show no sign of axial specialization (summarized by Miller and Ball 2005, Schierwater 2005), five ANTP-class genes have been recovered, but no Hox-related genes (summarized by Montiero et al. 2006) and this is still the case following sequencing of the genome (Schierwater et al. 2008, Srivastava et al. 2008). The only gene closely related to the Hox/ParaHox classes is *Trox-2* (Schierwater and Kuhn 1998), which groups with the cnidarian cnox2 and bilaterian Gsx-type sequences in phylogenetic analyses (Schierwater and Kuhn 1998, Finnerty and Martindale 1999, Gauchat et al. 2000, Hayward et al. 2001)

The other potentially informative group, which all studies locate near the base of the Metazoa, is the ctenophores. A full ctenophore genome sequence is not yet available, but ca 15,000 ESTs are now available on GenBank and Pang and Martindale (2008) have recently surveyed the homeobox genes of the ctenophore *Mnemiopsis leidyi*, using a degenerate PCR strategy, and found no Hox-like genes. As in sponges, the most Hox-like genes in *Mnemiopsis* were NK-like subclass homeobox genes, leading the authors to suggest that Hox genes may have evolved from this group.

Thus, although there is considerable confusion as to the branching order of phyla at the base of the metazoan tree (Dunn et al. 2008, Dellaporta et al. 2006, Srivastava et al. 2008), it is now clear that all of the groups that have been proposed as basal to the Cnidaria almost certainly lack Hox genes. This places the Cnidaria in a key position as the most basal extant phylum with Hox genes. Before turning to this much investigated phylum there is one other group that should be considered, namely the acoels. A number of phylogenetic studies (e.g. Philippe et al. 2007) support the idea that acoels are not true flatworms (Platyhelminthes) and may have diverged from the bilaterian stem after cnidarians but before the protostome/deuterostome split (Ruiz-Trillo et al. 2002, Telford et al. 2003). Cook et al. (2004) and Baguna and Riutort (2004) have investigated the Hox complement of phylogenetically divergent acoels and analyses of the (admittedly incomplete) sequences available suggest the presence of anterior, group 3, intermediate and posterior Hox types. Unfortunately, although the Hox genes of acoels continue to attract attention (eg. Jiminez-Guri et al. 2006, Deutsch 2008) no information appears to be yet available concerning possible linkages between these genes or on their expression.

Potential regulators of Hox genes include the miRNA mir-10 (Aboobaker et al. 2005, Woltering and Durston 2006). However, although there are miRANs in both the sponge *Amphimedon* and the anthozoan

Nematostella neither genome encodes a min-10 homolog (Grimson et al. 2008). Interestingly, acoels appear to have a miRNA repertoire that is intermediate in complexity between cnidarians (represented by *Nematostella*) and bilaterians (Sempere et al. 2007).

The Cnidarian "Hox" Gene Complement

The Cnidaria is by far the most extensively investigated lower metazoan phylum in terms of the Hox-like gene complement, and in the bilaterian/ cnidarian common ancestor the precedents of Hox genes, if not the Hox genes proper, are clearly represented for the first time. Much of the debate over the presence/absence of Hox genes in cnidarians is due to a lack of agreed definitions. For example, must the ancestors of a Hox gene once have belonged to a cluster of similar genes that pattern the body axis? and What constitutes a cluster? Below we will attempt to summarise what is presently agreed and disagreed. Hox (or Hox-like) genes have been identified in a range of cnidarians (summarized by Gauchat et al. 2000, Ryan et al. 2006, 2007) but the analyses of Kamm et al. (2006) and Chourrout et al. (2006) cast serious doubts on the idea that a Hox cluster, in the sense of two or more linked genes that are orthologous with distinct HOX classes of bilaterians, existed at the time of the cnidarian/ bilaterian divergence. Phylogenetic analyses indicate that most cnidarians have anterior Hox-like genes and that several other cnidarians, including the hydrozoan *Eleutheria*, have a second type that is related to both the posterior Hox and Cdx classes according to Kamm et al. (2006), or is a true posterior Hox gene according to Ryan et al (2007). All other presently characterized cnidarian Hox-like genes post-date the bilaterian divergence. In *Nematostella*, three paralogous Hox-like genes are linked to the *eve/ anthox6* pair, but this "cluster" arose since the bilaterian divergence, and is therefore unrelated to the Hox clusters of bilaterians. Moreover, the expression patterns of the cnidarian Hox-like genes are highly divergent and do not fit the expectation, based on the true Hox genes of bilaterians, of relatively conserved expression across the phylum (Fig. 3), a phenomenon that Ryan et al. (2007) explain away as being due to great differences in morphology and the long separation of the cnidarian Classes. *anthox6*, the only *Nematostella* gene that all investigators appear to accept as a bona-fide member of a Hox orthology group, is expressed in a narrow stripe in the pharynx (Finnerty et al. 2004). The authors of all of these papers would presumably agree that available data are consistent with the hypothesis that the cnidarian/bilaterian common ancestor (Ureumetazoa) is likely to have had two or three genes (Hox precursors; see above) that may have later given rise to the Hox cluster in Urbilateria. The two views are essentially that either a Hox system was present in the Ureumetazoan

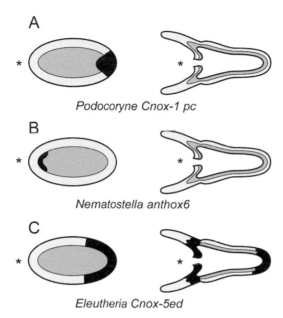

Fig. 3. Orthologous Hox-like genes are expressed in highly divergent patterns across the Cnidaria. The diversity of these expression patterns contrasts with the high degree of similarity in the expression patterns of true Hox genes across the Bilateria. **(A)** *Podocoryne Cnox-1pc* is expressed in ectoderm and endoderm at the aboral end of the planula and is not expressed in the polyp. **(B)** *Nematostella anthox6* expression is restricted to the oral endoderm of the planula and not expressed in the polyp. **(C)** *Eleutheria Cnox-5ed* is expressed in the aboral ectoderm of the planula and in ectoderm and endoderm at either end of the polyp. Expression is shown in black. Drawn from Kamm et al. (2006).

and it has subsequently diverged almost beyond recognition (Ryan et al. 2007) or such a system evolved in Urbilateria after the divergence of the Cnidaria, so that what we see today in the latter reflects a pre-Hox state (Kamm et al. 2006, Chourrout et al. 2006). Arguments can be made for either scenario, but regardless it is clear that the definition of a "Hox" gene has itself degenerated, almost beyond recognition, when applied in the context of the lower Metazoa. Progress in understanding the origins of the bilaterian Hox clusters appears to require not just new data, but also general agreement on a definition of Hox that is appropriate for "lower" animals.

Cnidarian Homologs of *Drosophila* Head-Gap Genes

Given the equivocal nature of the "Hox" data, what other genes might be good markers for the cnidarian equivalent of the anterior pole? Two of the most universal anterior markers are homeobox genes corresponding to the Drosophila head-gap genes empty-spiracles (ems) and orthodenticle (otd), which have conserved roles in anterior patterning from insects to mammals (reviewed by Lichtneckert and Reichert 2005). Whereas the Hox data are equivocal, unambiguous homologs of ems and otd have been cloned from several cnidarians (Mokady et al. 1998, Müller et al. 1999, Smith et al. 1999, de Jong et al. 2006, Mazza et al. 2007).

In *Acropora*, the ems gene *emx-Am* is expressed in the aboral two thirds of the planula larva (Fig. 4A), most obviously in presumed neurons (de Jong et al. 2006), but also more generally throughout the ectoderm; this is reminiscent of *empty-spiracles* function in *Drosophila*, which has roles in patterning both the anterior ectoderm and the brain (Dalton et al. 1989, Hartmann et al. 2001). While expression of this head-gap gene at the aboral end of the primary axis (i.e. the 'anterior' end with respect to swimming direction) suggests correspondence between aboral and anterior, an *Acropora* homolog of another 'definitive' anterior marker, otd/Otx, is expressed at the opposite end of the primary axis (Fig 4B). *otxA-Am* is expressed initially around the blastopore (which later becomes the oral pore), and later in an undefined cell type at the base of the ectoderm; these cells are distributed along the axis but only those in the oral half of the planula express *otxA-Am* (de Jong et al. 2006). A second otx gene is present in *Acropora* that is first expressed in presumptive endoderm at the start of gastrulation (Hayward et al. unpublished) but whole later expression pattern remains unknown. Three otx genes, OTX-A, OTX-B, and OTX-C, have recently been described from *Nematostella* by Mazza et al. (2007). OTX-A corresponds to the sequence designated Otx2Nv by de Jong et al. (2006), OTX-B was previously undescribed, and OTX-C to Otx1Nv.

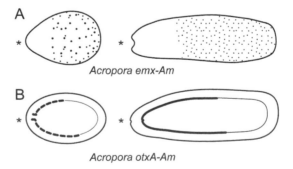

Acropora emx-Am

Acropora otxA-Am

Fig. 4. Apparently contradictory expression patterns of "anterior" genes in the coral *Acropora*. *emx-Am* and *otx-Am*, which are members of gene families widely considered to be diagnostic anterior genes, are expressed at opposite ends of the *Acropora* planula larva. Direct comparisons are complicated by the fact that both genes have been independently duplicated within the Cnidaria. Even if paralogs of one or both of these genes are expressed at opposite ends of the axis, however, the ancestral pattern will be equivocal. Externally visible expression is shown for *emx-Am*, while *otxA-Am* is shown in section. Drawn from de Jong et al. (2006).

According to Mazza et al. (2007) none of the *Nematostella* genes is strictly orthologous to either of the *Acropora* genes. Neither of the coral genes has the conserved WSP motif downstream of the homeodomain which Mazza et al. (2007) describe as being found in many Otx proteins and which is present in *Nematostella* OTX-B. All three of the *Nematostella* genes have similar expression patterns, appearing first in the future endoderm as it gastrulates, later in restricted regions of the endoderm at both ends of the planula, and finally in the ectoderm of the tentacles once they have formed. Given that these various "head genes" can be expressed at either or both ends of these developing cnidarians it is difficult to justify using them as anterior markers.

The "Directive" Axis of Cnidarians

Expression data for genes of the Dpp/BMP2/4 type in *Acropora* (Hayward et al. 2002 and unpublished) and *Nematostella* (Finnerty et al. 2004) provides one of the most convincing lines of evidence that cnidarians (or, at least, anthozoans) have more than one axis. In flies and vertebrates Dpp/BMP4 plays a key role in D/V patterning; on one side of the D/V axis the activity of this anti-neurogenic protein is antagonized by sog/chordin, allowing the nerve cord to develop in both groups. During gastrulation of both *Acropora* (Hayward et al. 2002) and *Nematostella* (Finnerty et al. 2004), *Dpp/BMP4* is expressed asymmetrically in one quadrant of the blastopore lip. In planulae of *Nematostella* (Finnerty et al. 2004) *Dpp/BMP4* is also asymmetrically expressed in the pharyngeal ectoderm. Combined with expression data for a number of *Nematostella* "Hox" genes, which show at least three distinct domains of expression along the primary axis (Finnerty et al. 2004), these intriguing similarities have been used to argue for a direct correspondence of cnidarian "bilaterality" with that of members of the Bilateria, underlain by common molecular mechanisms— so that, as in bilaterians, cnidarian "Hox" genes pattern along the primary axis (therefore O/A = A/P) and Dpp is involved in patterning the second axis (therefore the directive axis = D/V). More recently many of the same authors have established the three-dimensional patterns of the expression of these genes in more detail (Matus, Pang et al. 2006) and have proposed that they may be involved in the patterning of the nervous system.

Although expression data for *Dpp/BMP4* (see above) provided much cause for speculation concerning correspondence of cnidarian and bilaterian axes, more extensive investigation of the expression of TGFβ ligands, antagonists and receptors (Rentzsch et al. 2006, Matus, Pang et al. 2006, Matus, Thomsen et al. 2006) considerably complicates the situation. As outlined above, in bilaterians Dpp/BMP4 activity is antagonized on the opposite side of the D/V axis by short gastrulation (sog) in *Drosophila* or its homolog chordin in vertebrates, allowing neuroectoderm to develop on the sog/chordin side. In *Nematostella*, however, *chordin* is expressed on the same side of the directive axis as *dpp*, in a partially overlapping manner (Rentzsch et al. 2006, Matus, Thomsen et al. 2006). One of the two *Nematostella* homologs of another TGFβ antagonist, *noggin*, is also expressed specifically on the *dpp/chordin* side of the directive axis (Fig. 5B) (Matus, Pang et al. 2006). On the opposite side, the *Nematostella* homolog of yet another class of TGFβ antagonist, *gremlin*, and another TGFβ ligand, *gdf5-like* are again expressed in complex and semi-overlapping patterns (Rentzsch et al. 2006). It should be noted here that the differing expression patterns reported by Matus, Pang, et al. (2006) and Rentzsch et al. (2006) for *Nematostella gremlin* are the result of working on different paralogs of the gene. The gene described by the former authors is expressed at a time and in a manner that precludes a role in axis formation and therefore is not discussed here. Analyses in heterologous systems imply that these molecules function in *Nematostella* as do their orthologs in flies and vertebrates—for example, both NvChordin and NvGremlin can antagonize the ability of NvDpp mRNA to ventralize zebrafish (Rentzsch et al. 2006). What this means is that, "Taken together, the expression pattern of *NvGrm* is compatible with the generation of two endodermal BMP-like activity gradients: as an antagonist of *NvGDF-5-like* along the oral-aboral axis and as an antagonist of *NvDpp* along the directive axis (Rentzsch et al. 2006)".

In addition to components of TGFβ signalling systems, the Martindale laboratory has catalogued a number of other genes that are differentially expressed in the so-called directive axis in *Nematostella*. As outlined in the caption to Fig. 5, the most obvious restriction of the expression domains of the Hox-like genes *anthox1a*, *anthox7* and *anthox8* is in the directive axis (Finnerty et al. 2004, Ryan et al. 2007)) and these genes are expressed on the *gdf5-like* + *gremlin* side (Rentzsch et al. 2006, Matus, Pang et al. 2006). In addition, homologs of *netrin* and *goosecoid* (*gsc*) are clearly differentially expressed in the second axis (Matus, Pang et al. 2006). The significance of the *netrin* data is unclear, as this is an extensive family of axon-guidance molecules in bilaterians, but the *gsc* data are intriguing given that this is a key player in the vertebrate organizer—the small group of cells on the blastopore lip that directs the development of the embryonic body plan. The *Nematostella goosecoid* gene NvGsc is not expressed during gastrulation, but in the planula is expressed asymmetrically with respect to the directive axis - two bands of expression are seen, the stronger being on the same side as *anthox1a/7/8*, *gdf5-like* and *gremlin*, but a less extensive stripe is seen on the opposite side (i.e. where *dpp*, *chordin* and *noggin* expression is seen).

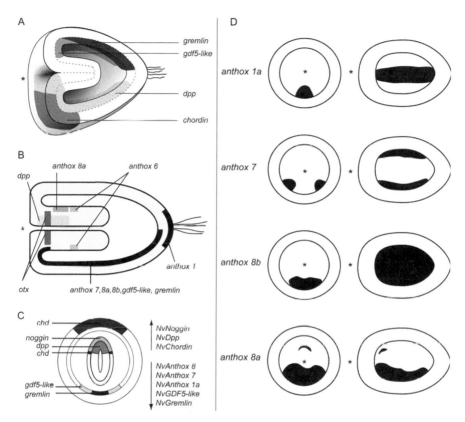

Fig. 5. Expression of *Nematostella* homologs of key axial patterning genes. Because gene expression patterns are constantly changing, static diagrams are inadequate but do serve to illustrate overall patterns of message distribution. **(A)** In contrast to the situation in bilaterians, *dpp* and *chordin* are expressed on the same side of the directive axis and in gradients both along the oral-aboral axis and at right angles to it, as shown in this diagram redrawn from Rentzsch et al. (2006). **(B)** Expression domains of some genes potentially involved in axis determination or providing axial markers (redrawn from Matus, Pang et al. 2006). The expression of many additional genes, belonging to a broader Hox-like class, has recently been described (Ryan et al. 2007), but has been omitted in the interests of clarity. **(C)** Transverse section of (B) at the level of the pharynx showing the distribution of differentially expressed messages in the directive axis (redrawn from Matus, Pang et al, 2006 with *gremlin* distribution estimated from Rentzsch et al. 2006). **(D-G)** Expression patterns of four *Nematostella* Hox-like genes along the directive (left column) and oral/aboral axes (right column). The position of the oral pore is shown by an asterisk. These expression patterns suggest the possibility of functions in patterning in the directive axis, rather than, or in addition to, the oral/aboral axis (drawn from Ryan et al. 2007).

Cnidarian Homologs of *Drosophila* 'Columnar' Genes

Another suite of genes that are potentially informative in terms of the correspondence of axes are those that in *Drosophila* are known as the 'columnar' genes, which include three interacting homeobox genes that subdivide the neuroectoderm into three columns in the D/V axis. In vertebrates, orthologous genes are expressed in strikingly similar patterns, leading to this being considered an evolutionarily conserved system (reviewed in Chan and Jan 1999, Cornell and von Ohlen 2000). As in the case of the head-gap genes, unambiguous members of each of the columnar homeobox gene classes are also present in cnidarians. If the directive axis of cnidarians does correspond to the D/V axis of bilaterians, then one might expect differential expression of these genes in the directive axis, but this is not the case. Instead, in *Acropora* expression of each of these is restricted along the primary (O/A) axis, but the patterns do not fit expectations based on a simple correspondence of O/A and dorsal/ventral axes (de Jong et al. 2006, Ball et al. 2007).

Should Differences be Ignored?

While apparent similarities in expression patterns are generally interpreted as reflecting conservation of function, discrepancies between cnidarians and bilaterians are often ignored because of the assumption that these are consequences of the depth of the divergence. Should we really ignore differences?

It is clear that the gene complement of Ureumetazoa was surprisingly complex and did not differ greatly from those of modern chordates (Kortschak et al. 2003, Technau et al. 2005, Nichols et al. 2006, Putnam et al. 2007). Anthozoan cnidarians appear to have maintained much of this ancestral complexity, and we should not therefore be surprised that *Nematostella* and *Acropora* have clear homologs of many of the genes that play key roles in patterning both major axes of bilaterians. Many cnidarians are not radially symmetrical, and among anthozoans a kind of bilaterality is most clearly and convincingly seen. The big question is whether the axes of cnidarians and bilaterians are homologous and, while the assumption has often been that this is the case, the presently available data are equivocal. While it has been argued that the expression data for Hox-like genes and *dpp* support a direct correspondence of the cnidarian O/A with the bilaterian A/P axis, and the bilaterian D/V axis with a second cnidarian axis most often known as the 'directive' axis (e.g. Finnerty et al. 2004), this simple interpretation is made less certain by more detailed analyses of these and other key genes.

In terms of the correspondence of cnidarian and bilaterian axes, at the present time, one can essentially pick gene expression data to support any favoured evolutionary scenario, but such exercises require the turning of a blind eye to other data that do not fit (or flatly contradict) expectations. For example, the most obvious restriction of the expression of *anthox1a*, *anthox7*, *anthox8a* and *anthox8b*, and that which seems to us more likely to serve a patterning function, is in the directive axis (Fig 5D).

If preconceptions are discarded, the presently available data suggest a more intriguing possibility—that cnidarians might reflect a separate experiment (or, perhaps, a separate set of experiments) in axial patterning. Starting with much the same genetic repertoire as Urbilateria but lacking a canonical Hox cluster, they have independently achieved a tremendous diversity of body plans—witness, for example, the spectacular morphologies of siphonophores and octocorals. Independent solutions to the problem of how to pattern along the primary axis may include deployment of Wnts (Kusserow et al. 2005, Miller et al. 2005, Guder et al. 2006, Lee et al. 2006) and homologs of both the head gap and columnar genes of bilaterians (de Jong et al. 2006).

Martindale (2005) presented an attractive general theory in which cnidarians fit neatly into a general animal plan. In the long run this may turn out to be true when we have more data on the genes of various cnidarian groups and their expression and better understand how these have changed in the course of cnidarian evolution. Indeed, since this paper was written two papers have appeared experimentally demonstrating a role for β-catenin stabilization and wnt signalling in axis determination in the hydrozoan, *Clytia* (Momose and Houliston 2007, Momose et al. 2008). These papers, which were consistent with earlier work in *Nematostella* (Wickramanyake et al. 2003) but apparently inconsistent with classic papers apparently demonstrating post oogenesis axis determination by the site of first cleavage initiation (summarised by Goldstein and Freeman (1997)) seem to bring the Cnidaria more closely into line with other animals, and seem to make it more likely that oral/aboral corresponds to anterior/posterior, while leaving the polarity of this axis unclear. Nevertheless, at present some of the data discussed here appear inconsistent with such a theory. We are only beginning to uncover the complexity of gene duplication and gene loss within the Cnidaria and some of the present inconsistencies may well be due to comparisons of paralogs (e.g the two *Nematostella gremlins* cited above). Further uncertainty is added by the lack of information on the distribution of proteins, rather than mRNAs, for the genes considered here. Thus, apparent inconsistencies may disappear as more work is carried out on cnidarian genes and genomes. Nevertheless in our present state of knowledge it seems premature to try to force cnidarians into the same mould as all other animals.

References

Aboobaker, A.A. and P. Tomancak, N. Patel, G.M. Rubin, and E.C. Lai. 2005. *Drosophila* microRNAs exhibit diverse spatial expression patterns during embryonic development. Proc. Natl. Acad. Sci. USA 102: 18017–18022.

Adamska, M. and S.M. Degnan, K.M. Green, M. Adamski, A. Craigie, C. Larroux, and B.M. Degnan. 2007. Wnt and TGF-beta expression in the sponge *Amphimedon queenslandica* and the origin of metazoan embryonic patterning. PLoS ONE 2 e1031.

Akam, M. 1995. Hox genes and the evolution of diverse body plans. Philos. Trans. R. Soc. Lond. B Biol. Sci. 349: 313–319.

Baguna, J. and M. Riutort. 2004. The dawn of bilaterian animals: the case of acoelomorph flatworms. Bioessays 26: 1046–1057.

Ball, E.E. and D.M. de Jong, B. Schierwater, C. Shinzato, D.C. Hayward, and D.J. Miller. 2007. Implications of cnidarian gene expression patterns for the origins of bilaterality—is the glass half full or half empty? Int. Comp. Biol. 47: 701–711.

Boero, F. and C. Gravili, P. Pagliara, S. Pariano, J. Bouillon, and V. Schmid. 1998. The cnidarian premises of metazoan evolution: from triploblasty, to coelom formation, to metamery. Ital. J. Zool. 65: 5–9.

Bütschli, O. 1884. Bemerkungen zur Gastraeatheorie. Morph. Jahrb. 9: 415–427.

Chan, Y.M. and Y.N. Jan. 1999. Conservation of neurogenic genes and mechanisms. Curr. Opin. Neurobiol. 9: 582–588.

Chourrout, D. and F. Delsuc, P. Chourrout, R.B. Edvardsen, F. Rentzsch, E. Renfer, M.F. Jensen, B. Zhu, P. de Jong, R.E. Steele, and U. Technau. 2006. Minimal ProtoHox cluster inferred from bilaterian and cnidarian Hox complements. Nature 442: 684–687.

Conway Morris, S. 2006. Darwin's dilemma: the realities of the Cambrian 'explosion'. Philos. Trans. R. Soc. Lond. B Biol. Sci. 361: 1069–1083.

Cook, C.E. and E. Jimenez, M. Akam, and E. Salo. 2004. The Hox gene complement of acoel flatworms, a basal bilaterian clade. Evol. Dev. 6: 154–163.

Cornell, R.A. and T.V. Ohlen. 2000. Vnd/nkx, ind/gsh, and msh/msx: conserved regulators of dorsoventral neural patterning? Curr. Opin. Neurobiol. 10: 63–71.

Dalton, D. and R. Chadwick, and W. McGinnis. 1989. Expression and embryonic function of empty spiracles: a *Drosophila* homeo box gene with two patterning functions on the anterior-posterior axis of the embryo. Genes Dev. 3: 1940–1956.

de Jong, D.M. and N.R. Hislop, D.C. Hayward, J.S. Reece-Hoyes, P.C. Pontynen, E.E. Ball, and D.J. Miller. 2006. Components of both major axial patterning systems of the Bilateria are differentially expressed along the primary axis of a 'radiate' animal, the anthozoan cnidarian *Acropora millepora*. Dev. Biol. 298: 632–643.

Degnan, B. and S. Leys, and C. Larroux. 2005. Sponge development and antiquity of animal pattern formation. Int. Comp. Biol. 45: 335–341.

Dellaporta, S.L. and A. Xu, S. Sagasser, W. Jakob, M.A. Moreno, L.W. Buss, and B. Schierwater. 2006. Mitochondrial genome of *Trichoplax adhaerens* supports placozoa as the basal lower metazoan phylum. Proc. Natl. Acad. Sci. USA 103: 8751–8756.

Deutsch, J. 2008. Do acoels climb up the "Scale of Beings"? Evol. Dev. 10: 135–140.

Duboule, D. 2007. The rise and fall of Hox gene clusters. Development 134: 2549–2560.

Dunn, C.W. and A. Hejnol, D.Q. Matus, K. Pang, W.E. Browne, S.A. Smith, E. Seaver, G.W. Rouse, M. Obst, G.D. Edgecombe, M.V. Sorensen, S.H. Haddock, A. Schmidt-Rhaesa, A. Okusu, R.M. Kristensen, W.C. Wheeler, M.Q. Martindale, and G. Giribet. 2008. Broad phylogenomic sampling improves resolution of the animal tree of life. Nature 452: 745–749.

Finnerty, J.R. and M.Q. Martindale. 1999. Ancient origins of axial patterning genes: Hox genes and ParaHox genes in the Cnidaria. Evol. Dev. 1: 16–23.

Finnerty, J.R. and K. Pang, P. Burton, D. Paulson, and M.Q. Martindale. 2004. Origins of bilateral symmetry: Hox and dpp expression in a sea anemone. Science 304: 1335–1337.

Gauchat, D. and F. Mazet, C. Berney, M. Schummer, S. Kreger, J. Pawlowski, and B. Galliot. 2000. Evolution of Antp-class genes and differential expression of Hydra Hox/paraHox genes in anterior patterning. Proc. Natl. Acad. Sci. USA 97: 4493–4498.

Goldstein, B. and G. Freeman. 1997. Axis specification in animal development. Bioessays 19: 105–116.

Grimmelikhuijzen, C. and J. Westfall. The nervous systems of cnidarians. pp. 7–24. *In:* O. Breidbach and W. Kutsch. [eds] 1995. The Nervous Systems of Invertebrates: An Evolutionary and Comparative Approach. Birkhauser, Basel, Switzerland.

Grimson, A. and M. Srivastava, B. Fahey, B.J. Woodcroft, H.R. Chiang, N. King, B.M. Degnan, D.S. Rokhsar and D.P. Bartel. 2008. Early origins and evolution of micro RNAs and Piwi-interacting RNAs in animals. Nature 455: 1193–1197.

Gröger, H. and V. Schmid. 2001. Larval development in Cnidaria: a connection to Bilateria? Genesis 29: 110–114.

Guder, C. and I. Philipp, T. Lengfeld, H. Watanabe, B. Hobmayer, and T.W. Holstein. 2006. The Wnt code: cnidarians signal the way. Oncogene 25: 7450–7460.

Haeckel, E. 1874. The gastraea-theory, the phylogenetic classification of the animal kinxgdom and the homology of the germ-lamellae. Quart. J. Microsc. Sci. 14: 142–165.

Haeckel, E. 1875. Die Gastrula und die Entfuchung der Thiere. Jena Z. Naturwiss. 9: 402–508.

Hagadorn, J. and R.H.J. Dott, and D. Damrow. 2002. Stranded on a late Cambrian shoreline: Medusae from central Wisconsin. Geology 30: 147–150.

Hartmann, B. and H. Reichert, and U. Walldorf. 2001. Interaction of gap genes in the *Drosophila* head: tailless regulates expression of empty spiracles in early embryonic patterning and brain development. Mech. Dev. 109: 161–172.

Hayward, D.C. and J. Catmull, J.S. Reece-Hoyes, H. Berghammer, H. Dodd, S.J. Hann, D.J. Miller, and E.E. Ball. 2001. Gene structure and larval expression of cnox-2Am from the coral *Acropora millepora*. Dev. Genes Evol. 211: 10–19.

Hayward, D.C. and G. Samuel, P.C. Pontynen, J. Catmull, R. Saint, D.J. Miller, and E.E. Ball. 2002. Localized expression of a dpp/BMP2/4 ortholog in a coral embryo. Proc. Natl. Acad. Sci. USA 99: 8106–8111.

Hyman, L. 1940. The Invertebrates: Protozoa through Ctenophora. Metazoa of the Tissue Grade of Construction—The Radiate Phyla—Phylum Cnidaria. McGraw-Hill Book Company Inc, New York.

Jimenez-Guri, E. and J. Paps, J. Garcia-Fernandez, and E. Salo. 2006. Hox and ParaHox genes in Nemertodermatida, a basal bilaterian clade. Int. J. Dev. Biol. 50: 675–679.

Kamm, K. and B. Schierwater, W. Jakob, S.L. Dellaporta, and D.J. Miller. 2006. Axial patterning and diversification in the cnidaria predate the Hox system. Curr. Biol. 16: 920–926.

Kmita, M. and D. Duboule. 2003. Organizing axes in time and space; 25 years of colinear tinkering. Science 301: 331–333.

Kortschak, R.D. and G. Samuel, R. Saint, and D.J. Miller. 2003. EST analysis of the cnidarian *Acropora millepora* reveals extensive gene loss and rapid sequence divergence in the model invertebrates. Curr. Biol. 13: 2190–2195.

Kusserow, A. and K. Pang, C. Sturm, M. Hrouda, J. Lentfer, H.A. Schmidt, U. Technau, A. von Haeseler, B. Hobmayer, M.Q. Martindale, and T.W. Holstein. 2005. Unexpected complexity of the Wnt gene family in a sea anemone. Nature 433: 156–160.

Larroux, C. and B. Fahey, D. Liubicich, V.F. Hinman, M. Gauthier, M. Gongora, K. Green, G. Worheide, S.P. Leys, and B.M. Degnan. 2006. Developmental expression of transcription factor genes in a demosponge: insights into the origin of metazoan multicellularity. Evol. Dev. 8: 150–173.

Larroux, C. and B. Fahey, S.M. Degnan, M. Adamski, D.S. Rokhsar, and B.M. Degnan. 2007. The NK homeobox gene cluster predates the origin of Hox genes. Curr. Biol. 17: 706–710.

Lee, P. and K. Pang, D. Matus, and M. Martindale. 2006. A WNT of things to come: Evolution of Wnt signaling and polarity in cnidarians. Seminars in Cell & Developmental Biology 17: 157–167.

Lévi, C. Gastrulation and larval phylogeny in sponges. pp. 375–382. *In:* E.C. Dougherty. [ed.] 1963. The Lower Metazoa. Univ. of Calif. Press, Berkeley and Los Angeles.

Lichtneckert, R. and H. Reichert. 2005. Insights into the urbilaterian brain: conserved genetic patterning mechanisms in insect and vertebrate brain development. Heredity 94: 465–477.

Manuel, M. and Y. Le Parco. 2000. Homeobox gene diversification in the calcareous sponge, *Sycon raphanus*. Mol. Phylogenet. Evol. 17: 97–107.

Martin, G. 1987. Nomenclature for homeobox-containing genes. Nature 325: 21–22.

Martindale, M.Q. 2005. The evolution of metazoan axial properties. Nat. Rev. Genet. 6: 917–927.

Masuda-Nakagawa, L. and H. Groer, B. Aerne, and V. Schmid. 2000. The HOX-like gene Cnox2-Pc is expressed at the anterior region in all life cycle stages of the jellyfish *Podocoryne carnea*. Dev. Genes Evol. 210: 151–156.

Matus, D.Q. and K. Pang, H. Marlow, C.W. Dunn, G.H. Thomsen, and M.Q. Martindale. 2006. Molecular evidence for deep evolutionary roots of bilaterality in animal development. Proc. Natl. Acad. Sci. USA 103: 11195–11200.

Matus, D.Q. and G.H. Thomsen, and M.Q. Martindale. 2006. Dorso-ventral genes are asymmetrically expressed and involved in germ-layer demarcation during cnidarian gastrulation. Curr. Biol. 16: 499–505.

Mazza, M.E. and K. Pang, M.Q. Martindale, and J.R. Finnerty. 2007. Genomic organization, gene structure, and developmental expression of three clustered otx genes in the sea anemone *Nematostella vectensis*. J. Exp. Zoolog. B Mol. Dev. Evol. 308: 494–506.

Miller, D.J. and A. Miles. 1993. Homeobox genes and the zootype. Nature 365: 215–216.

Miller, D.J. and E.E. Ball, and U. Technau. 2005. Cnidarians and ancestral genetic complexity in the animal kingdom. Trends Genet. 21: 536–539.

Miller, D.J. and E.E. Ball. 2005. Animal evolution: the enigmatic phylum placozoa revisited. Curr. Biol. 15: R26–28.

Miller, D.J. and E.E. Ball. 2008. Cryptic complexity captured: the *Nematostella* genome reveals its secrets. Trends Genet. 24: 1–4.

Mokady, O. and M.H. Dick, D. Lackschewitz, B. Schierwater, and L.W. Buss. 1998. Over one-half billion years of head conservation? Expression of an ems class gene in *Hydractinia symbiolongicarpus* (Cnidaria: Hydrozoa). Proc. Natl. Acad. Sci. USA 95: 3673–3678.

Momose, T. and R. Derelle, and E. Houliston. 2008. A maternally localised Wnt ligand required for axial patterning in the cnidarian *Clytia hemisphaerica*. Development 135: 2105–2113.

Momose, T. and E. Houliston. 2007. Two oppositely localised frizzled RNAs as axis determinants in a cnidarian embryo. PLoS Biol. 5: e70.

Monteiro, A.S. and B. Schierwater, S.L. Dellaporta, and P.W. Holland. 2006. A low diversity of ANTP class homeobox genes in Placozoa. Evol. Dev. 8: 174–182.

Müller, P. and N. Yanze, V. Schmid, and J. Spring. 1999. The homeobox gene Otx of the jellyfish *Podocoryne carnea*: role of a head gene in striated muscle and evolution. Dev. Biol. 216: 582–594.

Narbonne, G. 2005. The Ediacara biota: Neoproterozoic origin of animals and their ecosystems. Annu. Rev. Earth Planet Sci. 33: 421–442.

Nichols, S.A. and W. Dirks, J.S. Pearse, and N. King. 2006. Early evolution of animal cell signaling and adhesion genes. Proc. Natl. Acad. Sci. USA 103: 12451–12456.

Pang, K. and M.Q. Martindale. 2008. Developmental expression of homeobox genes in the ctenophore *Mnemiopsis leidyi*. Dev. Genes Evol. 218: 307–319.

Pantin, C. Homology, analogy and chemical identity in the Cnidaria. pp.1–15. *In:* W.J. Rees. [ed.] 1966. The Cnidaria and their Evolution. Academic Press, New York.

Philippe, H. and H. Brinkmann, P. Martinez, M. Riutort, and J. Baguna. 2007. Acoel flatworms are not platyhelminthes: evidence from phylogenomics. PLoS ONE 2: e717.

Pickerill, R. 1982. Cambrian medusoids from the St John Group, southern New Brunswick. Canadian Geological Survey Paper 82-1B: 71-76.

Putnam, N.H. and M. Srivastava, U. Hellsten, B. Dirks, J. Chapman, A. Salamov, A. Terry, H. Shapiro, E. Lindquist, V.V. Kapitonov, J. Jurka, G. Genikhovich, I.V. Grigoriev, S.M. Lucas, R.E. Steele, J.R. Finnerty, U. Technau, M.Q. Martindale, and D.S. Rokhsar. 2007. Sea anemone genome reveals ancestral eumetazoan gene repertoire and genomic organization. Science 317: 86-94.

Rentzsch, F. and R. Anton, M. Saina, M. Hammerschmidt, T.W. Holstein, and U. Technau. 2006. Asymmetric expression of the BMP antagonists chordin and gremlin in the sea anemone *Nematostella vectensis*: implications for the evolution of axial patterning. Dev. Biol. 296: 375-387.

Rieger, R.M. and P. Ladurner, B. Hobmayer. 2005. A clue to the origin of the Bilateria? Science 307: 353c-355c.

Ruddle, F.H. and J.L. Bartels, K.L. Bentley, C. Kappen, M.T. Murtha, and J.W. Pendleton. 1994. Evolution of Hox genes. Annu. Rev. Genet. 28: 423-442.

Ruiz-Trillo, I. and J. Paps, M. Loukota, C. Ribera, U. Jondelius, J. Baguna, and M. Riutort. 2002. A phylogenetic analysis of myosin heavy chain type II sequences corroborates that Acoela and Nemertodermatida are basal bilaterians. Proc. Natl. Acad. Sci. USA 99: 11246-11251.

Russell-Hunter, W. 1979. A Life of Invertebrates. Macmillan, New York.

Ryan, J.F. and P.M. Burton, M.E. Mazza, G.K. Kwong, J.C. Mullikin, and J.R. Finnerty. 2006. The cnidarian-bilaterian ancestor possessed at least 56 homeoboxes: evidence from the starlet sea anemone, *Nematostella vectensis*. Genome Biol. 7: R64.

Ryan, J.F. and M.E. Mazza, K. Pang, D.Q. Matus, A.D. Baxevanis, M.Q. Martindale, and J.R. Finnerty. 2007. Pre-bilaterian origins of the Hox cluster and the Hox code: evidence from the sea anemone, *Nematostella vectensis*. PLoS ONE 2: e153.

Salvini-Plawen, L. and H. Splechtna. 1978. On the origin and evolution of the lower metazoa. Z. f. Zool. Systematik Evolutionsforschung 16: 40-88.

Schierwater, B. and K. Kuhn. 1998. Homology of Hox genes and the zootype concept in early metazoan evolution. Mol. Phylogenet. Evol. 9: 375-381.

Schierwater, B. 2005. My favorite animal, *Trichoplax adhaerens*. Bioessays 27: 1294-1302.

Schierwater, B. and K. Kamm, M. Srivastava, D. Rokhsar, R.D. Rosengarten, and S.L. Dellaporta. 2008. The Early ANTP Gene Repertoire: Insights from the Placozoan Genome. PLoS ONE 3: e2457.

Scott, M.P. 1992. Vertebrate homeobox gene nomenclature. Cell 71: 551-553.

Seimiya, M. and H. Ishiguro, K. Miura, Y. Watanabe, and Y. Kurosawa. 1994. Homeobox-containing genes in the most primitive metazoa, the sponges. Eur. J. Biochem. 221: 219-225.

Seipel, K. and V. Schmid. 2005. Evolution of striated muscle: jellyfish and the origin of triploblasty. Dev. Biol. 282: 14-26.

Sempere, L.F. and P. Martinez, C. Cole, J. Baguna, and K.J. Peterson. 2007. Phylogenetic distribution of microRNAs supports the basal position of acoel flatworms and the polyphyly of Platyhelminthes. Evol. Dev. 9: 409-415.

Seo, H.C. and R.B. Edvardsen, A.D. Maeland, M. Bjordal, M.F. Jensen, A. Hansen, M. Flaat, J. Weissenbach, H. Lehrach, P. Wincker, R. Reinhardt, and D. Chourrout. 2004. Hox cluster disintegration with persistent anteroposterior order of expression in *Oikopleura dioica*. Nature 431: 67-71.

Smith, K.M. and L. Gee, I.L. Blitz, and H.R. Bode. 1999. CnOtx, a member of the Otx gene family, has a role in cell movement in hydra. Dev. Biol. 212: 392-404.

Srivastava, M. and E. Begovic, J. Chapman, N. Putnam, U. Hellsten, T. Kawashima, A. Kuo, T. Mitros, M. Carpenter, A. Signorovic, M. Moreno, K. Kamm, H. Shapiro, I. Grigoriev, L. Buss, B. Schierwater, S. Dellaporta, and D. Rokhsar. 2008. The Trichoplax Genome and the Enigmatic Nature of Placozoans. Nature 454: 955-960.

Technau, U. and S. Rudd, P. Maxwell, P.M. Gordon, M. Saina, L.C. Grasso, D.C. Hayward, C.W. Sensen, R. Saint, T.W. Holstein, E.E. Ball, and D.J. Miller. 2005. Maintenance of ancestral complexity and non-metazoan genes in two basal cnidarians. Trends Genet. 21: 633–639.

Telford, M.J. and A.E. Lockyer, C. Cartwright-Finch, and D.T. Littlewood. 2003. Combined large and small subunit ribosomal RNA phylogenies support a basal position of the acoelomorph flatworms. Proc. Biol. Sci. 270: 1077–1083.

Wikramanayake, A.H. and M. Hong, P.N. Lee, K. Pang, C.A. Byrum, J.M. Bince, R. Xu, and M.Q. Martindale. 2003. An ancient role for nuclear beta-catenin in the evolution of axial polarity and germ layer segregation. Nature 426: 446–450.

Willmer, P. 1990. Invertebrate Relationships: Patterns in Animal Evolution. Cambridge University Press, Cambridge.

Woltering, J.M. and A.J. Durston. 2006. The zebrafish hoxDb cluster has been reduced to a single microRNA. Nature Genetics 38: 601–602.

Chapter 10

Key Transitions During Animal Phototransduction Evolution: Co-duplication as a Null Hypothesis for the Evolutionary Origins of Novel Traits

Todd H. Oakley and *David C. Plachetzki*

Introduction

> *"Novelties come from previously unseen association of old material.*
> *To create is to recombine".*
>
> — Jacob (1977)
>
> *"Gene duplication emerged as the major force of evolution".*
>
> — Ohno (1970)

Two different but interdependent processes likely contribute to the majority of major evolutionary transitions and each will leave different signatures that are distinguishable in comparative analyses. Some components of novel traits have their mutational origins in duplication, subsequently followed by differential divergence in both form and function. Gene duplications are perhaps the best studied duplication events, but "duplication" (either by copying or by fission) happens at other levels of biological organization, including protein domains, groups of interacting proteins, chromosomes,

Ecology Evolution and Marine Biology, University of California-Santa Barbara Santa Barbara. CA 93106; (805) 893-4715.
E-mail: oakley@lifesci.ucsb.edu

genomes, cells, organs, castes, species, and ecosystems. Other novel traits originate as new combinations of existing traits, a process sometimes termed bricolage, or tinkering (Jacob 1977). As with duplication, recombinational novelties occur at multiple levels of biological organization: domains fuse to form new proteins, proteins gain new interactions with other proteins, cells/species merge as in the endosymbiotic origin of eukaryotes, and new ecological interactions emerge by dispersal of species. Recognizing and distinguishing duplication, and co-option requires a historical or phylogenetic perspective.

This historical or "tree thinking" perspective (O'Hara 1997), which focuses on duplication and co-option, has not always been the dominant mode of thinking about the origin of novelties. Many evolutionists have instead suggested linear histories for the origin of complex traits, where trait evolution is viewed as a linear and gradual progression, often from simple to complex phenotype, and driven by natural selection. For example, Salvini-Plawen and Mayr (1977) constructed linear histories of eye evolution, which they termed "morphological sequences of differentiation" by collecting examples of eyes of differing complexity from closely related species (Fig. 1a). Additionally, Nilsson and Pelger (1994) constructed a linear conceptual model for the gradual origin of lens eyes (Fig. 1b). These linear histories are important for understanding how and how quickly natural selection might incrementally mold complexity; yet at the same time linear thinking discourages questions about the origins of novelty. Under linear models, all that is needed to proceed in a gradual and linear fashion from simple to complex traits is nondescript heritable variation and selection. As such, heritable variation—which is abundant in nature—was taken for granted and the focus was placed almost exclusively on selection. When considering linear, gradual models, little attention was therefore paid to how variations originate or to whether certain types of variation (e.g. duplication or co-option) are more common fuel for evolution. In contrast, the possibility of identifying the different types of variation that have been involved in the macroevolutionary origins of novelties (key transitions) is precisely what a historical, tree thinking perspective can add.

Duplication/divergence and co-option in the origin of novelties

Certainly, many novelties originate by duplication and subsequent differential divergence. In fact, duplication and divergence has become widely accepted as the primary means of origin of new genes (Taylor and Raes 2004; Zhang 2003). Here, the mutational mechanisms causing duplication are well understood, and include retrotransposition, segmental and genome duplications (Hurles 2004). But the original mutational events

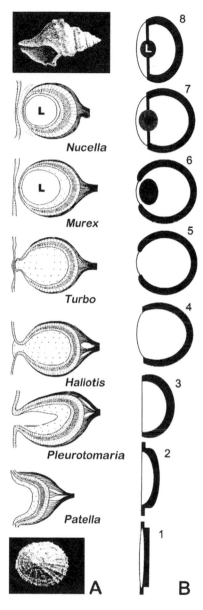

Fig. 1. Linear models of eye evolution. **(A)** (Salvini-Plawen and Mayr 1977) provided several "morphological series of differentiation", such as the gastropod eyes depicted here. They viewed eye evolution as a gradual and linear transformation from simple to complex eyes, as illustrated by these living gastropod eyes. **(B)** (Nilsson and Pelger 1994) constructed a conceptual model of the evolution of lens eyes that allowed them to estimate the time necessary to evolve such an eye. This model assumes a gradual and linear progression from simple to complex eye. Figure taken from (Kutschera and Niklas 2004) with permission.

are only part of the story, the other part being the subsequent differential divergence of the newly arisen gene copies. Divergence often occurs in both form and function of genes, and involves natural selection on one or both copies or neutral processes that fix complementary loss of gene functions (Zhang 2003). Besides genes, duplication may occur also at other levels of biological organization. For example, co-expressed gene networks can be duplicated (Conant and Wolfe 2006). Here, the mechanism of duplication may often be whole genome duplications, unless all genes in the network are copied independently or co-expressed genes are genomically adjacent allowing segmental duplication of the network. In yeast, differential divergence of duplicated co-expression networks has occurred by differential loss of network interactions (Conant and Wolfe 2006). Duplication also occurs in tissues and organs. As a second example, serial homologs like vertebrae, teeth, segments, limbs, and eyes and other nervous system traits may be considered duplicates (Arthur et al. 1999; Minelli 2000; Oakley 2003; Oakley and Rivera 2008). The mechanistic origins of duplicated organs are not as well understood as duplications that occur at the genetic level. One likely mechanism is fission of a formerly homogeneous field of cells into multiple fields, a developmental mechanism involved in the duplication of arthropod eyes (Friedrich 2006; Rivera and Oakley 2008).

In addition to duplication plus divergence, evolutionary novelties often originate as new combinations of existing genes, structures, or species. For example, many new genes originate as novel combinations of existing protein domains. As one of many examples, the nervous system specific transcription factor gene *Pax-6* originated as a fusion of a *Paired* domain and a *Homeodomain* (Catmull et al. 1998). The mechanism of domain fusion likely involves the duplication of those domains, thus illustrating one way that duplication and co-option are inter-related, especially when considering together multiple levels of biological organization. New combinations of existing elements also occur at other levels of biological organization. Gene interaction networks gain new interactions with existing genes (Olson 2006). In addition, whole gene networks are often co-opted to function in new contexts (e.g. Gompel et al. 2005). Mechanistically, this often occurs by changes in gene regulation. In laboratory experiments, changes in regulation of transcription factor genes that lead to new sites of expression can cause ectopic production of organs like eyes (Halder et al. 1995), a process that has been termed heterotopy (Haeckel 1866; West-Eberhard 2003). Interestingly, a co-option event at the level of the developmental regulatory network that leads to ectopic organ expression can be considered a mechanism for the duplication of organs, illustrating another relationship between co-option and duplication. Namely, co-option at one level may be viewed as duplication at another level of biological organization. Finally, co-option is a process that may act at the species level. Fusion events of existing species may have led to novel species, for example by endosymbiosis (but see Kutschera and Niklas 2005; Margulis and Sagan 2002; Thompson 1987).

Although analogous processes of duplication and co-option generalize across numerous levels of biological organization, we focus our attention here on protein interaction networks. In protein networks —like other biological levels—duplication-divergence and co-option lead to different patterns that can be distinguished by phylogenetic analyses (Abouheif 1999; Geeta 2003; Serb and Oakley 2005). Despite the focus here on networks, the general approach can be easily translated to other levels of biological organization. Before examining the evolutionary origins of specific phototransduction networks below, we will outline the phylogenetic patterns in protein networks expected to result from origins of novelty by duplication and co-option.

Phylogenetic patterns following duplication and co-option

If two protein networks originated by duplication of an ancestral network, then multiple components of the descendent networks will show patterns of co-duplication. In other words, simultaneous duplication of multiple genes of an ancestral network is a pattern consistent with the origin of the descendent networks by duplication (Fig. 2a). In contrast, the components of a novel protein network that evolved by co-option will have originated by gene duplication events that occurred at different times (Fig. 2b). Co-duplication and co-option need not be considered discrete alternatives. Instead, some genes of a network may co-duplicate, whereas others may be co-opted, an intermediate situation that will often be true, especially when examining the evolution of particular networks at multiple time scales, and when examining networks at increasing spatial scales by increasing the number of interactions considered.

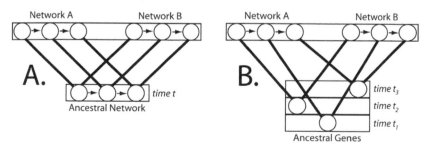

Fig. 2. Co-duplication and co-option can be understood in terms of the phylogenetic interval within which the interacting elements originate. **(A)** Co-duplication and divergence is characterized by a shared temporal interval of element originations. **(B)** Co-option occurs when interactions between elements are de-coupled from their temporal origins.

Differentiating co-duplication from co-option requires determining the relative timing of duplication events for the genes involved in the network. Co-duplication may be viewed as a null hypothesis, which can be rejected by different hypothesis-testing procedures, in favor of the alternative hypothesis of co-option. A primary means of testing gene co-duplication is by reconciled tree analysis (RTA), the comparison of gene phylogenies to a species phylogeny (Akerborg et al. 2009; Goodman et al. 1979; Page and Charleston 1998). RTA determines the phylogenetic interval of gene duplication events, with upper bounds at one speciation event and lower bounds at another speciation event. For example, if two sister species possess duplicate genes and an outgroup species possesses only one such gene, we might infer that a gene duplication event occurred prior to the origin of the sister species, but after their divergence with the outgroup. Algorithmic approaches are available for more complicated situations (Akerborg et al. 2009; Page and Charleston 1997; Page and Charleston 1998). If duplication events giving rise to the origins of genes in a protein network map to different nodes on a species phylogeny, the null hypothesis of co-duplication is rejected in favor of network assembly by recombination of protein interactions (co-option). Another means of testing a co-duplication hypothesis is by explicitly estimating divergence times of the nodes leading to the network genes, for example by relaxed molecular clock analysis (Britton et al. 2002; Sanderson 2002; Thorne and Kishino 2002) to compare origination times of gene duplication events in a gene tree. Such an analysis may offer an additional test after RTA, especially in situations where RTA maps co-duplication events to the same large phylogenetic interval due to paucity of sampling or due to extinction of intermediate species. Conceptually analogous statistical tests for co-speciation in host and parasite phylogenies also have been described (Huelsenbeck et al. 2000).

Once the evolutionary patterns of co-duplication and co-option have been established for a protein network, additional questions can be addressed. To what extent do co-duplicated genes respond to natural selection as a unit of interacting proteins versus as individual genes? Are rates of evolution in co-duplicated genes concordant? Does one descendent network evolve at a different rate than the other? Are duplicated networks sub-functionalized (Conant and Wolfe 2006)? Can we determine the cis-regulatory changes responsible for co-option events? Do rates of evolution change after co-option? Does selection act to allow co-opted genes to specialize to their new roles? Next, we propose that metazoan phototransduction will provide a valuable model for investigating co-duplication, co-option, and the evolutionary outcomes of these novelty-generating mechanisms.

Tree Thinking and the Evolution of Animal Phototransduction Cascades

Multiple different signaling pathways enable much of life to respond to and interact with the light environment (Bjorn 2008; Spudich et al. 2000). The dominant signaling pathways found in animals for detecting light are phototransduction cascades initiated by opsins, members of the G protein-coupled receptor (GPCR) class of proteins. Variations of opsin pathways mediate a wide range of acuity for sensing light, from simple photosensitivity to vision (Land and Nilsson 2002). In this section we explore the origin, evolution and diversification of animal phototransduction cascades with the purpose of highlighting instances where duplication and co-option may have contributed to key transitions in the evolution of these pathways. Namely, we will focus on two examples. First, we review evidence that the different transduction pathways of vertebrate rod and cones cells originated primarily by duplication: multiple phototransduction proteins differentially expressed in rods and cones duplicated early in the history of vertebrates (Hisatomi and Tokunaga 2002; Nordström et al. 2004; Serb and Oakley 2005). Second, we suggest that the evolutionary origins of two primary transduction pathways of animals (named ciliary and rhabdomeric) involved extensive co-option.

Duplication and the origin of vertebrate rod and cone phototransduction cascades

The dual presence of rods and cones allow vertebrate retinas to overcome an inherent trade-off between acuity and sensitivity. Whereas rods are specialized for sensitivity in dim light (scotopic vision), cones are specialized for acuity in bright light (photopic vision). This duplicity theory (Schultze 1866) was first supported by the observation of two morphological classes of vertebrate photoreceptor cells and is now universally accepted, supported with additional physiological and molecular data. How did the novelty of dual retinas originate during evolution? In this section, we will first discuss a hypothesis framed in linear thinking. Next, we will take a tree thinking perspective and find that we are largely unable to reject a null hypothesis of co-duplication in the origin of rod and cone phototransduction pathways. Establishing duplication of pathways requires two things. First, the elements of the pathways must differ. Second, the origin of the elements of the pathway must be coincident in time. We consider this temporal coincidence to be a null hypothesis that could be rejected by various means, and whose acceptance is subject to considerations about the power of the tests employed.

One perspective on the origin of dual retinas is that a subset of one cell type gradually transformed into another cell type. In a timeless classic, *The Vertebrate Eye and its Adaptive Radiation* (Walls 1942), Gordon Walls invoked one such linear model of transformation from primitive to derived stating "They [rods] were derived quite simply from cones by the enlargement of the outer segment and by an increase in the number of visual cells connected to each nerve cell" (Walls 1942). That vertebrate cones are more ancient than rods is in fact supported by current knowledge that rod opsins are derived within paraphyletic cone opsin clades (Okano et al. 1992). However, Walls' astute ascertainment of ancestry, entrenched in the linear mode of thinking, tells us little about the types of evolutionary processes or the types of variation that could have given rise to the novelty of vertebrate rod cells. In fact, if rods and cones evolved by some means other than gradual transformation, the linear framework would obscure this phenomenon.

Another perspective is that rods and cones represent evolutionary duplicates or paralogs: Just as genes duplicate within lineages, so too may cell types duplicate within lineages. Such hypotheses of duplication are supported by the similarity of components. The similarity of amino acid components of proteins supports duplication of the proteins. In an analogous way, the similarity of expressed protein components of cell types may support duplication of the cell types (Arendt 2008). Although the entire repertoire of expressed proteins in rods and cones is not yet known, we can begin to address a duplication hypothesis for the cell types by investigating the phototransduction pathways of the cell types. Rod and cone cells use similar but different phototransduction pathways, which may have originated by duplication.

More specifically, rods express one set of parologous cell-type-specific phototransduction proteins and cone cells express another (Hisatomi and Tokunaga 2002; Nordström et al. 2004). Rod- and cone-specific paralogues include those of opsin, the G protein α subunit (Gα), phosphodiesterase (PDE), cyclic-nucleotide gated ion channels (CNGs) and arrestin (Nordström et al. 2004) (Fig. 3). From a tree thinking perspective, we can ask if these data indicate duplication or co-option in the origin of rod and cone phototransduction. If the rod phototransduction pathway is a duplicate of the cone pathway, then most components of those pathways should have duplicated simultaneously. In contrast, if one pathway originated largely by co-option, for example by utilizing other existing proteins that gained new expression patterns, then the components should have originated by duplication at different times. The co-duplication of phototransduction components can be treated as a null hypothesis that could be rejected in favor of co-option.

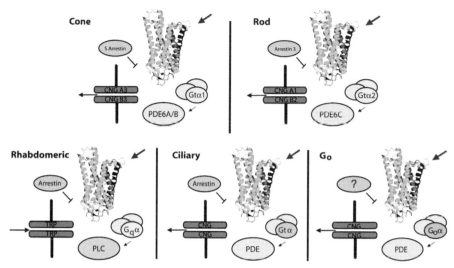

Fig. 3. Simplified phototransduction cascades of the rod and cone photoreceptors of the vertebrate retina and the more general rhabdomeric, ciliary and Go signaling pathways found in animals. The rod and cone phototransduction cascades are members of the ciliary class. The phototransduction cascade is initiated by light (arrow) ultimately leading to the efflux (left-pointing arrow) or influx (right-pointing arrow) of K+ and Na+ ions. Rod-specific and cone-specific paralogs of Gtα, PDE, CNG, and arrestin are known to have arisen by segmental genome duplications (Nordström et al. 2004). PDE, phosphodiesterase; Gα, G α-subunit of the G protein; CNG, cyclic-nucleotide-gated ion channel; TRP, transient-receptor-potential ion channel.

For rod and cone phototransduction pathways, the null hypothesis of co-duplication currently cannot be rejected, but accepting a null hypothesis is always subject to the power that the available tests have to reject. Paralogs for each rod and cone specific gene originated through large-scale segmental duplication of the genome (Nordström et al. 2004). RTA indicates that each of these duplications pre-dates the origin of gnathostomes (jawed vertebrates) and post-dates the split of cephalochordates/urochordates and vertebrates (Nordström et al. 2004, Fig. 4a). As such, the duplication of multiple genes of the pathways maps to the same phylogenetic interval, consistent with co-duplication. However, this phylogenetic interval is large and encompasses branches leading to two extant groups—agnathans [assumed monophyletic (Takezaki et al. 2003), but see Lamb et al. 2007] and condrichthyians—that can be used to further refine the timing of the gene duplications. If agnathans and condrichthyians both possess rod and cone paralogs of multiple phototransduction genes, then the phylogenetic interval—thus inferred to be pre-vertebrate—for these duplications would be reduced, providing a more powerful test of the co-duplication null hypothesis.

By potentially narrowing the phylogenetic interval of inferred gene duplication events, agnathans represent a particularly valuable group to more powerfully test hypotheses that rod and cone phototransduction genes co-duplicated. As of 2009, an assembled genome at 5.9x coverage is available for the lamprey *Petromyzon marinus*. This genomic data can be used to determine the timing of gene duplications, and *Petromyzon* is further amenable to examination of gene expression (Muradov et al. 2008) in different photoreceptor cell types called long and short photoreceptors. Do long and short receptors of the *Petromyzon* retina possess paralagous phototransduction pathways, homologous to cone and rod receptors? If so, the phylogenetic interval for co-duplication would be narrowed substantially. Testing this hypothesis requires examining individual phototransduction genes.

Even before the *Petromyzon* genome project, evidence surfaced that differentiated rod and cone phototransduction cascades were present at the origin of vertebrates. Namely, multiple opsin genes are known from agnathans. The most in-depth phylogenetic treatment of the problem of basal vertebrate opsin evolution concluded that rod—and cone-class opsins are present in agnathans, and that rod-opsins branch from within paraphyletic cone opsin clades (Collin 2006; Pisani et al. 2006). From this, and assuming agnathan monophyly, we can conclude that the duplication of the gene that gave rise to the rod-specific opsin class took place prior to the last common ancestor of living vertebrates.

More recently, additional phototransduction genes have been studied in *Petromyzon* and both are consistent with duplication before the origin of vertebrates, but since the divergence of tunicates and vertebrates. An example is $G\alpha_t$ (transducin). Muradov et al. (2008) found two G_t paralogs in *Petromyzon*, and found using immunofluorescent antibody probes that one gene is expressed in short photoreceptors and the other is expressed in long photoreceptors. Phylogenetic analysis suggested that the short-expressed G_t from lamprey is orthologous to rod G_t proteins of gnathostomes. The phylogenetic position of the long-expressed G_t is somewhat equivocal based on amino acid similarity alone, but minimizing implied gene duplication events among similarly likely gene trees might place the G_t-long expressed gene as an ortholog of gnathostome cones. These considerations place the duplication of rod and cone-specific transducins before the origin of vertebrates, and suggest homology of lamprey-short and gnathostome-rod photoreceptors, and homology of lamprey-long with gnathostome-cone receptors. Similar to transducin, the inhibitory (gamma) subunit of phosphodiesterase 6 (PDE6) supports the hypothesis of co-duplication. This gene is duplicated in *Petromyzon*, with one paralog expressed in short and the other in long photoreceptor cells (Muradov et al. 2007).

Despite the concordance of duplicative patterns in opsin, transducin and PDE6-gamma, one *Petromyzon* gene is not duplicated, even though it is duplicated in gnathostomes. The catalytic subunit of PDE6 has rod and cone-specific paralogs in gnathostomes, but Muradov et al (Muradov et al. 2007) found only a single copy in lamprey. Their phylogenetic analysis using neighbor-joining found the lamprey gene in a clade with cone-specific PDE6 suggesting a possible loss or failure to detect a rod-paralog in lamprey. However, our Maximum Likelihood re-analysis (not shown) gives high support for the lamprey gene as the sister to all rod and cone-specific genes. These results reject co-duplication for the PDE6 catalytic sub-unit prior to the origin of vertebrates. It will be interesting to learn whether PDE6 was duplicated before or after the divergence of chondrichthyes and other gnathostomes.

Figure 5 summarizes current knowledge about the timing of origin of rod-specific gene expression. Some components have not yet been studied in agnathans like *Petromyzon*, but the assembled genome sequence makes this a feasible project. As such, the origins of opsin kinase, arrestin, GCAP, GC, G-protein beta and gamma subunits, and CNG alpha and beta, have been narrowed only to a broad phylogentic interval. Recoverin and RGS9 duplicated in the same large interval, but these genes lack specific expression in rods. RGS9 is expressed in both rods and cones, and its sister gene RGS11 is expressed in retinal neurons. In addition, one known exception to co-duplication is that catalytic PDE6 duplicated within the vertebrate lineage, Petromyzon has only one copy (Muradov et al. 2007).

Nevertheless, based on current knowledge, it is clear that multiple components of rod phototransduction duplicated in the same phylogenetic interval, suggesting that the pathways themselves can largely be considered duplicates. In many cases, the duplicated genes are known to part of paralagons, chromosomal regions with similar gene order that were most likely duplicated early in vertebrate history (Larhammar et al. 2009). This indicates that the mutational mechanism for the origin of the dual retina largely involved whole-genome or block duplications (Nordström et al. 2004).

The realization that rod and cone phototransduction pathways largely represent evolutionary duplicates suggests at least two areas for future work. First, the mode of differentiation of the duplicate pathways may be examined. Was natural selection involved in the differentiation? To what extent does selection act on the entire biochemical pathway as compared to the individual genes? At least one interesting study has been conducted that bears on these questions. (Carleton et al. 2005) examined opsin genes and found that there are no amino acids that uniquely define the rod opsin clade. This result indicates that the origin of the rod pathway probably was not accompanied by a change in the biochemical function of opsin.

As such, if selection acted to differentiate all genes in the rod pathway including opsin, it may have involved changes in rod opsin expression and not its biochemical interaction with other proteins. Another possibility is that selection acted on other components of the pathway, but was not involved in the fixation of the duplicated rod opsin.

A second area for future work and understanding is to extend our interpretation to the next level of biological organization above the pathway, to the entire cell type. How have the dual processes of co-duplication plus divergence and co-option contributed to the evolution of rod and cone cells? In order to answer this question we must be able to link the cell (morphological) phenotype with the physiological phenotype in an evolutionary explanation (Fig. 4a). In the case of rods and cones, we can begin to make some preliminary connections. It has been shown that the transcription factor Nr2e3 serves as a switch for the terminal differentiation and maintenance of rod cells in the vertebrate retina (Chen et al. 2005; Peng et al. 2005). Comparisons between wild-type and Nr2e3 mutants have revealed

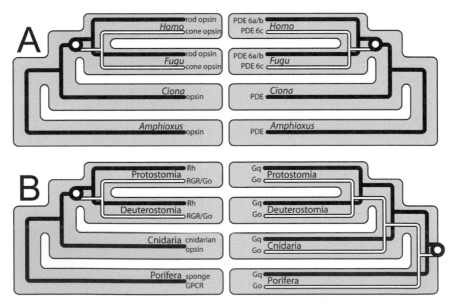

Fig. 4. The co-duplication and co-option modes of variation can be understood using tree-thinking. **(A)** The evolution of vertebrate rod and cone cells may be an example of co-duplication and divergence. Rod (black) and cone (white) opsins duplicated within the same phylogenetic interval as their PDE-binding partners (PDE6C, white; PDE6AB, black) (Nordström et al. 2004). **(B)** The origin of the animal phototransduction cascade itself involved co-option. The origin of opsin must have occurred after G-protein α-subunits from existing GPCR pathways had diversified (Suga et al. 1999). The G-protein α subunit involved in cnidarian opsin signaling remains unknown. PDE, phosphodiesterase; Rh, rhabdomeric opsin; RGR/Go, RGR/Go, opsin; GPCR, G-protein-coupled receptor.

that this transcription factor up-regulates rod-specific phototransduction genes in rod transcriptional contexts and at the same time down-regulates cone-specific genes in cone-specific contexts. In the course of development, vertebrate rod and cone cells—together with other retinal cell types—are derived from a pool of multi-potent progenitor cells (Turner and Cepko 1987; Wetts and Fraser 1988). At some point prior to the evolution of rod and cone cells Nr2e3 must have been co-opted to serve a role in either the inhibition or activation of each phototransduction cascade.

Co-option and the major animal phototransduction networks

Animals possess two major classes of photoreceptor cell types, ciliary and rhabdomeric. Like rods and cones (which are both ciliary photoreceptor

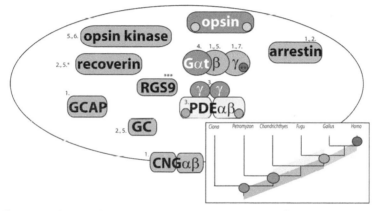

Fig. 5. The timing of origin of components of rod phototransduction. Phototransduction genes are illustrated as symbols that are colored according to when a gene duplication event led to paralogs expressed separately in rods and cones. Red are mammal-specific gene duplications, orange are tetrapod-specific, yellow duplicated between the common ancestor of *Petromyzon*+mammals and *Fugu*+mammals, green are vertebrate-specific (assuming monophyly of hagfish+*Petromyzon*), grey are genes duplicated between the ancestor of *Ciona*+mammals and *Fugu*+mammals. Genes colored in grey are targets for study in *Petromyzon*, genes colored in yellow are targets for study in cartilaginous fishes. References for gene trees are indicated with numbers next to each gene family's symbol: 1. (Nordström et al. 2004); 2. (Hisatomi and Tokunaga 2002); 3. (Muradov et al. 2007); 4. (Muradov et al. 2008); 5. (Larhammar et al. 2009); 6. (Yasutaka et al. 2006); 7. (Chen et al. 2007). Small orange circles in PDE indicate timing of origin of alpha and beta PDE, both expressed within rods. Small pink circles in opsin indicates that multiple recent gene duplications have occurred within vertebrates, usually having specialized for different wavelength sensitivity in cones. *Recoverin is not known to have rod and cone-specific expression. **Mammal-specific duplication present and expressed in various tissues. *** RGS9 is expressed in both rods and cones, but at different levels. The gene is unique to vertebrates. Its sister gene family is RGS11, which is expressed in retinal neurons. Our unpublished phylogenetic analyses indicates that RGS9 and RGS11 are sister paralog, and that the purple urchin has a single copy that is sister to the RGRS9/11 clade. Agnathans or condrycthyes have not been investigated. Therefore, RGS9 and RGS11 duplicated within deuterostomes, before the common ancestor of teleosts and humans.

Color image of this figure appears in the color plate section at the end of the book.

cells), the two major animal classes were first distinguished by morphological and physiological differences (Eakin 1963). The presence of these major classes has been further supported by molecular data (Arendt 2003). In this section, we first briefly discuss a linear perspective on the origins of the major photoreceptor cell types. We next take a tree thinking perspective on their origins, and address the null hypothesis of duplication. As we will discuss below, unlike rod and cone origins, co-duplication can be clearly rejected in favor of co-option in the origins of ciliary and rhabdomeric cell types.

One perspective on the origins of ciliary and rhabdomeric cells is a gradual linear model. Salvini-Plawen and Mayr (1977) in a landmark paper, include a steady infusion of linear explanations throughout their analysis. Based on morphological data available at the time (Eakin 1979), these authors provided a scheme of linear modifications that are possible in ciliary and rhabdomeric photoreceptor cell types in the course of evolution. Their scheme for ciliary photoreceptors portrays 14 different morphologies from disparate metazoan phyla, each derived by transitioning from a common prototypical ancestor. Their explanation for diversity among ciliary photoreceptors is displayed as a carousel of varying photoreceptor morphologies revolving around a central ancestral state, each separated by a single linear transition (Fig 9, Salvini-Plawen and Mayr 1977).

Another perspective on animal photoreceptor origins is a tree thinking perspective. Instead of treating animal photoreceptor cell type as a single morphological unit and providing descriptive accounts of possible modes of transitions between types, we can examine separately the individual components of the cells, and quantify the timing of their origins. Although the entire suite of expressed proteins in photocells is unknown, especially in non-model organisms, we can begin by investigating the phototransduction pathways, which tend to correlate with different morphological cell types. A null hypothesis of duplication for the origin of ciliary and rhabdomeric cells has been described previously (Plachetzki et al. 2005). As we will next describe, ciliary and rhabdomeric cell origins reject this null hypothesis of duplication in favor of co-option, providing a counter-example to duplicated rod and cone phototransduction.

Many similarities exist in the signaling networks of phototransduction in all animals studied to date. In bilaterian animals, three primary phototransduction cascades are often distinguished—two of these cascades tend to correlate with ciliary and rhabdomeric morphological cell classes (Arendt 2003). All three pathways are based on canonical GPCR signaling, yet they differ in important ways, including the sub-class of the opsin protein that initiates each cascade. In once class, photoreceptors with ciliary morphology (such as the rod and cone cells in the example above) generally utilize ciliary opsins, $G_t\alpha$ and PDE in cell signaling. Ciliary phototransduction leads to the closing of CNG ion channel proteins, the reduction of cation concentration and the hyperpolarization of the cell. Second, the rhabdomeric phototransduction cascade (such as that present in insect ommatidia) is also initiated by a class-specific opsin paralog, but it utilizes $G_q\alpha$ and phospholipase C (PLC) in cell signaling. Upon activation of the rhabdomeric phototransduction cascade, transient receptor potential (TRP) ion channels open allowing the influx of cations and the depolarization of the cell.

An additional class of phototransduction in bilaterian animals has been proposed based on the presence of a third clade of opsin proteins known from mollusks, cephalochordates and vertebrates that utilize $G_o\alpha$ in cell signaling. This phototransduction pathway has been studied in the dual retina mantle eye of the scallop *Patinopecten* (Gomez and Nasi 2000; Gomez and Nasi 1995; Kojima et al. 1997), the parietal eye of the reptile *Uta* (Su et al. 2006) and from the cephalochordate *Branchiostoma* (Koyanagi et al. 2002). These few investigations have indicated many unanswered questions. For instance, and similar to the ciliary pathway, the G_o mediated pathway in both the vertebrate parietal eye and the scallop retina manifest their physiological signal through CNG ion channels. However, unlike existing data from ciliary and rhabdomeric phototransduction cascades, it would appear that G_o phototransduction displays some evolutionary plasticity in the physiological outcome it specifies. As a first example, in the vertebrate parietal eye, G_o signaling leads to a depolarization of the cell but in the scallop this pathway causes hyperpolarization (Su et al. 2006). As another example of G_o plasticity during evolution, some members of the G_o class of opsins appear to have lost their role in phototransduction altogether and instead play important enzymatic functions as photoisomerases (Chen et al. 2001; Shen et al. 1994), enzymes that regenerate the light reactive chromophore that binds with opsin.

If ciliary, rhabdomeric, and G_o phototransduction cascades originated by duplication, we would expect to find a simultaneous origin by gene duplication of the four major components: opsin, G-protein, PLC/PDE, and ion channel. Testing this co-duplication null hypothesis requires dating the origin of each component separately.

To begin testing the co-duplication hypothesis, we have recently estimated dates for the origin of opsin clades using RTA (Plachetzki et al. 2007). Early opsin history is challenging to reconstruct, and other subsequent authors have obtained different results using different data sets, including cnidarian opsins from *Tripedalia* (Kozmik et al. 2008), *Cladonema* (Suga et al. 2008) and *Carybdea* (Koyanagi et al. 2008). Our analyses indicate that opsin phylogeny is strongly influenced and often destabilized by the inclusion of non-opsin outgroups (Plachetzki et al. 2007).

Future studies of opsin history will benefit from explicit incorporation of species tree information when estimating gene tree (Akerborg et al. 2009) and perhaps from sequence evolution models specific to opsins or GPCR's and intron patterns.

Opsins of the three major clades (ciliary, rhabdomeric, G_o) were already known to predate bilaterians (Arendt 2003). However, no analyses had included opsins from early branching animals, thus precluding a specific upper bound for opsin origins. Our study reported opsins discovered by screening trace genome sequence data from the cnidarians *Hydra magnipapillata* and *Nematostella vectensis* and the poriferan *Amphimedon* (Plachetzki et al. 2007), allowing us to date the origins of the three opsin clades. This study had three major results. First, we identified a new class of opsin genes present so far only in cnidarians that we call cnidops. Public databases also contain a ctenophore opsin gene that is a member of this clade (unpublished observation). Second, we uncovered ciliary opsins from the cnidarian *Nematostella*. Third, although our search did uncover opsin-like GPCR's, we were unable to find any opsins in the genome of the sponge *Amphimedon*. Based on the best supported phylogeny, and using RTA allowed us to conclude that the ciliary class of opsin genes has an ancient origin in the Eumetazoan ancestor of cnidarians + bilaterian animals, while the rhabdomeric and G_o sub-families are bilaterian innovations with a single sister group of opsins in cnidarians. Using this timing for the origins of major opsin clades, we can test for co-duplication of opsin and the other components of the signaling cascade.

If G-proteins co-duplicated with opsins at the origin of the three animal phototransduction cascades, they should map by RTA to the same phylogenetic interval as the opsins. As we shall discuss, co-duplication of G-proteins and opsins can be rejected. Assuming co-duplication, the opsin results above allow a specific prediction for the timing of origin of the other transduction components. Under co-duplication, $G_o\alpha$ and $G_q\alpha$ (the G_o and rhabdomeric G-proteins) would be the result of a bilaterian-specific duplication and $G_t\alpha$ (the ciliary G-protein) would date to a duplication preceding the Eumetazoan ancestor. In contrast to this prediction, the individual G-protein classes pre-date animals, thus rejecting the co-duplication hypothesis. A previous survey (Suga et al. 1999) identified the full complement of $G\alpha$ paralogues found in animal phototransduction from the demosponge *Ephydatia fluviatilis*, indicating pre-animal origins. Therefore, existing $G\alpha$ proteins gained new interactions with opsin proteins during animal evolution, and furthermore, co-option, not co-duplication was responsible for the origins of the interactions between opsins and G-proteins in the first ciliary, rhabdomeric and G_o pathways. Bolstering the hypothesis of co-option even farther is evidence that the cnidarian *Tripedalia* uses $G_s\alpha$ in rhopalial phototransduction (Koyanagi et al. 2008).

This indicates that opsins signal to at least four ancient G-protein families $(G_t G_s G_q G_o)$, requiring multiple shifts in protein-protein interactions during animal evolution.

A similar approach to G-proteins can be taken to test the hypothesis of co-duplication between opsins with PLC/PDE and opsins with ion channels. Here, co-option at some point in phototransduction evolution is evident immediately as the ciliary cascade uses PDE and the rhabomeric cascade uses PLC, two non-homologous genes with ancient histories. Co-option is also evident in the origin of rhabdomeric cascade, the only pathway known to use TRP ion channels. However, despite the differences in how these signaling networks manifest their physiological effects, CNG ion channels are known to play a role in both ciliary and G_o phototransduction networks. In addition, we have found, based on gene expression and behavioral/pharmacological experiments, that CNG is used for phototransduction in the cnidarian *Hydra*, which only has cnidops opsins (Plachetzki et al. 2010). CNG thus represents a shared, and perhaps conserved, component of cnidops, ciliary and G_o pathways, networks that otherwise changed through opsin duplication events.

Summary

Tree thinking provides a general framework for understanding the types of variation present at the origins of novel traits. Understanding these elements is important if we are to construct more complete evolutionary narratives and a more expansive model for the evolutionary process. In some ways, the linear approach has been an important heuristic for understanding the directionality and rate of evolution, but the linear mode falls short of illuminating processes in evolution that have given rise to novel traits. This is precisely where tree thinking approaches excel. We have used the evolution of phototransduction cascades as a model protein interaction network to explore the utility of tree thinking in explaining some of the key transitions in the evolution animal photoreception. This approach is also useful for analyzing traits from other levels of biological organization above the molecular level (Oakley and Rivera 2008). Not only can the evolution of elements at a single level be better understood using tree thinking, but interactions between levels of organization can also be addressed. One important outcome of expanding our understanding of the dual novelty-generating processes of duplication plus divergence and co-option to other levels of biological organization will be to gain insight into how varying levels interact in evolutionary processes (Buss 1987). For instance, can gene duplications provide a selective context for cell type duplications to occur? Can co-option by a given cell type of an alternative signaling or developmental pathway reinforce the evolution of novel tissues, such as vertebrate or mollusk retinas? Much of current evolutionary biology seeks to address the "novelty problem". We argue that the tree thinking framework can better facilitate our understanding of both the patterns and the processes of evolution.

References

Abouheif, E. 1999. Establishing homology criteria for regulatory gene networks: prospects and challenges. Novartis Found Symp. 222: 207–221.

Akerborg, O. and B. Sennblad, L. Arvestad, and J. Lagergren. 2009. Simultaneous Bayesian gene tree reconstruction and reconciliation analysis. Proc. Natl. Acad. Sci. USA 106: 5714–5719.

Arendt, D. 2003. Evolution of eyes and photoreceptor cell types. Intl. J. Develop. Biol. 47: 563–571.

Arendt, D. 2008. The evolution of cell types in animals: emerging principles from molecular studies. Nat. Rev. Genet. (In Press).

Arthur, W. and T. Jowett, and A. Panchen. 1999. Segments, limbs, homology, and co-option. Evol. Dev. 1: 74–76.

Bjorn, L.-O. 2008. Photobiology: The Science of Life and Light. Springer, New York.

Britton, T. and B. Oxelman, A. Vinnersten, and K. Bremer. 2002. Phylogenetic dating with confidence intervals using mean path lengths. Mol. Phylogenet. Evol. 24: 58–65.

Buss, L.W. 1987. The Evolution of Individuality. Princeton University Press, Princeton, NJ.

Carleton, K.L. and T.C. Spady, and R.H. Cote. 2005. Rod and cone opsin families differ in spectral tuning domains but not signal transducing domains as judged by saturated evolutionary trace analysis. J. Mol. Evol. 61: 75–89.

Catmull, J. and D.C. Hayward, N.E. McIntyre, J.S. Reece-Hoyes, R. Mastro, P. Callaerts, E.E. Ball, and D.J. Miller. 1998. Pax-6 origins—implications from the structure of two coral Pax genes. Develop. Genes and Evol. 208: 352–356.

Chen, P. and W. Hao, L. Rife, X.P. Wang, D. Shen, J. Chen, T. Ogden, G.B. Van Boemel, L. Wu, M. Yang, and H.K. Fong. 2001. A photic visual cycle of rhodopsin regeneration is dependent on Rgr. Nat. Genet. 28: 256–260.

Chen, J. and A. Rattner, and J. Nathans. 2005. The rod photoreceptor-specific nuclear receptor Nr2e3 represses transcription of multiple cone-specific genes. J. Neurosci. 25: 118–229.

Chen, H. and T. Leung, K.E. Giger, A.M. Stauffer, J.E. Humbert, S. Sinha, E.J. Horstick, C.A. Hansen, and J.D. Robishaw. 2007. Expression of the G protein [gamma]T1 subunit during zebrafish development. Gene Expr. Patterns 7: 574–583.

Collin, S.T.A. 2006. Response to Pisani et al. Current Biology (2006) 16::R320–R320. Curr. Biol.
16: R320–R320.

Conant, G.C. and K. H. Wolfe. 2006. Functional partitioning of yeast co-expression networks after genome duplication. Plos Biol 4: 545–554.

Eakin, R.M. Lines of evolution of photoreceptors. pp. 393–423. In: T.A. Mazia. [ed.] 1963. General Physiology of Cell Specification. McGraw-Hill, New York.

Eakin, R. 1979. Evolutionary significance of photoreceptors in retrospect. Am. Zool. 19: 647–653.

Friedrich, M. 2006. Continuity versus split and reconstitution: exploring the molecular developmental corollaries of insect eye primordium evolution. Dev. Biol. 299: 310–329.

Geeta, R. 2003. Structure trees and species trees: what they say about morphological development and evolution. Evol. Dev. 5: 609–621.

Gomez, M. and E. Nasi. 2000. Light transduction in invertebrate hyperpolarizing photoreceptors: possible involvement of a Go-regulated guanylate cyclase. J. Neurosci. 20: 254–262.

Gomez, M.D. and E. Nasi. 1995. Activation of light-dependent K+ channels in ciliary invertebrate photoreceptors involves Cgmp but not the Ip3/Ca²⁺ cascade. Neuron 15: 607–618.

Gompel, N. and B. Prud'homme, P.J. Wittkopp, V.A. Kassner, and S.B. Carroll. 2005. Chance caught on the wing: cis-regulatory evolution and the origin of pigment patterns in *Drosophila*. Nature 433: 481–487.

Goodman, M. and J. Czelusniak, G. Moore, and A. Romero. 1979. Fitting the gene lineage into its species lineage, a parsimony strategy illustrated by cladograms. Syst. Zool. 28: 132–163.

Haeckel, E. 1866. Generelle Morphologie der Organismen: Allegmeine Grudzuge der organischen Formen-Wissenschaft, mecanisch begrundet durch die von Charles Darwin reformite Descendez-Theorie, 2 volumes. Beorg Reimer, Berlin.

Halder, G. and P. Callaerts, and W.J. Gehring. 1995. Inducion of ectopic eyes by targeted expression of the *eyeless* gene in *Drosophila*. Science 267: 1788–1792.

Hisatomi, O. and F. Tokunaga. 2002. Molecular evolution of proteins involved in vertebrate phototransduction. Comp. Biochem. Phys. B 133: 509–522.

Huelsenbeck, J. P. and B. Rannala, and B. Larget. 2000. A Bayesian framework for the analysis of cospeciation. Evolution Int. J. Org. Evolution 54: 352–364.

Hurles, M. 2004. Gene duplication: the genomic trade in spare parts. Plos Biol 2: E206.

Jacob, F. 1977. Evolution and tinkering. Science 196: 1161–1166.

Kojima, D. and A. Terakita, T. Ishikawa, Y. Tsukahara, A. Maeda, and Y. Shichida. 1997. A novel Go-mediated phototransduction cascade in scallop visual cells. J. Biol. Chem. 272: 22979–22982.

Koyanagi, M. and A. Terakita, K. Kubokawa, and Y. Shichida. 2002. Amphioxus homologe of Go-coupled rhodopsin and peropsin having 11-cis- and all-trans-retinals as their chromophores. Febs. Lett. 531: 525–528.

Koyanagi, M. and K. Takano, H. Tsukamoto, K. Ohtsu, F. Tokunaga, and A. Terakita. 2008. Jellyfish vision starts with cAMP signaling mediated by opsin-G(s) cascade. Proc. Natl. Acad. Sci. USA 105: 15576–15580.

Kozmik, Z. and J. Ruzickova, K. Jonasova, Y. Matsumoto, P. Vopalensky, I. Kozmikova, H. Strnad, S. Kawamura, J. Piatigorsky, V. Paces, and C. Vlcek. 2008. Assembly of the cnidarian camera-type eye from vertebrate-like components. Proc. Natl. Acad. of Sci. USA 105: 8989–8993.

Kutschera, U. and K.J. Niklas. 2004. The modern theory of biological evolution: an expanded synthesis. Naturwissenschaften 91: 255–276.

Kutschera, U. and K.J. Niklas. 2005. Endosymbiosis, cell evolution, and speciation. Theory in Biosciences 124: 1–24.

Lamb, T.D. and S.P. Collin, and E. N. Pugh. 2007. Evolution of the vertebrate eye: opsins, photoreceptors, retina and eye cup. Nat. Rev. Neurosci. 8: 960–975.

Land, M.F. and D.-E. Nilsson. 2002. Animal Eyes. Oxford University Press, Oxford.

Larhammar, D. and K. Nordstrom, and T.A. Larsson. 2009. Evolution of vertebrate rod and cone phototransduction genes. Phil. Trans. Roy. Soc. 364: 2867–2880.

Margulis, L. and D. Sagan. 2002. Acquiring Genomes: A Theory of the Origins of Species. Basic Books, New York.

Minelli, A. 2000. Limbs and tail as evolutionarily diverging duplicates of the main body axis. Evol. Dev. 2: 157–165.

Muradov, H. and K.K. Boyd, V. Kerov, and N.O. Artemyev. 2007. PDE6 in lamprey *Petromyzon marinus*: implications for the evolution of the visual effector in vertebrates. Biochemistry 46: 9992–10000.

Muradov, H. and V. Kerov, K.K. Boyd, and N.O. Artemyev. 2008. Unique transducins expressed in long and short photoreceptors of lamprey *Petromyzon marinus*. Vision Research 48: 2302–2308.

Nilsson, D.E. and S. Pelger. 1994. A pessimistic estimate of the time required for an eye to evolve. Phil. Trans. Roy. Soc. B 256: 53–58.

Nordström, K. and T.A. Larsson, and D. Larhammar. 2004. Extensive duplications of phototransduction genes in early vertebrate evolution correlate with block (chromosome) duplications. Genomics 83: 852–872.

O'Hara, R.J. 1997. Population thinking and tree thinking in systematics. Zoologica Scripta 26: 323–329.

Oakley, T.H. 2003. The eye as a replicating and diverging, modular developmental unit. TREE 18: 623–627.

Oakley, T.H. and A.S. Rivera. 2008. Genomics and the evolutionary origins of nervous system complexity. Curr. Opin. Genet. Dev. 18: 479–492.

Ohno, S. 1970. Evolution by Gene Duplication. Springer-Verlag, New York.

Okano, T. and D. Kojima, Y. Fukada, Y. Shichida, and T. Yoshizawa. 1992. Primary structures of chicken cone visual pigments: vertebrate rhodopsins have evolved out of cone visual pigments. Proc. Natl. Acad. Sci. USA 89: 5932–5936.

Olson, E.N. 2006. Gene regulatory networks in the evolution and development of the heart. Science 313: 1922–1927.

Page, R.D.M. and M.A. Charleston. 1997. From gene to organismal phylogeny: reconciled trees and the gene tree/species tree problem. Mol. Phylogenet. Evol. 7: 231–240.

Page, R.D.M. and M.A. Charleston. 1998. Trees within trees: phylogeny and historical associations. TREE 13: 356–359.

Peng, G.H. and O. Ahmad, F. Ahmad, J. Liu, and S. Chen. 2005. The photoreceptor-specific nuclear receptor Nr2e3 interacts with Crx and exerts opposing effects on the transcription of rod versus cone genes. Hum. Mol. Genet. 14: 747–764.

Pisani, D. and S.M. Mohun, S.R. Harris, J.O. McInerney, and M. Wilkinson. 2006. Molecular evidence for dim-light vision in the last common ancestor of the vertebrates. Curr. Biol. 16: R318–9; author reply R320.

Plachetzki, D.C. and J.M. Serb, and T.H. Oakley. 2005. New insights into the evolutionary history of photoreceptor cells. TREE 20: 465–467.

Plachetzki, D.C. and T.H. Oakley. 2007. Key transitions during animal eye evolution: Novelty, tree thinking, co-option and co-duplication. Int. Comp. Biol.: doi:10.1093/icb/icm050.

Plachetzki, D.C. and B.M. Degnan, and T.H. Oakley. 2007. The Origins of Novel Protein Interactions during Animal Opsin Evolution. PLoS ONE 2: e1054.

Plachetzki, D.C. and C.R. Fong, and T.H. Oakley. 2010. Evidence that a cyclic nucleotide gated (CNG) ion channel functioned in the ancestral animal phototransduction cascade. Proc. R. Soc. Lond. B: InReview.

Rivera, A.S. and T.H. Oakley. 2008. Ontogeny of sexual dimorphism via tissue duplication in an ostracod (Crustacea). Evolution and Development.

Salvini-Plawen, L.V. and E. Mayr. 1977. On the Evolution of Photoreceptors and Eyes. Plenum Press, New York.

Sanderson, M.J. 2002. Estimating absolute rates of molecular evolution and divergence times: A penalized likelihood approach. Mol. Phy. Evol. 19: 101–109.

Schultze, M. 1866. Anatomie und Physiologie der Retina. Achiv Fur Mikroskopisches Anatomie 2: 175–286.

Serb, J.M. and T.H. Oakley. 2005. Hierarchical phylogenetics as a quantitative analytical framework for Evolutionary Developmental Biology. Bioessays 27.

Shen, D. and M. Jiang, W. Hao, L. Tao, M. Salazar, and H. K. Fong. 1994. A human opsin-related gene that encodes a retinaldehyde-binding protein. Biochemistry 33: 13117–13125.

Spudich, J.L. and C.S. Yang, K.H. Jung, and E.N. Spudich. 2000. Retinylidene proteins: structures and functions from archaea to humans. Annu. Rev. Cell. Dev. Biol. 16: 365–392.

Su, C.Y. and D.G. Luo, A. Terakita, Y. Shichida, H.W. Liao, M.A. Kazmi, T.P. Sakmar, and K.W. Yau. 2006. Parietal-eye phototransduction components and their potential evolutionary implications. Science 311: 1617–1621.

Suga, H. and M. Koyanagi, D. Hoshiyama, K. Ono, N. Iwabe, K. Kuma, and T. Miyata. 1999. Extensive gene duplication in the early evolution of animals before the parazoan-eumetazoan split demonstrated by G proteins and protein tyrosine kinases from sponge and hydra. J. Mol. Evol. 48: 646–653.

Suga, H. and V. Schmid, and W.J. Gehring. 2008. Evolution and functional diversity of jellyfish opsins. Curr. Biol. 18: 51–55.

Takezaki, N. and F. Figueroa, Z. Zaleska-Rutczynska, and J. Klein. 2003. Molecular phylogeny of early vertebrates: monophyly of the agnathans as revealed by sequences of 35 genes. Mol. Biol. Evol. 20: 287–292.

Taylor, J.S. and J. Raes. 2004. Duplication and divergence: the evolution of new genes and old ideas. Ann. Rev. Gen. 38: 615–643.

Thompson, J. 1987. Symbiont-induced speciation. Biol. J. Linn. Soc. 32: 385–393.

Thorne, J.L. and H. Kishino. 2002. Divergence time and evolutionary rate estimation with multilocus data. Syst. Biol. 51: 689–702.

Turner, D.L. and C.L. Cepko. 1987. A common progenitor for neurons and glia persists in rat retina late in development. Nature 328: 131–136.

Walls, G.L. 1942. The Vertebrate Eye and Its Adaptive Radiation. Cranbrook Press., Bloomfield Hills, MI.

West-Eberhard, M.J. 2003. Developmental Plasticity and Evolution. Oxford University Press, Oxford.

Wetts, R. and S.E. Fraser. 1988. Multipotent precursors can give rise to all major cell types of the frog retina. Science 239: 1142–1145.

Yasutaka, W. and S. Junichi, O. Toshiyuki, and F. Yoshitaka. 2006. GRK1 and GRK7: Unique cellular distribution and widely different activities of opsin phosphorylation in the zebrafish rods and cones.J. Neurochem. 98: 824–837.

Zhang, J.Z. 2003. Evolution by gene duplication: an update. TREE 18: 292–298.

Chapter 11

Vertebrate Hox Genes and Specializations in Mammals

Claudia Kappen

Introduction

The role of Hox genes in patterning of vertebrate body plans has been well documented (Capecchi 1996). Since the discovery of homeotic genes in the fruit fly Drosophila (Lewis 1978), it has been held that changes in body plan would also involve Hox genes. By mutation analyses in the mouse, changes in Hox gene expression have been shown to cause changes in body plan, most prominently in the skeletal system (Wellik 2009).

Linking body plans to Hox genes also implies that differences between body plans, such as between two particular species, could be attributed to altered Hox gene function (Chen et al. 2005, Behringer 2009). Yet, proof for this hypothesis in an evolutionary context, has been sparse to date (Sears et al. 2007, Galis et al. 2005). Many comparative studies have been published that document similarities and differences in the patterns and timing of Hox gene expression between vertebrate species (Burke et al. 1995), but whether such differences in expression are causative for, or a reflection of, anatomical differences is difficult to discern (Ray and Capecchi 2008, Sakamoto et al. 2009). This would require reciprocal manipulations in vertebrate genomes, when such techniques are not available for most species.

Even less is known about the role of Hox genes in tissues and organs that can be considered evolutionary novelties within the vertebrates, such as the turtle carapace (Gilbert et al. 2001, Ohya et al. 2005), or the

Department of Maternal Biology, Pennington Biomedical Research Center, 6400 Perkins Road, Baton Rouge, LA 70808.
E-mail: Claudia.Kappen@pbrc.edu

characteristic reproductive organs of female mammals. This manuscript reviews the current knowledge on Hox gene function in mammary gland, placenta, and the female reproductive tract of mammals.

Clustered Organization of Hox Genes

The Hox genes encode transcription factors that are thought to regulate gene expression through DNA binding of the conserved homeodomain (Gehring et al. 1994). This may involve the formation of complexes with other transcription factors—some of them also homeodomain containing —such as the Pbx or Meis proteins (Pöpperl et al. 1995). The formation of complexes appears to modulate target recognition by the homeodomain, suggesting that the complement of factors present in a given cell will have important bearing on the repertoire of Hox target genes expressed in that cell (Saleh et al. 2000). Nevertheless, mutant studies in the mouse have demonstrated unique and overlapping functions for Hox genes in development of the nervous system, the axial skeleton, the limbs, the craniofacial region, all internal organs that have been investigated, and the skin. Which specific Hox gene affects which tissue is regulated through the expression patterns of Hox genes, which, in as yet poorly understood fashion are coupled to their genomic clustered organization (Duboule and Morata 1994, Deschamps 2007, Herault et al. 1999).

Hox genes are organized in clusters in many species; except for fish (Kuraku and Meyer 2009), vertebrates have four clusters, typically on different chromosomes. The clustered organization of Hox genes in the mouse is shown in Fig. 1. Based on sequence similarity, cognate (paralogous) groups can be recognized in which the closest relatives are located in comparable positions on different clusters. Thirteen such groups exist in vertebrates, although on some clusters, members of a given group may be absent. Also, while similarities between members of a given cognate group may extend beyond the homeodomain, each gene encodes a distinct protein. Overall, the colinearity and conservation between the four clusters was interpreted early on as evidence for duplications of entire clusters if not genomes (Kappen et al. 1989, Schughart et al. 1989).

However, not only are Hox gene coding sequences highly conserved among vertebrates, non-coding sequences along the clusters also bear recognizable conservation between distant vertebrate species (Shashikant et al. 2007). Such conserved non-coding elements are thought to be involved in regulation of expression of the Hox genes. Indeed, Hox genes are expressed—and function—in distinct, yet overlapping domains along the anterior-posterior axis in the hindbrain and spinal cord and the axial mesoderm, and the proximal-distal axis in the limbs (Fig. 1). The region-specific expression of each Hox genes is accomplished through specific regulatory elements that also control the timing of expression (Juan and Ruddle 2003, Gerard et al. 1997). However, to date, no specific elements have been identified that are responsible for Hox gene expression in female reproductive tissues.

THE VERTEBRATE *HOX* GENE SYSTEM

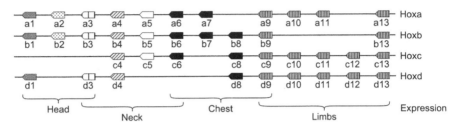

Fig. 1. The vertebrate Hox system.

Legend: The location of Hox genes on clusters is shown schematically; no attempt has been made to depict actual distances. Orientation of arrows indicates direction of transcription.

Hox Genes in Mammalian Female Reproductive Tissues

Mammary gland

Soon after reports of Hox gene expression in the male reproductive system (Lindsey and Wilkinson 1996), the female reproductive organs were investigated (Taylor et al. 1997). In normal mammary gland as well as in neoplastic tissues and breast cancer cell lines, Hox gene expression has been reported (Friedmann 1994, Cantile et al. 2003, Srebow et al. 1998, Chen and Capecchi 1999, Garcia-Gasca and Spyropoulos 2000, Garin et al. 2006, Raman et al. 2000). However, little is known about the functional role of Hox genes in mammary development or etiology of breast cancer. Mouse mutagenesis has established a role for Hoxa5, Hoxc6 and the Hox9 paralogous genes in mammary development (Chen and Capecchi 1999, Garcia-Gasca and Spyropoulos 2000, Garin et al. 2006). Mice homozygous for a loss-of-function allele of Hoxc6 have underdeveloped mammary glands in the thoracic region, with a lack of epithelial cells by adulthood. The glands also do not respond properly to pregnancy hormones (Garcia-Gasca and Spyropoulos 2000), suggesting that Hoxc6 has a role in mammary expansion and involution. This phenotype is more pronounced in the anterior domain of Hoxc6 expression, probably due to redundancy with other Hox genes in more posterior regions. Mammary gland development in response to pregnancy is also impaired in mice that carry mutations for three of the four Hox9 paralogs, Hoxa9, Hoxb9 and Hoxd9 (Chen and Capecchi 1999), resulting in insufficient milk production (Duboule 1999).

Hoxb6 mutants also appear to have defective mammary glands as no milk is found in the stomach of pups born to Hoxb6-deficient dams (Kappen, unpublished). The same phenotype was observed in Hoxa5 mutants (Garin et al. 2006), and was traced back to cell proliferation defects. Even though only a few Hox genes have thus been investigated, it emerges that there may be regional specificities as well as selective roles in specific cell types within the mammary gland. It is interesting to note that Hox gene expression in the mammary gland is responsive to sex/pregnancy hormones (Chen and Capecchi 1999, Garica-Gasca and Spyropoulos 2000, Garin et al. 2006), a feature shared with Hox genes in the uterus.

Female reproductive tract

In analogy to the Hox code for regions of the skeletal system (see Fig. 1), a Hox code has been proposed for the female reproductive tract (Eun Kwon and Taylor 2004). This is based upon data that indicate highest levels of expression for Hoxa9 in oviduct, Hoxa10 and Hoxa11 in uterus, and Hoxa13 in vagina (Block et al. 2000). Hoxa9 mutants are fertile, indicating that Hoxa9 is either not required or functionally redundant with other Hox genes in the oviduct; even triple Hox9 mutants are fertile (Chen and Capecchi 1999). In contrast, Hoxa10 mutants display transformations of the rostral reproductive tract (Benson et al. 1996) into oviduct-like identity, while the more posterior regions are less affected. Hoxa10 appears to regulate both proliferation of stromal cells and epithelial cell morphogenesis in uterus, and this may contribute to receptivity of uterus at the time of blastocyst implantation (Bagot et al. 2001). Defective implantation in Hoxa10 mutants has been interpreted to be due to stromal cell deficiencies (Lim et al. 1999), possibly associated with abnormal cell proliferation and differentiation (Yue et al. 2005). Wnt4 expression was absent and Sfrp4 expression aberrantly broader in stromal cells on the day of implantation (Daikoku et al. 2004). Hoxa11 mutants also display an abnormal reproductive tract (Gendron et al. 1997) with deficient development of stromal and decidual cells at early stages of gestation. Hoxa11 mutants also lack a critical burst of LIF production from glandular cells at gestation day 4.5 (Gendron et al. 1997). Analysis of compound heterozygotes for Hoxa10 and Hoxa11 mutant alleles indicates that both genes interact in regulating the responsiveness of uterine tissue to implantation (Branfond 2000). In Wnt7a null mutants shorter and thinner uterine horns were found, associated with loss of Hoxa10 and Hoxa11, indicating that Wnt7a maintains their expression (Miller and Sassoon 1998). When Hoxa13 was expressed from the Hoxa11 locus, the uterus was transformed into tissue similar to cervix and vagina, where Hoxa13 is normally expressed (Zhao and Potter 2001). These results indicate that Hoxa11 and Hoxa13 cannot functionally substitute for each other (as in the male reproductive tract, for example). More important from an evolutionary perspective, however, is that the phenotypes of the Hox mutants provide evidence for a functional Hox code in the mammalian female reproductive tract.

In humans, mutations in Hox genes have not been reported for reproductive tract abnormalities, such as congenital absence of uterus and vagina (Burel et al. 2006). Also, the expression patterns of Hoxc and Hoxd cluster genes in human endometrium differ from those for Hoxa cluster genes (Akbas and Taylor 2004): Hoxc10, Hoxc11, Hoxd10, and Hoxd11 are all expressed in proliferating endometrium, but they do not appear to be stimulated by hormones. Hoxa10 regulates stromal cell proliferation in response to progesterone (Lim et al. 1999), and Hoxa11 expression is induced by progesterone in stroma, while it is repressed by progesterone in myometrial cells (Cermik et al. 2001). Similarly, exogenous steroids affect Hox gene expression during development of the reproductive tract itself: Diethylstilbestrol (DES), applied to embryos between embryonic days 9 and 16, shifted Hox gene expression more posteriorly, associated with more anterior morphology of reproductive tract (Block et al. 2000). The greatest changes were seen for Hoxa9, Hoxa10 and Hoxa11, and were maintained into adulthood. Altered Hoxa10 expression in offspring of treated females was induced by several compounds, suggesting that Hoxa10 is a common target in reproductive tract (Smith and Taylor 2007). It is interesting to note that steroid regulation of Hoxa10 and Hoxa11 appears to involve non-canonical response elements (Smith and Taylor 2007, Dabtary and Taylor 2006).

Placenta

Much less is known about Hox genes in the placenta. The placenta develops upon implantation of the blastocyst into the endometrium. The trophectoderm cells then become trophoblast giant cells that border the maternal decidua. A second wave of trophoblast generation (Hemberger 2008) then produces the different types of trophoblasts found in the placenta, including the secondary trophoblast giant cells, spongiotrophoblasts, syncytiotrophoblasts and labyrinthine trophoblasts. Numerous other cell types are found associated with placenta at various stages of development, including natural killer cells, macrophages, pericytes and other hematopoietic cells (Georgiades et al. 2002). While the morphology of placentae differs between species (Rawn and Cross 2008, Benirschke 1983), Fig. 2 shows a typical histological section and a simplified graphic representation of midgestation placenta in the mouse. As the placenta matures, distinct layers of different cell types are apparent, and based on layer-specific gene expression patterns, it has been suggested that the different cell types have distinct progenitors (Simmons et al. 2008).

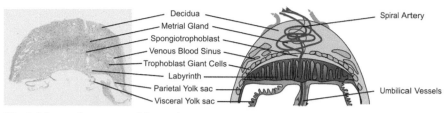

Decidua
Metrial Gland
Spongiotrophoblast
Venous Blood Sinus
Trophoblast Giant Cells
Labyrinth
Parietal Yolk sac
Visceral Yolk sac

Spiral Artery

Umbilical Vessels

Fig. 2. Mouse placenta at midgestation.

Legend: Left panel shows a section from paraffin embedded mouse placenta isolated at gestation day 10.5, stained with the Periodic acid Schiff method, counterstained with methyl green.

Color image of this figure appears in the color plate section at the end of the book.

Expression in placenta has been reported for a number of Hox genes: Hoxb6, Hoxc6, and Hoxa11 were detected in human trophoblastic cell lines (Zhang et al. 2002) and in human placenta at term (Amesse et al. 2003). More divergent homeobox-containing genes Hlx, Msx2 and Meox2 are also expressed in human placenta (Quinn et al. 1997). Dlx4, Msx2 and Meox2 were found in the cytotrophoblast layer in first trimester placenta, and in syncytiotrophoblasts in term placenta (Quinn et al. 1998, 2000). Hoxc4, Hoxc5 and Hoxc6 were found expressed in chorionic villi (Oudejans et al. 1990) where Hlx is also expressed (Rajaraman et al. 2008). In the mouse, Hoxc9 is expressed in giant trophoblast cells at gestational days 12 and 19, but not at earlier time points (E8); it was also detected in luminal epithelial cells in uterus at gestational day 12 (E12) (Murasawa et al. 2000). Dlx3 is expressed in mouse placenta, but is restricted to the labyrinth and not detected in trophoblast giant cells (Berghorn et al. 2005); its expression diminishes towards later developmental stages (E15.5). Thus, expression of homeobox-containing genes in the placenta is temporally dynamic and may be cell-type specific. However, comprehensive surveys have not been performed yet, and for many of these genes, the developmental sequence of expression has not been established. Cdx2 is the earliest homeobox gene to date detected in extra-embryonic tissue: it is expressed in the trophectoderm cells of blastocysts, but is absent from the inner cell mass; at later stages, Cdx2 is expressed in glycogen-positive cells of ectoplacental cone, and in spongiotrophoblasts (Beck et al. 1995). Thus, Cdx2 is one of the earliest markers for the progenitors of cells in the trophoblast lineage.

We recently performed microarray gene expression profiling of murine placenta at midgestation (Salbaum et al. manuscript in preparation) and analyzed the results with specific focus on Hox genes and homeobox-containing genes. Hybridization signals were detected for many Hox genes (Fig. 3); the highest levels of expression involve posterior Hox genes of the Hox9, Hox10 and Hox11 cognate groups. It is noteworthy that more anterior Hox genes were also detected, such as Hoxa1, Hoxb2, Hoxb3, Hoxd3, and Hoxd4 and Hoxb6. However, we currently do not

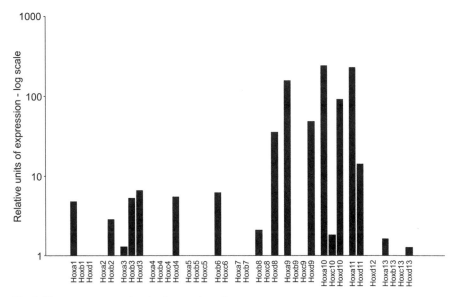

Fig. 3. Hox gene expression in the murine placenta.

Placenta was isolated from pregnancies of normal FVB mice at midgestation (embryonic day 10.5), and subjected to Affymetrix microarray analysis as described elsewhere (Salbaum et al., in preparation). Three separate microarrays were analyzed and the data were averaged and expressed in arbitrary units as the fold-change over the level of the lowest expressed gene, Hoxc8. Note that the y-axis has a logarithmic scale. No signal was detected for Hoxb1, Hoxd1, Hoxa2, Hoxa4, Hoxb4, Hoxc4, Hoxa5, Hoxb5, Hoxc5, Hoxc6, Hoxa7, Hoxb7, Hoxb9, Hoxc9, Hoxd12, Hoxb13, Hoxc13 with the probes present on the microarray. Probes for some Hox genes were not present on the array.

The placenta at this stage contains maternal decidual cells, maternal blood, embryonic cells, including embryonic blood. Thus, the results depicted are the summation of expression for the mixture of these cell types.

have information which specific cell types express these genes; it is possible that some genes could be restricted to hematopoietic cells present in the maternal or fetal vessels of the placenta. Similarly, Fig. 4 shows the expression of divergent homeobox-containing genes for which the highest hybridization signals were found. Pbx4 gave only a weak signal, consistent with the literature reporting that Pbx4 expression is restricted to the testis (Wagner et al. 2001). But other Hox-cofactors of the Pbx and Pknox classes are present. Most notably, we also detect expression of Rhox genes (MacLean et al. 2005) with highest expression for Rhox5, Rhox6 and Rhox9, consistent with previous findings (MacLean et al. 2005). However, not all Rhox genes were represented on the microarrays. The Rhox genes are a group of clustered X-lined genes that is subject to rapid evolution (Wang and Zhang 2006, MacLean et al. 2006). It will therefore be particularly interesting to investigate their role in mammalian reproductive tissues.

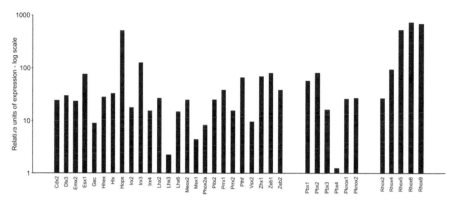

Fig. 4. Expression of other homeobox genes in mouse placenta.
The results are derived from the same experiment as depicted in Figure 2 and also were normalized to Hoxc8. No signal was detected for Cdx4, Emx1, Lhx4, Pou5f1 (Oct3/4), Rhox13 at this gestational time point with the probes present on the array.

Genetic assays have demonstrated that Hoxa13 is required during formation of the placental labyrinth (Shaut et al. 2008). Hoxa13 is first expressed in allantoic bud mesoderm, and later in allantois and chorionic mesoderm. It was also found expressed in the vessel walls of the umbilical arteries, and in endothelial cells in the labyrinth region, as demonstrated by PECAM-1 staining. As a consequence of Hoxa13-deficiency, the size of the labyrinth remains small, and vascular branching is impaired, leading to embryonic lethality by E15.5 at the latest. Expression of endothelial cells was altered, as well as expression or pro-vascular genes, indicating that defective specification of endothelial cells in the labyrinth vasculature leads to defective placental development (Shaut et al. 2008). This was interpreted as a cooption of Hox gene function in specialized structures, analogous to the role of Hoxc13 in hair follicle development (Godwin and Capecchi 1998, Augulewitsch 2003).

Of interest in this regard will be the characterization of Hox gene expression in mutants for the polycomb group gene Eed (Wang et al. 2002). Eed is required for stable repression of Hox genes in the skeleton, but the placenta is also affected in eed mutants. While labyrinth and spongiotrophoblasts are present, placental lactogen 1 expression was found greatly reduced, demonstrating that eed function is required in secondary trophoblast cells. However, it is unknown at present, whether the defects are associated with altered Hox gene expression, such as de-repression.

A vascularization defect in the labyrinth was also found in the Esx1 mutant (Li and Behringer 1998). Esx1 is expressed in the endodermal layer of the visceral yolk sac, but not in mesodermal cells, and in labyrinth trophoblasts. The labyrinth appears to be correctly specified in the Esx1 mutants at first, but by E14.5 disorganization of the normally distinct boundary between spongiotrophoblasts and labyrinthine layer was obvious, and accumulation of glycogen-filled cells was observed. Precursors of the syncytiotrophoblasts were absent from the fetal capillaries, providing evidence for defective vascularization.

The divergent homeobox gene Dlx3 is also involved in labyrinth development, where it is normally expressed in labyrinthine trophoblasts (Han et al. 2007). In Dlx3 mutants, the labyrinth does not expand, possibly because of insufficient production of placental growth factor. Although the specification of placental cell types does not appear to be affected by Dlx3 deficiency (Morasso et al. 1999), aberrant gene expression was found in spongiotrophoblasts. Esx1 expression was also dramatically reduced, and the placenta did not establish an effective circulatory system in the mutants.

Two other homeobox genes that were reported to be expressed in trophoblasts are Lhx3 and Lhx4 (Tian et al. 2008). Although Lhx4 is expressed in spongiotrophoblasts, this did not produce a signal in our microarray assay. Lhx4 mutants also do not have a placental phenotype and neither do Lhx3 mutants (Tian et al. 2008). Analyses of compound mutants in which both deficiencies were combined revealed some placentae with abnormal labyrinth structure. However, because no consistent interference with transport or expression of putative target genes, such as Gnhrh was detected, the authors conclude that these genes are likely dispensable for placental development and function (Tian et al. 2008).

The earliest effects on placental development were seen in mutants for Cdx2 (Chawengsaksophak et al. 2004). Homozygous mutants do not implant effectively, due to failure of development of the trophectoderm (Deb et al. 2006). Absence of trophectoderm development was also seen in Tead4 mutants (Nishioka et al. 2008), which lack Cdx2 expression. This also prevented generation of trophoblasts giant cells from such embryos in culture. The trophectoderm defect can be rescued by tetraploid reconstitution (Chawengsaksophak 2004), but the allantois remained underdeveloped and yolk sac vascularization was abnormal. Placental defects were also found in compound heterozygotes for Cdx2 and Cdx4 (van Nes et al. 2005). Although Cdx4 deficiency by itself is asymptomatic in placenta, reducing the gene dosage of both genes resulted in labyrinth defects. These could be traced back to incomplete fusion of allantois with the chorion and a subsequent defect in branching morphogenesis within the labyrinth. Taken together, the results indicate that Cdx2 and Cdx4 interact in placental development, particularly in the labyrinth. Because the Cdx genes are known regulators of Hox gene expression in the posterior region of developing embryos, these findings suggest that

Cdx gene deficiencies may also result in deregulated Hox gene expression in the placenta.

Placenta-Specific Regulation of Hox Gene Expression

Nothing is known about molecular mechanisms responsible for Hox gene expression in the placenta. Numerous enhancers have been identified that control region-specific pattern of Hox gene expression, but how they have evolved in vertebrates has only been studied for a few examples, such as the Hoxc8 enhancer in the axial skeleton (Schughart et al. 1989, Shashikant et al. 1995, Belting et al. 1998, Anand et al. 2003), or the Hoxd13 (Ray and Capecchi 2008) and Prx1(Cretekos et al. 2008) loci in the limb.

The common theme from these investigations was that subtle differences in the activity of Hox gene enhancers may explain differences in expression patterns that are associated with phenotypic differences in the features of specific structures. However, sorting out which enhancer activities are causative for development of a different structure and which activities simply reflect the anatomical difference between two species still faces technical limitations. The situation is even less clear for cases where new features have emerged during evolution. Information on enhancers for female reproductive tissues is currently lacking.

In fact, very few regulatory elements have been identified to date that drive cell-type-specific gene expression in the placenta. An enhancer for the rat Placental lactogen gene II was defined by transfection into choriocarcinoma and pituitary cells, but there is no evidence available whether this element is active in vivo (Sun and Duckworth 1999). AP2γ was identified as a regulator of this enhancer (Ozturk et al. 2006). However, since AP2α is expressed throughout the placenta, unidentified cell-type specific elements will also be required to restrict Placental lactogen II expression to specific cell types.

The human Cyp19 gene contains tissue-specific promoters upstream of tissue-specific first exons, with the placental module 100,000 bp upstream of the second exon. Deletion studies in transgenic mice were used to narrow the active element within a 501 bp fragment (Kamat et al. 2005). This fragment was sufficient to drive reporter gene expression in the labyrinthine layer, and a 254 bp shorter portion was also active in trophoblast giant cells. This suggests that a negative regulatory element is present in the larger fragment that suppresses activity in trophoblast cells, while the labyrinth-specific enhancer would be present within the residual 246 bp. Deletion of the element by another 45 bp abolished enhancer activity completely. However, the mouse ortholog is not expressed in placenta, and thus, the functional relevance of this observation remains unclear. In fact, the few genes known as placenta-specific genes to date are not present in all placental mammals (Rawn and Cross 2008), suggesting that divergent regulatory mechanisms may be involved.

Concluding Summary

In summary, very few studies have addressed the roles of Hox genes in evolutionary novelties that were invented during vertebrate evolution, such as the mammary gland or the placenta. Our own data indicate that a selected subset of Hox genes is expressed in midgestation placenta, among them those that have been previously described in the mammalian uterus. We also detect expression of divergent homeobox genes, including genes known to be expressed in placenta. We also detect genes for which placental expression has not heretofore been reported, suggesting that they could play a role in placental biology. The different expression levels of the various Hox genes also suggest that there could be heterogeneity in timing or cell type distribution of Hox gene expression in the placenta. This is supported by the emerging evidence from the mammary gland (see above).

While a function for Hox genes within the murine uterus has been clearly established through in vivo mutagenesis, evidence for a role in placenta is more elusive. Over 100 genes have been identified that play a role in various aspects of placental development (Rossant and Cross 2001), but up to now, Hox genes are not listed among them (Watson and Cross 2005). However, upstream regulators of Hox genes, such as eed and Cdx2 are required for placental development, and hence it is plausible to speculate that their targets, the Hox genes, also may be involved. As discussed above, Hoxa13 is the only gene that has been shown to result in placental defects when mutated (Shaut et al. 2008). The paucity of reported placental phenotypes in Hox mutants could be an indicator for redundancy, where impaired placental development or function would only become evident with multiple deficiencies. Alternatively, it is also possible that subtle phenotypes exist that have so far escaped attention due to the strong focus on the role of Hox genes in embryonic patterning. For example, abnormalities in the hematopoietic system are compatible with survival to adulthood in mutants for Hoxc8 (Takeshita et al. 1993), Hoxb6 (Kappen 2000), Hoxb3 (Ko et al. 2007), Hoxb4 (Bijl et al. 2006) and the complete absence of Hoxb1 through Hoxb9 (Bijl et al. 2006). Similarly, functional roles for Hox genes in regionalizing the vascular system are just beginning to be reported (Pruett et al. 2008).

It is interesting to note that in mammary gland and uterus, Hox genes are responsive to hormonal regulation, a property not reported to date for other sites of Hox gene expression, such as the skeletal system. This indicates that (a) either this aspect is underinvestigated, or that (b) this source of input into the Hox system is a new feature that co-evolved with the appearance of the mammary gland, uterus and placenta. Then, one would predict that Hox genes contain specific hormone-responsive elements that are targets for regulation by hormone receptors, although, in analogy to Hoxa10 and Hoxa11 (see above), these could be non-canonical motifs. Nothing is known about the regulatory mechanisms that control expression of Hox genes in placental cells. Considering the insights that have emerged from studies of Hox enhancers in the embryo, a few predictions can be made:

1) Given the evolutionary conservation of Hox enhancers within vertebrates, we can postulate that sequences with activity in placental cell types should be conserved between the genomes of mammals, and would likely be absent from non-mammalian genomes.

2) The regulatory elements are likely separate from already identified promoters and enhancers. This would be consistent with the unique properties of placental cells. However, since few cell lines exist that adequately represent the different placental cell types, a screen for placental enhancer might be complicated. Transgenic experiments will be required anyway, since so little is known about placental enhancers, and cell type-specificity will need to be ascertained in vivo.

3) In analogy to the situation in the embryo, it is likely that initiation and maintenance of Hox gene expression could be regulated by different transcription factors. In this regard, it is interesting to note that Cdx proteins are downregulated as the placenta matures, and the eed regulator similarly is only active in early placentation. This may argue that additional elements exist that function to maintain Hox gene expression in placental cell types at later stages.

4) Also in analogy to the timing of Hox gene expression in the embryo, the more anterior Hox genes might be activated earlier and more posterior Hox genes later. If Hox genes similarly affect cell proliferation in placenta as they do in the embryo, one would also expect higher levels of expression in proliferating cells, and hence, the early stages of placenta formation.

5) Once placental enhancers have been identified, the evolutionary trajectory of these sequences can be reconstructed to long before the emergence of mammals, and thus provide a model for regulatory innovation. A focus of such studies on the placenta has a distinct advantage over other vertebrate tissues: Many other body parts, even though quite distinct between species, such as the bird wing, bat wing, and the limb, are homologous to each other, and under common constraints since the regulatory networks in their formation have been established. This likely also applies to placenta in mammals, although it has been hypothesized that co-option (Rawn and Cross 2008), gene duplication and divergence have further

diversified placental development to suit specific physiological needs (Knox and Baker 2008). Indeed, gonadotropin expression is only found in placenta of primates (Maston and Ruvolo 2002). Then, two scenarios can be envisioned: either the existing enhancers are being reused/co-opted for regulation of expression in placental cells, or placenta-specific enhancers co-evolved with the placental cell types. Both possibilities are testable at the level of sequence comparisons and in experimental systems. Thus, on a background of otherwise conserved functions in common features in vertebrate body patterning, mammal-specific features, such as the placenta or the mammary gland, provide particularly suitable models to further our understanding of evolutionary innovations.

Acknowledgements

I thank Dr. J. Michael Salbaum for the placenta schematic and assistance with the annotation of microarrays, Loula Burton for help in assembling the references, and the Peggy M. Pennington Cole endowment for financial support.

References

Akbas, G.E. and H.S. Taylor. 2004. HOXC and HOXD gene expression in human endometrium: lack of redundancy with HOXA paralogs. Biol. Reprod. 70(1): 39–45.

Amesse, L.S. and R. Moulton, Y.M. Zhang, and T. Pfaff-Amesse. 2003. Expression of HOX gene products in normal and abnormal trophoblastic tissue. Gynecol. Oncol. 90(3): 512–518.

Anand, S. and W.C. Wang, D.R. Powell, S.A. Bolanowski, J. Zhang, C. Ledje, A.B. Pawashe, C.T. Amemiya, and C.S. Shashikant. 2003. Divergence of Hoxc8 early enhancer parallels diverged axial morphologies between mammals and fishes. Proc. Natl. Acad. Sci. USA 100(26): 15666–15669.

Awgulewitsch, A. 2003. Hox in hair growth and development. Naturwissenschaften 90(5): 193–211.

Bagot, C.N. and H.J. Kliman, and H.S. Taylor. 2001. Maternal Hoxa10 is required for pinopod formation in the development of mouse uterine receptivity to embryo implantation. Dev. Dyn. 222(3): 538–544.

Beck, F. and T. Erler, A. Russell, and R. James. 1995. Expression of Cdx-2 in the mouse embryo and placenta: possible role in patterning of the extra-embryonic membranes. Dev. Dyn. 204(3): 219–227.

Behringer, R.R. and J.J. Rasweiler, C.H. Chen, and C.J. Cretekos. 2009. Genetic regulation of mammalian diversity. cold spring harb. symp. Quant. Biol. [Epub Dec.22].

Belting, H.G. and C.S. Shashikant, and F.H. Ruddle. 1998. Modification of expression and cis-regulation of Hoxc8 in the evolution of diverged axial morphology. Proc. Natl. Acad. Sci. USA 95(5): 2355–2360.

Benirschke, K. 1983. Placentation. J. Exp. Zool. 228(2): 385–389.

Benson, G.V. and H. Lim, B.C. Paria, I. Satokata, S.K. Dey, and R.L. Maas. 1996. Mechanisms of reduced fertility in Hoxa-10 mutant mice: uterine homeosis and loss of maternal Hoxa-10 expression. Development 122(9): 2687–2696.

Berghorn, K.A. and P.A. Clark, B. Encarnacion, C.J. Deregis, J.K. Folger, M.I. Morasso, M.J. Soares, M.W. Wolfe, and M.S. Roberson. 2005. Developmental expression of the homeobox protein Distal-less 3 and its relationship to progesterone production in mouse placenta. J. Endocrinol. 186(2): 315–323.

Bijl, J. and A. Thompson, R. Ramirez-Solis, J. Krosl, D.G. Grier, H.J. Lawrence, and G. Sauvageau. 2006. Analysis of HSC activity and compensatory Hox gene expression profile in Hoxb cluster mutant fetal liver cells. Blood 108(1): 116–122.

Block, K. and A. Kardana, P. Igarashi, and H.S. Taylor. 2000. In utero diethylstilbestrol (DES) exposure alters Hox gene expression in the developing mullerian system. FASEB J. 14(9): 1101–1108.

Branford, W.W. and G.V. Benson, L. Ma, R.L. Maas, and S.S. Potter. 2000. Characterization of Hoxa-10/Hoxa-11 transheterozygotes reveals functional redundancy and regulatory interactions. Dev. Biol. 224(2): 373–387.

Burel, A. and T. Mouchel, S. Odent, F. Tiker, B. Knebelmann, I. Pellerin, and D. Guerrier. 2006. Role of HOXA7 to HOXA13 and PBX1 genes in various forms of MRKH syndrome (congenital absence of uterus and vagina). J. Negat. Results Biomed. 5: 4.

Burke, A.C. and C.E. Nelson, B.A. Morgan, and C. Tabin. 1995. Hox genes and the evolution of vertebrate axial morphology. Development 121(2): 333–346.

Cantile, M. and G. Pettinato, A. Procino, I. Feliciello, L. Cindolo, and C. Cillo. 2003. In vivo expression of the whole HOX gene network in human breast cancer. Eur. J. Cancer 39(2): 257–264.

Capecchi, M.R. 1996. Function of homeobox genes in skeletal development. Ann. N.Y. Acad. Sci. 785: 34–37.

Cermik, D. and M. Karaca, and H.S. Taylor. 2001. HOXA10 expression is repressed by progesterone in the myometrium: differential tissue-specific regulation of HOX gene expression in the reproductive tract. J. Clin. Endocrinol. Metab. 86(7): 3387–3392.

Chawengsaksophak, K. and W. de Graaff, J. Rossant, J. Deschamps, and F. Beck. 2004. Cdx2 is essential for axial elongation in mouse development. Proc. Natl. Acad. Sci. USA 101(20): 7641–7645.

Chen, F. and M.R. Capecchi. 1999. Paralogous mouse Hox genes, Hoxa9, Hoxb9, and Hoxd9, function together to control development of the mammary gland in response to pregnancy. Proc. Natl. Acad. Sci. USA 96(2): 541–546.

Chen, C.H. and C.J. Cretekos, J.J. Rasweiler, and R.R. Behringer. 2005. Hoxd13 expression in the developing limbs of the short-tailed fruit bat, *Carollia perspicillata*. Evol. Dev. 7(2): 130–141.

Cretekos, C.J. and Y. Wang, E.D. Green, J.F. Martin, J.J. Rasweiler, and R.R. Behringer. 2008. Regulatory divergence modifies limb length between mammals. Genes Dev. 22(2): 141–151.

Daftary, G.S. and H.S. Taylor. 2006. Endocrine regulation of HOX genes. Endocr. Rev. 27(4): 331–355.

Daikoku, T. and H. Song, Y. Guo, A. Riesewijk, S. Mosselman, S.K. Das, and S.K. Dey. 2004. Uterine Msx-1 and Wnt4 signaling becomes aberrant in mice with the loss of leukemia inhibitory factor or Hoxa-10: evidence for a novel cytokine-homeobox-Wnt signaling in implantation. Mol. Endocrinol. 18(5): 1238–1250.

Deb, K. and M. Sivaguru, H.Y. Yong, and R.M. Roberts. 2006. Cdx2 gene expression and trophectoderm lineage specification in mouse embryos. Science 311(5763): 992–996.

Deschamps, J. 2007. Ancestral and recently recruited global control of the Hox genes in development. Curr. Opin. Genet. Dev. 17(5): 422–427.

Duboule, D. and G. Morata. 1994. Colinearity and functional hierarchy among genes of the homeotic complexes. Trends in Genetics 10: 358–364.

Duboule, D. 1999. No milk today (my Hox have gone away). Proc. Natl. Acad. Sci. USA 96(2): 322–323.

Eun Kwon, H. and H.S. Taylor. 2004. The role of HOX genes in human implantation. Ann. NY. Acad. Sci. 1034: 1–18.

Friedmann, Y. and C.A. Daniel, P. Strickland, and C.W. Daniel. 1994. HOX genes in normal and neoplastic mouse mammary gland. Cancer Research 54(22): 5981–5985.

Galis, F. and M. Kundrat, and J.A. Metz. 2005. Hox genes, digit identities and the theropod/bird transition. J. Exp. Zool. B. Mol. Dev. Evol. 304(3): 198–205.

Garcia-Gasca, A. and D.D. Spyropoulos. 2000. Differential mammary morphogenesis along the anteroposterior axis in Hoxc6 gene targeted mice. Dev. Dyn. 219(2): 261–276.

Garin, E. and M. Lemieux, Y. Coulombe, G.W. Robinson, and L. Jeannotte. 2006. Stromal Hoxa5 function controls the growth and differentiation of mammary alveolar epithelium. Dev. Dyn. 235(7): 1858–1871.

Gehring, W.J. and Y.Q. Qian, M. Billeter, K. Furukubo-Tokunaga, A.F. Schier, D. Resendez-Perez, M. Affolter, G. Otting, and K. Wüthrich. 1994. Homeodomain-DNA Recognition. Cell 78: 211–223.

Gendron, R.L. and H. Paradis, H.M. Hsieh-Li, D.W. Lee, S.S. Potter, and E. Markoff. 1997. Abnormal uterine stromal and glandular function associated with maternal reproductive defects in Hoxa-11 null mice. Biol. Reprod. 56(5): 1097–1105.

Georgiades, P. and A.C. Ferguson-Smith, and G.J. Burton. 2002. Comparative developmental anatomy of the murine and human definitive placentae. Placenta 23(1): 3–19.

Gerard, M. and J. Zakany, and D. Duboule. 1997. Interspecies exchange of a Hoxd enhancer in vivo induces premature transcription and anterior shift of the sacrum. Dev. Biol. 190(1): 32–40.

Gilbert, S.F. and G.A. Loredo, A. Brukman, and A.C. Burke. 2001. Morphogenesis of the turtle shell: the development of a novel structure in tetrapod evolution. Evol. Dev. 3(2): 47–58.

Godwin, A.R. and M.R. Capecchi. 1998. Hoxc13 mutant mice lack external hair. Genes Dev. 12(1): 11–20.

Han, L. and M. Dias Figueiredo, K.A. Berghorn, T.N. Iwata, P.A. Clark-Campbell, I.C. Welsh, W. Wang, T.P. O'brien, D.M. Lin, and M.S. Roberson. 2007. Analysis of the gene regulatory program induced by the homeobox transcription factor distal-less 3 in mouse placenta. Endocrinology 148(3): 1246–1254.

Hemberger, M. 2008. IFPA award in placentology lecture—characteristics and significance of trophoblast giant cells. Placenta 29 Suppl. A: S4–9.

Herault, Y. and J. Beckers, M. Gerard, and D. Duboule. 1999. Hox gene expression in limbs: colinearity by opposite regulatory controls. Dev. Biol. 208(1): 157–165.

Juan, A.H. and F.H. Ruddle. 2003. Enhancer timing of Hox gene expression: deletion of the endogenous Hoxc8 early enhancer. Development 130(20): 4823–4834.

Kamat, A. and M.E. Smith, J.M. Shelton, J.A. Richardson, and C.R. Mendelson. 2005. Genomic regions that mediate placental cell-specific and developmental regulation of human Cyp19 (aromatase) gene expression in transgenic mice. Endocrinology 146(5): 2481–2488.

Kappen, C. and K. Schughart, and F.H. Ruddle. 1989. Two steps in the evolution of Antennapedia-class vertebrate homeobox genes. Proc. Natl. Acad. Sci. USA 86(14): 5459–5463.

Kappen, C. 2000. Disruption of the homeobox gene Hoxb-6 results in increased numbers of early erythrocyte progenitors. Am. J. Hematol. 65: 111–118.

Knox, K. and J.C. Baker. 2008. Genomic evolution of the placenta using co-option and duplication and divergence. Genome Res. 18(5): 695–705.

Ko, K.H. and Q.L. Lam, M. Zhang, C.K. Wong, C.K. Lo, M. Kahmeyer-Gabbe, W.H. Tsang, S.L. Tsang, L.C. Chan, M.H. Sham, and L. Lu. 2007. Hoxb3 deficiency impairs B lymphopoiesis in mouse bone marrow. Exp. Hematol. 35(3): 465–475.

Kuraku, S. and A. Meyer. 2009. The evolution and maintenance of Hox gene clusters in vertebrates and the teleost-specific genome duplication. Int. J. Dev. Biol. 53(5–6): 765–773.

Lewis, E.B. 1978. A gene complex controlling segmentation in *Drosophila*. Nature 276(5688): 565–570.

Li, Y. and R.R. Behringer. 1998. Esx1 is an X-chromosome-imprinted regulator of placental development and fetal growth. Nat. Genet. 20(3): 309–311.

Lim, H. and L. Ma, W.G. Ma, R.L. Maas, and S.K. Dey. 1999. Hoxa-10 regulates uterine stromal cell responsiveness to progesterone during implantation and decidualization in the mouse. Mol. Endocrinol. 13(6): 1005–1017.

Lindsey, S. and M.F. Wilkinson. 1996. Homeobox genes and male reproductive development. J. Assist. Reprod. Genet. 13(2): 182–192.

MacLean, J.A. 2nd and M.A. Chen, C.M. Wayne, S.R. Bruce, M. Rao, M.L. Meistrich, C. Macleod, and M.F. Wilkinson. 2005. Rhox: a new homeobox gene cluster. Cell 120(3): 369–382.

MacLean, J.A. 2nd and D. Lorenzetti, Z. Hu, W.J. Salerno, J. Miller, and M.F. Wilkinson. 2006. Rhox homeobox gene cluster: recent duplication of three family members. Genesis 44(3): 122–129.

Maston, G.A. and M. Ruvolo. 2002. Chorionic gonadotropin has a recent origin within primates and an evolutionary history of selection. Mol. Biol. Evol. 19(3): 320–335.

Miller, C. and D.A. Sassoon. 1998. Wnt-7a maintains appropriate uterine patterning during the development of the mouse female reproductive tract. Development 125(16): 3201–3211.

Morasso, M.I. and A. Grinberg, G. Robinson, T.D. Sargent, and K.A. Mahon. 1999. Placental failure in mice lacking the homeobox gene Dlx3. Proc. Natl. Acad. Sci. USA 96(1): 162–167.

Murasawa, H. and R. Takashima, K. Yamanouchi, H. Tojo, and C. Tachi. 2000. Comparative analysis of HOXC-9 gene expression in murine hemochorial and caprine synepitheliochorial placentae by in situ hybridization. Anat. Rec. 259(4): 383–394.

Nishioka, N. and S. Yamamoto, H. Kiyonari, H. Sato, A. Sawada, M. Ota, K. Nakao, and H. Sasaki. 2008. Tead4 is required for specification of trophectoderm in pre-implantation mouse embryos. Mech. Dev. 125(3–4): 270–283.

Ohya, Y.K. and S. Kuraku, and S. Kuratani. 2005. Hox code in embryos of Chinese soft-shelled turtle *Pelodiscus sinensis* correlates with the evolutionary innovation in the turtle. J. Exp. Zool. B. Mol. Dev. Evol. 304(2): 107–118.

Oudejans, C.B. and M. Pannese, A. Simeone, C.J. Meijer, and E. Boncinelli. 1990. The three most downstream genes of the Hox-3 cluster are expressed in human extraembryonic tissues including trophoblast of androgenetic origin. Development 108(3): 471–477.

Ozturk, A. and L.J. Donald, L. Li, H.W. Duckworth, and M.L. Duckworth. 2006. Proteomic identification of AP2 gamma as a rat placental lactogen II trophoblast cell-specific enhancer binding protein. Endocrinology 147(9): 4319–4329.

Pöpperl, H. and M. Bienz, M. Studer, S.K. Chan, S. Aparicio, S. Brenner, R.S. Mann, and R. Krumlauf. 1995. Segmental expression of Hoxb-1 is controlled by a highly conserved autoregulatory loop dependent upon exd/pbx. Cell. 81(7): 1031–1042.

Pruett, N.D. and R.P. Visconti, D.F. Jacobs, D. Scholz, T. McQuinn, J.P. Sundberg, and A. Awgulewitsch. 2008. Evidence for Hox-specified positional identities in adult vasculature. BMC Dev. Biol. 8: 93.

Quinn, L.M. and B.V. Johnson, J. Nicholl, G.R. Sutherland, and B. Kalionis. 1997. Isolation and identification of homeobox genes from the human placenta including a novel member of the Distal-less family, DLX4. Gene 187(1): 55–61.

Quinn, L.M. and S.E. Latham, and B. Kalionis. 1998. A distal-less class homeobox gene, DLX4, is a candidate for regulating epithelial-mesenchymal cell interactions in the human placenta. Placenta 19(1): 87–93.

Quinn, L.M. and S.E. Latham, and B. Kalionis. 2000. The homeobox genes MSX2 and MOX2 are candidates for regulating epithelial-mesenchymal cell interactions in the human placenta. Placenta 21 Suppl. A: S50–4.

Rajaraman, G. and P. Murthi, L. Quinn, S.P. Brennecke, and B. Kalionis. 2008. Homeodomain protein HLX is expressed primarily in cytotrophoblast cell types in the early pregnancy human placenta. Reprod. Fertil. Dev. 20(3): 357–367.

Raman, V. and A. Tamori, M. Vali, K. Zeller, D. Korz, and S. Sukumar. 2000. HOXA5 regulates expression of the progesterone receptor. J. Biol. Chem. 275(34): 26551–26555.

Rawn, S.M. and J.C. Cross. 2008. The evolution, regulation, and function of placenta-specific genes. Annu. Rev. Cell. Dev. Biol. 24: 159–181.

Ray, R. and M. Capecchi. 2008. An examination of the Chiropteran HoxD locus from an evolutionary perspective. Evol. Dev. 10(6): 657–670.

Rossant, J. and J.C. Cross. 2001. Placental development: lessons from mouse mutants. Nat. Rev. Genet. 2(7): 538–548.

Sakamoto, K. and K. Onimaru, K. Munakata, N. Suda, M. Tamura, H. Ochi, and M. Tanaka. 2009. Heterochronic shift in Hox-mediated activation of sonic hedgehog leads to morphological changes during fin development. PLoS ONE 4(4): e5121.

Saleh, M. and I. Rambaldi, X.J. Yang, and M.S. Featherstone. 2000. Cell signaling switches HOX-PBX complexes from repressors to activators of transcription mediated by histone deacetylases and histone acetyltransferases. Mol. Cell. Biol. 20(22): 8623–8633.

Schughart, K. and C. Kappen, and F.H. Ruddle. 1989. Duplication of large genomic regions during the evolution of vertebrate homeobox genes. Proc. Natl. Acad. Sci. USA 86(18): 7067–7071.

Sears, K.E. and R.R. Behringer, J.J. Rasweiler, and L.A. Niswander. 2007. The evolutionary and developmental basis of parallel reduction in mammalian zeugopod elements. Am. Nat. 169(1): 105–117.

Shashikant, C.S. and C.J. Bieberich, H.-G. Belting, J.C.H. Wang, M.A. Borbely, and F.H. Ruddle. 1995. Regulation of Hoxc-8 during mouse embryonic development: identification and characterization of critical elements involved in early neural tube expression. Development 121: 4339–4347.

Shashikant, C.S. and S.A. Bolanowsky, S. Anand, and S.M. Anderson. 2007. Comparison of diverged Hoxc8 early enhancer activities reveals modification of regulatory interactions at conserved cis-acting elements. J. Exp. Zool. B. Mol. Dev. Evol. 308(3): 242–249.

Shaut, C.A. and D.R. Keene, L.K. Sorensen, D.Y. Li, and H.S. Stadler. 2008. HOXA13 Is essential for placental vascular patterning and labyrinth endothelial specification. PLoS Genet. 4(5): e1000073.

Simmons, D.G. and D.R. Natale, V. Begay, M. Hughes, A. Leutz, and J.C. Cross. 2008. Early patterning of the chorion leads to the trilaminar trophoblast cell structure in the placental labyrinth. Development 135(12): 2083–2091.

Smith, C.C. and H.S. Taylor. 2007. Xenoestrogen exposure imprints expression of genes (Hoxa10) required for normal uterine development. FASEB J. 21(1): 239–246.

Srebrow, A. and Y. Friedmann, A. Ravanpay, C.W. Daniel, and M.J. Bissell. 1998. Expression of Hoxa-1 and Hoxb-7 is regulated by extracellular matrix-dependent signals in mammary epithelial cells. J. Cell. Biochem. 69(4): 377–391.

Sun, Y. and M.L. Duckworth. 1999. Identification of a placental-specific enhancer in the rat placental lactogen II gene that contains binding sites for members of the Ets and AP-1 (activator protein 1) families of transcription factors. Mol. Endocrinol. 13(3): 385–399.

Takeshita, K. and J.A. Bollekens, N. Hijiya, M. Ratajczak, F.H. Ruddle, and A.M. Gewirtz. 1993. A homeobox gene of the Antennapedia class is required for human adult erythropoiesis. Proc. Natl. Acad. Sci. USA 90(8): 3535–3538.

Taylor, H.S. and G.B. Vanden Heuvel, and P. Igarashi. 1997. A conserved Hox axis in the mouse and human female reproductive system: late establishment and persistent adult expression of the Hoxa cluster genes. Biol. Reprod. 57(6): 1338–1345.

Tian, G. and U. Singh, Y. Yu, B.S. Ellsworth, M. Hemberger, R. Geyer, M.D. Stewart, R.R. Behringer, and R. Fundele. 2008. Expression and function of the LIM homeobox containing genes Lhx3 and Lhx4 in the mouse placenta. Dev. Dyn. 237(5): 1517–1525.

van Nes, J. and W. de Graaff, F. Lebrin, M. Gerhard, F. Beck, J. Deschamps. 2006. The Cdx4 mutation affects axial development and reveals an essential role of Cdx genes in the ontogenesis of the placental labyrinth in mice. Development. Feb; 133(3): 419–428.

Wagner, K. and A. Mincheva, B. Korn, P. Lichter, and H. Popperl. 2001. Pbx4, a new Pbx family member on mouse chromosome 8, is expressed during spermatogenesis. Mech. Dev. 103(1–2): 127–131.

Wang, J. and J. Mager, E. Schnedier, and T. Magnuson. 2002. The mouse PcG gene eed is required for Hox gene repression and extraembryonic development. Mamm. Genome. 13(9): 493–503.

Wang, X. and J. Zhang. 2006. Remarkable expansions of an X-linked reproductive homeobox gene cluster in rodent evolution. Genomics 88(1): 34–43.

Watson, E.D. and J.C. Cross. 2005. Development of structures and transport functions in the mouse placenta. Physiology (Bethesda) 20: 180–193.

Wellik, D.M. 2009. Hox genes and vertebrate axial pattern. Curr. Top. Dev. Biol. 88: 257–278.

Yue, L. and T. Daikoku, X. Hou, M. Li, H. Wang, H. Nojima, S.K. Dey, and S.K. Das. 2005. Cyclin G1 and cyclin G2 are expressed in the periimplantation mouse uterus in a cell-specific and progesterone-dependent manner: evidence for aberrant regulation with Hoxa-10 deficiency. Endocrinology 146(5): 2424–2433.

Zhang, Y.M. and B. Xu, N. Rote, L. Peterson, and L.S. Amesse. 2002. Expression of homeobox gene transcripts in trophoblastic cells. Am. J. Obstet. Gynecol. 187(1): 24–32.

Zhao, Y. and S.S. Potter. 2001. Functional specificity of the Hoxa13 homeobox. Development 128(16): 3197–3207.

Section 3

Pattern and Process at the Base of the Metazoan Tree of Life

Chapter 12

Field Biology of Placozoans (*Trichoplax*): Distribution, Diversity, Biotic Interactions

Vicki Buchsbaum Pearse[1]* and *Oliver Voigt*[2]

Introduction

The goal of this review is to highlight what little is known, and point to the bulk of what is yet to be learned, about the natural history of placozoans in the field—in order to stimulate a broader search for placozoans and a fuller exploration of their distribution, diversity, and all other aspects of their enigmatic lives. The documented geographic distribution of placozoans lies mostly in the nearshore, warm, marine waters of the tropics and subtropics. Although placozoans have long been viewed as benthic organisms, they can be more readily collected from the water column, well above the sea bottom. The full life-history of placozoans is unknown, including the nature of this abundant pelagic phase and all details of sexual reproduction and development. We note observations on the biota associated with placozoans in field collections, in particular the other regular members of the microcommunity in which placozoans occur on our collecting plates and on some factors influencing this assemblage. Among the animals found are some potential predators against which placozoans appear to be defended. Yet to be uncovered is the full breadth of

[1]Long Marine Laboratory, Institute of Marine Sciences, University of California, Santa Cruz, CA 95060, USA.
E-mail: vpearse@ucsc.edu, vpearse@gmail.com
[2]Department of Earth- and Environmental Sciences & GeoBioCenterLMU, Ludwig-Maximilians-Universität München Richard-Wagner-Str. 10, 80333 München, Germany.

diversity in this phylum, certainly under-represented by its single named species. We summarize here the distributions for known haplotypes; report specimens from a cool-temperate coast, including a new haplotype; and review the evidence that many more surprises almost surely await discovery. We also describe some methods for collecting and handling these small, fragile animals.

As the simplest of living metazoans, *Trichoplax adhaerens* Schulze, 1883, phylum Placozoa, holds a special fascination for biologists (Fig. 1). Most phylogenetic analyses, both morphological and DNA-based, place it among the so-called diploblasts , but its position within the animal tree remains controversial (e.g., Halanych 2004; Dellaporta et al. 2006; Cartwright and Collins 2007; Lavrov 2007, 2011; Schierwater and DeSalle 2007; Signorovitch et al. 2007 ; Hejnol et al. 2009; Philippe et al. 2009; Mallatt et al. 2010; Pick et al. 2010). Studies on laboratory-cultured animals have provided substantial background on structure and ultrastructure (reviewed by Grell and Ruthmann 1991) and on cell and molecular biology (e.g., quantitative DNA measurements, Ruthmann 1977; evidence of a single Hox gene, Schierwater and Kuhn 1998; documentation of a circular mitochondrial (mt) genome, Ender and Schierwater 2003; complete

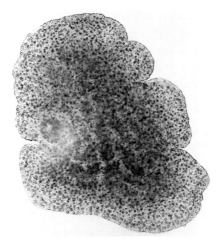

Fig. 1. Placozoan. W. Samoa. (K.J. Marschall). This individual displays the plasticity and asymmetry of shape typical of larger placozoans; small ones are roughly circular in outline. Placozoans are the simplest metazoans known. They look like thin microscopic pancakes, commonly less than a millimeter across and only about 20 µm thick. The nearly transparent body consists of upper and lower epithelia, both flagellated and enclosing between them a syncytial network of contractile cells. Distributed throughout the upper epithelium are the so-called shiny spheres, lipid-filled vesicles; in this photo, they are most visible around the margin of the animal. The histology is extremely simple, with only a few cell types and no basal lamina or other extracellular matrix. The specialized cells, tissues, or organ systems typical of most animals (digestive, circulatory, excretory, neurosensory, and muscular) are also absent.

mt genome sequences, Dellaporta et al. 2006, Signorovitch et al. 2007; expression of developmental genes, Martinelli and Spring 2003, Jakob et al. 2004, Hadrys et al. 2005). Reviews of research on *Trichoplax* (Syed and Schierwater 2002; Miller and Ball 2005; Schierwater 2005; Pearse and Voigt 2007) will need frequent updating, as growing interest in these animals accelerates the rate of publications. *Trichoplax* has joined the elite list of model animals whose entire genome has been sequenced (Srivastava et al. 2008).

Thus, the rocket science is well launched. In contrast, almost no one has bothered to glance down at the launch pad: little is known of the most basic biology of the living animals in the field—their biogeography, habitat, behavior, biotic interactions, or other aspects of their lives. Grell (e.g., 1981) pioneered the first outlines of their distribution by extracting placozoans from samples of seaweeds and other potential substrates collected from several tropical and subtropical sites. Decades later, we have still barely begun to document the rich taxonomic diversity that living placozoans promise (Voigt et al. 2004; Signorovitch et al. 2006; Eitel and Schierwater 2010). The goal of this chapter is to update our recent review (Pearse and Voigt 2007). We hope to stimulate a broader search for placozoans; fuller exploration of their distribution and diversity; and better understanding of all aspects of their biology.

Since first working with these animals in Hawaii in 1986, VBP has continued to sample for placozoans at many locales, in marine laboratory facilities and in the sea, in order to further document where they are found. Both authors have contributed samples, including two new haplotypes, to the genetic and phylogenetic analysis presented here. Additional information about the field biology of placozoans, mostly fortuitous, was collected as opportunities arose. Also described are techniques for collecting and handling these minute, delicate organisms.

Collection and Observation

VBP collected placozoans on standard glass microscope slides (25 x 76 mm) placed in the sea or in laboratory seawater systems, usually in racks made from plastic slide boxes, with most of the top and bottom cut out to allow water to circulate through them. The slides were spaced ~10–15 mm apart. The top and bottom of the rack were then tied together with nylon monofilament line or cable ties, and the rack was suspended from any handy support (e.g., a coral branch, mangrove prop, or dock) (Fig. 2). Racks were oriented so that the surfaces of the slides were either parallel or perpendicular to the surface of the water. Some racks were placed directly on the substrate, with the slides roughly perpendicular to the surface of the water.

Fig. 2. Racks of glass microscope slides hanging on a coral reef, Sesoko Island, Okinawa, Japan. Slides suspended in the water capture more placozoans than do slide racks placed on the substrate (see text, Benthos versus water column).

Single microscope slides were sometimes suspended in the water (weighted with a small rock by means of a rubber band slipped around the slide) or placed directly on the substrate. Being less conspicuous, these were less likely to attract tampering.

When the slides were retrieved, typically after 10 to 15 days, the racks of slides were placed into securely lidded plastic boxes filled with seawater *in situ*, so that the slides would be exposed neither to air nor to seawater from another location or depth. Loose slides were placed into racks and treated the same way. Returned to the laboratory, the slides were kept and examined in the same water, at room temperature, within hours or no more than 1–2 days.

To examine a slide for placozoans, it was quickly transferred to a Petri dish containing enough seawater to cover the slide. The slide was supported on small bits of rock, so that while the upper surface was being examined, organisms on the lower surface were protected from rubbing against the bottom of the dish. Illumination from a mirror or light below the dish was adjusted to optimal angle and intensity, and both sides of the slide were scanned systematically under a dissecting microscope. Placozoans were counted; other organisms were recorded as present.

Factors affecting the likelihood of success in capturing placozoans under various conditions are noted below (see Macrohabitat and Microhabitat).

Strengths and Limitations of Collection Methods

The first published finding of placozoans on glass microscope slides suspended in the sea was by Sudzuki (1977) in Sagami Bay, Japan. This method was used from the start of studies by VBP in 1986, and others have followed suit (e.g., Maruyama 2004; Signorovitch et al. 2006), because placozoans are almost invisible on any substrate except clear glass. Sampling with glass slides is also a relatively benign process, causing little or no disturbance to the benthic habitat; there is some by-catch, but it is minor on an oceanic scale.

Capture of a placozoan on a slide establishes its presence at a known location and depth, provided that the slide is kept in seawater from the original site or in adequately filtered seawater. However, because settlement may have occurred at any time after the slide was placed, the time cannot be known precisely. Likewise, the number of placozoans observed on a slide depends on where and for how long a slide is exposed and results from some combination of settlement and fission, as well as possible loss of individuals over the period of exposure and increasing difficulty in seeing placozoans as the slide becomes overgrown with other biota. Thus, only semi-quantitative sampling has been achieved.

Assuming that two or more slides bearing placozoans represent independent settlement events, reporting the number of positive slides, as well as the number of placozoans on each, adds information (e.g., Maruyama 2004). The alternative of estimating biomass (versus number) is problematic (Pearse 1989). Another method is to collect pieces of rock or shell and shake them in a plastic bag filled with seawater (Maruyama 2004). Although faster, this process further obscures the number of placozoans present, because many become broken into fragments. However, one is assured of direct sampling of the benthic phase (provided the seawater is well filtered), whereas the stage of the life-history that settles on glass slides, either suspended or placed on the bottom, is uncertain.

Also, as with any small, easily transported planktonic organism, the location of the source population on the benthos is difficult to determine. Thus, although more than one clade of placozoans can often be found together in a single sample (Voigt et al. 2004; Signorovitch et al. 2006), interpretations of sympatry are complicated by the limitations of the sampling method. Placozoans swimming or drifting in the water column settle on the suspended slides. However, two or more placozoans settling on a single slide may have originated from benthic individuals in quite separate and different microhabitats. Lacking any definition of "an ecologically relevant spatial scale" (Signorovitch et al. 2006), conclusions of sympatry may not apply to the adjacent benthic habitat, but strictly, only to sympatry in the water column.

Distribution

Biogeography

Sites where placozoans are known to have been collected are summarized on a global map (Fig. 3, see also Eitel and Schierwater 2010.). The annotated list below includes published records [P], previously unpublished observations of either of this review's authors [A], and personal communications from other biologists [C]. Reported haplotypes of the mitochondrial 16S large subunit ribosomal RNA gene (16S rDNA) are given in bold (using the numbering of Voigt et al. 2004 and of this publication). This list includes several additions to the one previously published (Pearse and Voigt 2007). Given the ubiquity of placozoans, however, and the extremely limited extent of collecting achieved to date, this collection of sites must not be considered even close to comprehensive. On the contrary, it will certainly be outdated even by the time this volume is published and increasingly so with time.

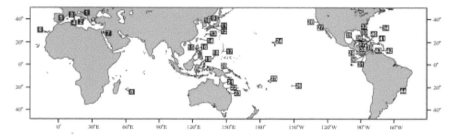

Fig. 3. Geographic distribution of reported occurrences of placozoans. The numbers correspond to those of the sites listed immediately below.

Mediterranean and Eastern Atlantic

(1) Gulf of Trieste, Adriatic Sea. Schulze 1883; Stiasny 1903. Schulze's type locality for the species *Trichoplax adhaerens*. [P]

(2) Gulf of Naples, Tyrrhenian Sea. Monticelli 1893, 1896. Monticelli's type locality for the species Treptoplax reptans Monticelli 1896, currently not recognized. [P]

(3) Orbetello Lagoon, Tyrrhenian Sea. Tomassetti et al. 2005 [P] **H2,5**

(4) Tunisia. Eitel and Schierwater 2010 [P]

(5) Almuñecar coast (Granada), Spain. Ocaña and Ibáñez 2006. [P]

(6) Tenerife, Canary Islands. Eitel and Schierwater 2010 [P]

Roscoff, Brittany, France. Ivanov 1973 [P]

Red Sea and Indian Ocean

(7) Eilat, Red Sea. Grell 1971 [P] **H1**
(8) La Reúnion. N. Gravier-Bonnet 2001 [C] Pacific
(9) Presumed from Sea of Japan. Ivanov et al. 1980 [P] (similar to **H1**)
(10) Oki Islands, Sea of Japan. Y.K. Maruyama [C]
(11) Shimoda, southcentral Honshu, Japan. Sudzuki 1977 [P]
(12) Shirahama, south Honshu, Japan. V. Pearse 1989 [A], Maruyama 2004 [P]
(13) Okinawa, Ryukyu Islands, Japan. V. Pearse 1989 [A], Pearse et al. 1994 [P]
(14) Iriomote, Ryukyu Islands, Japan. V. Pearse 1989 [A]
(15) Hong Kong. V. Pearse 2004 [A]
(16) Zambales, Philippines, from material transported to Monterey Bay Aquarium, Monterey, California. V. Pearse 1992 [A]
(17) Guam. Univ. Guam Marine Lab. V. Pearse 1988 [A], Voigt et al. 2004 [P] **H8**
(18) Palau. V. Pearse 1988 [A]
(19) North of Manado, northeast Sulawesi Indonesia. V. Pearse 1994 [A]
(20) Madang, Papua New Guinea. V. Pearse 1988 [A]
(21) Lizard Island, Great Barrier Reef, northeastern Australia. O. Voigt 2006 [A] **LIZ** (similar to **H9**)
(22) Orpheus Island, Great Barrier Reef, northeastern Australia. V. Pearse 1988 [A]
(23) Heron Island, Great Barrier Reef, northeastern Australia. A.T. Newberry 1988 [C]
(24) Oahu, Hawaii. Kewalo Marine Lab. M.G. Hadfield ~1978 [C], V. Pearse 1989 [P]. E. Gaidos 2007: **H1,2,4,6** [C]
(25) Western Samoa. K.J. Marschall, films, 1970, and cited in Grell 1980 [P], K.J. Marschall and V.Pearse 1989 [A]
(26) Moorea, French Polynesia. V. Pearse 1989 [A]
(27) Probably from southern California. V. Pearse 2006 [A] **H11**
(28) Monterey Bay, California. V. Pearse 2007 [A] **H2,17**
(29) Achotines Laboratory, Azuero Peninsula, Panama. Voigt et al. 2004 [P] **H4**
(30) Isla Iguana, Azuero Peninsula, Panama. Voigt et al. 2004 [P] **H6,8**
(31) Naos Island Laboratories, Panama. Voigt et al. 2004 [P] **H4,7**

Caribbean and Western Atlantic

(32) Southeast Atlantic coast of US: North Carolina, Grell 1980 (R.M. Rieger, personal communication); South Carolina, Klauser 1982 [P]
(33) Southeast Atlantic coast of US: Florida. V. Pearse 2006 [A]

(34) Bermuda. Grell 1980 [P]. T. Syed, S. Sagasser 2001 [C], Signorovitch et al. 2006 [P] **H2,4,9,10**
(35) Puerto Morelos, Quintana Roo, Mexico. Grell and López-Ochoterena 1988 [P]
(36) Carrie Bow Cay and Twin Cays, Belize. Signorovitch et al. 2006. [P] **H2-4,6-8**
(37) Roatan, Honduras. V. Pearse 1985 [A]
(38) Bocas del Toro, Panama. Voigt et al. 2004; Signorovitch et al. 2006 [P] **H1-4,8**
(39) Galeta, Panama. V. Pearse 1997 [A]
(40) Discovery Bay, Jamaica. Signorovitch et al. 2006 [P] **H1,8**
(41) Western Puerto Rico. J.M. Ward 2007 [C]
(42) Grenada. Signorovitch et al. 2006 [P] **H6**
(43) Cubagua Island/Margarita Island, Venezuela. Voigt et al. 2004 [P] **H4**
(44) São Sebastião Channel, São Paulo State, Brazil. Morandini et al. 2006 [P]

Macrohabitat and Microhabitat

VBP found placozoans in abundance on coral reefs and in full-salinity mangroves. Results can be locally patchy, both positive and negative samples being obtained from a single site. Nonetheless, consistently negative samples or smaller yields in certain types of habitats or conditions point to factors restricting placozoan distribution. No systematic survey has been attempted but examples of findings are briefly summarized here (see also Pearse 1988).

Seasonality/Temperature

Placozoans were found on slides in samples taken at various tropical sites and all months of the year. However, no long-term surveys of placozoans at a single location have been done at tropical latitudes. At one subtropical site, Sesoko Island, Okinawa, Japan, where sea surface temperatures vary seasonally (means 20–28°C; Loya et al. 2001), collections were made at a site close to the Sesoko Marine Laboratory from March to August. In March and April, when sea temperatures at Sesoko average ~21–22°C, all samples were negative; only during May–August, when temperatures average ~24–28°C, were placozoans found on slides. Few sites from temperate latitudes have been examined; some data suggest possible seasonality in abundance at Shirahama, Japan (Maruyama 2004). A 1992 survey by VBP in Monterey Bay, central California, where annual sea surface temperatures range ~11–18°C, failed to yield any specimens of placozoans. In 2007, however, slides retrieved from Monterey's harbor yielded a number of specimens belonging to two haplotypes: H2, widely distributed in warmer waters also (Fig. 4), and a new haplotype, H17 (for haplotypes H12-16, see Eitel and Schierwater 2010). Sea surface temperatures during the period when the slides were deployed (September–October) ranged ~13–17°C (measured at Hopkins Marine Station of Stanford University, 2 km from the sampling site). Another example of placozoans occurring under cool temperate conditions (11–14°C, October) was reported by Tomasetti et al. (2005) from the Orbetello Lagoon (western Mediterranean), and those specimens likewise included members of H2 and of a new haplotype, H5, not yet found elsewhere. Although we can now expect to find placozoans readily at tropical and subtropical temperatures, we

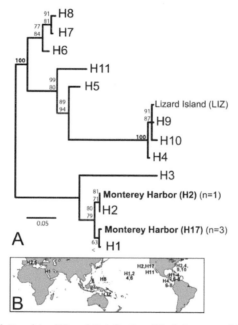

Fig. 4. Phylogenetic relationships (A) and distribution (B) of placozoan 16S rDNA haplotypes. **(A)** Phylogenetic tree (MrBayes), support values >50 are shown at the corresponding nodes; Bayesian clade credibility values (posterior probabilities x100) are shown above bootstrap values (from 500 replicates. Accession numbers from GenBank sequences are H1: AY652522; H2: AY652523; H2 (Monterey Bay): EU700987; H3: AY652524; H4: AY652525; H9 (=H4-2): DQ389828+DQ389763; H10 (=H4-3): DQ389825+DQ389760; H5: AY652526; H6: AY652527; H7: AY652528; H8: AY652529; LIZ: EF421454, H11: EF421455; H17: EU700986. **(B)** Geographic distribution of known occurrences of haplotypes. Specific locations are listed under Biogeography. H7 comprises, besides the points shown, a single specimen of uncertain origin, probably Indo-Pacific, isolated from an aquarium. See text under Molecular analysis and haplotype distribution for notes on the origins of H11 and H17. Unpublished data for Hawaii courtesy of J.M. Ward and E. Gaidos. See also Eitel and Schierwater (2010) for additional haplotypes and haplotype distributions.

must evidently begin to expand our search into cooler waters, both for relatively widespread, apparently more eurythermal haplotypes such as H2 and for the possibility of specifically cool-adapted lineages, such as H5 and H17 might represent.

A search was made specifically for placozoans, using the same methods, but no placozoans were found at two sites illustrating extreme cold temperature conditions: $-1.6°C$ in McMurdo Sound, Antarctica (Pearse and Pearse 1991) and ~3°C in the Monterey Canyon, central California, ~1000–3000 m depth.

Salinity

In laboratory trials at the Christensen Research Inst. (CRI), Madang, Papua New Guinea, exposure to seawater of reduced salinity rapidly killed placozoans: exposure for 1 h to seawater diluted to 75% or 0.5 h in seawater diluted to 50% proved 100% lethal. After only 8 min or 1 min, respectively, in the test solutions, the edges of the animals began to curl, and the animals detached from the glass and became motionless and unresponsive to touch. Placed again into full salinity seawater (32.5‰), some individuals showed improvement, but all died within <24 h. This intolerance for seawater of reduced salinity was evident in field collections, e.g., nearshore, shallow, previously positive sites near CRI were negative after rains. In contrast, placozoans have been found to tolerate elevated salinities in the lab, up to 40–50‰ (A. Signorovitch and L. Buss, pers. comm., 2007).

Currents and wave surge

In a narrow channel with strong tidal flow, samples were consistently small or negative, relative to samples from the mangrove pond and fringing reef that the channel connected (Kranket Island, Madang, Papua New Guinea). Samples from reef flats with strong currents likewise failed to yield placozoans (Guam, Western Samoa), even in places where placozoans were recovered from nearby protected inlets (Cook Bay and Opunohu Bay, Moorea, French Polynesia). See also Associated biota, below.

Sandy bottoms

Almost all samples were negative in seagrass beds (Madang, Papua New Guinea; Apia, Western Samoa) and other sandy habitats. For example, at Orpheus Island, Great Barrier Reef, Australia, placozoans were found on slides placed on a fringing coral reef, but not at a site only ~200 m away where slides were suspended over a sandy bottom. However, placozoans were abundant as meiofauna in intertidal sand of Tamare Beach on

Moorea, French Polynesia (R. Hochberg, pers. comm., 2009). We know of no other reports of placozoans in sand, and this type of habitat should be further explored.

Benthos versus water column

Although placozoans have long been viewed as benthic organisms, they were regularly collected on glass slides suspended in the water column well above the bottom (Fig. 2). Moreover, when paired samples were compared (free slides or slide-racks placed on the substrate versus others hung in shallow water at the same site 20–60 cm above the bottom), those suspended in the water column bore significantly higher numbers of placozoans. For example, for 12 sets of 8 slides each in Madang, Papua New Guinea, the number of placozoans was significantly greater on slides suspended above the bottom versus slides on the bottom (CRI dock, mean ± S.D. was 8.1 ± 3.4 for slides suspended above the bottom, 0.23 ± 0.32 for slides on the bottom, t-test, p=0.02; Kranket Island, mean ± S.D. was 13.9 ± 3.1 for slides suspended above the bottom, 3.4 ± 2.0 for slides on the bottom, t-test, p<0.01). Comparison of VBP's midsummer results in Shirahama, Japan, compared to those of Maruyama (2004) likewise indicate a strong effect of sampling off the bottom versus on the bottom. VBP found large numbers of placozoans on slide racks suspended in the water column in July, comparable to yields during the fall peaks documented by Maruyama (2004, his fig. 3), whereas his slide racks, always placed on the substrate, captured very few or no specimens in July.

Substrate orientation

At the CRI dock, Madang, Papua New Guinea, placozoans were counted on two sets of 8 slides each, hung in the sea for 17 days about 20 cm off the bottom so that their surfaces were horizontal (parallel to the water surface) versus vertical (perpendicular to the water surface). The number of placozoans per slide was not statistically different (mean ± S.D. was 5.9 ± 3.4 for a set of 8 horizontal slides, 5.0 ± 3.3 for a set of 8 vertical slides, t-test, p=0.61). In contrast, on horizontal slides, the number of placozoans was significantly greater on lower versus upper surfaces (mean ± S.D. was 5.5 ± 0.96 for lower surfaces, 2.0 ± 0.93 for upper surfaces for 3 sets of slides, t-test, p=0.01). Similar results were obtained from a trial at Sesoko Island, Okinawa, Japan (mean ± S.D. was 3.1 ± 1.6 for lower surfaces, 0.17 ± 0.13 for upper surfaces, for 4 sets of slides, t-test, p=0.01).

Remarks: Biogeography and Habitats

Given the limitations of collection methods, we can nonetheless draw some broad conclusions about the distribution of placozoans in terms of biogeography and at other scales. First, they have been widely documented in tropical and subtropical nearshore marine habitats, especially coral reefs and mangroves, as well as, more surprisingly, in a few cool-temperate locations. VBP seldom recovered them near sandy habitats, and this was also the most common experience of Signorovitch et al. (2006), even when hard substrates positive for placozoans were nearby. Although this might suggest a measurable limit on the distance that placozoans can travel between adjacent habitats, the number of observations is few and cannot be evaluated without information on the direction of currents and other confounding influences. Protected areas yield more placozoans than exposed sites with strong currents or wave-surge. We would expect to find fewer placozoans in areas of freshwater runoff, as near river mouths, as much because of the quantities of sediment and soft bottoms as because of their intolerance of reduced salinities.

Although whole biogeographic regions, such as the Indian Ocean, remain scantly sampled on the map of placozoan distribution, sampling in those waters will certainly yield additional positive results. Indeed, any further success in warm marine waters will be unsurprising, though all additions to the known diversity will surely be welcome. In contrast, our expectations of finding placozoans in deep or polar waters remain minimal. Only recently (Pearse and Voigt 2007), we predicted the same negative potential for coasts characterized by cold upwelling. Our recent recovery of placozoans from the cool temperate waters of Monterey in central California, a coast that experiences regular cold upwelling and has been fairly intensively studied, was unexpected. Such coasts should now be considered as a potentially promising focus as we pursue a more extended global search for placozoans. To our knowledge, no one has looked for placozoans in the open ocean. Where collections are from laboratory tanks or commercial or personal tropical aquariums rather than directly from the sea, some uncertainty about the origins of the specimens is inevitable because placozoans can easily be introduced with exotic materials.

Placozoans exhibit an abundant nearshore pelagic phase and can be efficiently collected from the water column, well above the sea bottom. In paired samples, slide racks hung suspended in the water column yielded more placozoans than did those placed directly on the substrate. The latter collect more silt and may be subject to disturbance from other benthic animals, or perhaps the pelagic phase settling on the slides is less numerous close to the bottom than within the water column. Compared to freestream flow in the middle of the water column, flow along the bottom is expected to be slower and therefore to supply fewer settlers per unit time; arriving settlers, however, probably attach more easily in slower flow. Thus, no simple conclusion emerges from the hydrodynamics.

The larger number of placozoans found on the lower surface of slides might relate to the greater amount of silt and ultraviolet radiation on the upper surface. Ultraviolet may be very important in restricting the microhabitats of these organisms and should be examined. (Placozoans have been observed to react strongly when exposed to ultraviolet radiation from laboratory sources; their sensitivity under field conditions has never been tested). Serpulids and some other organisms also appeared to settle preferentially on or migrate to the lower surface, perhaps responding to the same factors. Alternatively, placozoans might be responding directly to other biota. A study of temporal succession in this microcommunity could provide some answers. No difference was observed between upper and lower surfaces under laboratory conditions (e.g., Pearse 1989) and placozoans settled preferentially on a biofilmed surface, whether upper or lower. Thus, this differential is likely related to secondary factors present only in the field rather than directly to the orientation of the substrate.

Relations to Other Organisms

Associated biota

At some time or other, small representatives of most phyla of sessile invertebrates, as well as a great variety of protists and algae, have been seen on VBP's glass slides together with placozoans. Nonetheless, a few particular organisms often dominate the microcommunity that develops on glass slides and occur together with placozoans with sufficient regularity that one may almost predict placozoans by the presence of these other associates, in particular several kinds of sessile ciliates: solitary and colonial vorticellids as well as folliculinids; spirorbid and other serpulid polychaetes; and, in smaller numbers, free-living loxosomatid kamptozoans (entoprocts). Especially remarkable was the repeated finding of this same assemblage, though not necessarily represented by the same species, together with placozoans, at a variety of sites extending across the tropical Pacific and into the Caribbean Sea. Occasionally, the typical assemblage would be present except for placozoans; even more rarely, placozoans were found in the absence of their usual associates.

The association may be roughly illustrated by two semi-quantitative vertical transects (Table 1). At an exposed, wave-swept coral pinnacle, Tripod Reef, near CRI Madang, Papua New Guinea, slides were placed at intervals from near the surface down to 20 m in depth. Along this transect, placozoans were found only at the three deepest stations, 18–20 m, largely below the influence of wave-surge, and the associated microcommunity displayed the same distribution. In another vertical transect in a quiet part of Opunohu Bay, Moorea, French Polynesia, placozoans and the associated microcommunity were present in comparable abundance throughout the range in depth sampled, down to 16 m (Table 1).

On a smaller scale, the association between serpulids and placozoans, was seen even in their distribution on each single glass slide. For example, in Madang, Papua New Guinea, data for placozoans (see Substrate orientation, above) paralleled that for small spirorbids: on horizontal slides, the number of spirorbids (diameter 200–500 µm) was significantly greater on lower versus upper surfaces (mean/slide ± S.D. was 11.4 ± 4.1 for lower surfaces, 2.9 ± 1.6 for upper surfaces for a set of 9 slides, t-test, $p<<0.001$).

Table 1. Two vertical transects, on Tripod Reef, Madang, Papua New Guinea from 3–20 m, and in Opunohu Bay, Moorea, French Polynesia from 3–16 m. Semi-quantitative estimates of the abundance of placozoans and three other commonly associated organisms were made from glass slides placed down the reef slope at intervals of several meters. The upper part of Tripod Reef experiences strong wave surge, whereas the reef at Opunohu Bay is in a quiet, protected situation. Plac = placozoans; Vort = colonial vorticellids (Ciliata); Foll = folliculinids (Ciliata); Spir = spirorbids (Polychaeta). (–) denotes absence; (+, ++, +++) denotes increasing abundance.

Depth (m)	Madang				Moorea			
	Plac	Vort	Foll	Spir	Plac	Vort	Foll	Spir
3	–	+	–	+	+	++	–	–
6	–	–	–	+	+++	++	–	+
8	–	–	–	++				
9					+	++	+	+
10	–	–	–	++				
12	–	–	–	++	++	++	+	+
14	–	+	–	+++	+	++	+	+
16	+	+	–	+++	+	++	+	+
18	+	+	+	+++				
20	+	+++	+	+++				

Diet

Placozoans grew and multiplied on glass slides bearing a varied assemblage of small organisms that had settled on slides suspended in the sea; typically, we were unable to discern what they were eating. In Madang, Papua New Guinea, some slides that bore an abundant film of green algae supported unusually large placozoans. In Western Samoa, conspicuous feeding tracks were visible in the thin, red, algal film growing on aquarium glass, and at the end of each track was a pink placozoan.

Encounters with Potential Predators and Other Organisms

Despite the diversity of protists and animals occurring together with placozoans on the glass slides, and numerous interactions, not a single instance of predation was seen, either on or by placozoans. In several cases, however, potential predators were observed to recoil after contact with placozoans or reject them as food. At Orpheus Island, Great Barrier Reef, Australia, a small snail was watched as it headed straight for a placozoan, clearing a swath with its radula as it went. Just before reaching the placozoan, the snail extended one tentacle anteriorly, touched the placozoan, recoiled abruptly, turned, and proceeded in another direction. Similarly, at Madang, Papua New Guinea, a small rhabdocoel flatworm crawling on a slide among placozoans was observed contacting a placozoan, recoiling vigorously, and changing its direction of movement only to encounter another placozoan and repeat this behavior several times. A swimming placozoan caught in the feeding current of a small sabellid tubeworm was passed down the food-groove to the mouth, where it was rejected, flicked out into the water, and then caught again; this process was repeated six times, until the placozoan finally contacted the substrate, adhered, and crept away, apparently unharmed. Various ciliates, tiny nematodes, and other small animals sometimes crawled around or under placozoans without displaying any reaction or evoking any visible response from the placozoans. Ciliates were observed to gather around wounded or degenerating placozoans (Pearse 1989).

In none of these instances did the other organisms appear to be harmed by the placozoans. In contrast, in laboratory trials carried out by Jackson and Buss (2009), placozoans that were dropped onto the tentacles of three species of hydroids caused paralysis. After transferring the placozoans from the tentacles to the mouth, the polyp became immobile and unresponsive. By 24 h after contact with a single placozoan, some polyps had recovered, but exposing a polyp to placozoans two days in a row was fatal. Only small placozoans fed to large polyps were ever swallowed; most placozoans, or fragments of them, were able to crawl away.

Remarks: Relations to Other Organisms

Associated biota

Studies of the assemblage of microbiota that includes placozoans share many common problems, a major one being the difficulty of determining the natural substrates of these organisms that we see settling together on glass slides. Attempts to examine preferred placozoan substrates by collecting seaweeds, corals, and other possible habitats, and placing glass slides among them, typically failed because equal numbers of placozoans settled in the no-substrate controls, an early hint of the presence of placozoans in the water. Minute, solitary loxosomatids were thought to live exclusively as commensals on other organisms, almost certainly because they were observed only incidentally by biologists whose interest was in their hosts. Using glass slides as settling plates, however, T. Iseto has described numerous new species of free-living loxosomatids (e.g., Iseto 2002), and in the course of his sampling has, not coincidentally, also captured placozoans (T. Iseto, pers. comm., 2004).

No specific interactions were ever observed by VBP between placozoans and the members of the associated microcommunity, and their relationship must be presumed to consist of no more than a tendency to settle under a shared set of conditions, although it remains possible that the settlement of one positively influences that of another. Their patterns of abundance, similar among the taxa but differing between vertical transects at two quite different sites (Table 1), appear most easily explained by their common response to differences in wave action (see Associated biota, above).

Diet

In the laboratory, placozoans grow on a diet of cryptomonads, other algal unicells, or heat-killed nauplii of *Artemia* (Grell 1983; Grell and López-Ochoterena 1988) or on commercial food for aquarium filter-feeders (M.G. Hadfield and A. Carwile, pers. comm., 1985) or for fish (Y.K. Maruyama, pers.comm., 2004). Kept on clean glass slides versus biofilmed slides, they grew significantly only on the latter (Pearse 1989). They have been seen to capture and eat ciliates in a culture dish (Klauser 1982). Wild placozoans are probably opportunistic grazers and scavengers on organic detritus and on algae and bacteria in biofilms covering a diversity of substrates.

Potential predators

The sole published report of predation on placozoans involved the opisthobranch *Rhodope* (Riedl 1959). K.J. Marschall told VBP of seeing a small nemertean apparently preying on placozoans in an aquarium in Western Samoa (pers. comm., 1973). We have never observed any animal to prey upon healthy specimens; on the contrary, we report several instances of potential predators reacting to contact with placozoans as they would to a strongly noxious substance. We suggest that the most logical site of such a defensive substance is the so-called shiny spheres that lie throughout the upper epithelium (see, e.g., fig. 2 of Rassat and Ruthmann 1979). The high lipid content of these osmiophilic vesicles has been confirmed; such lipid deposits normally serve for nutrient storage, and in placozoans they do become more numerous when the animals are well fed (Rassat and Ruthmann 1979). A more likely and better protected location for nutrient storage, however, would be in the lower, nutritive epithelium. Instead, these large lipid-rich vesicles are situated at the surface of the animal between the upper epithelial cells, where they are freely exposed to contact and appear to be readily expelled to the outside (Rassat and Ruthmann 1979, their SEM fig. 4, and observations of live animals by VBP)—a most perverse fate for nutrient reserves. Thus, it seems more plausible that the large investment represented by the abundant shiny spheres of placozoans serves for the defense of these soft-bodied morsels. Some support for this idea has been provided by the experiments of Jackson and Buss (2009), who fed placozoans to hydroids: although intact placozoans were not usually swallowed, those previously induced to expel many of their shiny spheres were ingested far more often, and those polyps that swallowed placozoans subsequently died. Analysis of the chemical contents of the shiny spheres awaits advances in microchemistry. The developmental history of these vesicles, which appear to lie within anucleate cells (Grell and Ruthmann 1991), also remains to be investigated.

Diversity

Collection and sequencing

For genetic analysis, placozoans were sampled by washing animals from the wall of a tank (containing only local reef fishes and seawater) at Lizard Island Research Station, Australia, February 2006 (n = 17) or by settlement on glass slides (a) in a holding-tank in the Tuna Research and Conservation Center (TRCC) at Hopkins Marine Station of Stanford University on Monterey Bay, California, November 2006 (n = 2), (b) exposed in Monterey Harbor for almost 5 weeks, September-October 2007 (n = 4). Single placozoans were first transferred from the slides, through a wash of clean seawater, into 1-ml microcentrifuge tubes. Because these animals are so small, thin, and fragile, as well as extremely sticky (the species name *adhaerens* is indeed apt), they were first loosened from the glass using gentle jets from a Pasteur pipette, then drawn into the pipette and quickly expelled into a clean depression slide or watch glass, followed by a second transfer into a microcentrifuge tube, in a minimal amount of seawater. About 1 ml of ethanol was added to the tube, which was then closed and sealed with Parafilm®. Animals along with the residual precipitated salt from the seawater were pelleted by centrifugation; the pellet was air-dried at room temperature and 50 μl of sterile ultra-pure water containing 5% Chelex® 100 (sodium form, Sigma-Aldrich, http://www.sigmaaldrich.com) and 20 μl of 10 mg/ml proteinase K (Sigma-Aldrich, http://www.sigmaaldrich.com) were added. Alternatively, the animals were directly transferred into a tube containing the Chelex® solution and the proteinase K.

All tubes were then treated as described by Voigt et al. (2004). They were stored at –20°C and 1–3 μl of supernatant was used in PCR to amplify a fragment of 16S rDNA (primers: fw: 5'-GCCTGCCCARTGRTTGTA-3'; rv:-5'- GGTCGCAAACATCGTCA-3'; program: 95°/5 min, 37x[95°C/30s; 50°C/30S; 72°C/1 min]). The PCR products were sequenced in both directions with our PCR primers, using chemistry and equipment as described by Dohrmann et al. (2006). For the samples from Lizard Island, cycle sequencing reactions were modified to sequence over a G-C rich partition with stable secondary structure by adding an initial denaturation step (96°C/10 min) and containing 5% DMSO and 0.1 μl BIOTAQ™-DNA-polymerase (5 u/μl; Bioline, http://www.bioline.com). Sequences submitted to GenBank (http://www.ncbi.nlm.nih.gov) have accession numbers EF421454 (LIZ, Lizard Island), EF421455 (H11, TRCC), EU700987 (H2, Monterey Harbor), EU700986 (H17, Monterey Harbor).

Molecular Analysis and Haplotype Distribution

We have chosen a fragment of 16S rDNA as a molecular marker for our phylogenetic analyses, including additional sequences from GenBank (for accession numbers, see caption of Fig. 4A). This marker is known to yield good phylogenetic resolution, but also shows considerable polymorphism in length among haplotypes (Voigt et al. 2004), hampering alignment of all sequences in three regions within our analysis. Because some differences between haplotypes (e.g., H1-H2; H7-H8) appear in these regions, we locally blocked the alignment to include the maximum of information by introducing gaps for non-alignable haplotypes. With this method, otherwise unsupported or poorly supported relationships of similar, yet not identical haplotypes could be resolved (H6, H7, H8), while the tree topology was otherwise unaffected. In all phylogenetic analyses, we applied a GTR + G model of nucleotide evolution suggested by the Akaike information criterion with the software modeltest 3.7 (Posada and Crandall 1998); likelihood values were calculated in PAUP* 4.0b10 (Swofford 1998). A maximum likelihood (ML) analysis including bootstrap resampling was carried out with in PAUP* 4.0b10 (Swofford 1998). A Bayesian analysis was carried out with MrBayes 3.1.2 (Huelsenbeck and Ronquist 2001; Ronquist and Huelsenbeck 2003), performing a Markov chain Monte Carlo analysis with two runs (8 chains each) for 10,000,000 generations, with sample frequency set to 100 and temperature for the heating chains set to 0.2. From the sampled trees, we discarded the first 25% (25,000 trees) as burn-in. In the absence of a suitable outgroup that would result in a well-supported rooting with our marker (see Voigt et al. 2004), midpoint rooting was applied to display the tree in Fig. 4A.

Sequences from the 2 TRCC samples were identical to each other, and sequences from the 17 samples from Lizard Island were also identical to each other. In the latter, a stretch in the middle of the fragment, flanked by several Cs on the one side and several Gs on the other side, caused problems in sequencing similar to those reported by Signorovitch et al. (2006). Even after applying the modified cycle sequencing protocol, sequence reads from both sides overlapped only slightly in the critical region. The new Australian haplotype is identical to the H9 haplotype, reported (as H4-2) from Bermuda by Signorovitch et al. (2006), in parts where information exists for H9. Our sequence, however, covers a larger part of the gene and contains sequence of the critical region not available for the H9 haplotype; thus, it is not clear if our new sequence is identical or shows base exchanges in these regions. Therefore, we do not use the label H9 for this haplotype, but refer to it as LIZ (Fig. 4) to avoid the introduction of another, possibly redundant haplotype number. In contrast, from the two TRCC samples, we can report a new, relatively divergent haplotype, designated H11, and another new haplotype, differing only by two base pairs from H1, designated H17.

The most likely origin of the TRCC specimens is southern California or northern Baja California, as the tank contained Bluefin Tuna (*Thunnus thynnus*) caught in that region. However, this tank is maintained on central California (Monterey Bay) seawater (filtered through sand filters of nominal pore size ~20 µm), and has received in the past water from other tanks in the facility which held fishes from Hawaii; both of these eastern Pacific regions are thus other possible sources of these specimens. In contrast, the Monterey Harbor specimens were collected directly from the sea. Although they may possibly have been carried on a boat from some other part of the world, the same is true of specimens collected almost anywhere at this stage of shipping history. In any case, if they were only recently introduced, the animals were apparently able to establish themselves under the local temperature regime, possibly in the short term (summer of 2007), possibly longer-term but during a temperature peak in some recent period such as the ENSO year of 1998.

Our phylogenetic analyses yielded trees that differ only in the occurrence of a polytomy in the Bayesian analysis (clade H9/H10/LIZ). All other clades find high support by bootstrap or clade credibility values (Fig. 4A). According to our resulting trees and the applied rooting, the newly reported haplotype H11 is the sister group to the clade (H5, H4, H9, H10, LIZ), with high support. The newly reported haplotype H17 falls tightly within the H1-H2 clade.

Until distinguishing morphological characters are discovered, placozoan lineages can be recognized only by DNA analysis with various sequence markers. The 16S rDNA fragment we used identifies at least 12 different haplotypes, including the newest one reported here. Our scant knowledge about the geographic distribution of placozoan haplotypes is summarized in Fig. 4B. The known richness of haplotypes in a region is undoubtedly correlated with sampling effort: the Caribbean and Sargasso Sea, where sampling for molecular work has been most intensive (especially by Signorovitch et al. 2006), together currently have the highest number of haplotypes (9 of the 12 known). Thus, we predict that more sampling will yield not only new haplotypes but also an increase in the documented distributions of haplotypes. Clearly, genetic distances in Placozoa are not at all correlated with geographic distances. Several (in the case of LIZ, presumably) identical haplotypes exhibit a large distributional range: H1: Caribbean, Central Pacific (Hawaii), Red Sea; H2: Sargasso Sea (Bermuda), Caribbean, Central California, Central Pacific (Hawaii), Mediterranean; H4: Sargasso Sea, Caribbean, Eastern Pacific (Panama), Central Pacific (Hawaii) and H6: Caribbean, Eastern Pacific (Panama), and Central Pacific (Hawaii); H8: Caribbean, Eastern Pacific (Panama), Central Pacific (Guam); H9/LIZ: Sargasso Sea, Western Pacific (Australia). At the same time, distantly related haplotypes can occur within the same region, as in the Caribbean (7 haplotypes), on the Pacific coast of Panama (4 haplotypes) or Bermuda (4 haplotypes) or Hawaii (4 haplotypes), or in the Mediterranean (2 haplotypes). Even the hints of geographical pattern mentioned by Voigt et al. (2004) are, with additional data, now almost entirely moot. Signorovitch et al. (2006) in their extensive sampling did not find haplotypes H5 or H17 in the Caribbean/Sargasso; these are examples of haplotypes found to date only in cool temperate waters and each at a single site (see Seasonality/Temperature), but with extended exploration in higher latitudes, these or other haplotypes potentially restricted to lower temperatures could turn out to have global distributions also. In any case, it appears increasingly likely that at least most placozoan lineages are distributed worldwide. These are minute animals with a high capacity for dispersal, easily intermingled through natural agencies—being carried in currents, on the surfaces of other animals, or on floating objects. So we may be observing species whose diversification depends on isolation in microhabitats: as mentioned, sympatry of placozoans in the plankton does not necessarily apply to the sexually reproducing forms that we assume are on the benthos. The swarmers observed by Thiemann and Ruthmann (1991) lived only about one week in culture before settling; however, benthic-phase placozoans probably survive for longer periods on floating debris.

An alternative, perhaps more likely hypothesis for the wide distribution of lineages is transfer by human activities, which have probably at least accelerated the mixing of lineages and contributed to the global patchwork of distributions we see today. For example, in Panama, where identical haplotypes occur on either side of the isthmus, gene flow is most parsimoniously explained by transport through the Panama Canal in ballast water, as placozoans' sensitivity to reduced salinities would otherwise probably prevent their surviving passage through the Canal's lakes. The ability of placozoans to proliferate rapidly through fission, fragmentation, and budding (e.g., Schulze 1891; Grell and Ruthmann 1991) would further facilitate invasive events. Thus, even if placozoan lineages were once geographically isolated, it may now be impossible to uncover their history.

Taxonomic Diversity

The taxonomic status of clades discovered by DNA analysis has not yet been addressed. The extremely simple organization of placozoans offers few specific traits to systematists, and to date, we lack morphological or other characters that can be used to distinguish placozoan lineages. Potentially useful characters include the birefringent granules, which may be present or absent (Pearse et al. 1994), and the conditions for culture, which differ for specimens from various sources. For example, Grell and López-Ochoterena (1988) cultured placozoans from Quintana Roo, Mexico on a green alga instead of the cryptomonad used for the Red Sea strain, and Grell (pers. comm.) was unable to culture a strain from the Mediterranean. Electron microscopy may provide ultrastructural characters; to date, only Grell's strain from the Red Sea has been examined in any detail, although hints of differences have been suggested by Ivanov et al. (1980) and by Klauser and Ruppert (1981).

The question remains, how such cryptic, yet distinct, taxa are to be handled by systematists, because all evidence indicates that at least the more distant clades in Fig. 4A are indeed different biological species, if not genera or higher-level taxa; the level of divergence between placozoan lineages in the nuclear 18S-rRNA gene is similar to or higher than the levels reported between species of other diploblast phyla, or even between genera or families (Voigt et al. 2004). Moreover, Signorovitch et al. (2007) have published three additional mt genomes from placozoans having the 16S rDNA haplotypes H3, H4, and H8 (Signorovitch et al. 2006). Compared to the already published mt genome of a H1 haplotype (Dellaporta et al. 2006), several gene rearrangements were observed (e.g., large inversions), accompanied by remarkable divergence in length—the mt genome size varies from 32.7 to 43.1 kb (Signorovitch et al. 2007). Such rearrangements of genes and extraordinary differences in length in the mt genome have not been reported within any metazoan species. All these findings of unexpected genetic diversity support the view that Placozoa comprises at least several deep clades that are reproductively isolated and highly divergent, comparable to species, if not genera or higher taxa of other phyla (for evidence of sexual reproduction, see Grell 1972; Signorovitch et al. 2005). The species boundaries between different lineages remain to be recovered. 16S rRNA data are not sufficient to detect reproductive isolation and a species definition on genetic distance alone is biologically not satisfactory. A much better approach to define species would be large scale genetic multi-marker studies, quantifying the amount of gene-flow and uncovering reproductive isolation between lineages. An additional problem with 16S rRNA trees is that the position of the root remains uncertain. Therefore the tree still lacks any orientation in time. Strictly speaking, even the definition of clades (necessary to establish a hierarchical taxonomic scheme) is impossible in an unrooted phylogenetic tree, as any group of haplotypes containing the root becomes obsolete.

Given the lack of both biogeography and other distinguishing characters, the eventual description and naming of placozoan species will challenge the usual taxonomic conventions. Type locality will often be meaningless, and place-based names equally so. The name *Trichoplax adhaerens* Schulze, 1883 might best attach, not to a specimen from the Adriatic Sea, but to the durable laboratory strain from the Red Sea, which K.G. Grell long ago established in culture and which survives today. This strain continues as the basis of most of our knowledge of the phylum Placozoa Grell 1971, including all of what we know about fine structure.

Life History

Regarded by their discoverer, F.E. Schulze, as solely benthic organisms, placozoans inspired O. Bütschli (1884) to propose the placula theory of metazoan origins from a holobenthic ancestor. Always a minority view, in the shadow of Haeckel's powerful tradition, the idea of a benthic ancestor has most recently been reproposed and championed by Degnan and Degnan (2006), now from the perspective of sponge development and the well-described pelagobenthic life histories of sponges. Framing the processes of gametogenesis, embryogenesis, and metamorphosis as the essence of a pelagobenthic cycle, these authors argue persuasively that a pelagic phase would arise simply and inevitably from a benthic adult ancestor, practicing sex, whereas the reverse requires multiple reinvention of a sexual benthic phase. If only Bütschli had known what we now know, his placula might not have been so easily dismissed: placozoans have sex (Grell 1972; Signorovitch et al. 2005) and they also have a pelagic phase that is abundant in the water column in warm seas around the world.

The incompletely known life history of placozoans thus presents a significant gap in our understanding, not only of *Trichoplax*, but of the framework of metazoan history. For placozoans, not only do gametogenesis, embryogenesis, and metamorphosis remain undescribed, but also meiosis, sperm, and fertilization. Although field studies have revealed tropical waters teeming with pelagic placozoans, the nature of what is settling on glass slides remains a puzzle: small fragments of the benthic phase, or budded swarmers (Thiemann and Ruthmann 1991), or sexually produced larvae? Like most invertebrate larvae, they require the development of a biofilm on glass slides in order to settle (Pearse 1989), but this fact alone cannot begin to distinguish between possible larvae and various asexual products. Free-swimming placozoans in a variety of shapes have been observed (Fig. 5); although the size of these seems most consistent with fragments, the shape has varied from gastrula-like to flat, very much as Thiemann and Ruthmann (1991) described and illustrated hollow swarmers preparing to settle. If VBP was observing swarmers, then this stage is not an artifact of laboratory culture, an alternative judiciously discussed by Grell and Ruthmann (1991). The diameter of the smallest individuals seen on slides is ~120 μm, similar to that reported for both eggs (Grell 1972) and for swarmers (Thiemann and Ruthmann 1991). Such small specimens are typically circular in outline initially, grow rapidly, and can begin to fission within a few days (Fig. 6). Eggs were never observed by VBP or by A.Y. Signorovitch (pers. comm., 2006) in placozoans collected on slides from the field; the field conditions required for sexual reproduction are completely unknown.

Fig. 5. Drawing of swimming placozoans, estimated at no more than 200 μm diameter, seen in dishes in the laboratory; placozoans are too small to be observed directly in the sea. It may possibly be forms such as these that settle on glass slides, but their developmental history is unknown. They could be small fragments of benthic-phase placozoans; a stage of the swarmers described by Thiemann and Ruthmann (1991); larvae developed from sexually produced eggs; or some other phase of the incompletely known life history. (Drawing by J. Keller and C. Patton, based on observations by VBP).

We may be able to induce sex reliably in the laboratory, however, by discovering and reproducing the conditions under which production of eggs occurs (Grell 1972), e.g., by reducing food or increasing cadmium (or other heavy metals), as has been found effective in hydrozoans such as *Laomedea* (as *Campanularia*) *flexuosa* (Stebbing 1980) and *Eleutheria dichotoma* (Schierwater and Hadrys 1998). We might then uncover suitable conditions for development, which so far has invariably ceased after a few cleavages (Grell 1972; Signorovitch et al. 2005). Given that no typical animal sperm have ever been documented in placozoans, the eggs so far observed may have been unfertilized or failed to undergo maturation; perhaps the Red Sea strain in culture is female, as Grell once suggested (pers. comm., 1992). Other missing requirements might be straightforward

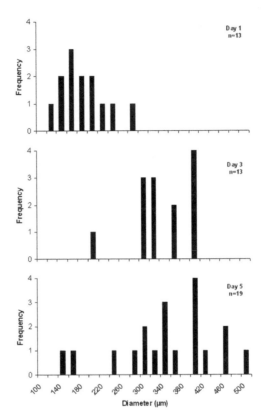

Fig. 6. Growth dynamics of a cohort of placozoans at the University of Guam Marine Laboratory. Sizes measured as diameter, plotted in 20-μm increments. On Day 1, one day after preconditioned slides were placed into the outflow of the lab's seawater system, a cohort of uniformly small placozoans had settled; mean diameter ± S.D. was 183 ± 45 μm. On Day 3, the same number of animals were present, but they had grown significantly (t-test, p<<0.001), more than tripled in mean size: 337 ± 58 μm. By Day 5, mean size had leveled (t-test, p=0.76, 345 μm ± 96 μm), while an increase in total number, and the reappearance of individuals in small size classes, showed that some animals had already begun to fission while others continued to grow.

modifications of culture conditions, or might conceivably be as complex as a host organism in which a parasitic placozoan phase normally completes its development. This speculative possibility has been suggested by the regularity with which placozoans turn up in fish mariculture facilities (observations by VBP; Tomassetti et al. 2005). As parasitism often leads to secondary reduction, it could explain the extreme simplicity of placozoans. The coincidence of placozoans with fishes, however, is far more likely a simple result of enrichment in a protected environment. Another remote possibility is that the evolution of placozoans involved mutations in their equivalent of stem cells (see fig. 5 in Schierwater 2005), comparable to interstitial cells or archeocytes, simplifying the histology and modifying other aspects of normal development. While these ideas are purely speculative, they are put forward here to challenge the assumption that the normal cellular processes of sexual reproduction known for almost all other animals must necessarily occur in the familiar benthic phase of placozoans and have somehow merely been overlooked.

Finally, even the benthic development of placozoans is still incompletely documented. The smallest individuals are roughly circular, and they take on increasing irregular shapes as they grow, from somewhat ameba-like (Fig. 1) to the long stringy forms (see, e.g., fig. 1 of Schulze 1891) commonly seen in older laboratory cultures or after placozoans have been growing and multiplying on aquarium glass for some time. The occasional spontaneous development of a ring-shaped form with a hole in the center has so far been recorded, to our knowledge, only in Western Samoa. As documented in film by K.J. Marschall 1970, the ring subsequently breaks through, producing a long, stringy shape, and fragments generated from both free ends crawl away as small individuals. A ring-shaped placozoan observed by VBP in 1989 in W. Samoa, however, reverted to a normal form by closing up the central hole. Yet to be understood is whether the plastic spectrum of planar shapes, and of asexual proliferation (Thiemann and Ruthmann 1991), reflects developmental stages or environmental conditions or taxonomic diversity.

Concluding Remarks

Placozoans, like sponges, lack a digestive cavity and nervous system, and like most sponges (except homoscleromorphs, see e.g., Boury-Esnault et al. 2003; Nichols et al. 2006), also lack a basal lamina. Yet the body plans and ways of life of the adults of these two groups could hardly be more different: the sessile sponges grow to large sizes and filter-feed whereas the nearly microscopic placozoans wander freely over the benthos as active grazers or scavengers. In this respect, placozoans are more similar to acoels, which likewise lack a basal lamina and digestive cavity (Rieger et al. 1991), and to larval sponges. Because placozoans are such appealing models as ancestral metazoans, we are naturally avid to know where they fit into animal phylogeny. At the same time, in order to understand them as functional, living organisms, to be compared with others and understood in an ecological context, we need to learn more of the facts of their basic biology. We hope that the observations collected here might stimulate others to seek further understanding of placozoans in their natural haunts.

Acknowledgments

VBP owes special thanks to John S. Pearse and Devon E. Pearse for their loyal field assistance and other support throughout her studies. A grant from the Christensen Foundation to VBP and NSF INT-8817807 to JSP made much of this field work possible. For support of the molecular work reported here, we thank Gert Wörheide and the German Research Foundation (DFG, Project No. W0896/3). The Gesellschaft für wissenschaftliche Datenverarbeitung Göttingen (GWDG) is acknowledged for providing computational resources for our Bayesian analysis. Especially warm thanks are due to all our many hosts and host institutions.

VBP carried out field work at: Seto Marine Laboratory, Univ. of Kyoto, Shirahama, Japan. Sesoko Marine Laboratory, Univ. of the Ryukyus, Okinawa, Japan. Iriomote Marine Research Station, Tokai University, southwest Ryukyu Islands, Japan. Coastal Marine Laboratory, Hong Kong Univ. of Science & Technology. Univ. of Guam Marine Laboratory, Mangilao, Guam. Palau Mariculture Demonstration Center (Micronesian Mariculture Development Center), Palau. Christensen Research Institute, Madang, Papua New Guinea. Orpheus Island Research Station, James Cook University, Great Barrier Reef, northeastern Australia. Kewalo Marine Laboratory, Univ. of Hawaii, Oahu, Hawaii. Laboratory of the late Karl J. Marschall, Apia, Western Samoa. Gump Field Station, Univ. of California, Berkeley, Moorea, French Polynesia. Tuna Research and Conservation Center, Hopkins Marine Station, Stanford University, Pacific Grove, California. Monterey Bay Aquarium, Monterey, and Monterey Bay Aquarium Research Institute, Moss Landing, California. Long Marine Laboratory, Univ. of California, Santa Cruz, California. Achotines Laboratory, Azuero Peninsula, Pacific coast of Panama. Smithsonian Tropical Research Institute, Panama: Naos Island Laboratories on the Pacific coast; Bocas del Toro Research Station and Galeta Marine Laboratory on the Atlantic coast. Rosenstiel School of Marine and Atmospheric Science, University of Miami. Roatan Institute of Marine Biology, Roatan, Honduras. Museo Marino de Margarita, Boca del Rio, Nueva Esparta, Venezuela. McMurdo Station, Antarctica.

O. Voigt carried out field work at: Lizard Island Research Station, Australian Museum, Great Barrier Reef, northeastern Australia.

Finally, we both offer thanks to Bernd Schierwater and Rob DeSalle for their encouragement of our work on placozoans; for inviting VBP to participate in the symposium on "Key Transitions in Animal Evolution," January 2007, Phoenix, Arizona, annual meeting of the Society for Integrative and Comparative Biology and the American Microscopical Society; and for organizing the present book. This chapter is an update of our review arising from that symposium, originally published in the journal Integrative and Comparative Biology, 2007, volume 47, number 5 and largely reprinted here, by permission.

References

Boury-Esnault, N. and A. Ereskovsky, C. Bézac, and D. Tokina. 2003. Larval development in the Homoscleromorpha (Porifera, Demospongiae). Invertebr. Biol. 122(3): 187–202.

Bütschli, O. 1884. Bemerkungen zur Gastraea-Theorie. Morph. Jahrb. 9: 415–427.

Cartwright, P. and A. Collins. 2007. Fossils and phylogenies: integrating multiple lines of evidence to investigate the origin of early major metazoan lineages. Integr. Comp. Biol. 47: 744–751.

Degnan, S.M. and B.M. Degnan. 2006. The origin of the pelagobenthic metazoan life cycle: what's sex got to do with it? Integr. Comp. Biol. 46(6): 683–690.

Dellaporta, S. and A. Xu, S. Sagasser, W. Jakob, M.A. Moreno, L. Buss, and B. Schierwater. 2006. Mitochondrial genome of *Trichoplax adhaerens* supports Placozoa as the basal lower metazoan phylum. Proc. Natl. Acad. Sci. USA 103(32): 8751–8756.

Dohrmann, M. and O. Voigt, D. Erpenbeck, and G. Wörheide. 2006. Non-monophyly of most supraspecific taxa of calcareous sponges (Porifera, Calcarea) revealed by increased taxon sampling and partitioned Bayesian analysis of ribosomal DNA. Mol. Phylogenet. Evol. 40(3): 830–843.

Ender, A. and B. Schierwater. 2003. Placozoa are not derived cnidarians: evidence from molecular morphology. Mol. Biol. Evol. 20(1): 130–134.

Eitel, M and B. Schierwater. 2010. The phylogeography of the Placozoa suggests a taxon-rich phylum in tropical and subtropical waters. Mol. Ecol. 19(11): 2315–2327.

Grell, K.G. 1971. *Trichoplax adhaerens*, F. E. Schulze und die Entstehung der Metazoen. Naturw. Rundschau 24: 160–161.

Grell, K.G. 1972. Eibildung und Furchung von *Trichoplax adhaerens* F.E. Schulze (Placozoa). Z. Morph. Tiere. 73: 297–314.

Grell, K.G. Stamm Placozoa. pp. 247–250. *In:* K.G. Grell, H.-E. Gruner and E.F. Kilian [eds.] 1980 Lehrbuch der Speziellen Zoologie, Band I: Wirbellose Tiere, Teil 1: Einführung, Protozoa, Placozoa, Porifera. VEB Gustav Fisher Verlag, Jena.

Grell, K.G. 1981. *Trichoplax adhaerens* and the origin of Metazoa. *In:* Origine dei Grandi Phyla dei Metazoi. Acc. Naz. Lincei, Convegno Intern. pp. 107–121.

Grell, K.G. 1983. Ein neues Kulturfahren für *Trichoplax adhaerens* F.E. Schulze. Z Naturforsch 38c: 1072.

Grell, K.G. and E. López-Ochoterena. 1988. A new record of *Trichoplax adhaerens* F.E. Schulze (phylum Placozoa) in the Mexican Caribbean Sea. Anales del instituto de ciencias del mar y limnología, Unversidad Nacional Automa de Mexcio 14: 255–256.

Grell, K.G. and A. Ruthmann. Placozoa. pp. 13–27. *In:* F.W. Harrison and J.A. Westfall [eds.] 1991. Microscopic Anatomy of Invertebrates, vol. 2. Placozoa, Porifera, Cnidaria, and Ctenophora. Wiley-Liss, New York.

Hadrys, T. and R. DeSalle, S. Sagasser, N. Fischer, and B. Schierwater. 2005. The *Trichoplax* PaxB gene: a putative Proto-PaxA/B/C gene predating the origin of nerve and sensory cells. Mol. Biol. Evol. 22(7): 1569–1578.

Halanych, K.M. 2004. The new view of animal phylogeny. Ann. Rev. Ecol. Evol. Syst. 35: 229–256.

Hejnol, A. and M. Obst, A. Stamatakis, M. Ott, G.W. Rouse, G.D. Edgecombe, P. Martinez, J. Baguñà, X. Bailly, U. Jondelius, M. Wiens, W.E.G. Müller, E. Seaver, W.C. Wheeler, M.Q. Martindale, G. Giribet, and C.W. Dunn. 2009. Assessing the root of bilaterian animals with scalable phylogenomic methods. Proc. Royal Society B, December 2009 276(1677): 4261–4270.

Huelsenbeck, J.P. and F. Ronquist. 2001. MRBAYES: Bayesian inference of phylogenetic trees. 17(8): 754–755.

Iseto, T. 2002. *Loxocorone*, a new genus of the family Loxosomatidae (Entoprocta: Solitaria), with descriptions of two new *Loxomitra* (sensu stricto) and a new *Loxocorone* from Okinawa, the Ryukyu Archipelago, Japan. Zoolog. Sci. 19(3): 359–367.

Ivanov, A.V. 1973. *Trichoplax adhaerens*, a phagocytal animal. Zool. Zh. 52(8): 1117–1131.

Ivanov, D.L. and V.V. Malakhov, A.B. Tsetlin. 1980. A new finding of primitive multicellular organism *Trichoplax* sp. Zool. Zh. 59(12): 1765–1767.

Jackson, A.M. and L.W. Buss. 2009. Shiny spheres of placozoans (Trichoplax) function in anti-predator defense. Invertebr. Biol. 128(3): 205–212.

Jakob, W. and S. Sagasser, S. Dellaporta, P. Holland, K. Kuhn, and B. Schierwater. 2004. The Trox-2 Hox/ParaHox gene of *Trichoplax* (Placozoa) marks an epithelial boundary. Dev. Genes Evol. 214(4): 170–175.

Klauser, M.D. and E.E. Ruppert. 1981. Non-flagellar motility in the phylum Placozoa: ultrastructural analysis of the terminal web of *Trichoplax adhaerens*. Am. Zool. 21(4): 1002 (Abst.)

Klauser, M.D. 1982. An ultrastructural and experimental study of locomotion in *Trichoplax adhaerens* (Placozoa). Unpubl. thesis, Clemson Univ., Clemson, SC.

Lavrov, D.V. 2007. Key transitions in animal evolution: a mitochondrial DNA perspective. Integr. Comp. Biol. 47: 734–743.

Lavrov, D.V. 2011. Key transitions in animal evolution: a mitochondrial DNA perspective. Chap. 3, pp. 35–54 *In:* B. Schierwater and R. DeSalle, eds. Key Transitions in Animal Evolution. Science Publishers & CRC Press.

Loya, Y. and K. Sakai, K. Yamazato, Y. Nakano, H. Sambali, and R. van Woesik. 2001. Coral bleaching: the winners and the losers. Ecol. Lett. 4: 122–131.

Mallatt, J. and C.W. Craig, and M.J. Yoder. 2010. Nearly complete rRNA genes assembled from across the metazoan animals: Effects of more taxa, a structure-based alignment, and paired-sites evolutionary models on phylogeny reconstruction. Mol. Phylogenet. Evol. 55(1): 1–17.

Martinelli, C. and J. Spring. 2003. Distinct expression patterns of the two T-box homologues Brachyury and Tbx2/3 in the placozoan *Trichoplax adhaerens*. Dev. Genes Evol. 213(10): 492–499.

Maruyama, Y.K. 2004. Occurrence in the field of a long-term, year-round, stable population of placozoans. Biol. Bull. 206: 55–60.

Miller, D. and E. Ball. 2005. Animal evolution: the enigmatic phylum Placozoa revisited. Curr. Biol. 15(1): R26–28.

Monticelli, F.S. 1893. *Treptoplax reptans* n.g., n.sp. Atti dell' Academia dei Lincei, Rendiconti (5)II:39–40.

Monticelli, F.S. 1896. Adelotacta zoologica. 2. *Treptoplax reptans* Montic. Mitt. Zool. Stat. Neapel 12: 444–462.

Morandini, A.C. and S.N. Stampar, and F.L. da Silveira. 2006. *Trichoplax* from marine cultures in Brazil—first record of the phylum Placozoa in the South Atlantic Ocean. Zool. Anz. 254(2): 127–129.

Nichols, S.A. and W. Dirks, J.S. Pearse, and N. King. 2006. Early evolution of animal cell signaling and adhesion genes. Proc. Natl. Acad. Sci. 103(33): 12451–12456.

Ocaña, A. and A. Ibáñez. 2006. A new record of Placozoa from the Mediterranean Sea. Belg. J. Zool. 136: 255–256.

Pearse, V.B. 1988. Field biology of placozoans, August—October 1988. Unpublished report to the Christensen Research Institute, Madang, Papua New Guinea. (Available as pdf from VBP.).

Pearse, V.B. 1989. Growth and behavior of *Trichoplax adhaerens*: first record of the phylum Placozoa in Hawaii. Pac. Sci. 43(2): 117–121.

Pearse, V.B. and J.S. Pearse. 1991. Year–long settling plate study yields no antarctic placozoans, and surprisingly little else. Antarctic J. US 26: 149–150.

Pearse, V.B. and T. Uehara, and R.L. Miller. 1994. Birefringent granules in placozoans (*Trichoplax adhaerens*). Trans. Am. Micr. Soc. 113(3): 385–389.

Pearse, V.B. and O. Voigt. 2007. Field biology of placozoans (*Trichoplax*): distribution, diversity, biotic interactions. Integr. Comp. Biol. 47: 677–692.

Philippe, H. and R. Derelle, P. Lopez, K. Pick, C. Borchiellini, N. Boury-Esnault, J. Vacelet, E. Renard, E. Houliston, E. Quéinnec, C. Da Silva, P. Wincker, H. Le Guyader, S. Leys, D.J. Jackson, F. Schreiber, D. Erpenbeck, B. Morgenstern, G. Wörheide, and M. Manuel. 2009. Phylogenomics revives traditional views on deep animal relationships. Current Biol. 19(8): 706–712.

Pick, K. S. and H. Philippe, F. Schreiber, D. Erpenbeck, D.J. Jackson, P. Wrede, M. Wiens, A. Alié, B. Morgenstern, M. Manuel, and G. Wörheide. 2010. Improved phylogenomic taxon sampling noticeably affects non-bilaterian relationships. Mol. Biol. Evol., in press: doi:10.1093/molbev/msq089

Posada, D. and K.A. Crandall. 1998. Modeltest: testing the model of DNA substitution. Bioinformatics 14(9): 817–818.

Rassat, J. and A. Ruthmann. 1979. *Trichoplax adhaerens* F.E. Schulze (Placozoa) in the scanning electron microscope. Zoomorphol. 93: 59–72.

Riedl, R. 1959. Beiträge zur Kenntnis der *Rhodope veranii,* Teil I. Geschichte und Biologie. Zool. Anz. 163: 107–122.

Rieger, R.M. and S. Tyler, J.P.S. Smith III, G.E. Rieger. Platyhelminthes: Turbellaria. pp. 7–140. *In:* F.W. Harriosn and B.J. Bogitsh [ed.] 1991. Microscopic anatomy of invertebrates, v.3. Wiley-Liss, New York.

Ronquist, F. and J.P. Huelsenbeck. 2003. MRBAYES 3. Bioinformatics 19: 475–481.

Ruthmann, A. 1977. Cell differentiation, DNA content, and chromosomes of *Trichoplax adhaerens* F.E. Schulze. Cytobiologie 15: 58–64.

Schierwater, B. and H. Hadrys. 1998a. Environmental factors and metagenesis in the hydroid *Eleutheria dichotoma.* Invertebr. Reprod. Dev. 34(2–3): 139–148.

Schierwater, B. and K. Kuhn. 1998b. Homology of Hox genes and the zootype concept in early metazoan evolution. Mol. Phyl. Evol. 9: 375–381.

Schierwater, B. 2005. My favorite animal, *Trichoplax adhaerens.* BioEssays 27: 1294–1302.

Schierwater, B. and R. DeSalle. 2007. Can we ever identify the Urmetazoan? Integr. Comp. Biol. 47: 670–676.

Schulze, F.E. 1883. *Trichoplax adhaerens,* nov. gen., nov. spec. Zool. Anz. 6: 92–97.

Schulze, F.E. 1891. Über *Trichoplax adhaerens.* Abhandlungen der königl. preuss. Akad. der Wissenschaften pp. 1–23.

Signorovitch, A.Y. and S.L. Dellaporta, and L.W. Buss. 2005. Molecular signatures for sex in the Placozoa. Proc. Natl. Acad. Sci. USA 102(43): 15518–15522.

Signorovitch, A.Y. and S.L. Dellaporta, and L.W. Buss. 2006. Caribbean placozoan phylogeography. Biol. Bull. 211: 149–156.

Signorovitch, A.Y. and L.W. Buss, and S.L. Dellaporta. 2007. Comparative genomics of large mitochondria in placozoans. PLoS 3(1): e13.

Srivastava, M. and E. Begovic, J. Chapman, N.H. Putnam, U. Hellsten, T. Kawashima, A. Kuo, T. Mitros, A. Salamov, M.L. Carpenter, A.Y. Signorovitch, M.A. Moreno, K. Kamm, J. Grimwood, J. Schmutz, H. Shapiro, I.V. Grigoriev, L.W. Buss, B. Schierwater, S.L. Dellaporta, and D.S. Rokhsar. 2008. The *Trichoplax* genome and the nature of placozoans. Nature 454(7207): 955–960.

Stebbing, A.R.D. Increase in gonozooid frequency as a response to stress in *Campanularia flexuosa.* pp. 27–42. *In:* P. Tardent and R. Tardent [eds.] 1980. Development and Cellular Biology of Coelenterates. Elsevier, New York.

Stiasny, G. 1903. Einige histologische Details über *Trichoplax adhaerens.* Zeitschr. wiss. Zool. 75: 430–436.

Sudzuki, M. 1977. Microscopical marine animals scarcely known from Japan. II. Occurrence of *Trichoplax* (Placozoa) in Shimoda. Proc. Jap. Soc. Syst. Zool. No. 13:1–3.

Swofford, D.L. 1998. PAUP*. Phylogenetic analysis using parsimony (*and other methods) Version 4. Sinauer Associates, Sunderland, Massachusetts.

Syed, T. and B. Schierwater. 2002. *Trichoplax adhaerens*: discovered as a missing link, forgotten as a hydrozoan, re-discovered as a key to metazoan evolution. Vie et Milieu 52(4): 177–188.

Thiemann, M. and A. Ruthmann. 1991. Alternative modes of asexual reproduction in *Trichoplax adhaerens* (Placozoa). Zoomorphol. 110: 165–174.

Tomassetti, P. and O. Voigt, A.G. Collins, S. Porrello, and V.B. Pearse, B. Schierwater. 2005. Placozoans (*Trichoplax adhaerens* Schulze, 1883) in the Mediterranean Sea. Meiofauna Marina 14: 5–7.

Voigt, O. and A.G. Collins, V.B. Pearse, J.S. Pearse, A. Ender, H. Hadrys, and B. Schierwater. 2004. Placozoa—no longer a phylum of one. Current. Biol. 14(22): R944–R945.

Chapter 13

Trichoplax and Placozoa: One of the Crucial Keys to Understanding Metazoan Evolution

Bernd Schierwater,[1,2] Michael Eitel,[1] Hans-Jürgen Osigus,[1] Karolin von der Chevallerie,[1] Tjard Bergmann,[1] Heike Hadrys,[1,3] Maria Cramm,[1] Laura Heck,[1] Wolfgang Jakob,[1] Michael R. Lang,[1] Rob DeSalle[2]

Trichoplax adhaerens—Historical Background

The discovery of *Trichoplax adhaerens*: In 1883, the German zoologist Franz Eilhard Schulze published a short communication on the description of a new species, *Trichoplax adhaerens* (the "sticky hairy plate"; Greek trich=hair, plax=plate); a flattened, crawling marine animal of up to a few millimeters in size (Schulze 1883). Schulze found these organisms settling on the glass sides of seawater aquaria at the University of Graz (Austria), recognizing their amoeba-like movements and continual shape changes. These were new features for metazoan animals.

Schulze's histological analysis of *Trichoplax adhaerens*, based on microtome sections and various staining procedures, revealed a three-layered sandwich organization of the animal, with morphologically different upper and lower epithelia (Fig. 1). The epithelia enclose an inner, connective-tissue-like union of cells. Summarizing his first results, Schulze

[1]ITZ, Ecology & Evolution, TiHo Hannover, Bünteweg 17d, D-30559 Hannover, Germany.
E-mail: bernd.schierwater@ecolevol.de
[2]American Museum of Natural History, New York, Sackler Institute for Comparative Genomics, 79 St. at Central Park West, New York, NY 10024, USA.
E-mail: desalle@amnh.org
[3]Department of Ecology and Evolutionary Biology, Yale University, New Haven, CT, 06520-8104, USA.

(1883) concluded that *Trichoplax adhaerens* did not fit into any of the bauplan patterns of sponges, coelenterates (ctenophores and cnidarians), or the vermiform phyla. Consequently, he assumed that *Trichoplax adhaerens* was an isolated, basal offshoot close to the root of the metazoan phylogenetic tree (see Schierwater 2005 for overview and refs.).

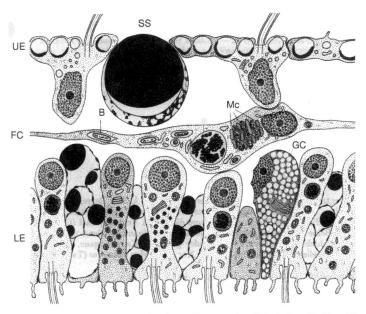

Fig. 1. Schematic cross section of *Trichoplax adhaerens* (modified after Grell and Ruthmann 1991). UE = upper epithelium, LE = lower epithelium, FC = contractile fiber cell, GC = gland cell, SS = shiny sphere, Mc = mitochondrial complex, B = (endosymbiotic?) bacterium in endoplasmic reticulum. Note that the interspace between fiber cells and epithelia is free of ECM and that a basal lamina is missing (from Syed and Schierwater 2002a).

Schulze's original description of *Trichoplax adhaerens* soon sparked debate on the hypothetical first metazoan ("urmetazoa" or "archimetazoa") between Haeckel, Lankester, Metschnikoff, and other zoologists (for overview see Gruner 1993, Syed and Schierwater 2002a). Only one year after Schulze's original description of *Trichoplax adhaerens*, O Bütschli (University of Heidelberg) published an improved version of Haeckel's "gastrea hypothesis" (Bütschli 1884). While Haeckel's "gastrea" (Haeckel 1874), a hypothetical spherical, pelagic organism, invaginates from a pelagic "blastaea" at its posterior pole, Bütschli tried to derive the gastraea from a flat, benthic-vagile ancestor, the hypothetical "placula". According to Bütschli, the first metazoans emerged after colonial flagellates (Protozoa) fused into a benthic, single layered organism with ciliary locomotion. From this stage, the two-layered "placula" developed with an upper "epithelium" and a lower "epithelium". Gradual invagination of the lower "epithelium" layer led to a benthic gastraea-like animal. This lower "epithelium" invagination finally led to closed gastric cavities or through-guts, as was already described in Haeckel's model (Syed and Schierwater 2002a, 2002b).

Bütschli argued that the three-layered *Trichoplax adhaerens* is a comparatively derived organism, still mirroring the two-layered placula's mode of life. It is important to note that both Schulze and Bütschli agreed in interpreting the upper epithelium of *Trichoplax adhaerens* as an ectoderm homolog and the lower epithelium as an entoderm homolog. The question then arose whether the interior fiber cell complex of *Trichoplax adhaerens* is a mesoderm homolog or not. Both authors hesitated to interpret the fiber cells as a mesoderm homolog because this inference would have implied a close affinity of *Trichoplax adhaerens* to the triploblastic phyla (the Bilateria). Bütschli (1884) therefore saw in the fiber cell layer an analogy to mesodermal structures of triploblasts. Schulze (1883) pointed out that observations on the ontogeny of *Trichoplax adhaerens* would be required to solve this question. Both researchers were aware of the principal counter-hypothesis that *Trichoplax adhaerens* might be a secondarily simplified organism, as this view had already been proposed for other groups of organisms such as the parasitic mesozoans. The latter alternative has always been regarded as relatively unlikely as there was not the slightest evidence for parasitism in *Trichoplax adhaerens*. After the first morphological descriptions and resulting phylogenetic interpretations it was expected that elucidating the ontogeny and the life cycle of *Trichoplax adhaerens* would be the next crucial step in resolving the phylogenetic position of the Placozoa (Schierwater 2005).

The hydrozoan interpretation: In 1890, one year before Schulze's detailed *Trichoplax adhaerens* description, FC Noll from the Senckenberg Museum in Frankfurt reported observations on the animal's normal mode of vegetative reproduction, which is binary fission of the entire body into two individuals (Fig. 2A, B). Noll wrongly suggested the presence of otoliths in large *Trichoplax adhaerens* specimens and thus a close relationship to acoel turbellarians. This idea was supported by L von Graff, an expert on Acoela. Schulze (1891) remained skeptical about this interpretation, mainly because of the arrangement of the inner contractile cells of the animal (which do not resemble a myoepithelium), and also because of the lack of any fixed axes of symmetry. At this time, most zoologists agreed with Schulze (1883) that the existence of a functional layer of contractile cells in *Trichoplax* leads to the rejection of a close relationship to either coelenterates or sponges. Through the end of the 19th century a close relationship of *Trichoplax adhaerens* to acoel turbellarians or mesozoans was discussed. After FS Monticelli (1893, 1896) described another *Trichoplax*-like animal,

Treptoplax reptans, both forms were united as Mesenchyma, in reference to the fiber cells, and grouped within the Mesozoa (Delage and Herouard 1899), a phylum that had already become "a dumping ground for a host of multicellular but presumed nonmetazoan organisms" (Brusca and Brusca 1990). True metazoan phyla were seen as showing an invaginating gastrula stage during embryogenesis (e.g. Neresheimer 1912), although this definition had been intensely debated from the very beginning (c.f. Syed and Schierwater 2002a).

The question of the complete life cycle of *Trichoplax adhaerens* initially yielded a most surprising—and completely wrong—answer when the German zoologist T Krumbach observed these animals in a seawater aquarium that was settled by sexual medusae of the hydrozoan *Eleutheria*

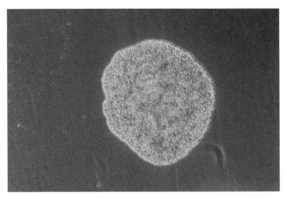

Fig. 2A. Photograph of a live *Trichoplax adhaerens*. This specimen belongs to the so-called "Grell clone" and is about 2 mm in diameter.
Color image of this figure appears in the color plate section at the end of the book.

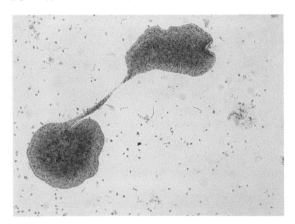

Fig. 2B. Dividing *Trichoplax adhaerens*. During binary fission of the mother individual two genetically identical daughter individuals are generated.

krohni (Krumbach 1907). As in medusae of other *Eleutheria* species, for example *E. dichotoma*, the eggs of *E. krohni* develop in a brood pouch, which eventually opens to release well-developed planula larvae (Hauenschild 1956, Schierwater 1989, Hadrys et al. 1990). Krumbach (1907) reported that he found *Trichoplax adhaerens* individuals in exactly those positions where *Eleutheria* planulae had settled before. Krumbach was convinced that *Trichoplax adhaerens* was a deformed larva of *E. krohni*, although he never observed the supposed metamorphosis. Surely Krumbach's interpretations were influenced by speculations of other authors, who thought of *Trichoplax adhaerens* as an "abnormal" organism that was unable to complete its life cycle under culture conditions (e.g. Ehlers 1887: 497). Although rejected by Schulze (1891), this interpretation persisted in zoological textbooks of that time (e.g. Lankester 1901: 158). Krumbach's 1907 publication led to a corresponding statement on *Trichoplax adhaerens* in the first installment of Bütschli's "Vorlesungen über vergleichende Anatomie" (1910) and was cited as fact in a reference book (Neresheimer 1912: 827). However, the hydrozoan interpretation was soon criticized by Schubotz (1912) and Schulze (1914). Schubotz compared the histological organization of *E. krohni* planulae and *Trichoplax adhaerens*. He noted that the ectoderm of the planulae already contains nematocysts, which would have to vanish during any transformation into a *Trichoplax adhaerens*. Schulze completed Schubotz's argumentation by mentioning some special features of the inner fiber cell layer of *Trichoplax adhaerens*. Note that Schulze's article of 1914 was the last publication on *Trichoplax adhaerens* in a zoological journal for more than half a century (Syed and Schierwater 2002a).

How could an exciting animal like *Trichoplax adhaerens* be pushed out of scientific research by this "shaky" larva hypothesis? It is an amazing fact that the completely unsupported larva-hypothesis remained in German, French, and Anglo-American textbooks for decades. After World War One, the first German encyclopedia on animal phyla was prepared by the zoologist W. Kuekenthal, who died one year before the first volumes were published in 1923 (also, FE Schulze died in 1921). The editor who finished the volumes was Thilo Krumbach, a supporter of the larva-hypothesis. Through his hands, volume one of the "Handbuch der Zoologie" contained the Protozoa, Porifera, Coelenterata, and Mesozoa. *Trichoplax adhaerens* is briefly mentioned in the chapter "Hydroida" (H Broch, Oslo) and the chapter "Mesozoa" (M Hartmann, Berlin-Dahlem). Both authors interpreted *Trichoplax adhaerens* as a transformed planula of *E. krohni*. Hartmann cites Krumbach's paper from 1907 but ignores the replies of Schubotz (1912) and Schulze (1914). It was through this single, pivotal circumstance that Krumbach's larva-hypothesis became widely accepted. In her influential "Invertebrates", LH Hyman (1940) also cites only Krumbach's paper (p. 247) and ignores the other data: "... *Trichoplax adhaerens* and *Treptoplax*, which have the construction of planulae, were found actually to be modified planulae of Hydroidea" (p. 243). PP Grassés "Taité de zoologie IV" (1961: 694) mentions *Trichoplax adhaerens* in a similar way (Syed and Schierwater 2002a).

The rediscovery of *Trichoplax adhaerens*: It is often said that *Trichoplax adhaerens* was rediscovered when the German protozoologist KG Grell (University of Tübingen) found this animal in an algal sample from the Red Sea in 1969. Although it is true that the first electron-microscopical examinations by Grell were decisive for the final falsification of Krumbach's larva-hypothesis, at this time the animal had already found its way back into scientific circles.

In July 1961, the cell biologist W Kuhl (University of Frankfurt) found *Trichoplax adhaerens* in a seawater aquarium containing organisms from the Mediterranean Sea. Although Kuhl's research on *Trichoplax adhaerens* concentrated on locomotion and regeneration, he and co-workers clearly stated that they had never observed any connection between *Trichoplax adhaerens* and hydrozoans in the aquarium (Kuhl and Kuhl 1963, 1966). At about the same time, *Trichoplax adhaerens* was also cultured in Moscow, where it inspired Russian researchers to reinforce E Metschnikoff's phagocytella-hypothesis of metazoan evolution (Ivanov 1968, 1973, 1988).

When Grell (Grell 1971b, 1972, Grell and Benwitz 1974) discovered oogenesis and cleavage processes (after mixing *Trichoplax adhaerens* clones from different locations) it became clear that *Trichoplax adhaerens* specimens in culture represent an adult stage (Fig. 2A). Unfortunately, the embryos regularly died after reaching the 64-cell-stage (cf. Ruthmann et al. 1981, Grell 1984b), and the further development of *Trichoplax adhaerens* remained unknown at that time. However, Grell's meticulous research provided sufficient support for placing *Trichoplax adhaerens* in a new phylum, the "Placozoa" (Grell 1971a). The new phylum was named after Bütschli's placula hypothesis, and consequently the Russian researchers around AV Ivanov responded by proposing a phylum "Phagocytellozoa" for *Trichoplax adhaerens* (Ivanov 1973). Grell's phylum Placozoa has survived to this day and has marked the rediscovery of *Trichoplax adhaerens* (Schierwater et al. 2009a).

German and Russian research groups focused mainly on morphological descriptions, while researchers in the US began fieldwork on placozoans (close to nothing was known about the biology of *Trichoplax adhaerens* in its natural habitat). It quickly became clear that *Trichoplax adhaerens* could be found worldwide in the littoral of subtropical and tropical regions (Pearse 1989a). *Treptoplax reptans* Monticelli 1893 has never been found again, and its existence must be doubted.

Detailed electron-microscopical studies by KG Grell and colleagues in Tübingen and A Ruthmann and colleagues in Bochum confirmed and extended Schulze's (1883, 1891) classical descriptions of *Trichoplax adhaerens* (Fig. 1). No basal lamina could be found in *Trichoplax adhaerens*, and the interspace between the fiber cells and epithelia was found to be free of any collagenous ECM (Grell and Benwitz 1971, 1981). For the fiber cells, a syncytial (Buchholz and Ruthmann 1995) and wide-meshed organization (instead of a compact mass (Stiasny 1903) was described. Interestingly, von Graff (1891) and Stiasny (1903) had described unicellular algae in the cell bodies of the fiber cells and interpreted them as symbiotic or commensal zooxanthelles. Wenderoth (1986) found that algae and other food particles adhere to the slime layer of the upper epithelium and are subsequently phagocytized by the inner fiber cells. Food particles must be pulled through gaps of the upper epithelium, and Wenderoth (1986) called this unique mode of feeding "transepithelial cytophagy". Thus the incorporated algae are prey. However, there may also be endosymbionts present in *Trichoplax adhaerens* as bacteria were regularly found in the endoplasmatic reticulum of the fiber cells (Grell and Benwitz 1971). The ability for transepithelial cytophagy indicates a relatively loose arrangement of the epithelia (Syed and Schierwater 2002b).

Only two types of epithelial cell-cell connections are present in *Trichoplax adhaerens*, belt and septate desmosomes (Ruthmann et al. 1986, Ruthmann 2000). Connections between the epithelia and the fiber cells remain unknown. It seems likely that these connections are continually rearranged. Studies on isolated fiber cells (Thiemann and Ruthmann 1989) revealed their ability to form cytoplasmatic extensions by microtubule assembly. Those extensions are probably mediated by an actinomyosin system (Ruthmann 2000). Since isolated fiber cells live for hours in seawater, Grell and Ruthmann (1991) suggested that the interspace between epithelia and fiber cells may not be very different from the seawater medium. The loose arrangement of the epithelia and the lack of an underlying basal lamina support this idea.

Trichoplax adhaerens, the "Archimetazoon": With the beginning of the molecular revolution in biology, molecular phylogenetic approaches have been used in an attempt to unravel the phylogenetic position of the Placozoa. Ironically, some of the modern molecular analyses moved the phylogenetic view on *Trichoplax adhaerens* back to the turn of the century, and Krumbachs larva hypothesis almost was rejuvenated in a "phylogenetic" version. Based on DNA sequence analyses of ribosomal genes it was proposed that the phylogenetic position of *Trichoplax adhaerens* could fall within or very close to Cnidaria, and therefore placozoans could potentially be derived from a neotenic planula larva (e.g. Bridge 1994, Aleshin et al. 1995, Collins

1998, Aleshin and Petrov 2002). The use of new molecular markers, such as mtDNA and rRNA molecular morphology, however, showed that placozoans are not very closely related to Cnidaria (Odorico and Miller 1997, Ender and Schierwater 2003). For example, the secondary structure of the 16S rRNA molecule is substantially more complex in *Trichoplax adhaerens* than in any cnidarians (Ender and Schierwater 2003).

In 2001 the Human Frontier Science Program awarded a Research Grant to the Antp Superclass Gene Consortium in order to clarify the early evolution of Antp-type genes in basal metazoans and to develop *Trichoplax adhaerens* as a model system for research in development and evolution. Together with Peter Holland (Oxford) and Stephen Dellaporta (Yale), Bernd Schierwater (Hannover) explored *Trichoplax adhaerens* from different perspectives. Between 2001 and 2005 several new Antp-type genes were isolated, the existence of a single Hox/ParaHox gene was verified and comparative functional characters for several Antp superclass genes were analyzed. This research has yielded some surprising results. For example, the putative Proto-Hox gene *Trox-2* (Schierwater and Kuhn 1998, Kuhn et al. 1999, Sagasser and Schierwater 2002, Jakob et al. 2004) and the *Pax B* (Hadrys et al. 2005) gene are expressed in a region where the upper and lower epithelia meet, and where yet undescribed pluripotent cells are suspected. The *T-box* (Martinelli and Spring 2003) and *Not* (Martinelli and Spring 2004) gene also seem to be expressed in the same area. Especially noteworthy was the first usage of RNAi experiments on *Trichoplax adhaerens* in Jakob et al. (2004), showing that inhibition of *Trox-2* causes complete cessation of growth and binary fission. These investigations led to the conclusion that gene regulation even in an organism with a simple bauplan such as *Trichoplax adhaerens* is much more complex than hitherto expected.

***Trichoplax adhaerens*, a cryptic species:** In 2004 Aleoshin and colleagues (Aleoshin et al. 2004) compared *T. spec* (possibly Japanese Sea) with *T. adhaerens* (Red Sea) and considered *Trichoplax adhaerens* to be monotypic. Quite contrary Voigt et al. (2004) proclaimed in the same year that Placozoa might no longer be a "phylum of one". Placozoans can be found in warm, shallow, marine environments around the world (Pearse 1989b), and all observed individuals fit the general morphological description of *Trichoplax adhaerens*. Their analyses, however, showed that the phylum Placozoa is significantly more diverse than previously thought. By using four different molecular markers (18S rDNA & 28S rDNA; ITS1 & 2; 5,8S rDNA; 16S rDNA) extensive genetic variation was revealed (Voigt et al. 2004). For example from only 31 placozoans sampled around the world, they obtained eight different haplotypes of mitochondrial 16S rDNA, displaying length variation of up to 145 bp, a level far exceeding that documented for any metazoan species or genus. Most recently the number of putative placozoan species was estimated to be in the dozens or hundreds with deep phylogenetic branches (Eitel and Schierwater 2010). While sexual reproduction has not been observed in culture, putative oocyte formation in degenerating animals is routinely seen (Grell and Benwitz 1974). These large cells have been observed to undergo cleavage up to a 128-cell stage before degenerating (Eitel and Schierwater 2010). Recent DNA polymorphism analysis also provided evidence for sexual reproduction within the Placozoa (Signorovitch et al. 2005).

Placozoa, the closest link to Urmetazoa: New insights into the riddle called Placozoa was provided by the complete mapping of the mitochondrial genome of *Trichoplax adhaerens* by Dellaporta et al. in 2006 (Dellaporta et al. 2006; see Box 1). Mitochondrial genomes of multicellular animals are typically 15- to 24-kb circular molecules that encode a nearly identical set of 12–14 proteins for oxidative phosphorylation and 24–25 structural RNAs (16S rRNA, 12S rRNA, and tRNAs). These genomes lack significant intragenic spacers and are generally without introns. The unicellular choanoflagellate, *Monosiga brevicollis*, has mtDNA that is nearly four times larger (76,568 bp) than the typical animal mtDNA genome and encodes 55 different genes, often separated by large intragenic spacer regions, including two genes interrupted by introns (Burger et al. 2003). In comparison, the *Trichoplax adhaerens* mitochondrion contains the largest known metazoan mtDNA genome at 43,079 bp, more than twice the size of the typical metazoan mtDNA. The mitochondrion's size is due to numerous intragenic spacers, several introns and ORFs of unknown function, and protein-coding regions that are generally larger than those found in other animals. Not only does the *Trichoplax adhaerens* mtDNA have characteristics of the mitochondrial genomes of known metazoan outgroups, such as chytrid fungi and choanoflagellates, but, more importantly, it shares derived features unique to the Metazoa (Dellaporta et al. 2006). Phylogenetic analyses of mitochondrial proteins provide strong support for the placement of the phylum Placozoa at the root of the Metazoa (Schierwater et al. 2009a, 2009b).

In 2007 Signorovitch et al. supplemented three additional placozoan mitochondrial genomes representing three highly divergent clades. By adding these data they showed that the large *Trichoplax adhaerens* mtDNA is a shared feature among members of the phylum Placozoa and not a uniquely derived condition. All three mitochondrial genomes were found to be very large, 32 to 37 kb, circular molecules, having the typical 12 respiratory chain genes, 24 tRNAs, and the large and small ribosomal RNA's (*rnS* and *rnL*). Each placozoan strain sequenced in their study was unique in its mitochondrial genome content and structure, supporting the theory that *Trichoplax adhaerens* is a cryptic species of unknown diversity (Eitel and Schierwater 2010, Voigt et al. 2004, see Fig. 3). A phylogenetic comparison of these complete placozoan mitochondrial genome sequences to other phyla further supported the placement of Placozoa as a basal lower metazoan phylum and provided evidence for the ancestral animal mtDNA condition (Signorovitch et al. 2007).

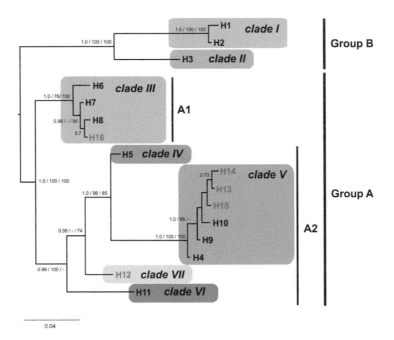

Fig. 3. 16S haplotype cladogram of all known placozoan lineages. The cladogram shows a distinctive hierarchical arrangement independent of the tree-building algorithm applied. Haplotype numbers (H) refer to strains listed in Table 1. Numbers beside nodes are from left to right: Baysian posterior probabilities, Maximum likelihood and Maximum Parsimony bootstrap support. Values below 70% are marked with '-'. Two main groups ('A' and 'B') are found within the Placozoa probably representing higher taxonomic units. Within group 'A' two subgroups ('A1' and 'A2') are clearly distinguishable. Red labeling marks formerly undescribed haplotypes. (from Eitel and Schierwater 2010).

Color image of this figure appears in the color plate section at the end of the book.

Appraising the common features among placozoan mtDNA genomes, the first hypotheses about the ancestral condition of the mitochondrial genome of all animals have been made. For example, the data suggested that, contrary to conventional wisdom, the common ancestor of all animals actually possessed large, non-compact mitochondria, owing to the fact that both Placozoa and their closest non-animal relative, Monosiga (Lang et al. 2002), have distinctively large mitochondrial genomes.

Compaction of the mitochondrial genome likely occurred secondarily after the emergence of higher metazoans. The data support the hypothesis that loss of ribosomal protein genes from the mitochondrial genome is a metazoan synapomorphy, as no animal mitochondrial genome sequenced so far, including the large mtDNAs of placozoans described here, has identified sequences coding for ribosomal proteins (Signorovitch et al. 2007).

In a large multinational collaboration the sequencing and analysis of the ~98 Mb nuclear genome of the placozoan *Trichoplax adhaerens* was published (Srivastava et al. 2008). The compact genome shows remarkable complexity, including conserved gene content, gene structure and synteny relative to human and other eumetazoan genomes. Despite the absence of any known developmental program and only a modest number of cell types, the *Trichoplax adhaerens* genome encodes a rich array of transcription factors and signaling genes that are typically associated with embryogenesis and cell fate specification in eumetazoans, as well as other genes that are consistent with cryptic patterning of cells, unobserved life history stages and/or complex execution of biological processes such as fission and embryonic development in these enigmatic creatures.

Biased by the fact that Placozoa harbor the largest known mitochondrial genome in animal history and only a small nuclear genome as reported in Srivastava et al. (2008), it was claimed that sponges and not Placozoa are the earliest diverging animals. Their hypothesis is based on phylogenetic analyses of a large number of nuclear genes supporting the identification of placozoans as a basal eumetazoan lineage that diverged before the separation of cnidarians and bilaterians but after the divergence of demosponges from other animals. It is unfortunate Srivastava et al. (Srivastava et al. 2008) could not include a comprehensive taxon sampling.

Although this result disagrees with results from mitochondrial DNA analysis (Dellaporta et al. 2006, Haen et al. 2007, Signorovitch et al. 2007) all of the analyses are complicated by long branch lengths (that is, unusually high amounts of amino acid divergence and hence potential convergence of taxa as a result of attraction of the long branches to each other) found in bilaterian mitochondrial peptides relative to their basal metazoan orthologues (Lavrov et al. 2005, Haen et al. 2007). Peptides encoded by the nuclear genome have no notable differences in amino acid substitution levels between basal metazoans and bilaterians, suggesting that their proposed phylogeny on the basis of nuclear genes is less susceptible to long branch attraction artefacts.

To keep it even more interesting Dunn et al. stated in the same year (Dunn et al. 2008) that ctenophores might be the earliest diverging extant multicellular animals. Dunn et al. used a comparable large dataset by utilizing Expressed Sequence Tag (EST) data, but ironically they omitted the Placozoa. The Dunn et al. paper was supported by Miller and Ball (Miller and Ball 2008). Despite the more complex morphology of Ctenophora compared to sponges and Placozoa it was argued that the latter two are secondary simplified animals or that there is an undiscovered morphologically complex life stage for Placozoa while the complex morphology of ctenophores has arisen independently from that of other metazoans. Recent publications on this and other subjects related to the phylogeny of early metazoans, demonstrate that while the battlefield might have changed from morphology to phylogenetic analysis the debate about the evolutionary relationships at the base of the Metazoa is livelier than ever before. Any tree supporting Ctenophora as most basal phylum contradicts all principles of comparative zoology and calls for a revisit of the dataset and analysis. It is very unlikely that Placozoa are secondarily reduced animals given that their mitochondrial genome (dimension and content) is the best link between Protists and higher Metazoa discovered so far and that the morphology cannot be explained by reduction (Syed and Schierwater 2002b). Particular attention is required when interpreting trees, as incomplete data, inadequate sampling or the wrong choice of outgroups, to name a few examples, can easily corrupt the output of the analysis.

A concatenated analysis using over 30 genes and anatomy was published by Schierwater et al. in 2009 (Schierwater et al. 2009c). For the first time all relevant taxa regarding the Urmetazoon Hypothesis were included and morphological features were combined with molecular sequence data to produce a phylogenetic hypothesis. This extensive study is consistent with the general view that Bilateria and Diploblasts (Placozoa, Porifera, Cnidaria and Ctenophora) are sister groups with the choanoflagellate Monosiga basal to these taxa with high jackknife values and Bayesian posteriors. Notably, Placozoa are robustly observed as the most basal diploblastic group. Porifera, Bilateria and Fungi all form strong monophyletic groups. The four Cnidaria classes together with the Ctenophora form a monophyletic group, the "Coelenterata". Within the Cnidaria, the generally accepted basal position of the anthozoans is also recovered by this analysis. Both choanoflagellates and Placozoa are strongly excluded from a Porifera-Coelenterata monophyletic group. However none of the three recent scenarios, Placozoa basal (Schierwater et al. 2009c), Porifera basal (Philippe et al. 2009), Ctenophora basal (Dunn et al. 2008) provide unambiguous support (Siddall 2009).

As if the origin of Metazoa was not controversial enough as it is, a forgotten theory was revitalized by Mikhailov et al. (Mikhailov et al. 2009). They argued that in addition to Haeckel's Gastraea theory (Haeckel 1874) and Bütschli's Placula theory (Bütschli 1884) new evidence and insights into gene regulation and cell development supported a completely different scenario, the Synzoospore theory (Zakhvatkin 1949, Sachwatkin 1956, Zakhvatkin 2008). This theory envisions the metazoan ancestor as a protist with a complex life cycle that includes monotonously dividing trophic cells (or cellular aggregates), hypertrophic growth of gametes, and their subsequent palintomic cleavage producing non-feeding dispersal zoospores. The transition to multicellularity occurs with (i) integration of trophic cells into a differentiated colonial body and (ii) integration of zoospores into the uniform synzoospore, the primary lecitotrophic dispersal larva of the animals. This theory is supported by observations that genes that typically control metazoan development, cell differentiation, cell-to-cell adhesion, and cell-to-matrix adhesion are found in various unicellular relatives of the Metazoa, suggesting that cell differentiation might have preceded the emergence of multicellularity, and not vice versa.

More recently new evidence for the Placula theory (Fig. 4) was presented by Schierwater et al. (Schierwater et al. 2009a) in a larger analysis of 72 taxa to reinforce the inference obtained from the smaller taxonomic sample (Schierwater et al. 2009c). The data clearly showed that Bilateria and Diploblasta are monophyletic taxa and sister groups to each other with robust bootstrap support for both parsimony and maximum likelihood analyses. The last common metazoan ancestor (LCMA) likely possessed a pre-nervous system with some kind of unspecialized proto-nerve cells. Placozoa and Porifera *cum grano salis* conserved this stage, while both Coelenterata and Bilateria developed specialized nerve cells from this stage. In the light of this finding the parallel invention of nerve cells, and consequently a nervous system, in Bilateria and Coelenterata is hardly problematic and not much more than a morphological and physiological specialization of preexisting proto-nerve cells (Fig. 5A, B).

Trichoplax adhaerens—General Biology

Morphology

T. adhaerens is 2–3 mm wide and approximately 15 µm high (Grell and Ruthmann 1991). Thiemann and Ruthmann 1990). Its bauplan is based on a sandwich organization. An upper and a lower epithelium surround a loose network (but not an epithelium) of so-called fiber cells (Schierwater 2005). Five cell types have been described in *Trichoplax*, upper and lower epithelial cells, gland cells within the lower, feeding epithelium, fiber cells sandwiched between the epithelia forming a syncytium and pluripotent stem or precursor cells, localized within the boundary area of the animal (Stiasny 1903, Grell and Benwitz 1971, Ivanov 1973, Grell 1974, Behrendt and Ruthmann 1986, Ruthmann, G. et al. 1986, Buchholz and Ruthmann 1995, Jakob et al. 2004). The cells of the upper epithelium are monociliated and able to excrete mucus (Ruthman 1996). Cells of the lower epithelium

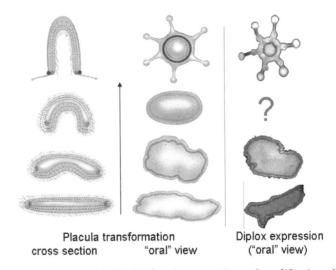

Placula transformation
cross section   "oral" view

Diplox expression
("oral" view)

Fig. 4. Modern Placula Hypothesis. Modern interpretation and modification of the placula hypothesis of metazoan origins. A non-symmetric and axis-lacking Placula bauplan transforms into a typical symmetric metazoan bauplan with a defined oral–aboral body axis. The Placula transforms from a primitive disk consisting of an upper and a lower epithelium (lower row) to a form with an external feeding cavity between its lower epithelium and the substrate (2nd row from bottom). For this the Placula lifts up the center of its body, as naturally seen in feeding *Trichoplax*. If this process is continued, the external feeding cavity increases (cross section, 3rd row) while at the same time the outer body shape changes from irregular to more circular (see oral views). Eventually the process results in a bauplan where the formerly upper epithelium of the Placula remains "outside" (and forms the ectoderm) and the formerly lower epithelium becomes "inside" (and forms the entoderm, upper row). This stage represents the basic bauplan of Cnidaria and Porifera. Three of the four transformation stages have living counterparts in form of resting *Trichoplax*, feeding *Trichoplax*, and cnidarian polyps and medusae (right column). From a developmental genetics point of view a single regulatory gene would be required to control separation between the lower and upper epithelium (three lower rows).We find this realized in *Trichoplax* in the form of the putative Proto-Hox/Para-Hox gene, *Trox-2*. More than one regulatory gene would be required to organize new head structures

originating from the ectoderm-entoderm boundary of the oral pole (upper row) in Cnidaria. Indeed (and quite intriguing), two cnidarian orthologs of the *Trox-2* gene, *Cnox-1* and *Cnox-3*, show exactly these hypothesized expression patterns (Diplox expression upper row; for simplicity only the ring for *Cnox-1* expression is shown). (from Schierwater et al. 2009a).

Color image of this figure appears in the color plate section at the end of the book.

are either clubbed without cilia or cylindrical and monociliated (Grell and Benwitz 1971, 1981). Between these, so-called gland cells of different number and size can be found (Schulze 1892, Grell and Benwitz 1971, Grell and Benwitz 1981). No organs, specialized nerve or muscular cells,

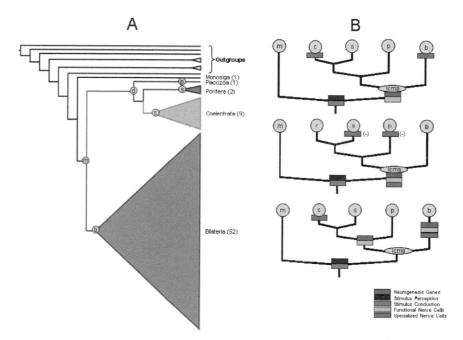

Fig. 5A. Phylogenetic tree with relationships within Bilateria, Coelenterata, and Porifera collapsed. Numbers in parentheses refer to number of species in each of these groups. To establish tree support measures we ran both maximum likelihood (ML) and maximum parsimony (MP) analyses. In the reporting of tree statistics we give the ML value first and the MP value second. Tree support statistics for the nodes marked by circles with letters inside are: (B) Bilateria 100/100, (C) Coelenterata100/82, (S) Porifera 100/100, (D) Diploblasta 100/99, (M) Metazoa 100/63; (P) Placozoa is a single taxon. Within the Bilateria - Deuterostomia 100/100, Protostomia 100/100. (from Schierwater et al. 2009a).

Fig. 5B. Phylogenetic scenarios for the evolution of nerve cells mapped onto the Diploblast-Bilateria Sister hypothesis. Five potential characters (represented by colored boxes in the figure) important in the evolution of nerve cells are mapped onto the Diploblast-Bilateria Sister. Most qualities of a nerve cell seem to have been present already in the last common metazoan ancestor (lcma in light blue). In the top figure we present the most parsimonious explanation for the evolution of these five characters (6 steps). Only the specialization of multi-functional proto-nerve cells into uni-functional nerve cells would have occurred in parallel in Bilateria and Coelenterata in the above scenario. The middle scenario is similar to the top only instead of hypothesizing independent gain of specialized nerve cells it hypothesizes independent loss of specialized nerve cells (7 steps). The bottom tree shows a highly unlikely scenario where the number of steps is nearly twice the top scenario (from Schierwater et al. 2009b).

Color image of this figure appears in the color plate section at the end of the book.

basal lamina, extracellular matrix or any kind of axis symmetry are present (Grell and Benwitz 1971, 1981, Grell 1984, Schierwater 2005). Body shape is irregular and changes constantly. Merely two epithelia are definable: a functional upper epithelium, facing the open water, and a lower (= feeding) epithelium, facing the substrate, (Schierwater 2005, see Fig. 1).

In laboratory culture the animal feeds down algae. Extracellular digestion is mediated by the clubbed cells of the lower epithelium followed by pinocytosis via the cylindrical cells (Grell and Ruthmann 1991). Additionally, algae und food particles from the upper epithelium are phagocytized by fiber cells between the two epithelia (transepithelial cytophagy; Wenderoth 1986). The fiber cells have comparatively large numbers of mitochondria which form complexes with vesicles. Within the cisterns of the endoplasmatic reticulum bacteria have been found, which could potentially be endosymbionts (Grell and Benwitz 1971, 1981). So far nothing is known about the relative importance of the different feeding modes or about the role of potentially symbiotic bacteria (Schierwater 2005). *Trichoplax* individuals glide across the substrate by action of the cilia of the lower epithelium or move in an amoeboid way by changing body shape. Body shape changes result from contractions and expansions of the fiber cell syncytium (Ruthmann et al. 1986).

Habitat & Ecology

Placozoa can be found in the littoral zone of all warm oceans, crawling over algae, pieces of coral and smooth stones and are distributed globally in tropical and subtropical waters (Pearse 1989a, Pearse 1989b, Schierwater 2005). There are reports about sources in the Red Sea (Grell 1974), the Mexican Caribbean Sea (Signorovitch et al. 2006), the Pacific Ocean (Pearse 1989a, Pearse 1989b), the Mediterranean Sea (Tomassetti et al. 2005a, Tomassetti et al. 2005b) and the Japanese Sea (Maruyama 2004). Molecular analyses of the genetic variation within the phylum Placozoa show distinct ecological niche separations between different clades strongly differing in their ecological needs (Eitel and Schierwater 2010; see Figs. 3 and 6). Besides that close to nothing is known about placozoan ecology.

Life Cycle & Reproduction

T. adhaerens is able to reproduce vegetatively as well as bisexually. Vegetative reproduction occurs by binary fission into two daughter individuals of similar size or by budding off of small spherical swarmers which are planktonic (Grell 1974, Grell 1984, Thiemann and Ruthmann 1990). The latter most likely are dispersal stages, which may float in the open water for up to a week (Thiemann and Ruthmann 1990). In the laboratory the animal mostly undergoes binary fission. Bisexual reproduction of *Trichoplax* is currently being analyzed. Male gametes have not been observed yet. Female gametes appear in small numbers in individual placozoans in the laboratory, mostly after a period of degeneration (Grell and Benwitz 1971, Grell and Benwitz 1974, Schierwater 2005). No embryonic development has been observed beyond early cleavage stages (64 cells) (Grell 1972, Grell 1984, Schierwater 2005). Yet genetic analyses show allelic variations and

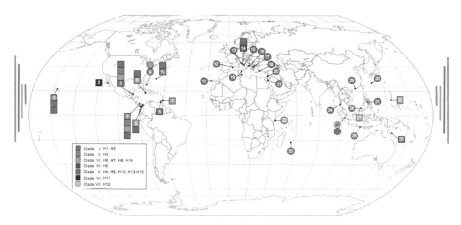

Fig. 6. Worldwide distribution of genetically characterized placozoan specimens. Aquarium samples (A.s.) with presumed origin are labeled with dashed lines. Note that several numbers combine multiple sampling sites. **1.** Oahu, Hawaii (US), **2.** Southern California (A.s., US), **3.** Caribbean coast of Belize, **4.** Caribbean coast of Panama, **5.** Pacific coast of Panama, **6.** Cubagua Island/Margarita Island (Venezuela), **7.** Grenada, **8.** Discovery Bay (Jamaica), **9.** Bahamas, **10.** Bermuda (GB), **11.** Tenerife, Canary Islands (Spain), **12.** Majorca, Balearic Islands (Spain), **13.** Castiglioncello (Italy), **14.** Orbetello Lagoon (Italy), **15.** San Felice Circeo (Italy), **16.** Otranto (Italy), **17.** Kateríni and Ormos Panagias (Greece), **18.** Bay of Turunç (Turkey), **19.** Gulf of Hammamet and near Zarzis (Tunisia), **20.** Caesarea (Israel), **21.** Elat (Israel), **22.** Mombasa (Kenya), **23.** Réunion (France), **24.** Laem Pakarang (Thailand), **25.** 'Indonesia' (A.s.), **26.** Bali (A.s), **27.** 'Indo-Pacific' (A.s.), **28.** Kota Kinabalu, Sabah (Malaysia), **29.** Hong Kong (China), **30.** Okinawa, Ryukyu Islands (Japan), **31.** Boracay (Philippines), **32.** Guam (US), **33.** Lizard Island, (NE Australia). (from Eitel and Schierwater 2010).

Color image of this figure appears in the color plate section at the end of the book.

evidence of genetic recombination, consistent with bisexual reproduction (Signorovitch et al. 2005). Thus, the life cycle of *T. adhaerens* is still not completely understood, but most likely the life cycle is simple and bisexual (Eitel et al. in prep.).

Trichoplax adhaerens—A Model System

The simplest known metazoan animals, the Placozoa, are a unique model system for investigations in the fields of evolutionary, developmental and cell biology (Miller and Ball 2005, Kiefer 2006, Schierwater et al. 2009a). Its simple morphology and genetic constitution, that seems to be the last living surrogate for the genetic constitution of the urmetazoon, renders it a perfect candidate for several research interests from evodevo to cell cycle studies up to cancer research. Its worldwide distribution could furthermore qualify Placozoa to be used as an indicator organism to determine marine pollution levels and to estimate the ecological impact of global warming.

Yeast has turned out to be a prime model system regarding cellular processes in unicellular organisms (e.g. Kerscher et al. 2004, Smith and Snyder 2006). Lower metazoans serve as a simple multicellular system which provides the simplest platform for analysis of biological processes unique to multicellular organisms such as cell-cell communication, differentiation or signaling. In contrast to systems like sponges or ctenophores, Placozoa are kept routinely in long-term laboratory culture under standardized conditions (for culturing details see Schierwater 2005). Furthermore, its vegetative reproduction via budding or fission facilitates the maintenance of clonal lineages and makes the organism putatively immortal (Grell 1971, Grell and Ruthmann 1991). *Trichoplax'* simple bauplan offers one of its greatest advantages over the known traditional model organisms. As it consists of merely five somatic cell types which form three distinct cell layers, experiments can be kept simple, well arranged and much cheaper. Whole mount experiments can be performed without any problems as its average size of 2 mm in diameter renders time consuming preparations in most cases unnecessary and additionally allows several *in vivo* applications which are not possible in more complex model organisms. Its translucent morphology furthermore facilitates to track single cells of interest in the microscope. As Placozoans can be considered a "crawling cell culture" they could potentially be used as animal models for chemical tests and drugs screens and could therefore become important for the pharmaceutical industry.

Trichoplax possesses the ability to quickly and effectively regenerate after injuries. After transection of the marginal tissue, the wound closes quickly when the outer margin starts to bend inwards. If one individual is cut into two, both parts can survive as independent organisms (Kuhl and Kuhl 1966, Schwartz 1984), and even if cohesive cell arrangement is destroyed by means of chemical treatment or a dissection net, the animal can regenerate a new individual from a few cells (Terwelp 1978, Ruthmann and Terwelp 1979). It is also possible to combine two genetically different individuals and thus create a chimeric individual which is viable and can be kept in laboratory culture (von der Chevallerie, unpublished data). The great regeneration capacity is likely based on very efficient repair mechanisms, and investigations in this field can provide new insights into the evolution of the immune system and wound healing mechanisms. Placozoa are also very resistant to irradiation with X-rays (von der

Chevallerie, unpublished data) which suggests the existence of a very efficient DNA repair system.

With the evolution of multicellularity came the requirement for several mechanisms controlling cellular homeostasis, regulating cell division processes and tracing the elimination of ineffective cells (Rokas 2008). The whole complexity of the genetic control of cellular homeostasis is far from being resolved (Tyson et al. 2002) which clearly demands further investigation. *Trichoplax adhaerens* is undoubtedly the most simple animal model system in which processes and mechanisms of cell proliferation and apoptosis can be studied.

In Evodevo research we consider the evolution of different animal bauplans to be one of the great mysteries of biology. Transcription factors of the homeodomain family play important roles during bauplan development, cell differentiation and cell proliferation (e.g. Holland and Garcia-Fernandez 1996, Hughes and Kaufman 2002). For example the study of a putative Hox gene in Placozoa has fueled some highly new ideas about Proto-bauplan evolution at the basis of Metazoa (DeSalle and Schierwater 2008, Schierwater et al. 2009a, 2009b, 2009c).

Trichoplax adhaerens—Some Final Comments on the Phylogenetic position of the Placozoa in the Metazoan Tree of Life

From their extensive morphological and embryonic studies F.E. Schulze (1891) and K.G. Grell (1971c) came to the same conclusion: The phylum Placozoa, with its yet only described species *Trichoplax adhaerens*, represents morphologically the simplest living animal and has "to be placed isolated at the lowest level of metazoan evolution" (Grell 1971c, author's translation). Although several studies are in favor of this view from a morphological standpoint (Grell 1971c, Schierwater 2005, Schulze 1883), others disagree, placing sponges as the closest relative of the 'Urmetazoon' (e.g. Nielsen 2001, Nielsen 2008; Fig. 7A). This view is mainly based on a presumed synapomorphic collar structure surrounding a flagellum shared among sponges and choanoflagellates. Several arguments have been discussed that either support or reject homology between these structures (Gonobobleva and Maldonado 2009, King 2004, Maldonado 2004, Rieger 1976, Willmer 1991). Some authors are in favor of a convergent evolution of collar structures and metazoan choanocytes (Maldonado 2004) or even claim that the choanoflagellates are derived sponges (Clark 1868, Kent 1878, Maldonado 2004).

Genomic techniques and associated algorithms to handle genetic information from different animals were used to decipher metazoan relationships from the very early 1990s. Early molecular studies were mainly based on ribosomal DNA (18S and 28S) because of their high conservation

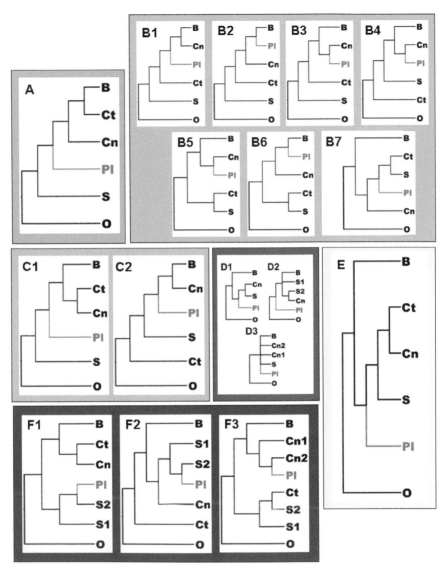

Fig. 7. An overview of published intra-relationships of the four diploblastic groups (Placozoa, Porifera, Cnidaria, Ctenophora) and their inter-relationship to the Bilateria. Shown are a few examples for each of the five character groups defined in Table 1: morphology **(A)**, ribosomal DNA **(B)**, nuclear encoded protein sequences **(C)**, mitochondrial encoded protein sequences **(D)** and combined data sources **(E)**. Placozoans have been placed at nearly every possible relationship to the other four groups, even within Porifera (F1, F2) and within Cnidaria (F3). Hence, a consensus on the phylogenetic placement of the Placozoa is still missing.

O = outgroup(s), S = Porifera, Pl = Placozoa, Cn = Cnidaria, Ct = Ctenophora, B = Bilateria.

Color image of this figure appears in the color plate section at the end of the book.

in certain regions making it easy to design primer sets likely to work well even in different animal phyla. These early studies significantly improved our knowledge on phylogenetic relationships among some—mostly bilaterian—groups (Lavrov 2007). But the relationships among very early branching metazoans—Placozoa, Porifera, Cnidaria and Ctenophora —still remained unresolved. To the authors' knowledge a total of 33 papers have been published in the last two decades using placozoan partial or complete 18S and/or 28S sequences for phylogenetic tree reconstructions. Mostly sponges have been placed as the earliest branching animals in these studies and nearly every possible relationships at the base of the Metazoa has been published based on 18S and 28S data (see Fig. 7B and Table 1 for an overview and references). Even the most modern phylogenetic reconstruction methods using complete 28S sequences from 197 taxa didn't resolve this problem showing early branching and paraphyletic sponges with one representative grouping together with a Ctenophore—a morphologically highly doubtable scenario, which is based on the analysis of a single marker gene (Mallatt et al. 2010). One has to note that most of the older 18S and 28S studies mentioned above and in Table 1 are based on limited taxon sampling and statistical methods that are now considered insufficient.

State of the art molecular phylogenetic approaches using highly adopted algorithms and high computing power were promising to overcome such problems as genetic information from hundreds to thousands of genes could be used to study metazoan evolution. Several approaches have been used to resolve the metazoan tree of life. Single gene amplification strategies or EST libraries with several thousand characters still resulted in different and sometimes highly contradictory trees (Fig. 7C and Table 1). Hardly any consensus can be found, but mostly the assumed linear evolution from simple (non-bilaterian or diploblastic) to complex (bilaterian or triploblast) organisms has been supported by these concatenated nuclear genes studies (for refs. see Table 1). This view is currently widely accepted. In most phylogenetic scenarios following this assumption sponges were found branching off first (Philippe et al. 2009, Srivastava et al. 2008) thus being the closest living relative to the Urmetazoon.

Very recently a new and—from a morphological perspective—highly debatable scenario has been proposed. In two studies Ctenophora were shown to be the earliest branching metazoans followed by Porifera, Placozoa and Cnidaria as sister taxa to bilateria (Dunn et al. 2008, Hejnol et al. 2009; Fig. 7 C2). These results suggest that the metazoan radiation started very early after the new level of organization (multicellularity) had been invented, rapidly leading to quite complex bauplans such as those of ctenophores. Based on these studies the simple bauplans of placozoans and sponges would have to be interpreted as derived features.

Another new and quite striking scenario of metazoan evolution has recently been supported based on concatenated data from several different sources: mitochondrial and nuclear DNA sequences as well as gross and molecular morphology (Schierwater et al. 2009a, Schierwater et al. 2009c; Fig. 7E). In this scenario diploblasts (non-bilateria *sensu stricto*) and tribloblasts (bilateria) are sister groups that share a common urmetazoan ancestor. Placozoans have a pivotal role in this scenario, as they are earliest branching off in the diploblast clade, sharing lots of features with the hypothesized 'placula' (Bütschli 1884) and thus possibly being the closest extant relative to the animals' ancestor. This diploblast-triploblast-sister scenario has already been proposed before based on concatenated analysis of mitochondrial protein-coding genes (Fig. 7D) and on 18S sequence data (Aleshin et al. 1995, Katayama et al. 1995, Winnepenninckx et al. 1995; Fig. 7 B7 and D1).

A continuous accumulation of new sequence information from diploblast animals will help to overcome the current problem of highly underrepresented genetic information from that key taxon. These and other sources of phylogenetic information will hopefully enable us to understand the currently controversial phylogenetic relationships among early branching metazoan animals.

Box 1. *Trichoplax adhaerens*—Comparative mitochondrial genomics in Placozoa

The analysis of complete mitochondrial genomes is a useful tool to decipher deeper relationships between the previously described placozoan haplotypes, which have been postulated based on 16S, 18S and 28S rDNA and ITS sequences (Voigt et al. 2004, Signorovitch et al. 2006, Eitel and Schierwater 2010). There are some advantages of using mitochondrial genomes for phylogenetic analyses. They are easier to handle than nuclear genomes due to their relatively small genome size between 15 kb and 24 kb in most basal metazoans. Additionally there is a higher number of mt DNA molecules in each cell and despite of some exceptions (e.g. Hellberg 2006) the nearly identical set of mitochondrial genes throughout Metazoa generally evolve faster than nuclear genes (for review see Lavrov 2007 and Lynch et al. 2006, and references therein).

The first mitochondrial genome of the only described placozoan species *Trichoplax adhaerens* (H1, clade I) was published in 2006 by Dellaporta and co-workers. With a size of over 43 kb it is the largest metazoan mt genome known so far with a mt genome size between the choanoflagellate *Monosiga brevicollis* (> 76 kb) (Burger et al. 2003) and the sponge *Suberites domuncula* (> 26 kb) (Lukić-Bilela et al. 2008). The circular

Table 1. Summary of published phylogenetic studies inferring metazoan relationships. The table comprises all references that included placozoan data. Shown are five character groups using different sources of information: ribosomal DNA sequences (ribo), morphological characters (morph), nuclear encoded protein sequences (nuclear), mitochondrial encoded protein sequences (mito) and information from combined sources (mixed). WGS = whole genome sequence, ESTs = expressed sequence tags, CP = cladistic parsimony, NJ = neighbor joining, MP = maximum parsimony, ML = maximum likelihood analysis , BA = bayesian analysis, cons = consensus, Min = minimum length, O = outgroup(s), S = Porifera (S1-S3 in case of paraphyly), Pl = Placozoa, Cn = Cnidaria (Cn1-Cn3), Ct = Ctenophora, B = Bilateria (B1–B3).

# in Fig. 2	reference	data source	marker(s)	method	missing taxa	tree topology	remarks
	Abouheif et al., 1998	ribo	18S	MP	-	(S,(Ct,(Pl,(Cn,B))))	rooted on sponges
	Aleshin et al., 1995	ribo	18S	ML	-	(O,(B,(((S1,(S2,Ct)),(Pl,Cn)))))	
	Aleshin et al., 1995	ribo	18S	NJ	-	(O,(S1,(S2,Ct),(B,(Pl,Cn)));)	
B4	Aleshin et al., 1998	ribo	18S	MP	-	(O,(Ct,(S,(B,(Pl,Cn)))))	
	Bass et al., 2007	ribo	18S	ML, BA	Ct	(O,(S1,(S2,(Cn,(Pl,B)))))	
	Berntson et al., 2001	ribo	18S	ML	B	(O,(S,(Ct,(Pl,Cn))))	
	Borchiellini et al., 2001	ribo	18S	MP	-	(S,(Ct,(B,(Pl,Cn))))	rooted on sponges
	Carranza et al., 1997	ribo	18S	ML	S	(O,(Ct,(Cn,(Pl,B))))	
	Cavalier-Smith & Chao, 1995	ribo	18S	ML	-	(O,((S1,(S2,Ct)),(B,(Pl,Cn))))	
	Cavalier-Smith & Chao, 2003	ribo	18S	ML	B	(O,(S,(Ct,(Cn1,(Cn2, (Pl,Cn3))))))	Placozoa within Cnidaria
	Collins, 1998	ribo	18S	CP	-	(O,((S1,(S2,Ct)),(Cn,(Pl,B))))	
	Collins, 1998	ribo	18S	ML, NJ	-	(O,((S1,(S2,Ct)),(Pl,(Cn,B));))	
	Collins, 2000	ribo	18S	MP	-	(S,(Ct,(Cn,(Pl,B))))	rooted on sponges
	Collins, 2002	ribo	18S	MP	-	(O,(S1,(S2,(Ct,(Pl,(Cn,B))));;))	
	Gerlach et al., 2007	ribo	18S	NJ	-	(O,((S1,Ct),(S2,Cn,(Pl,B))));	
	Glenner et al., 2004	ribo	18S	BA	-	(O,S,Ct,(B,(Pl,Cn)))	
	Katayama et al., 1995	ribo	18S	ML	-	(O,((S,(Pl,Ct)),(Cn,B)))	
B7	Katayama et al., 1995	ribo	18S	MP, NJ	-	(O,(B,Cn,(Pl,(S,Ct))))	
	Kim et al., 1999	ribo	18S	ML	-	(O,(S,(Ct,(Pl,Cn,B)))	
	Kober & Nichols, 2007	ribo	18S	MP, BA	-	(O,((S1,(S2,Ct)),((Cn, (Pl,((S,Ct),(Cn,B)))	
	Littlewood et al., 1998	ribo	18S	NJ	-	(Pl,((S,Ct),(Cn,B)))	unrooted tree
	Medina et al., 2003	ribo	18S	ML, MP, BA	-	(O,(S1,(S2,(S3,(B,(Pl,Cn))))))	
B1	Podar et al., 2001	ribo	18S	ML	-	(O,(S,(Ct,(Pl,(Cn,B))))))	
	Sidall et al., 1995	ribo	18S	MP	-	(O,((S1,(S2,Ct)),(B,(Pl,Cn)));;)	

Table 1. contd...

Table 1. contd...

# in Fig. 2	reference	data source	marker(s)	method	missing taxa	tree topology	remarks
B5	Smothers et al., 1994	ribo	18S	MP, NJ	-	(O,((S,Ct),(B,(Pl,Cn))))	
B3	Wainright et al., 1993	ribo	18S	ML	-	(O,(S,(Ct,(B,(Pl,Cn))))	
B2	Wallberg et al., 2004	ribo	18S	MP	-	(O,(S,(Cn,(Pl,B))))	
	Winnepenninckx et al., 1998	ribo	18S	NJ	-	(O,((S1,(S2,Ct),(Cn,(Pl,B))))	
	Zrzavy et al., 1998	ribo	18S	MP	-	(O,((S1,(S2,Ct),(Pl,(Cn1,(Cn2,B))))	
	Christen et al., 1991	ribo	28S	MP	-	(O,((S1,(S2,Pl),(B,(Cn,Ct))))	Placozoa within sponges
	Kober & Nichols, 2007	ribo	28S	MP, BA	-	(O,((S1,B1),(Pl,(Cn,(S2,Ct,B2))))	paraphyletic Bilateria
	Lafay et al., 1992	ribo	28S	ML, MP, NJ	-	(B,(S1,(S2,(Pl,(Ct,(S3,Cn))))))	unrooted tree
	Zrzavy & Hypsa, 2003	ribo	28S	MP	-	(S,(Ct,(B,(Pl,Cn)))	rooted on sponges
B6	Cartwright & Collins, 2007	ribo	18S, 28S	ML	-	(O,((S,Ct),(Cn,(Pl,B)))	
B3	Da Silva et al., 2007	ribo	18S, 28S	ML	-	(O,(S,(Ct,(B,(Pl,Cn))))	
	Mallatt et al., 2009	ribo	18S, 28S	ML	-	(O,(S1,((S2,Ct),(B,(Pl,Cn)))))	
	Mallatt et al., 2009	ribo	18S, 28S	BA	-	(O,(S1,(S2,(Ct,(B,(Pl,Cn)))))	
	Odorico & Miller, 1997	ribo	18S (3' end) to 28S (5' end)	ML	B	(S,Cn,(Pl,Ct))	unrooted tree
A	Glenner et al., 2004	morph	94 characters	BA	-	(O,(S,(Pl,(Cn,(Ct,B))))	
A	Nielsen et al., 1996	morph	61 characters	Min	-	(O,(S,(Pl,(Cn,(Ct,B))))	
	Nielsen, 2001	morph	64 chartacters	Min	Ct	(O,(S,(Pl,(Cn,B)))	
A	Peterson & Eernisse, 2001	morph	138 characters	MP	-	(O,(S,(Pl,(Cn,(Ct,B))))	
	Zrzavy et al., 1998	morph	276 characters	MP	-	(O,(S,(Pl,(Cn,(B1,(Ct,B2))))))	
	Hejnol et al., 2009	nuclear	1487 nc-encoded proteins (270,580 aa)	ML	-	(O,(Ct,(S1,(Pl,(S2,(Cn,B))))))	
C2	Hejnol et al., 2009	nuclear	150 nc-encoded proteins (??? aa)	ML	-	(O,(Ct,(S,(Pl,(Cn,B))))	
F2	Marletaz et al., 2008	nuclear	77 ribosomal proteins (WAG)	ML	-	(O,(Ct,(B,(Cn,(S1,(S2,Pl))))))	Placozoa within sponges
	Marletaz et al., 2008	nuclear (11,730 aa)	77 ribosomal proteins	BA (CAT)	-	(O,((S1,(S2,Pl)),(B,(Cn,Ct)))	(11,730 aa) Placozoa within sponges

C1	Philippe et al., 2009	nuclear (30,257 aa)	128 nc-encoded proteins (CAT)	BA	-	(O,(S,(Pl,(B,(Cn,Ct)))))
	Ruiz-Trillo et al., 2006	nuclear	EF-1, HSP-70, actin	ML	Ct	(O,(B1,((Pl,(S,(B2,Cn1))), (Cn2,B3)))) — paraphyletic Bilateria
D1	Sperling et al., 2009	nuclear	house keeping genes	BA (WAG, CAT)	Ct	(O,((S1,(S2,(S3,(Pl,(Cn,B))))))
	Srivastava et al., 2008	nuclear	104 nc-encoded proteins (6,783 aa)	ML, MP, BA	Ct	(O,(S,(Pl,(Cn,B))))
D3	Burger et al., 2009	mito	13 mt-encoded proteins (3,004 aa)	BA (CAT)	Ct	(O,(Pl,S,Cn1,Cn2,B)
D1	Dellaporta et al., 2006	mito	12 mt-encoded proteins (2,730 aa)	ML, BA	Ct	(O,(B,(Pl,(S,Cn))))
D1	Erpenbeck et al., 2007	mito	13 mt-encoded proteins (??? aa)	ML, BA	Ct	(O,(B,(Pl,(S,Cn))))
	Haen et al., 2007	mito	12 mt-encoded proteins (2,678 aa)	ML	Ct	(O,((Pl,(S1,Cn)),(S2,B)))
	Haen et al., 2007	mito	12 mt-encoded proteins (2,678 aa)	BA (CAT)	Ct	(O,((S,Cn),(Pl,B)))
D2	Lavrov et al., 2008	mito	14 mt-encoded proteins (2,701 aa)	BA cons	Ct	(O,(Pl,(S1,B,(S2,Cn))))
	Lavrov et al., 2008	mito	14 mt-encoded proteins (2,701 aa)	ML, BA (cpREV)	Ct	(O,((S1,B),(Pl,(S2,Cn))))
	Lavrov et al., 2008	mito	14 mt-encoded proteins (2,701 aa)	BA (CAT)	Ct	(O,(Pl,(B,(S,Cn))))
D1	Ruiz-Trillo et al., 2008	mito	13 mt-encoded proteins (2,619 aa)	BA (CAT)	Ct	(O,((B,Pl),(S,Cn)))
D1	Signorovitch et al., 2007	mito	12 mt-encoded proteins (2,553 aa)	ML, BA	Ct	(O,(B,(Pl,(S,Cn))))
D1	Wang & Lavrov, 2007	mito	12 mt-encoded proteins (2,812 aa)	ML, BA, NJ	Ct	(O,(B,(Pl,(S,Cn))))

Table 1. contd...

Table 1. contd...

# in Fig. 2	reference	data source	marker(s)	method	missing taxa	tree topology	remarks
	Wang & Lavrov, 2008	mito	14 mt-encoded proteins (2,558 aa)	BA (CAT)	Ct	(O,(B,(Pl,(S1,(Cn1,(Cn2,S2))))))	
	Glenner et al., 2004	mixed	18S, morph	MP	-	(O,(S1,(S2,(Ct,(B,(Pl,Cn))))))	
	Glenner et al., 2004	mixed	18S, morph	BA	-	(O,(S,(Ct,(B,(Pl,Cn))))	
	Nielsen, 2008	mixed	18S, morph	cons (review)	-	(O,(S1,(S2,(S3,(Pl,(Cn,(Ct,B))))))	
	Peterson & Eernisse, 2001	mixed	18S, morph	MP	-	(O,(S1,(S2,(Pl,(Cn,(Ct,B))))))	
	Sidall et al., 1995	mixed	18S, morph	MP	-	(O,((S1,(S2,Ct)),(B,(Cn1, (Cn2,Pl))))	Placzoa within Cnidaria
	Zrzavy et al., 1998	mixed	18S, morph	MP	-	(O,(S1,(S2,(Pl,(Cn,(Ct,B))))))	
	Bridge et al., 1995	mixed	18S, morph, mitochondrial structure	Min	B	(S,(Ct,(Pl,Cn)))	rooted on sponges
	Peterson & Eernisse, 2001	mixed	18S, morph, mitochondrial structure	MP	-	(O,(S1,(S2,(Ct,(S3,(Pl, (Cn,B))))))	
	Carr et al., 2008	mixed	tubA, hsp90, 18S, 28S	BA	B	(O,((S1,Ct),(Cn,(Pl,S2))))	
	Schierwater et al., 2009a	mixed	WGS, ESTs, mt, cDNA	BA	-	(O,(B,(Pl,(S,(Cn1,(Ct,Cn2))))))	
E	Schierwater et al., 2009b	mixed	WGS, ESTs, mt, cDNA,. morph, mol. morph (17,664 characters from 51 partitions)	ML, MP, BA	-	(O,(B,(Pl,(S,(Ct,Cn))))	
E	Schierwater et al., 2009c	mixed	WGS, ESTs, mt, cDNA, morph, mol. morph. (17,664 characters from 51 partitions)	ML, MP	-	(O,(B,(Pl,(S,(Ct,Cn))))	

genome encodes a typical set of mitochondrial genes (*nad1-6, nad4L, cox1-3, cytb, atp6, rnS and rnL*) —although the *atp8* gene is missing—and a full complement of tRNAs but lacks ribosomal protein genes. It contains eight additional open reading frames and several group I and group II introns which are untypical for animal mt DNA. A group I intron is found in *nad5* and a group II intron in *rnLb*. The latter gene shows an untypical split into three exons. Recent re-annotation results confirm the unique organization of the *cox1* gene in three parts including eight exons and seven introns encoded on both strands. Despite of the complex arrangement of the gene fragments the *cox1* gene does not seem to be a pseudogene based on EST data (Burger et al. 2009). The percentage of intergenic regions of the whole mt genome was previously reported as ~50% but with special regard to the additionally found introns a percentage of ~22% seems to be more accurate (Burger et al. 2009). Initial phylogenetic analysis approaches of concatenated mitochondrial proteins including 14 taxa led to a basal placement of *Trichoplax adhaerens* within a monophyletic diploblasts. Bilateria and diploblasts thereby show a sister clade relationship (Dellaporta et al. 2006). Nevertheless, the availability of only one placozoan representative underscored the necessity of collecting mitochondrial genome data from additional placozoan haplotypes to gain further insight into the evolution of mitochondrial genomes at the base of the Metazoan tree of life.

Consequently, the mitochondrial genomes of Placozoa sp. H3 (clade II), Placozoa sp. H4 (clade V) and Placozoa sp. H8 (clade III) were published half a year later (Signorovitch et al. 2007; please note that this placozoan nomenclature is based on the recommendation by Eitel and Schierwater 2010. Placozoa sp. H3, sp. H4 and sp. H8 are stored in Genbank under different names). With a genome size of ~32 kb (H8), ~36 kb (H3) and ~37 kb (H4) they are smaller than the *Trichoplax adhaerens* mitochondrial genome but still larger than all other known mt genomes from metazoan representatives. Genome structure comparison revealed some differences to *Trichoplax adhaerens*, for example in the number of open reading frames and introns. Gene arrangement also differs remarkably between the haplotypes which allowed the authors to arrange Placozoa sp. H4 and Placozoa sp. H8 together in group A and *Trichoplax adhaerens* and Placozoa sp. H3 in group B based on structural similarities. A more detailed comparison of the mt genomes of *Trichoplax adhaerens* and Placozoa sp. H3 (both from group B) reveals a ~20kb genome fragment translocation including *nad1* and the tRNA encoding valine. Additionally the previously described *nad5* intron of *Trichoplax adhaerens* is absent in Placozoa sp. H3. The genome organization differences between the members of the mitochondrial groups A and B are even more significant. For example, an inversion of a ~22kb mt genome fragment, a reorganization of the *cox1* gene in only seven exons, a loss of one *cox1* intron and a gain of an additional *rnLb* intron are found in the mt genome of Placozoa sp. H8 in comparison to the mt genome of *Trichoplax adhaerens*. A reorganization of the *cox1* exon/intron structure can also be observed in the mt genome of Placozoa sp. H4 and interestingly this placozoan representative is the only known haplotype lacking any *rnLb* introns. The phylogenetic analyses including concatenated mitochondrial protein data of the four different placozoan haplotypes and fourteen additional taxa show the same tree topology and position of the Placozoa within the basal Metazoa as in the previously described phylogenetic analysis by Dellaporta et al. in 2006 (see above). With special regard to the relationships within the Placozoa the phylogenetic analyses support the arrangement of the four haplotypes in two groups A and B in agreement with the results based on mitochondrial genome structure similarities. Nevertheless, the number of taxa included in this analysis was limited and so the conclusions with regard to the relationships between the basal metazoan phyla should be viewed as preliminary, although recent concatenated analyses including morphological, mitochondrial and nuclear data resulted in an identical tree topology (Schierwater et al. 2009c).

The increasing number of available mitochondrial genome data from basal metazoan lineages (especially from Demospongiae and Anthozoa) resulted in multiple approaches to clarify the arrangement of the Placozoa, Porifera, Cnidaria and Bilateria at the base of the metazoan tree of life. These studies using only concatenated mitochondrial protein data suggest widely varying phylogenetic scenarios, depending on the choice of taxa included, the definition of outgroups and especially the evolutionary model used in the analysis. Whereas some studies place Placozoa basal within diploblasts with diploblasts as a sister group to Bilateria (e.g. Erpenbeck et al. 2007, Wang and Lavrov 2007), other studies yield highly contradictory results regarding the position of Placozoa (e.g. Haen et al. 2007). A remarkable number of studies even completely failed to clarify the relations between Placozoa, Porifera, Cnidaria and Bilateria (e.g. Burger et al. 2009, Kayal and Lavrov 2008, Wang and Lavrov 2008). In most of these cases the accelerated evolution rate of especially bilaterian mitochondrial genomes results in long-branch attraction (LBA) artifacts within the trees (Felsenstein 1978, Hendy and Penny 1989), and even the use of the CAT model (Lartillot and Philippe 2004) which is optimized to reduce LBA artifacts does not completely resolve this problem (Burger et al. 2009). The exclusion of phylogenetically unstable taxa may avoid this problem but it appears questionable if the currently available molecular data and phylogenetic tools are sufficient to illuminate the relationships at this part of the metazoan tree of life in detail.

Nonetheless the analysis of mitochondrial genomes is a useful tool to gain insight into the relationships between different members of the same phylum as shown for the Placozoa. In this case mitochondrial genome analysis revealed the splitting of the investigated placozoan haplotypes into two highly divergent major groups A and B, indicating the existence of different placozoan species and higher order taxonomic units. Future analyses of placozoan haplotypes that are little characterized to date might expand this scenario and result in a detailed definition of an extant placozoan lineage which hitherto has been regarded as monotypic.

Box 2. *Trichoplax adhaerens*—Regulatory genes of the ANTP superclass in Placozoans

The ANTP superclass of homeodomain-containing transcription factors functions in a wide range of developmental processes and consists of the Hox/ParaHox, extended Hox, and NKL genes. The homeobox gene complement of basal metazoan phyla has been the subject of many studies, and further comparative analysis both within and between phyla promises to provide valuable insights into the evolution of ANTP superclass homeobox genes. Regulatory genes of the ANTP superclass have been noted to play a central role in the anterior-posterior patterning throughout triploblastic animal phylogeny (e.g. Gehring 1998). In cnidarians, they are involved in determination of aboral-oral bauplan structures (Jakob and Schierwater 2007).

Recent preliminary research has focused on analysis of the ANTP-like superclass of homeobox genes in Placozoans (e.g. Jakob et al. 2004, Hadrys et al. 2005, Monteiro et al. 2006, Schierwater et al. 2008). The *Trichoplax* genome encodes representatives of Hox/ParaHox-like, NKL, and extended Hox genes. All Bilateria possess Hox/ParaHox, NKL, and extended Hox genes (Garcia-Fernandez 2005a, 2005b) and many of them serve conserved functions, so this repertoire possibly mirrors the condition of a hypothetical cnidarian-bilaterian ancestor. The evolution of cnidarian and bilaterian ANTP gene repertoires can be deduced by a limited number of cis-duplications of NKL and extended Hox genes and the presence of a single ancestral "ProtoHox" gene, *Trox-2*. In this scenario a Proto-ANTP homeobox gene would have duplicated into the predecessors of Hox/ParaHox, extended Hox and NKL genes. Recent data from cnidarian ANTP gene repertoires and remnants of linkage conform to this scenario of cis-duplications (Chourrout et al. 2006, Kamm et al. 2006, Kamm and Schierwater 2006, Ryan et al. 2006, Ryan et al. 2007). Cnidaria possess almost complete repertoires of NKL and extended Hox gene families, but may have branched before a complete Hox system with true anterior, medial, and posterior determinants had evolved. Recent analyses of a poriferan genome, however, suggest that the origin of the ANTP superclass rather lies in true NKL genes because the poriferan *Amphimedon* lacks any Hox/ParaHox-like or extended Hox genes but possesses several clustered NKL genes (Larroux et al. 2007). Seven homeobox classes comprised of 37 homeobox genes, including 14 from the ANTP superclass, were identified in the genome of *Trichoplax adhaerens* (Schierwater et al. 2008).

The *Trichoplax* genome harbors only one Hox/ParaHox type gene, *Trox-2*, although the possibility could not be completely excluded that *Trichoplax* has lost other Hox/ParaHox genes during evolution. Jakob et al. 2004 suggested that the *Trox-2* gene, the Gsx-ortholog of higher animals, might represent an ancestral ProtoHox gene. *Trox-2* is expressed in a ring around the periphery of *Trichoplax*, in small cells located between the outer margins of the upper and lower epithelial cell layers. Inhibition of *Trox-2* function, either by uptake of morpholino antisense oligonucleotides or by RNA interference, causes complete cessation of growth and binary fission. Supposably, *Trox-2* functions within a so far unrecognized population of possibly multipotential peripheral stem cells that contribute to differentiated cells at the epithelial boundary of *Trichoplax* (Jakob et al. 2004).

Placozoans harbor two ANTP superclass homeobox genes of the extended Hox family, *Not* and *Mnx*. Whole-mount *in situ* hybridization revealed expression of *Mnx* around the periphery of growing *Trichoplax*, albeit with a more patchy distribution than *Trox-2* and *Dlx* (NKL class; Monteiro et al. 2006). The *Not* gene has a unique expression profile in *Trichoplax* and it is highly expressed in folds of intact animals and in the wounds of regenerating animals (Martinelli and Spring 2004). Furthermore, *Trichoplax* possesses another eight NKL genes (*Dlx*, 3x *NK2*, *NK5*, *NK6*, *Hex* and *Dbx/Hlx*). *In situ* expression studies in growing placozoans revealed that *Dlx* is expressed at the periphery of the animal. The spatial expression pattern of Placozoan *Dlx* is similar to that of *Trox-2* (Monteiro et al. 2006).

Besides the ANTP superclass, there are nine homeobox genes of the Paired superclass (PRD) present in the Placozoan genome (*Arx, Ebx/Arx-like, Pax3-like, PaxB, Prd/Pax-like, Pitx, Gsc,* and *Otp*). Only for *PaxB*, gene expression data are available (Hadrys et al. 2005). The *Trichoplax* PaxB gene contains a paired domain, an octapeptide, and a full-length homeodomain. Semiquantitative RT-PCR experiments revealed that *TriPaxB* is expressed in adults, i.e. in growing and vegetatively reproducing animals. Whole-mount *in situ* hybridization studies revealed expression in distinct cell patches along a ring region close to the outer edge of the animal body; expression signals found in smaller animals were weaker than those found in larger animals. Tissue sections revealed that *TriPaxB*-expressing cells are not epithelial cells but cells inside the animal. Furthermore, *Trichoplax* harbors two POU superclass genes (*Pou3, Pou4*), five LIM superclass genes (*LIM1, 1/5, 3/4, 2/9, Isl*) and four genes of the TALE superclass (*PBX/PBC, Pknox, Irx, Meis;* Schierwater et al. 2008).

Depending on the phylogenetic placement of Placozoa, two scenarios for ANTP gene evolution come to mind: (1) Porifera branching first would support the hypothesis of an Urmetazoan harboring a set of NKL genes only, which later in eumetazoans (all non-poriferan metazoans) diverged into the different ANTP gene families (c.f. Pollard and Holland 2000, Castro and Holland 2003). (2) Placozoa branching first among extant phyla supports a scenario in which the ancestral Urmetazoan harbored a set of NKL, Hox/ParaHox-like and extended Hox genes (cf. Pollard and Holland 2000, Castro and Holland 2003). In this view Porifera have lost Hox/ParaHox-related and extended Hox genes. This interpretation is congruent with a recent hypothesis based on the alternative interpretation of gene trees (Petersen and Sperling 2007). Both scenarios as well as all recent phylogenetic analyses are compatible with the hypothesis that the extant Placozoan genome most closely mirrors the ancestral eumetazoan genome from which Cnidarian and Bilaterian genomes have been derived (Srivastava et al. 2008).

References

Adoutte, A. and G. Balavoine, N. Lartillot, O. Lespinet, B. Prud'homme, and R. de Rosa. 2000. The new animal phylogeny: reliability and implications. Proc. Natl. Acad. Sci. USA 97(9): 4453–4456.

Aleshin, V.V. and N.S. Vladychenskaya, O.S Kedrova, I.A Milyutina, and N.B Petrov. 1995. Phylogeny of invertebrates deduced from 18S rRNA comparisons. Molecular Biology 29(6): 843–855.

Aleoshin, V.V. and N.B Petrov. 2002. Molecular evidence of regression in evolution of metazoa. Zh. Obshch. Biol. 63(3): 195–208.

Aleshin, V.V. and A.V. Konstantinova, M.A Nikitin, and I.L. Okshtein. 2004. On the genetic uniformity of the genus *Trichoplax* (Placozoa). Russian Journal of Genetics 40(12): 1423–1425.

Behrendt, G. and A. Ruthmann. 1986. "The cytoskeleton of the fiber cells of *Trichoplax adhaerens* (Placozoa)." Zoomorphology 106(2): 123–130.

Bridge, D. 1994. Phylogeny and life cycle evolution in the phylum cnidaria. Yale: Yale University.

Brusca, R. and G. Brusca. 1990. Invertebrates. Sunderland, Massachusetts: Sinauer Associates.

Buchholz, K. and A. Ruthmann. 1995. The mesenchyme-like layer of the fibre cells of *Trichoplax adhaerens*: A. syncytium. Z. Naturforsch [C] 50c(3-4): 282–285.

Burger, G. and L. Forget, Y. Zhu, M.W. Gray, and B.F. Lang. 2003. Unique mitochondrial genome architecture in unicellular relatives of animals. Proc. Natl. Acad. Sci. USA 100(3): 892–897.

Burger, G. and Y. Yan, P. Javadi, and B.F. Lang. 2009. Group I-intron trans-splicing and mRNA editing in the mitochondria of placozoan animals. Trends Genet. 16: 276–277.

Bütschli, O. 1884. Bemerkungen zur Gastraea-Theorie. Morph. Jahrb. 9: 415–427.

Castro, L.F. and P.W. Holland. 2003. Chromosomal mapping of ANTP class homeobox genes in amphioxus: piecing together ancestral genomes. Evol. Dev. 5: 459–65.

Chourrout, D. and F. Delsuc, P. Chourrout, R.B. Edvardsen, F. Rentzsch, E. Renfer, M.F. Jensen, B. Zhu, P. de Jong, R.E. Steele, and U. Technau. 2006. Minimal ProtoHox cluster inferred from bilaterian and cnidarian Hox complements. Nature 442(7103): 684–687.

Clark, H. 1868. On the Spongiae ciliatae as Infusoria flagellata, or observations on the structure, animality and relationship of Leucosolenia botryoides. Annals and Magazine of Natural History 4, 133–142, 188–215, 250–264.

A.G. Collins. 1998. Evaluating multiple alternative hypotheses for the origin of Bilateria: an analysis of 18S rRNA molecular evidence. Proc. Natl. Acad. Sci. USA 95(26): 15458–15463.

Delage, Y. and E. Herouard. 1899. Traité de Zoologie Concrète II: Classe Mesenchymiens-Mesenchymia. Mason, Paris (1): 9–12.

Dellaporta, S.L. and A. Xu, S. Sagasser, W. Jakob, M.A. Moreno, L.W. Buss, B. Schierwater. 2006. Mitochondrial genome of Trichoplax adhaerens supports Placozoa as the basal lower metazoan phylum. Proc. Natl. Acad. Sci. USA 103(23): 8751–8756.

DeSalle, R. and B. Schierwater. 2008. An even „newer" animal phylogeny. Bio. Essays 30: 1043–1047.

Dunn, C.W. and A. Hejnol, D.Q. Matus, K. Pang, W.E. Browne, S.A. Smith, E. Seaver, G.W. Rouse, M. Obst, G.D. Edgecombe, M.V. Sørensen, S.H. Haddock, A. Schmidt-Rhaesa, A. Okusu, R.M. Kristensen, W.C. Wheeler, M.Q. Martindale, and G. Giribet. 2008. Broad phylogenomic sampling improves resolution of the animal tree of life. Nature 452(7188): 745–U745.

Ehlers, E. 1887. Zur. Auffassung des Polyparium ambulans (Korotneff). Zeitschrift wiss Zool. (45): 491–498.

Eitel, M. and B. Schierwater. 2010. The phylogeography of the Placozoa suggests a taxon. rich. phylum in tropical and subtropical waters. Mol. Ecol. 19: 2315–2327.

Ender, A. and B. Schierwater. 2003. Placozoa are not derived cnidarians: evidence from molecular morphology. Mol. Biol. Evol. 20(1): 130–134.

Erpenbeck, D. and S. Duran, K. Rützler, V. Paul, J.N.A. Hooper, and G. Wörheide. 2007. Towards a DNA taxonomy of Caribbean demosponges: a gene tree reconstructed from partial mitochondrial CO1 gene sequences supports previous rDNA phylogenies and provides a new perspective on the systematics of Demospongiae. J. Mar. Bio. Soc. 87: 1563–1570.

Felsenstein, J. 1978. Number of Evolutionary Trees. Systematic Zoology 27(1): 27–33.

Garcia-Fernandez, J. 2005a. Hox, ParaHox, ProtoHox: facts and guesses. Heredity 94(2): 145–152.

Garcia-Fernandez, J. 2005b. The genesis and evolution of homeobox gene clusters. Nature Reviews Genetics 6(12): 881–892.

Gehring, W. J. 1998. Master Control Genes in Development and Evolution: The Homeobox Story. Yale University Press New Haven, USA pp. 236.

Gonobobleva, E. and M. Maldonado. 2009. Choanocyte ultrastructure in Halisarca dujardini (Demospongiae, Halisarcida). Journal of Morphology 270, 615–627.

Graff, Lv. 1891. Die Organisation der Turbellaria acoela. Leipzig: W. Engelmann.

Grasse, P.P. 1961. Traite de Zoologie, Anatomie Systematique, Biologie. Paris.

Grell, K.G. and G. Benwitz. 1971. Die Ultrastruktur von *Trichoplax adhaerens* F.E. Schulze. Cytobiologie 4: 216–240.

Grell, K.G. and. 1971a. *Trichoplax adhaerens* F.E. Schulze und die Entstehung der Metazoan. Naturw Rdsch. 24(4): 160–161.

Grell, K.G. 1971b. Embryonalentwicklung bei *Trichoplax adhaerens* F.E. Schulze. Naturwiss 58: 570.

Grell, K.G. 1971c. Über den Ursprung der Metazoan. Mikrokosmos 60, 97–102.

Grell, K.G. 1972. Eibildung und Furchung von *Trichoplax adhaerens* F.E. Schulze (Placozoa). Z. Morph. Tiere. 73: 297–314.

Grell, K.G. 1974. Vom Einzeller zum Vielzeller. Hundert Jahre Gastraea-Theorie. Biologie in unserer Zeit. 4(3): 65–71.

Grell, K.G. and G. Benwitz. 1974. Elektronenmikroskopische Beobachtungen über das Wachstum der Eizelle und die Bildung der "Befruchtungsmembran" von *Trichoplax adhaerens* F.E. Schulze (Placozoa). Z. Morph. Tiere. 79: 295–310.

Grell, K.G. and G. Benwitz. 1981. Ergänzende Untersuchungen zur. Ultrastruktur von *Trichoplax adhaerens* F.E. Schulze (Placozoa). Zoomorphology 98(1): 47–67.

Grell, K.G. 1984. Reproduction of Placozoa. Advances in Invertebrate Reproduction. W. Engels, Elsevier 3: 541–546.

Grell, K.G. and A. Ruthmann. 1991. Placozoa. Microscopic Anatomy of Invertebrates, Placozoa, Porifera, Cnidaria, and Ctenophora. F. W. Harrison, Westfall, J.A. New York, Wiley-Liss. Vol. 2: 13–28.

Gruner, H. 1993. Einführung, Protozoa, Placozoa, Porifera. A.K. Fischer, editor. Jena.

Hadrys, H. and B. Schierwater, W. Mrowka. 1990. The feeding behaviour of a semi-sessile hydromedusa and how it is affected by the mode of reproduction. Anim. Behav. 40: 935–944.

Hadrys, T. and R. DeSalle, S. Sagasser, N. Fischer, and B. Schierwater. 2005. The *Trichoplax* PaxB gene: a putative Proto-PaxA/B/C gene predating the origin of nerve and sensory cells. Mol. Biol. Evol. 22(7): 1569–1578.

Haeckel, E. 1874. Die Gastrea-Theorie, die phylogenetische Klassifikation des Tierreichs, und die Homologie der Keimblätter. Jen. Z. Naturwiss (8): 1–55.

Haen, K.M. and B.F. Lang, S.A. Pomponi, and D.V. Lavrov. 2007. Glass sponges and bilaterian animals share derived mitochondrial genomic features: a common ancestry or parallel evolution? Mol. Biol. Evol .24(7): 1518–1527.

Halanych, K.M. 2004. The new view of animal phylogeny. Annual Review of Ecology Evolution and Systematics 35: 229–256.

Hauenschild, C. 1956. Experimentelle Unersuchungen über die Enstehung asexueller Klone bei der Hydromeduse Eleutheria dichotoma. Z. Naturforsch(11b): 394–402.

Hejnol, A. and M. Obst, A. Stamatakis, M. Ott, G.W. Rouse, G.D. Edgecombe, P. Martinez, J. Baguñà, X. Bailly, U. Jondelius, M. Wiens, W.E. Müller, E. Seaver, W.C. Wheeler, M.Q. Martindale, G. Giribet, and C.W. Dunn. 2009. Assessing the root of bilaterian animals with scalable phylogenomic methods. Proceedings of the Royal Society B Biological Sciences 276: 4261–4270.

Hellberg, M.E. 2006. No variation and low synonymous substitution rates in coral mtDNA despite high nuclear variation. Bmc. Evolutionary Biology 6: 24.

Hendy, M.D. and D. Penny. 1989. A Framework for the Quantitative Study of Evolutionary Trees. Systematic Zoology 38(4): 297–309.

Hertel, J. and D. de Jong, M. Marz, D. Rose, H. Tafer, A. Tanzer, B. Schierwater, and P.F. Stadler. 2009. Non-coding RNA annotation of the genome of *Trichoplax adhaerens*. Nucl. Acids. Res. 37: 1602–1615.

Holl, P.W.H. and J. Garcia-Fernández. Hox genes and chordate evolution. Dev. Biol. 1996; 173: 382–395.

Hughes C.L. and T.C. Kaufman. 2002. Review: Hox genes and the evolution of the arthropod body plan. Evolution and Development 4 (6): 459–499.

Hyman, L. 1940. The invertebrates. Protozoa through Ctenophora. In: M.G. Hill, editor. New York.

Ivanov, A. 1988. On the early evolution of the Bilateria. Fortschr Zoologie (36): 349–352.

Ivanov, A.V. 1968. Die Entstehung der vielzelligen Tiere. Phylogenetische Betrachtungen. Leningrad: Nauka.

Ivanov, A.V. 1973. *Trichoplax adhaerens*, a phagocytella-like animal. Zoologiceskij Zurnal 52: 1117–1131.

Jakob, W. and S. Sagasser, S. Dellaporta, P. Holland, K. Kuhn, and B. Schierwater. 2004. The Trox-2 Hox/ParaHox gene of Trichoplax (Placozoa) marks an epithelial boundary. Dev. Genes Evol. 214(4): 170–175.

Jakob, W. and B. Schierwater. 2007. Changing hydrozoan bauplans by silencing Hox-like genes. Plos One 8: e694.

Kamm, K. and B. Schierwater, W. Jakob, and D. Miller. 2006. Axial patterning and diversification in the Cnidaria predate the Hox system. Curr. Biol. 16: 920–926.

Kamm, K. and B. Schierwater. 2006. Complexity of the Non-Hox ANTP Gene Complement in the Anthozoan *Nematostella vectensis*. Implications for the Evolution of the ANTP Superclass. J. Exp. Zool. Mol. Dev. Evol. 306B: 589– 596.

Katayama, T. and H. Wada, H. Furuya, N. Satoh, and M. Yamamoto. 1995. Phylogenetic position of the dicymid Mesozoa inferred from 18S rDNA sequences. Biological Bulletin 189, 81–90.

Kayal, E. and D.V. Lavrov. 2008. The mitochondrial genome of Hydra oligactis (Cnidaria, Hydrozoa) sheds new light on animal mtDNA evolution and cnidarian phylogeny. Gene. 410: 177–86.

Kent, S. 1878. Notes on the embryology of sponges. Annals and Magazine of Natural History 5, 139–156.

Kerscher, S. and L. Grgic, A. Garofano, and U. Brandt. 2004. Application of the yeast Yarrowia lipolytica as a model to analyse human pathogenic mutations in mitochondrial complex I (NADH:ubiquinone oxidoreductase). Bio. chim. Biophys. Acta. 1659, 197–205.

Kessel, M. and F. Schulze, M. Fibi, and P. Gruss. 1987. Primary structure and nuclear localization of a murine homeodomain protein. Proc. natn. Acad. Sci. USA 84: 5306–5310.

Kiefer, J.C. 2006. Emerging developmental model systems. Developmental Dynamics 235, 2895–2899.

King, N. 2004. The unicellular ancestry of animal development. Developmental Cell 7, 313–325.

Krumbach, T. 1907. *Trichoplax*, die umgewandelte Planula einer Hydramedusae. Zool. Anz. 31: 450–454.

Kuhl, W. and G. Kuhl. 1963. Bewegungsphysiologische Untersuchungen an *Trichoplax adhaerens* F.E. Schulze. Zool. Anz. Suppl. 26: 460–469.

Kuhl, W. and G. Kuhl. 1966. Untersuchungen über das Bewegungsverhalten von *Trichoplax adhaerens* F.E. Schulze. Zeitschr Ökolog u Morph. Tiere. 56: 417–435.

Kuhn, K. and B. Streit, B. Schierwater. 1999. Isolation of Hox genes from the scyphozoan Cassiopeia xamachana: implications for the early evolution of Hox genes. J. Exp. Zool. 285(1): 63–75.

Lang, B.F. and C. O'Kelly, T. Nerad, M.W. Gray, and G. Burger. 2002. The closest unicellular relatives of animals. Curr. Biol. 12(20): 1773–1778.

Lankester, E. 1901. Platyhelmia, Mesozoa and Nemertini. A treatise on Zoology London, A & C Black. IV.

Larroux, C. and B. Fahey, S.M. Degnan, M. Adamski, D.S. Rokhsar, and B.M. Degnan. 2007. The NK homeobox gene cluster predates the origin of Hox genes. Curr. Biol. 17(8): 706–710.

Lartillot, N. and H. Philippe. 2004. A Bayesian mixture model for across-site heterogeneities in the amino-acid replacement process. Molecular Biology and Evolution 21(6): 1095–1109.

Lavrov, D.V. and L. Forget, M. Kelly, and B.F. Lang. 2005. Mitochondrial genomes of two demosponges provide insights into an early stage of animal evolution. Molecular Biology and Evolution 22(5): 1231–1239.

Lavrov, D.V. 2007. Key transitions in animal evolution: a mitochondrial DNA perspective. Integrative and Comparative Biology 47(5): 734–743.

Lukic-Bilela, L. and D. Brandt, N. Pojskić, M. Wiens, V. Gamulin, W.E. Müller. 2008. Mitochondrial genome of Suberites domuncula: Palindromes and inverted repeats are abundant in non-coding regions. Gene 412(1-2): 1–11.

Lynch, M. and B. Koskella, S. Schaack. 2006. Mutation pressure and the evolution of organelle genomic architecture. Science 311(5768): 1727–1730.

Maldonado, M. 2004. Choanoflagellates, choanocytes, and animal multicellularity. Invertebrate Biology 123, 1–22.

Mallatt, J. and C.W. Craig, and M.J. Yoder. 2010. Nearly complete rRNA genes assembled from across the metazoan animals: Effects of more taxa, a structure-based alignment, and paired-sites evolutionary models on phylogeny reconstruction. Molecular Phylogenetics and Evolution 55, 1–17.

Martinelli, C. and J. Spring. 2003. Distinct expression patterns of the two T-box homologues Brachyury and Tbx2/3 in the placozoan *Trichoplax adhaerens*. Dev. Genes. Evol. 213(10): 492–499.

Martinelli, C. and J. Spring. 2004. Expression pattern of the homeobox gene Not in the basal metazoan *Trichoplax adhaerens*. Gene. Expr. Patterns 4(4): 443–447.

Maruyama, Y.K. 2004. Occurrence in the field of a long-term, year-round, stable population of placozoans. Biol. Bull. 206(1): 55–60.

Mikhailov, K.V. and A.V. Konstantinova, M.A. Nikitin, P.V. Troshin, L.Y. Rusin, V.A. Lyubetsky, Y.V. Panchin, A.P. Mylnikov, L.L. Moroz, S. Kumar, V.V. Aleoshin. 2009. The origin of Metazoa: a transition from temporal to spatial cell differentiation. Bioessays 31(7): 758–768.

Miller, D.J. and E.E. Ball. 2005. Animal evolution: the enigmatic phylum placozoa revisited. Curr. Biol. 15, R26–8.

Miller, D.J. and E.E. Ball. 2008. Animal Evolution: *Trichoplax*, Trees, and Taxonomic Turmoil. Current Biology 18(21): R1003–R1005.

Monteiro, A.S. and B. Schierwater, S. Dellaporta, and P.W.H. Holl. 2006. A low diversity of ANTP class homeobox genes in Placozoa. Evol. Dev. 8:174–182.

Monticelli, F.S. 1893. *Treptoplax reptans* n.g., n.sp. Atti dell´ Academia dei Lincei, Rendiconti (5)II: 39–40.

Monticelli, F.S. 1896. Adelotacta zoologica. 2. Treptoplax reptans Montic. Mitt. Zool. Stat. Neapel 12: 444–462.

Neresheimer, E. 1912. Mesozoen. Handwörterbuch der Naturwissenschaften. E. Korschelt. Jena, Gustav Fischer. VI: 817–829.

Nielsen, C. 2001. Animal Evolution: Interrelationships of the Living Phyla. 2nd Ed. Oxford University Press, Oxford.

Nielsen, C. 2008. Six major steps in animal evolution: are we derived sponge larvae? Evolution & Development 10, 241–257.

Noll, F. 1890. Über das Leben niederer Seetiere. 85–87 p.

Odorico, D.M. and D.J. Miller. 1997. Internal and external relationships of the Cnidaria: implications of primary and predicted secondary structure of the 5´-end of the 23S-like rDNA. Proc. R. Soc. Lond. B. Biol. Sci. 264(1378): 77–82.

Pearse, V. 1989a. Growth and behaviour of *Trichoplax adhaerens*: first record of the phylum placozoa in Hawaii. Pacific Science 43(2): 117–121.

Pearse, V. 1989b. Stalking the wild placozoan: biogeography and ecology of *Trichoplax* in the Pacific. American Zoologist Abstracts of 1989 Centennial Meeting of the American Society of Zoologists: 772.

Peterson, K.J. and E.A. Sperling. 2007. Poriferan ANTP genes: primitively simple or secondarily reduced? Evolution & Development 9(5): 405–408.

Philip, G.K. and C.J. Creevey, and J.O. McInerney. 2005. The Opisthokonta and the Ecdysozoa may not be clades: Stronger support for the grouping of plant and animal than for animal and fungi and stronger support for the Coelomata than Ecdysozoa. Molecular Biology and Evolution 22(5): 1175–1184.

Philippe, H. and N. Lartillot, and H. Brinkmann. 2005. Multigene analyses of bilaterian animals corroborate the monophyly of Ecdysozoa, Lophotrochozoa, and Protostomia. Molecular Biology and Evolution 22(5): 1246–1253.

Philippe, H. and R. Derelle, P. Lopez, K. Pick, C. Borchiellini, N. Boury-Esnault, J. Vacelet, E. Renard, E. Houliston, E. Quéinnec, C. Da Silva, P. Wincker, H. Le Guyader, S. Leys, D.J. Jackson, F. Schreiber, D. Erpenbeck, B. Morgenstern, G. Wörheide, and M. Manuel. 2009.Phylogenomics Revives Traditional Views on Deep Animal Relationships. Curr. Biol. 19: 706–712.

Pollard, S.L. and P.W. Holland. 2000. Evidence for 14 homeobox gene clusters in human genome ancestry. Curr. Biol. 10(17): 1059–62.

Rieger, R.M. 1976. Monociliated epidermal cells in Gastrotricha: Significance for concepts of early metazoan evolution. Zeitschrift für Zoologische Systematik und Evolutionsforschung 14, 198–226.

Rokas, A. 2008. The origins of multicellularity and the early history of the genetic toolkit for animal development. Annu. Rev. Genet. 42, 235–51.

Ruthman, A. 1996. Placozoa. Spezielle Zoologie Teil 1: Einzeller und Wirbellose Tiere. Stuttgart, Jena, New York, Gustav Fisher Verlag: 121–124.

Ruthmann, A. 2000. Evolution und die Vielfalt des Lebens. Aachen: Shaker Verlag. pp. 466.

Ruthmann, A.G.B. and R. Wahl. 1986. The ventral epithelium of *Trichoplax adhaerens* (Placozoa). Zoomorphology 106(2): 115–122.

Ruthmann, A. and U. Terwelp. 1979. Disaggregation and reaggregation of cells of the primitive metazoan *Trichoplax adhaerens*. Differentiation 13, 185–198.

Ruthmann, A. and K.G. Grell, G. Benwitz. 1981. DNA-content and fragmentation of the egg-nucleus of Trichoplax adhaerens. Z Naturforsch [C] 60: 564–567.

Ryan, J.F. and P.M. Burton, M.E. Mazza, G.K. Kwong, J.C. Mullikin, and J.R. Finnerty. 2006. The cnidarian-bilaterian ancestor possessed at least 56 homeoboxes: evidence from the starlet sea anemone, Nematostella vectensis. Genome Biology 7(7): R64

Ryan, J.F. and A.D. Baxevanis. 2007. Hox, Wnt, and the evolution of the primary body axis: insights from the early-divergent phyla. Biology Direct 2.

Sachwatkin, A.A. 1956. Vergleichende Embryologie der niederen Wirbellosen: Ursprung und Gestaltungswege der individuellen Entwicklung der Vielzeller. Berlin, VEB Deutscher Verlag der Wissenschaften: 401.

Sagasser, S. and B. Schierwater. 2002. Expression analysis of Antennapedia-superclass gene Trox-2 in the placozoon *Trichoplax adhaerens*. Zoology, Abstract Proceedings, DZG-Tagung 2002 in Halle Suppl. V: 14 (95.11).

Schierwater, B. 1989. Allometric changes during growth and reproduction in Eleutheria dichotoma (Hydrozoa, Athecata) and the problem of estimating body size in a microscopic animal. J. Morphol (200): 255–267.

Schierwater, B. and K. Kuhn. 1998. Homology of Hox genes and the zootype concept in early metazoan evolution. Mol. Phylogenet. Evol. 9(3): 375–381.

Schierwater, B. 2005. My favorite animal, *Trichoplax adhaerens*. Bioessays 27, 1294–1302.

Schierwater, B. and K. Kamm, M. Srivastava, D. Rokhsar, R.D. Rosengarten, and S.I.. Dellaporta. 2008. The Early ANTP Gene Repertoire: Insights from the Placozoan Genome. PLoS ONE 3: e2457.

Schierwater, B. and D. de Jong, and R. DeSalle. 2009a. Placozoa and the evolution of Metazoa and intrasomatic cell differentiation. International Journal of Biochemistry & Cell Biology 41: 370–379.

Schierwater, B. and S.O. Kolokotronis, M. Eitel, and R. DeSalle. 2009b. The Diploblast-Bilateria sister hypothesis: parallel evolution of a nervous systems in animals. Communicative and Integrative Biology 2(5): 1–3.

Schierwater, B. and M. Eitel, W. Jakob, H.J. Osigus, H. Hadrys, S.L. Dellaporta, S.O. Kolokotronis, and DeSalle, R. 2009c. Concatenated Analysis Sheds Light on Early Metazoan Evolution and Fuels a Modern "Urmetazoon" Hypothesis. Plos Biology 7: 36–44.

Schubotz, H. 1912. Ist *Trichoplax* die umgewandelte Planula einer Hydromeduse? Zool. Anz. 39: 582–585.

Schulze, F.E. 1883. *Trichoplax adhaerens*, nov. gen. nov. spec. Zool. Anz. 6: 92–97.

Schulze, F.E. 1891. Über *Trichoplax adhaerens*. In: Abhandlungen der Königlichen Preuss. Akademie der Wissenschaften zu Berlin. (ed. Reimer G), pp. 1–23. Verlag der königlichen Akademie der Wissenschaften, Berlin.

Schulze, F.E. 1892. Über *Trichoplax adhaerens*. In Abhandlungen der Königlichen Preuss. Akademie der Wissenschaften zu Berlin. pp. 1–23. Berlin: Verlag der königlichen Akademie der Wissenschaften.

Schulze, F.E. 1914. Einige kritische Bemerkungen zu neuerern Mitteilungen über *Trichoplax*. Zoolog. Anz. 64: 33–35.

Schwartz, V. 1984. The radial polar pattern of differentiation in *Trichoplax adhaerens* F.E. Schulze (Placozoa). Z. Naturforsch. 39c, 818–832.

Scott, M.P. and J.W. Tamkun, G.W. Hartzell. 1989. The structure and function of the homeodomain. Biochim. Biophys. Acta 989: 25–48.

Siddall, M.E. 2009. Unringing a bell: metazoan phylogenomics and the partition bootstrap Cladistics 25: 1–9.

Signorovitch, A.Y. and S.L. Dellaporta, and L.W. Buss. 2005. Molecular signatures for sex in the Placozoa. Proc. Natl. Acad. Sci. USA 102(43): 15518–15522.

Signorovitch, A.Y. and S.L. Dellaporta, and L.W. Buss. 2006. Caribbean placozoan phylogeography. Biol. Bull. 211: 149–56.

Signorovitch, A.Y. and L.W. Buss, S.L. Dellaporta. 2007. Comparative genomics of large mitochondria in placozoans. PLoS Genet 3(1): e13.

Smith, M.G. and M. Snyder. 2006. Yeast as a model for human disease. Curr. Protoc. Hum. Genet. Chapter 15, Unit 15 6.

Srivastava, M. and E. Begovic, J. Chapman, N.H. Putnam, U. Hellsten, T. Kawashima, A. Kuo, T. Mitros, M. Carpenter, A. Signorovich, M. Moreno, K. Kamm, H. Shapiro, I. Grigoriev, L. Buss, B. Schierwater, S. Dellaporta, and D. Rokhsar. 2008. The *Trichoplax* genome and the nature of placozoans. Nature 454(7207): 955–U919.

Stiasny, G. 1903. Einige histologische Details über *Trichoplax adhaerens*. Zeitschr. wiss. Zool. 75: 430–436.

Syed, T. and B. Schierwater. 2002a. *Trichoplax adhaerens*: discovered as a missing link, forgotten as a hydrozoan, re-discovered as a key to metazoan evolution. Vie Milieu 52: 177–187.

Syed, T. and B. Schierwater. 2002b. The Evolution of the Placozoa: A new morphological model. Senckenb Lethaea 82: 315–324.

Terwelp, W. 1978. Reaggregation von Trichoplaxzellen, (ed. Bochum: Ruhr-University Bochum.

Thiemann, M. and A. Ruthmann. 1989. Microfilaments and microtubules in isolated fiber cells of *Trichoplax adhaerens* (Placozoa). Zoomorphology 109(2): 89–96.

Thiemann, M. and A. Ruthmann. 1990. Spherical forms of *Trichoplax adhaerens*. Zoomorphology 110(1): 37–45.

Thorley, J.L. and M. Wilkinson. 1999. Testing the phylogenetic stability of early tetrapods. Journal of Theoretical Biology 200(3): 343–344.

Tomassetti, P. and M. Balsamo, L. Guidi, L. Pierboni, V. Pearse, A.G. Collins, B. Schierwater. 2005a. Discovering again Placozoa in the Mediterranean Sea. National Meeting of the Italian Zoologists, Rome.

Tomassetti, P. and O. Voigt, A.G. Collins, S. Porrello, V.B. Pearse, B. Schierwater. 2005b. Placozoans (Trichoplax adhaerens Schulze 1883) in the Mediterranean sea. Meiofauna Marina 14: 5–7.

Tyson, J.J. and A. Csikasz-Nagy, and B, Novak. 2002. The dynamics of cell cycle regulation. Bioessays 24, 1095–109.

Voigt, O. and A.G. Collins, V.B. Pearse, J.S. Pearse, H. Hadrys, A. Ender, and B. Schierwater. 2004. Placozoa—no longer a phylum of one. Curr Biol 14(22): R944–945.

Wang, X. and D.V. Lavrov. Mitochondrial genome of the Homoscleromorph *Oscarella carmela* (Porifera, Demospongiae) reveals unexpected complexity in the common ancestor of sponges and other animals. Mol. Biol. Evol. 2007. 24: 363–73.

Wang, X. and D.V. Lavrov. 2008. Seventeen new complete mtDNA sequences reveal extensive mitochondrial genome evolution within the Demospongiae. PLoS One 3(7): e2723.

Wenderoth, H. 1986. Transepithelial cytophagy by *Trichoplax adhaerens* F.E. Schulze (Placozoa) feeding on yeast. Z. Naturforsch [C] 41c(3): 343–347.

Willmer, P. 1991. Invertebrate Relationships. Patterns in Animal Evolution. Cambridge: Cambridge University Press. p. 385.

Winnepenninckx, B. and T. Backeljau, L.Y. Mackey, J.M. Brooks, R. De Wachter, S. Kumar, and J.R. Garey. 1995. 18S rRNA data indicate that Aschelminthes are polyphyletic in origin and consist of at least three distinct clades. Mol Biol Evol 12, 1132–1137.

Zakhvatkin, A.A. 1949. The comparative embryology of the low invertebrates. Sources and method of the origin of metazoan development. Moscow, Soviet Science: 395.

Zakhvatkin, A.A. 2008. Generation continuity and integration. Zh. Obshch. Biol. (69): 243–263.

Chapter 14

A Food's-Eye View of Animal Transitions

Neil W. Blackstone

EPIGRAPH

"It would be a service to more than synthetic biology if we might now be permitted to dismiss the idea that life is a precise scientific concept."

Editorial, *Nature* (2007)

Introduction

While a precise definition of life may remain elusive, the elements of a description of life have been widely agreed on for some time, e.g. information, replication, energy, and so on (Schrödinger 1945, Morowitz 1992, Maynard Smith and Szathmáry 1999, de Duve 2002, Martin and Russell 2003). These common aspects of life have doubtless been honed by selection throughout life's history. Consider, for instance, one of the canonical features of living things—obtaining energy from the environment. No one would question that in the stem lineage leading to the last common ancestor of all life key innovations evolved in the basic cellular components used to convert environmental energy (e.g. food, light) into forms that were more useful to the cell. A number of shared primitive features of all modern cells, e.g. the Krebs cycle and electron transport chains, thus evolved. Perhaps less widely appreciated is that even in crown groups such as animals (see Valentine 2006 for discussion

---

Department of Biological Sciences, Northern Illinois University, DeKalb, Illinois 60115, USA.
E-mail: neilb@niu.edu

of terminology) novel mechanisms related to energy conversion continue to evolve and to profoundly shape diversity. While little in the way of cell-level innovation has taken place in animals (but see Bryant 1991), very significant steps have occurred to exploit the potential for heterotrophy inherent in multicellular Bauplans. A number of food-related adaptations have thus evolved. Some of these are structural, e.g. the animal mouth, gut, and trophic apparatus, while some required developing new ways of sensing and responding to environmental inputs, which are often food related. Many of these adaptations link old structures and pathways to new sources of environmental input and stimulus (Jacobs et al. 2007). Given the primacy of obtaining energy from the environment, strong selection continues to drive this process even in modern animals.

Nevertheless, multicellular animals represent a more complex target for selection than the first living things. Buss (1987: p. viii) provides a compelling rationale for why this is so:

"The history of life is a history of different units of selection. Novel selective scenarios dominate at times of transition between units of selection. Whereas the lower self-replicating unit was previously selected by the external environment alone, following the transition it became selected by traits expressed by the higher unit. Variants expressed in the lower unit influence not only the relative replication rate of the lower unit, but also that of the higher unit. The potential clearly exists for variants to have a synergistic effect (that is, to favor the replication of both the lower and the higher unit), or for conflicts to arise. The organization of any unit will come to reflect those synergisms between selection at the higher and the lower levels which permit the new unit to exploit new environments and those mechanisms which act to limit subsequent conflicts between the two units. This explicitly hierarchical perspective on evolution predicts that the myriad complexities of ontogeny, cell biology, and molecular genetics are ultimately penetrable in the context of an interplay of synergisms and conflicts between different units of selection".

Within the framework conceptualized by Buss, multicellular animals represent at least two levels of selection: the level of the cell and the level of the organism. For the majority of its history, all life was unicellular. The cells that had the most descendents were those cells that found a way to replicate regardless of the factors hindering such replication. As pointed out by Szathmáry and Wolpert (2003), the greatest obstacle to the evolution of multicellularity was thus "…the appropriate down-regulation of cell division…." Every animal must have a number of mechanisms to achieve this down-regulation. Yet the evolutionary tension between cells and organisms remains delicately balanced. As we humans unfortunately know from the effects of environmental carcinogens, new sources of variation and selection can upset this balance, allowing some cells to revert to the selfish replication that characterizes the unicellular state. In this "food's-eye view," it is suggested that early in the history of animals, adaptation to food-related selective pressures repeatedly perturbed the balance between these two levels of selection. New mechanisms consequently evolved to re-establish this balance. The combination of food-related adaptation and levels-of-selection innovation has thus been a potent contributor to several key transitions in animal evolution.

These general points will be illuminated by three related examples. The first—the transition from a colony of cells to a multicellular organism—occurred early in the stem lineage leading to animals. In this transition, pathways of metabolic signaling may have been co-opted into mediating conflicts between selection at the cell level and selection at the level of the organism (Blackstone 2000). The second transition—from multicellular animal to eumetazoan—occurred in the basal metazoans. The mouth and gut, food-related structural innovations, precipitated this transition by allowing large groups of cells to share a very similar metabolic environment. This not only reinforced the levels-of-selection innovations developed in the first transition but also allowed reorganization of metabolic signaling from unicellular to multicellular pathways. The final transition—from basal metazoan to bilaterian—occurred in the stem lineage of the bilaterians. In this stem lineage, the mouth, gut, and a battery of sense organs were linked to an active life style. Such animals were able to swiftly locate and exploit patchy resources. Mechanisms that allowed rapid and precise development may have evolved. These mechanisms once again perturbed the delicate balance of cell-level and organism-level selection in favor of the former. Derived features of bilaterians such as restricted somatic cell potency, a sequestered germ line, determinate growth, and "limited" clonality should be viewed in this context (Blackstone 2007). In each of the three transitions discussed below, it is suggested that food-related adaptations had an impact on the central selective conflict of multicellular organisms—between the cell and the individual. In each case a cascade of selective consequences resulted. Evolutionary responses to these consequences comprise a neglected aspect of these key evolutionary transitions.

Transition 1: From Group of Cells to Multicellular Organism

While attempts to identify the "urmetazoan" are fraught with challenges (DeSalle and Schierwater 2007), sufficient evidence does exist to characterize some aspects of the stem lineage of animals. Putative sister groups of metazoans such as choanoflagellates as well as the most basal animals (sponges and placozoans) consist of balls or disks of cells with cilia on the external surfaces. The earliest populations of the stem lineage of animals would likely also consist of such groups of externally ciliated cells. While this is a fairly obvious conclusion, some food-related implications entail (Blackstone 2000). For instance, metabolic gradients would seem almost inevitable in a ciliated ball or disk that is made up of a small population of cells. Simple geometry would produce at least two classes of cells: those on the surface (hereafter, "exterior cells") and those inside (hereafter, "interior cells"). Geometry would also dictate functional differences so that exterior cells could: (1) obtain food from the environment, (2) mediate gas exchange and excretion of wastes, and (3) move the colony by action of the cilia.

These functional differences could produce interior-to-exterior metabolic gradients in such an urmetazoan. If exterior cells obtained ample supplies of food, while interior cells were somewhat starved, there would be an oxidized-to-reduced gradient. On the other hand, if exterior cells obtained sufficient oxygen while interior cells were somewhat hypoxic, there would be a reduced-to-oxidized gradient. Similarly, if exterior cells were subject to intense metabolic demand, e.g. because of rapid ciliary action, there would also be a reduced-to-oxidized gradient. The direction and intensity of the metabolic gradient could effectively be used to adjust the differentiation of stem cells. A reduced-to-oxidized gradient might signal intense metabolic demand for ciliary action in exterior cells, or insufficient ventilation of interior cells, or both. Differentiating more stem cells into ciliated exterior cells would increase locomotion and ventilation so that the interior-to-exterior metabolic gradient diminished.

This sort of metabolic signaling remains common in modern animals (see Blackstone and Bridge 2005). When viewed in terms of levels-of-selection conflicts, however, there are likely more important, though less obvious, implications of metabolic signaling. Multicellular organisms have devised a number of mechanisms to limit selfish cellular replication (Grosberg and Strathmann 2007). Some of these mechanisms (e.g., a unicellular stage of the life cycle) align the evolutionary interests of the cell and the organism via kin selection. Other mechanisms (e.g., programmed cell death) coerce or punish defecting cells. Some cells inevitably manage to evade all of these mechanisms and reassert their primordial imperative for replication. The ultimate mechanism of organismal control over such unruly cells would be one in which these defecting cells suffer an automatic fitness cost. Metabolic signaling is one such mechanism. If the signals an organism uses to control cellular replication are the same as the signals a cell uses to control its own metabolism, then cells that are able to ignore these signals and carry out selfish replication will suffer from an inefficient energy metabolism. In turn, under most circumstances an inefficient metabolism translates into a fitness cost (Blackstone 2007).

This general case will be illuminated by, but not limited to, a particular example. Consider once again the urmetazoan, a simple colony of cells.

If the colony is subjected to fluctuating hypoxia, the most interior cells might suffer the highest levels of oxygen depletion. Under such conditions a variety of metabolic signals are emitted. For instance, the mitochondrial electron transport chain will become highly reduced and form reactive oxygen species (ROS, here broadly defined as partially reduced forms of diatomic oxygen such as superoxide and hydrogen peroxide). Much of the early research on ROS focused on the toxic effects of high concentrations of these molecules (Gilbert 2000). Nevertheless, it has become increasingly clear that ROS also play an important role in within- and between-cell signaling (Finkel 2001, Filomeni et al. 2005). Indeed, the risks of the toxic effects of hydrogen peroxide on one hand and the benefits of its signaling functions on the other are nicely summed up by the structure and function of peroxiredoxin proteins (Georgiou and Masip 2003). When H_2O_2 is reduced, the catalytic cysteine residue of the peroxiredoxin is oxidized. At low concentrations of peroxide, a disulfide bond will form with the resolving cysteine from the other subunit of the homodimer. Thioredoxin subsequently reduces the disulfide bond. On the other hand, at high concentrations of peroxide such as those typically found in a signaling pulse, there is a high probability that the oxidized cysteine will react with another H_2O_2 molecule before it reacts with the resolving cysteine. The higher oxidation products thus formed cannot be reduced by thioredoxins (Wood et al. 2003, Jönsson et al. 2008). The antioxidant function of the peroxiredoxin is then inactivated. The peroxide that is spared can subsequently function as a messenger in signaling pathways, such as those activated by protein tyrosine kinases. For instance, protein tyrosine phosphatase (PTP) 1B antagonizes the effects of these kinases, but PTP can be inactivated when the catalytic cysteine is oxidized by H_2O_2. Thus the activated pathway can proceed. These pathways, as well as peroxiredoxins and thioredoxins, seem to be evolutionarily ancient (Chan et al. 1994, Blackstone et al. 2004).

Returning to the colony of cells, the metabolic gradient induced by fluctuating hypoxia could plausibly affect cell differentiation via a PTP pathway. Once the phosphatase is inactivated by peroxide, the differentiation pathway proceeds. However a cell with a replacement substitution for cysteine in the PTP active site would not respond to the peroxide signal and instead would remain an undifferentiated stem cell. At the cell level, such a mutation would typically be selected for (Buss 1987, Michod 1999, 2003, Grosberg and Strathmann 2007). This is a well-worn argument, but now with a new twist: the peroxide signal, having inactivated peroxiredoxins, and not finding its usual PTP target may then damage other components of the cell before it is reduced to water. Thus a fitness cost is paid by the mutant cell (Fig. 1).

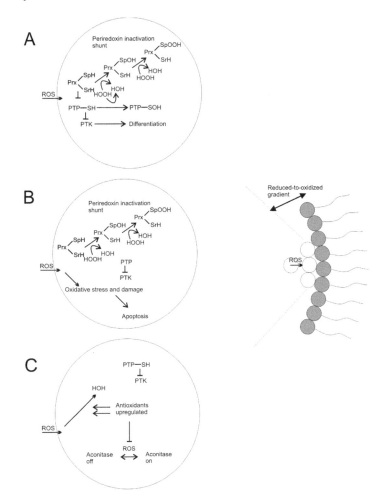

Fig. 1. A few representative examples suggest the sorts of metabolic signaling pathways that may have mediated levels of selection conflicts in an urmetazoan, which is represented as a small colony of cells. A reduced-to-oxidized gradient forms from interior to exterior cells. ROS released by the most interior cells induce different outcomes: **(A)** In the normal cell, the ROS pulse inactivates both peroxiredoxins (Prx) and protein tyrosine phosphatases (PTP). Protein tyrosine kinases (PTK) then activate pathways of differentiation. A flagellated somatic cells forms and the reduced-to-oxidized gradient is diminished. **(B)** In a passively defecting cell, the ROS pulse overwhelms the anti-oxidant function of peroxiredoxins (Prx), but the sensitivity of the PTP pathway to ROS has been lost through mutation. PTPs continue to antagonize PTKs, and terminal differentiation does not proceed. ROS damage other cellular components, ultimately leading to apoptosis. **(C)** In an actively defecting cell, the ROS pulse is entirely converted to water by a battery of upregulated anti-oxidant enzymes. PTPs continue to antagonize PTKs, and terminal differentiation does not proceed. A fitness cost is nevertheless inflicted because these upregulated anti-oxidants also affect the ROS that control the aconitase "rheostat," and inefficient and possibly debilitating metabolism results.

A plausible counter-argument is that the mutant cell need only invest in antioxidants to avoid much of this fitness cost. Upregulation of antioxidant pathways, however, typically depends on just the sort of thiol sensors present in PTP. Thus if the mutant cell deactivates these sensors, it will generally be unable to respond to increased ROS, even to the extent of mobilizing antioxidants. Moreover, cells require metabolic signals in general and ROS in particular to carry out their normal metabolism. For instance, Armstrong et al. (2004) provide a convincing hypothesis that crucial aspects of intermediary metabolism are directly regulated by ROS. Aconitase enzymes catalyse the reversible isomerization of citrate into isocitrate, an early step in the Krebs cycle. Mitochondrial aconitase is inhibited by the ROS produced by the electron transport chain. When aconitase is activated, citrate is channeled into the Krebs cycle, and NADH and $FADH_2$ carry the electrons from the oxidation of reduced carbon to the electron transport chain. If metabolic demand falters, or if less O_2 is available as the terminal electron acceptor, ROS formation by the electron transport chain increases. ROS then inactivates aconitase, and citrate is now channeled into storage as fat. Failure to inactivate aconitase will both waste substrate and increase ROS to more toxic levels. Mitochondria are thus complex organelles that cannot be operated efficiently without metabolic signals. A cell that upregulates antioxidants to avoid metabolic signals linked to differentiation pathways will nevertheless incur general fitness costs associated with inefficient, and possibly debilitating, metabolism (Fig. 1).

An alternative view of the Warburg effect thus emerges. This effect refers to the tendency of mammalian cancer cells to carry out glycolytic metabolism even in an aerobic environment (Garber 2006, Lane 2006). Mutational deactivation of metabolic signaling may require that ALL of the functions of mitochondria be halted. The result is an enormously inefficient metabolism, e.g. roughly an order of magnitude less ATP is produced per glucose molecule. Such cancer cells survive and replicate, however, to the extent that they can monopolize supplies of substrate, e.g., by proliferating the vascular tissue that provides them with these supplies. Indeed, this may explain why all human oncogenes and tumor-suppressor gene pathways have been implicated in angiogenesis, either directly or indirectly (Vogelstein and Kinzler 2004).

In summary, a crucial step in the emergence of the urmetazoan may have been linking metabolic signaling pathways, which date from the earliest life, to pathways mediating levels of selection conflicts between the cell and the individual. While many mechanisms mediate lower level conflicts in favor of the multicellular organism (Grosberg and Strathmann 2007), the most effective of these mechanisms are those that impose an automatic fitness cost on defecting cells. Metabolic signals provide just such mechanisms. Since these signals mimic the ones that the cell uses to control its own metabolism, ignoring these signals inflicts a heavy fitness cost on the cell and its descendents (Blackstone 2007). When confronted with external metabolic signals that direct it to differentiate, the cell that continues to replicate may be unable to manage its own metabolism. The Warburg effect, in which cancer cells shut down their mitochondrial metabolism, may be one manifestation of the connection between metabolic signals and selfish replication.

Transition 2: From Multicellular Animal to Eumetazoan

According to Martin and Russell (2003) "nothing about life is more conserved than compartmentalized redox chemistry." Thus signaling pathways mediated by the oxidation and reduction reactions of metabolism derive from the nature of life itself. Put another way, metabolic signaling appears to be a shared primitive character of all extant cells and organisms. Unicellular and simple multicellular organisms typically rely on metabolic signaling to integrate a variety of environmental inputs and to produce adaptive cell-level responses (Allen 1993). As described above, the stem lineage metazoans, perhaps in some ways comparable to modern sponges and placozoans, likely also relied on cell-level metabolic signaling and may have had few if any advantages when compared to other multicellular eukaryotes. In this context, perhaps the most important features derived in animal evolution were the mouth and gut (Blackstone 2007). For a heterotroph, the capacity to sequester and monopolize large amounts of substrate may have been a decisive advantage. The evolution of the mouth also may have had important implications with regard to metabolic signaling. First, the large amounts of substrate provided by the mouth may have allowed substantial increases in size. While these early animals were still relatively small, they were likely vastly larger than simple multicellular colonies. Second, a large quantity of substrate sequestered in a particular part of such an organism could easily produce large metabolic gradients extending between cells and even tissues. A number of cells, i.e., the cells lining the gut, would experience a similar environment and hence a similar metabolic state. At the same time, since concentrated food resources may be patchy in time and space, sedentary mouth-bearing animals in particular would be selected to respond to the changing availability of food (e.g., in terms of growth and differentiation). The mouth thus not only provided quantities of substrate to a particular subset of cells in the organism, but also required linking organism-level processes to the presence or absence of these quantities of substrate.

Consider the cells lining the gut of an individual of the stem lineage eumetazoan. These various endodermal cells—e.g. vacuole-rich cells, gland cells, ciliated cells—would likely be quiescent in between feedings. Assuming that the cells were still supplied with some substrate and not starving, they would exhibit relatively reduced metabolic states. When a food item was consumed, the digestive machinery would quickly be activated. Ciliated cells would agitate the material in the lumen of the gut, gland cells would secrete enzymes, and partly digested material would be taken up in food vacuoles and further broken down. The metabolic demand generated by these activities would shift the metabolic state of these cells in the direction of oxidation. As with the simple colony of cells described above, the resulting interior-to-exterior metabolic gradient could then be used to guide processes of cellular differentiation, except now the "exterior" cells would be those lining the lumen of the gut. Metabolic signals would continue to mediate levels-of-selection conflicts as described in Fig. 1.

Nevertheless, important differences exist between the first example and the present case. With a small colony of cells, there may be many instances when cells are individually affected by environmental stimuli (e.g. a single cell feeds, while the rest of the colony does not). This is much less likely with cells lining the lumen of the gut. Thus by providing a relatively uniform environment for many cells, the mouth and gut may have synchronized metabolic signaling pathways in many cells. By this view, collections of metabolically active cells that were strongly affected by mouth-related infusions of food may have begun emitting a disproportionate share of the signals that influence organism-level responses (Blackstone 2006a, Doolen et al. 2007). Two consequences are likely. First defecting cells could now be better regulated by these "group-level" signals. Second, organismal development itself could be guided by these large-scale metabolic signals. In the stem lineage eumetazoan, ciliated cells lining the gut may have functioned as "multicellular redox regulators." As pointed out by Jacobs et al. (2007), ciliated cells have many features in common with animal sensory cells in general. In some modern eumetazoans (e.g. corals), ciliated cells still play a primary role in circulating food (Gladfelter, 1983). Such ciliated cells may continue to play a role in metabolic signaling in these organisms (Fig. 2). On the other hand, many modern animals employ muscular contractions to circulate food. If consumption triggers contractions of muscular cells that drive the circulation of the semi-digested, fluidized food, it is then likely that the metabolic state of these cells might serve as an organism-wide signal related to feeding. These groups of cells—a muscular multicellular redox regulator—would be expected to be rich in mitochondria as well as responsive to organismal feeding in their metabolic signals. In particular, if the behavior of cells as a group determines their metabolic state, and thus the metabolic signals emitted (e.g. ROS), then multicellular redox regulation can be recognized (Blackstone 2006a, Doolen et al. 2007).

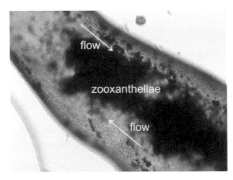

Fig. 2. A bright-field photomicrograph of a stolon of a stoloniferan octocoral growing on a glass cover slip (width of stolon ranges from 380 to 485 μm). Bidirectional gastrovascular flow (arrows) is driven by cilia. Distal flow occurs along the lower wall of the stolon, while proximal flow occurs along the upper wall. The dark mass in the central area of the stolon are numerous zooxanthellae, which have accumulated in an "eddy" between the proximal and distal flow. Because of their role in driving gastrovascular flow, ciliated cells lining the stolon may be the locus of metabolic signaling in these corals.

Studies of metabolic signaling have generally not focused on the possibility of multicellular redox signaling. Rather, with considerable success, these studies have focused on the cell level, elucidating the cellular and intracellular mechanisms of redox signaling and the genomic and proteomic targets of such signaling. The bias toward cell-level processes may reflect the model systems that are typically used in these studies. Clearly, unicellular organisms such as bacteria and yeast will not exhibit multicellular redox regulation (Georgellis et al. 2001). As well, plants do not obtain concentrated substrate as do animals with mouths (Karpinski et al. 1999, Pfannschmidt et al. 1999). Further, as discussed below bilaterians may have been released from many of the constraints of metabolic signaling when they began actively seeking out areas of concentrated substrate. On the other hand, some early-evolving non-bilaterian metazoans (animals with mouths and guts, but not heads) may provide the most promising group of extant animals to examine for evidence of multicellular redox regulation as well as other aspects of environmental signaling (Blackstone and Bridge 2005). Studies of cnidarians from an older literature support such an organism-level view of metabolic signaling (Blackstone 2006b).

Many cnidarians have relatively sedentary stages of the life cycle (e.g. polyps). Environmental heterogeneity requires plasticity in such sedentary organisms (Galloway and Etterson 2007). In such animals, this plasticity may be mediated by metabolic signaling. In symbiotic cnidarians, such signaling may primarily involve animal cells and zooxanthellae (see Fig. 2 and Furla et al. 2005, Perez and Weis 2006). Ciliated cells lining the lumen of the gut may also participate in metabolic signaling as described above. More derived cnidarians such as hydrozoans are more likely to use muscular contractions to circulate fluid in the gastrovascular system. The role of mitochondrion-rich epitheliomuscular cells is intriguing to consider in this context. These cells can be found in contractile regions of colonial hydroids (Fig. 3) and seem to function by pulling open a valve that connects a polyp to the stolon and hence to the rest of the colony (Doolen et al. 2007). The contractile function of these regions seems to be preserved even after polyp ablation. This view of basal metazoans suggests a level of complexity that approaches that of organ-level organization. Such a view is not unprecedented (Jacobs et al. 2007). In any event, cnidarian colonies show considerable plasticity in response to metabolic gradients. For instance, if a small area of polyps of a *Podocoryna carnea* colony is fed brine shrimp while the rest of the colony is starved, the fed area will develop a dense accumulation of polyps and stolons as compared to the rest of the colony. The overall pattern of colony development shows clear differences as compared to controls (Fig. 4). A variety of experiments involving pharmacological perturbations of metabolism suggest that mitochondrial redox state and ROS induce these changes in patterning (Blackstone 2001, 2003).

Transition 3: Basal Metazoan to Bilaterian

In competition for resources, animals with a mouth likely had a significant advantage compared to sponges, placozoans, and other multicellular protists. One way to employ this advantage is to use specialized cells (e.g. cnidocytes, colloblasts) to entangle and subdue prey that is then engulfed

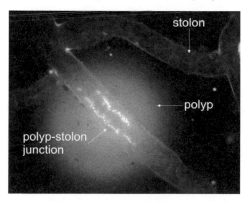

Fig. 3. A fluorescent micrograph of a colony of *Podocoryna carnea* growing on a glass cover slip (width of a stolon is 30–50 µm). Unlike corals, these hydroid colonies are entirely heterotrophic. Polyps feed and gastrovascular fluid is pumped throughout the stolons. Mitochondrion-rich epitheliomuscular cells are found at the polyp-stolon junction. Here these are visualized by H_2DCFDA, which typically fluoresces in the presence of intracellular peroxide. Each brightly fluorescent cell is ≈5 µm in diameter.

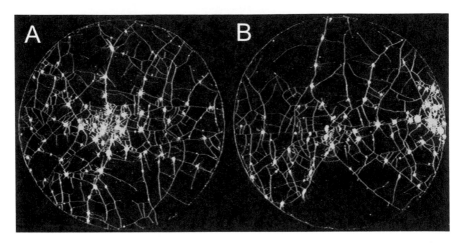

Fig. 4. *Podocoryna carnea* colonies growing on 15 mm diameter glass microscope cover slips. Polyps are bright and circular; stolons are darker and web-like. **(A)** A control colony was allowed to feed on brine shrimp in the normal fashion. Note the concentration of polyps and stolons in the center of the colony. **(B)** A differentially fed colony, genetically identical to colony in **(A)**. Only the small region on the extreme right of the colony was fed brine shrimp. This area shows a dense growth of polyps and stolons, while the rest of the colony shows relatively sparse growth (Blackstone 2001).

by the mouth. Early in the history of this mouth-bearing clade, however, another kind of animal evolved. Locating a battery of sensory equipment near the mouth and employing active movement, these first bilaterians were super consumers. Developing rapidly, dispersing widely, colonizing readily, these animals could efficiently locate areas of concentrated resources and rapidly exploit them. Slower developing organisms that lacked the capacity for directed movement were at a competitive disadvantage. Placozoans may have been strong competitors of these first bilaterians, since they too seem to have the capacity to locate and exploit concentrations of resources at least in aquariums (Fig. 5). Placozoans lack features crucial to general success in this competition, in particular a head and mouth, although their small size and simplicity may have allowed them to continue to succeed in some instances. The greater complexity of bilaterians, however, would impose a cost. Rapid development of a complex body plan could only be achieved by precise and unambiguous signaling mechanisms.

In this context, the limitations of metabolic signaling must be recognized. As perhaps is generally found with "generic" mechanisms (Newman 1994), metabolic signaling may be ambiguous under some circumstances. Consider the lack of oxygen and the lack of metabolic demand: both will shift the redox state of electron carriers in the direction of reduction. The metabolic signals will thus be similar even though the

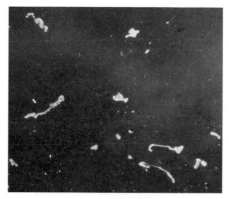

Fig. 5. *Trichoplax* individuals gathering in an area where ephemeral algae is growing in an aquarium. Larger individuals are several millimeters long. While placozoans lack several key features of the stem bilaterians, they may nonetheless exhibit some parallels in life history—rapid development and reproduction in resource-rich areas, followed by dispersal and colonization of other such areas. In the competition between placozoans and stem bilaterians, the head and mouth of the latter were likely decisive advantages.
Color image of this figure appears in the color plate section at the end of the book.

appropriate adaptive responses may be very different (e.g. suppress aerobic and activate anaerobic metabolism versus dormancy and/or sexual recombination). Because very different environmental stimuli may produce similar redox signals, the complexity of a signaling system based on metabolism may be limited (but see Blackstone 2008, Niethammer et al. 2008). Signaling systems involving protein-protein interactions may be considerably more precise. In rapidly developing bilaterians, selection may have favored these more precise signaling mechanisms. Limiting metabolic signaling, however, may have had costs. Because bioenergetic metabolism necessarily links external sources and sinks of electrons, monitoring these sources and sinks provides "honest" signals for controlling cell-level variation. Any mutation-based alteration of metabolism will perturb these signals (see Fig. 1 for hypothetical examples). Cancers are notorious for their altered metabolism (e.g. the Warburg effect—shutting down efficient oxidative phosphorylation and activating wasteful glycolysis—as well as overproduction of ROS—Garber 2006, Lane 2006). Because metabolic signaling is "honest" signaling, it provides effective policing of cancers and other mutation-based events that affect metabolism. Limiting metabolic signaling may thus have required other mechanisms to police cell-level variation. As a consequence, adults of the stem bilaterian may have exhibited diminished somatic cell potency, determinate growth, a sequestered germ line, and limited clonality (Blackstone 2007).

Bilaterian senescence evolved as a by-product of restricted somatic cell potency, itself a mechanism of cell policing required when metabolic signaling is limited. Senescence became a trade-off between the perils of cancer on one hand and the risks of too little self-renewal on the other (Tyner et al. 2002, but see Matheu et al. 2007, Radtke and Clevers 2005, Beausejour and Campisi 2006). Note that these statements are not in contradiction with evolutionary theories of aging (e.g. Rose 1991). Rather, a mechanism for an evolutionary view of bilaterian senescence is provided. By this view, the stem lineage bilaterian (in contrast to basal metazoans) evolved a life history (rapid development, dispersal, colonization, and reproduction) that in turn allowed senescence to evolve. The mechanism that actually led to senescence was restricted somatic cell potency, an innovation to police cell-level variation in the absence of metabolic signaling (Blackstone 2007).

Discussion

Obtaining energy from the environment is one of the canonical features of life. Considerable metabolic evolution occurred before the last common ancestor of modern life, and all cells share this legacy. Perhaps less widely appreciated, however, is that various forms of metabolic evolution have continued throughout the history of life. Indeed, some key transitions in animal evolution may have depended on just such evolution. In particular, in this food's-eye view of animal transitions, the suggestion is made that adaptation to food-related selective pressures repeatedly perturbed the balance between levels of selection. In animals, the principal levels-of-selection conflicts occur between cells and individuals. Numerous mechanisms have evolved to mediate these conflicts (Grosberg and Strathmann 2007). Nevertheless, in several animal transitions food-related adaptation triggered both synergisms and antagonisms between cell and organism-level selection. New mechanisms consequently evolved to re-establish this balance. The combination of food-related adaptation and levels-of-selection innovation has thus been a potent contributor to animal evolution. The interplay between these factors comprises a neglected aspect of the evolutionary history of animals.

In the transition from a colony of cells to a multicellular animal, pathways of metabolic signaling may have been co-opted into mediating conflicts between selection at the cell level and selection at the level of the organism (Blackstone 2000). These levels-of-selection conflicts may be particularly severe in animals in which selfishly replicating cells can easily become systemic (Buss 1987). Metabolic signaling can be a very effective mechanism for mediating these conflicts because defecting cells suffer an automatic fitness cost. If the signals an animal uses to control cellular replication are the same as the signals a cell uses to control its own metabolism, then cells that are able to ignore these signals and carry out selfish replication will suffer from an inefficient energy metabolism.

In turn, under most circumstances an inefficient metabolism translates into a fitness cost. Thus deploying metabolic signaling as a mechanism to keep selfish cells in check may have been a crucial event in animal history, allowing the multicellular organism to emerge as a full-fledged level of selection (Michod 1999, 2003).

In the transition from multicellular animal to eumetazoan, food-related structural innovations—the mouth and gut—arose. These innovations are likely responsible for much of the diversity and abundance of modern animals. For a heterotroph, the capacity to sequester and monopolize large quantities of substrate is an invaluable adaptation. Further, these innovations may have reinforced the utility of metabolic signaling in mediating levels-of-selection conflicts. With the origin of the gut, a number of cells could now share a very similar metabolic environment. Together, these cells could emit metabolic signals and thus better regulate defecting cells with this group-level censure. At the same time, adaptation of growth and development to food-related signals may have been influenced by this emerging group-level signaling. Pathways of metabolic signaling that heretofore functioned at the cell level may have been re-organized into pathways of "multicellular redox regulation". Organismal fitness may have achieved a dual benefit from greater control of defecting cells as well as a more integrated response to environmental food sources.

In the transition from basal metazoan to bilaterian, the mouth, gut, and a battery of sense organs were linked to an active life style. Such animals were able to swiftly locate and exploit patchy resources. Placozoans may have initially been strong competitors of these stem bilaterians, but the head and mouth of the latter likely were decisive advantages. On the other hand, the greater complexity of the bilaterian body plan in turn demanded greater precision than may have been possible with metabolic signaling. Alternative mechanisms that allowed rapid and precise development may have evolved. These mechanisms once again perturbed the delicate balance of cell-level and organism-level selection in favor of the former. Derived features of bilaterians such as restricted somatic cell potency, a sequestered germ line, and determinate growth should be viewed in this context (Blackstone 2007).

In each of the three transitions discussed above, it is suggested that food-related adaptations had an impact on the central selective conflict of multicellular organisms—between the cell and the individual. In each case a cascade of selective consequences resulted. Evolutionary responses to these consequences comprise a neglected aspect of these key evolutionary transitions. The power of a levels-of-selection approach to illuminate the history of animal life (Buss 1987) is further supported. Minimally, at least some small steps are suggested to understanding the "myriad complexities of ontogeny" in animals.

Acknowledgments

Many thanks to Bernd Schierwater and Rob DeSalle for all of their efforts in preparing this volume. The National Science Foundation (IBN-00-90580 and EF-05-31654) provided support for the research.

References

Allen, J.F. 1993. Control of gene expression by redox potential and the requirement for chloroplast and mitochondrial genomes. J. Theor. Biol. 165: 609–631.

Armstrong, J.S. and M. Whiteman, H. Yang, and D.P. Jones. 2004. The redox regulation of intermediary metabolism by a superoxide-aconitase rheostat. BioEssays 26: 894–900.

Beausejour, C.M. and J. Campisi. 2006. Balancing regeneration and cancer. Nature 443: 404–405.

Blackstone, N.W. 2000. Redox control and the evolution of multicellularity. BioEssays 22: 947–953.

Blackstone, N.W. 2001. Redox state, reactive oxygen species, and adaptive growth in colonial hydroids. J. Exp. Biol. 204: 1845–1853.

Blackstone, N.W. 2003. Redox signaling in the growth and development of colonial hydroids. J. Exp. Biol. 206: 651–658.

Blackstone, N.W. and K.S. Cherry, and S.L. Glockling. 2004. Structure and signaling in polyps of a colonial hydroid. Invert. Biol. 123: 43–53.

Blackstone, N.W. and D.M. Bridge. 2005. Model systems for environmental signaling. Integr. Comp. Biol. 45: 605–614.

Blackstone, N.W. 2006a. Multicellular redox regulation: integrating organismal biology and redox chemistry. BioEssays 28: 72–77.

Blackstone, N.W. 2006b. Charles Manning Child (1869–1954): the past, present, and future of metabolic signaling. J. Exp. Zool. (MDE) 306B: 1–7.

Blackstone, N.W. 2007. A food's-eye view of the transition from basal metazoans to bilaterians. Integr. Comp. Biol. 47: 724–733.

Blackstone, N.W. 2008. Metabolic gradients: a new system for old questions. Curr. Biol. 18: R351–R353.

Bryant, C. [ed.] 1991. Metazoan Life Without Oxygen. Chapman and Hall, New York, USA.

Buss, L.W. 1987. The Evolution of Individuality. Princeton University Press, Princeton, USA.

Chan, T.A. and C.A. Chu, K.A. Rauen, M. Kroiher, S.M. Tatarewicz, and R.E. Steele. 1994. Identification of a gene encoding a novel protein-tyrosine kinase containing SH2 domains and ankyrin-like repeats. Oncogene 9: 1253–1259.

de Duve, C. 2002. Life Evolving. Oxford University Press, Oxford, UK.

DeSalle, R. and B. Schierwater. 2007. Key transitions in animal evolution. Integr. Comp. Biol. 47: 667–669.

Doolen, J.F. and G.C. Geddes, and N.W. Blackstone. 2007. Multicellular redox regulation in an early-evolving animal treated with glutathione. Physiol. Biochem. Zool. 80(3): 317–325.

[Editorial] 2007. Meaning of 'life'. Nature 447: 1031–1032.

Filomeni, G. and G. Rotilio, and M.R. Ciriolo. 2005. Disulfide relays and phosphorylation cascades: partners in redox-mediated signaling pathways. Cell Death Differ. 12: 1555–1563.

Finkel, T. 2001. Reactive oxygen species and signal transduction. IUBMB Life 52: 3–6.

Furla, P. and D. Allemand, J.M. Shick, C. Ferrier-Pagès, S. Richier, A. Plantivaux, P.-L. Merle, and S. Tambutté. 2005. The symbiotic anthozoan: a physiological chimera between alga and animal. Integr. Comp. Biol. 45: 595–604.

Galloway, L.F. and J.R. Etterson. 2007. Transgenerational plasticity is adaptive in the wild. Science 318: 1134–1136.

Garber, K. 2006. Energy deregulation: licensing tumors to grow. Science 312: 1158–1159.

Georgellis, D. and O. Kwon, and E.C.C. Lin. 2001. Quinones as the redox signal for the arc two-component system of bacteria. Science 292: 2314–2316.

Georgiou, G. and L. Masip. 2003. An overoxidation journey with a return ticket. Science 300: 592–594.

Gilbert, D.L. Fifty years of radical ideas. pp. 1–14. *In*: C.C. Chiueh. [ed.] 2000. Reactive Oxygen Species: From Radiation to Molecular Biology. New York Academy of Sciences, Annals 899, New York, USA.

Gladfelter, E.H. 1983. Circulation of fluids in the gastrovascular system of the reef coral, *Acropora cervicornis*. Biol. Bull. 165: 619–636.

Grosberg, R.K. and R.R. Strathmann. 2007. The evolution of multicellularity: a minor major transition? Annu. Rev. Ecol. Evol. Syst. 38: 621–654.

Jacobs, D.K. and N. Nakanishi, D. Yuan, A. Camara, S.A. Nichols, and V. Hartenstein. 2007. Evolution of sensory structures in basal Metazoa. Integr. Comp. Biol. 47: 712–723.

Jönsson, T.J. and L.C. Johnson, and W.T. Lowther. 2008. Structure of the sulphiredoxin-peroxiredoxin complex reveals an essential repair embrace. Nature 451: 98–101.

Karpinski, S. and H. Reynold, B. Karpinska, G. Wingsle, G. Creissen, and P. Mullineaux. 1999. Systemic signaling and acclimation in response to excess excitation energy in *Arabidopsis*. Science 284: 654–657.

Lane, N. 2006. Power games. Nature 443: 901–903.

Martin, W. and M.J. Russell. 2003. On the origins of cells: a hypothesis for the evolutionary transitions from abiotic geochemistry to chemoautotrophic prokaryotes, and from prokaryotes to nucleated cells. Phil. Trans. Roy. Soc. Lond. B 358: 59–85.

Matheu, A. and A. Maraver, P. Klatt, I. Flores, I. Garcia-Cao, C. Borras, J.M. Flores, J. Vina, M.A. Blasco, and M. Serrano. 2007. Delayed ageing through damage protection by Arf/p53 pathway. Nature 448: 375–379.

Maynard Smith, J. and E. Szathmáry. 1999. The Orgins of Life. Oxford University Press, Oxford, UK.

Michod, R.E. 1999. Darwinian Dynamics. Princeton University Press, Princeton, USA.

Michod, R.E. 2003. Cooperation and conflict mediation during the origin of multicellularity. pp. 291–307. *In*: P. Hammerstein. [ed.] 2003. Genetic and Cultural Evolution of Cooperation. MIT Press, Cambridge, USA.

Morowitz, H.J. 1992. Beginnings of Cellular Life. Yale University Press, New Haven, USA.

Newman, S. 1994. Generic physical mechanisms of tissue morphogenesis: a common basis for development and evolution. J. Evol. Biol. 7: 467–488.

Niethammer, P. and H.Y. Kueh, and T. Mitchison. 2008. Spatial patterning of metabolism by mitochondria, oxygen, and energy sinks in a model cytoplasm. Curr. Biol. 586–591.

Perez, S. and V. Weis. 2006. Nitric oxide and cnidarian bleaching: an eviction notice mediates breakdown of symbiosis. J. Exp. Biol. 209: 2804–2810.

Pfannschmidt, T. and A. Nilsson, and J.F. Allen. 1999. Photosynthetic control of chloroplast gene expression. Nature 397: 625–628.

Radtke, F. and H. Clevers. 2005. Self-renewal and cancer of the gut: two sides of a coin. Science 307: 1904–1909.

Rose, M.R. 1991. Evolutionary Biology of Aging. Oxford University Press, Oxford, UK.

Schrödinger, E. 1945. What is Life? Cambridge University Press, Cambridge, UK.

Szathmáry, E. and L. Wolpert. The transition from single cells to multicellularity. pp. 271–290. *In*: P. Hammerstein. [ed.] 2003. Genetic and Cultural Evolution of Cooperation. MIT Press, Cambridge, USA.

Tyner, S.D. and S. Venkatachalam, J. Chol , S. Jones, N. Ghebranious, H. Igelmann, X. Lu, G. Soron, B. Cooper, C. Brayton, S.H. Park, T. Thompson, G. Karsenty, A. Bradley, and L.A. Donehower. 2002. p53 mutant mice that display early aging-associated phenotypes. Nature 415: 45–53.

Valentine, J.W. 2006. Ancestors and urbilateria. Evol. Dev. 8: 391–393.

Vogelstein, B. and K.W. Kinzler. 2004. Cancer genes and the pathways they control. Nature Med. 10: 789–799.

Wood, Z.A. and L.B. Poole, and P.A. Karplus. 2003. Peroxiredoxin evolution and the regulation of hydrogen peroxide signaling. Science 300: 650–653.

Chapter 15

Lost in Transition: The Biogeochemical Context of Animal Origins

Eric Gaidos

Introduction

The late Precambrian environment in which animals first emerged differed profoundly from that of the Phanerozoic in one key aspect—the incomplete oxidation of the oceans. Sulfidic deep water would have encroached into the mixed layer and photic zone, introducing both a toxin and reducing oxygen by the action of sulfide-oxidizing bacteria. The back-reaction of sulfide with oxygen in the oceanic mixed layer may have ultimately limited atmospheric oxygen to less than 4% of its present level. Animals would have been restricted to estuarine refugia where input of oxygenated freshwater held sulfide at bay. This biogeography may have profoundly impacted both the evolution and taphonomy of early animal life. Ultimately, sulfide and any attendant organic matter pool was swept from the oceans following a several episodes of intense glaciation and oxygen levels rose to Phanerozoic levels. Although there is no compelling evidence for pre-Ediacaran roots of the Metazoa, it is possible that oxygen levels may have permitted some sort of multicellular life immediately following the Lomagundi positive carbon isotope excursion ca. 2200 million years ago; after oxygen first appeared but before the deep oceans became intensely sulfidic. More work on this unique interval of Earth history is needed.

Department of Geology and Geophysics, University of Hawaii at Manoa, Honolulu, HI 96822 USA.
E-mail: gaidos@hawaii.edu

Life on the Cusp of Oxygen

Stephen Jay Gould regaled our collective imagination when he performed the thought experiment of rewinding the "tape of life" to illustrate the role of contingency in early animal evolution (Gould 1989). The fate of an entire phylum may have hinged on seemingly inconsequential events; entire groups were wiped out, while others survived, because of small changes in the environment or innocuous innovations. This controversial view has been challenged (Conway-Morris 2004), nevertheless the deep evolutionary history of the Metazoa involves a small number of events: obligate multicellularity and tissue-level organization evolved only once in the flagellated protists, although multicellularity evolved multiple times in the eukaryotes (King 2004), and all extant animal life (as far as we know) is represented by a modest number of phyla (36 described) and only five major groups (poriferans, placozoans, cnidarians, ctenophore, and bilaterians). Such statistics should have a sobering effect on our optimism for sleuthing the nature of events which led to the origin of the major animal lineages: If the five divergence events which led to the major animal groups occurred over a period between 580 millions of years before present (Ma)—the canonical but uncertain age of the fossil embryo-bearing Doushantuo phosphorite deposits (Kaufman 2005)—the appearance of bioturbated sediments at ca. 560 Ma (Droser et al. 2002) then the rate is one event per 4 million years. If the major lineages have much deeper roots, as some have suggested based on "molecular clock" calculations (Blair and Hedges 2005), then the interval between these events was 1–2 orders of magnitude longer. Thus we are considering extremely rare events in both geologic time and evolutionary history and it is not obvious that we can hope to understand them in the context of theories developed on the basis of more frequent, mundane events. Instead, they may represent extreme contingencies that occurred under particular conditions that no longer exist.

Proterozoic oceans, the habitat of early animal life, were probably very different from those that existed previously or since: They were present during a transition between the anoxic world of the Archean and the well-oxygenated realm of almost all of the Phanerozoic. Oxygen first appeared in geochemically significant quantities in the Paleoproterozoic around 2400 million years ago (Ma) (Bekker et al. 2004). Although surface waters may have been oxygenated, the chemistry of deep-sea sediments suggests deep ocean anoxia persisted at least through the last major Precambrian (Gaskiers) glaciation 580 Ma (Canfield et al. 2007) (Fig. 1). The level of Proterozoic atmospheric oxygen is poorly constrained but is thought to have been much lower than its present atmospheric level (1 PAL) (Kasting 1987). Sulfidic deep water masses may have developed after the appearance

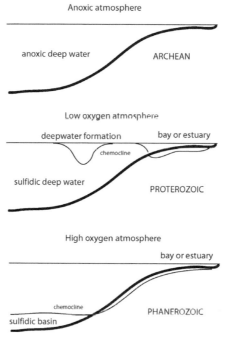

Fig. 1. A simple schematic of redox evolution of the ocean-atmosphere system and the appearance of oxygen oases in estuaries in the Proterozoic.

of atmospheric oxygen as pyritic sulfur weathered out of crustal rocks was oxidized to soluble sulfate, deposited in the oceans, and reduced to sulfide by bacterial anaerobic respiration in the deep ocean. Canfield proposed that sulfide titrated out ferrous soluble iron as insoluble iron sulfides such as pyrite, terminating banded iron formation around 1800 Ma (Canfield 1998). This scenario is supported by iron chemical and sulfur isotopic data indicating low sulfate and widespread anoxia (Shen et al. 2003, Arnold et al. 2004, Kah et al. 2004, Poulton et al. 2004). Once ferrous iron was overwhelmed by the input of sulfide, concentrations of the latter would have increased markedly. Sulfide-free but anoxic water masses may have existed near hydrothermal systems where high iron concentrations scavenged sulfide from the water column (Slack et al. 2007).

Oxygenated surface water and sulfidic deep water would not have been completely isolated. Upwelling and wind- and current-driving turbulent mixing would have introduced sulfide to the surface oceans and the photic zone (Fig. 1). At least one study found biomarkers for green and purple sulfur bacteria that use sulfide as the electron donor in anoxygenic photosynthesis, suggesting that sulfide had infiltrated the photic zone (Brocks et al. 2005). High stratigraphic resolution sulfur isotope data from the Ediacaran Doushantuo Formation in China (at or near wave base) show excursions in the isotopic composition of both pyrite and sulfates (McFadden et al. 2008) that could be interpreted in terms of incursions of sulfidic water.

The oxygen and sulfide contents of shallow water (the mixed layer, often corresponding very approximately to the photic zone) was controlled by diffusion or advection of oxygen from the atmosphere, upwelling of sulfide from deep water, net primary production, and sulfide oxidation. Kump et al. (2005) developed a simple criterion for the "collapse" of the chemocline and the appearance of sulfide in surface waters based on the stochiometry of the oxidation of sulfide:

$$HS^- + 2O_2 \rightarrow SO_4^{2-} + H^+$$

by balancing the mixing of oxygen from the atmosphere into surface waters and upwelling of sulfide from below

$$kK_HP_{O_2} = 2u[H_2S]$$

where k is piston velocity describing mixing of atmospheric oxygen into surface waters, K_H is the Henry's constant for oxygen, P_{O_2} is the partial pressure of atmospheric oxygen, u is an effective upwelling velocity, and [HS⁻] is the total sulfide concentration. Like Kump et al., I ignore production, but this may be significant under a low-oxygen Proterozoic atmosphere. I assume k = 1000 m yr⁻¹, and a maximum P_{O_2} of 5 percent PAL (Kasting 1987). The concentration of sulfide in the Proterozoic ocean is not well constrained, but if the lack of a $\Delta^{34}S$ values > 46 parts per thousand in pre-580 Ma sediments (Hurtgen et al. 2005) reflects suppression of bacterial sulfur disproportionation by high sulfide, then [HS⁻] must have exceeded 1 mM (Thamdrup et al. 1993). With these assumptions, upwelling rates exceeding 5 m yr⁻¹ would have caused sulfide to reach the surface. If atmospheric oxygen were slightly lower, or sulfide concentrations higher, the threshold falls below the modern global average of 4 m yr⁻¹ (Broeker and Peng 1982). In this case there is a near-global collapse of the chemocline and sulfide appears at the surface. This does not necessarily mean that oxygen is absent, as I shall show later.

Hydrogen sulfide at ≥ 100 ppm is intrinsically toxic to most animals due to reaction with the iron in the heme moiety of the mitochondrial cytochrome c oxidase. A few animals, such as those in hydrothermal vent ecosystems, have evolved adaptive detoxification mechanisms and may even produce ATP from sulfide oxidation (Grieshaber and Volkel 1998). One successful strategy has been symbioses with sulfide oxidizing bacteria in the gut or specialized organ and such a symbiotic metabolism has been proposed for some of the Edicaran fauna (Seilacher 1989, Dzik 2003). Clearly this will not work if oxygen, rather than sulfide, is the limiting reactant—not the case at deep sea hot springs on the present Earth. When sulfide is present, colorless sulfide oxidizing bacteria (SOB) can out-compete animals for available dissolved oxygen. Microaerophilic SOB can rapidly oxidize sulfide at oxygen concentrations as low as 1 µM (Zopfi et al. 2001), well below the threshold of animals (Mangum and Van Winkle 1973). Furthermore, this is well below the saturation level (13 µM) expected under an atmosphere with 5 percent PAL oxygen. This means that unproductive, sulfidic surface waters will be a sink for oxygen even under a hypoxic Proterozoic atmosphere. Of course, oxygen must be released somewhere into the atmosphere as a result of positive net primary productivity. Although sulfide and oxygen react spontaneously and sulfide-oxidizing bacteria will rapidly consume these reactants, some of each gas will escape to the atmosphere. Chemical sulfide reaction follows the rate law:

$$\frac{d[H_2S]}{dt} = k[H_2S][O_2]$$

with k ~5 mM^{-1} d^{-1} (Millero 2005). Reaction rates by SOB can exceed this by an order of magnitude or more, e.g. (Zopfi et al. 2001). Suppose that the availability of sulfide always exceeds that of oxygen and the exchange time between the mixed layer of the ocean and the atmosphere is 20 d, then the reaction time is shorter than the mixing time when dissolved oxygen exceed 10 µM, which is the saturation concentration at percent PAL (15°C). If oxygen concentration exceeds this, then chemical oxidation alone will consume this before oxygen it can reach the atmosphere; indeed, the oceans become a net sink for oxygen. If oxygen concentration is lower, then net primary production in the mixed layer is a source of oxygen to the atmosphere and O_2 will increase. Thus it is possible that this threshold (which depends on the activity of sulfide-oxidizing bacteria) actually sets the level of oxygen in the Proterozoic atmosphere. Of course, the mixed layer varied strongly over the global ocean while the atmospheric oxygen concentration did not; thus at any given PO$_2$ different regions of the ocean would be either sinks or sources of oxygen. Terrestrial microbial mats were also a source of oxygen.

Where might animals first thrive in the Proterozoic world, i.e., where would the highest dissolved oxygen concentrations be found? These would be environments with little or no exchange with the deep oceans, with minimal efflux of sulfide from sediments, maximum input of oxygenated surface waters, low temperatures maximizing oxygen solubility, and high photosynthetic activity. These conditions would be found in estuarine settings at high latitudes. In such coastal environments, influx of cold freshwater maximizes input of oxygen and dilutes sulfide fluxing from the sediment and marine sulfate available for the formation of sulfide (Fig. 1). Regions of deep water formation could also conceivably have high oxygen concentrations but these would be transient in nature. This is in contrast to many continental margins, sites of upwelling where high concentrations of sulfide would be introduced to shallower waters. Interestingly, a continental shelf offshore from an estuary (Lindstrom 1995) or a bay (Hagadorn 2002) have been suggested as possible settings for the lower Cambrian Chengjiang lagerstatte.

This scenario admits some interesting speculation. First, zones of high freshwater input may be sites of erosion, rather than deposition. This has obvious implications for the taphonomy of the earliest animal life. The "sudden" appearance of animal life in the fossil record could actually be an artifact of the oxygenation of shallow shelf waters and the dispersal of organisms formerly restricted to freshwater or brackish settings to sites where preservations was more favored. (Although some Ediacaran fauna were preserved in a deepwater setting (Narbonne 2005), these succeed the oxygenation of the ocean at 580 Ma.) Second, cold (near 0°C) water from melting glaciers at the end of major glaciations would have provided transiently enhanced input of oxygen to the oceans, expanding zones accessible to the existing fauna. This could explain evidence for animal herbivory immediately following the Marinoan glaciation 635 Ma (Butterfield 2007) and the appearance of the Ediacaran fauna after the Gaskiers glaciation 580 Ma (Narbonne 2005).

Third, the restriction of early metazoan life to isolated oxygenated oases, surrounded by an anoxic or hypoxic and sulfidic ocean, meant that mechanisms of wide dispersal available to Phanerozoic marine fauna were not possible. This would have lead to extreme genetic isolation and an "island biogeography," promoting divergence and radiation of many lineages, but also leaving many species vulnerable to extinction because they cannot easily disperse to neighboring habitats. This would presumably have had a dramatic effect on the topology of the metazoan tree. Finally, an estuarine setting makes specific predictions about the ecology of that earliest animal life. It must have been adapted to brackish or freshwater, and tolerant of large fluctuations in oxygen concentrations brought about by diurnal cycles of primary production and respiration by phototrophs, variable input of freshwater, and tidal- or storm-driven surges of sulfidic seawater. Early bacteriavorous metazoans could have been supported by microbial mats based on photosynthesis or oxidation of sulfide at the chemocline where oxygenated and sulfidic waters met. Perhaps limited dispersal was achieved by a lifecycle which included a sulfide-resistant cyst/egg phase.

Escape from Euxenia

What released the oceans from the grip of sulfide at the end of the Precambrian, causing the chemocline to creep into the sediment where, with the exception of the Permo-Triassic extinction interval and instances of deep ocean anoxia in the Cretaceous, it has remained since? Appeal to an increased oxygen content of the atmosphere is circular logic; as that increase could be a response to—not a driver of—the carbon oxygen cycles that balance net primary productivity with diagenesis in the sediments, or weathering of reduced C and S from uplifted sedimentary rocks. The standard explanations appeal to an *ad hoc* elevated rates of organic matter burial via animal activity (Logan et al. 1995), or *ad hoc* changes in phosphorus availability or changes in sediment mineralogy (Lenton and Watson 2004, Kennedy et al. 2006). A simpler explanation is that once readily labile sulfide disappeared from the oceans, higher oxygen levels were required to achieve the same level of consumption of the oxygen released by net primary productivity. Alternatively, changes in the relationship between total ecosystem aerobic respiration (which must balance NPP) and oxygen might have been responsible. There is accumulating isotopic evidence that the Proterozoic oceans contained a large pool of labile dissolved and/or suspended organic matter (Logan et al. 1995, Rothman et al. 2003, Jiang et al. 2007, McFadden et al. 2008). Gaidos et al. (2007) proposed that Proterozoic oxygen levels were held low by the respiration of the organic matter pool by aerobic heterotrophic bacteria, which can have K_m values of ~1 µM. Once the pool was removed, oxygen could rise (Fig. 2).

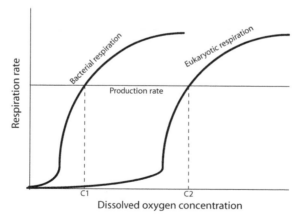

Fig. 2. Schematic of how the continued balance between net primary production and oxygen consumption—either by sulfide oxidation, aerobic bacterial respiration, or eukaryotic respiration, could have shifted the oxygen concentration from C1 to C2 (a higher level) at the end of the Proterozoic.

But what protected that organic matter pool from remineralization by anaerobic respiration, i.e. sulfate reduction? One possibility might be a high (>1 mM) concentration of sulfide, which could inhibit sulfate reduction. This is consistent with the absence of sulfur disproportionation suggested by the sulfur isotopic data. Sulfide itself could also be responsible for holding down oxygen levels between 1800 and 580 Ma, as discussed above. Once the sulfide was somehow depleted, then the organic matter pool was exposed to remineralization by sulfate-reducing bacteria. This could have happened by the Precambrian-Cambrian transition at 544 Ma, the location of a large negative excursion in the carbon isotope record and a brief episode of global anoxia (Schroder and Grotzinger 2007). Combined with subsequent mineralization of the organic matter by sulfate reduction would allow oxygen to rise. Depletion of sulfide is also consistent with the brief re-appearance of banded iron formation in the Neoproterozoic (Klein 2005). But what depleted sulfide? We might appeal to changes in ocean circulation that mix oxygen with sulfide. Another explanation was that weathering of sulfur from continental rocks was attenuated during episodes of cold climate in the late Precambrian (Canfield and Raiswell 1999). Input of sulfur into the ocean would already have been lowered by the lower rate of weathering necessary to compensate falling rates of volcanism and tectonic metamorphism with time, and perhaps changes in the sulfur content of crustal rocks.

How old are the Metazoa?

Persistent reports that the major metazoan lineages have deep Proterozoic roots (Runnegar 1982, Wray et al. 1996, Blair and Hedges 2005) have several possible explanations. One of the most likely is that the assumption of a molecular "clock" is grossly incorrect. There are many weak points in this house of cards, but now we also know of a serious crack in the foundation of the molecular clock hypothesis, i.e., the postulate that most changes are "nearly neutral" and thus are not subject to population size effects (Kimura 1968) . This assumption has found to be incorrect in an increasing number of taxa. One of the most recent and spectacular example has been illuminated by an analysis of 12 genomes of *Drosophila* species. This showed a correlation between recombination rates and substitution rates (Begun et al. 2007) that can most easily be explained by selection—probably through linkage ("hitchhiking") to a small fraction of loci on which selection is actually taking place (Andolfatto 2007). Some of its earliest proponents have declared the nearly-neutral theory "dead" (Ohta 2005). One collateral casualty of the demise of the nearly-neutral theory might be the molecular clock. Empirically, it might be claimed that a linear correlation between elapsed time and divergence exists amongst some taxa, but there is now no viable theory to expect this. Instead, we might expect clock rates to strongly co-vary with adaptive sweeps and speciation events.

Rates also might change with time in a global way. Stanley (2007) argued that background rates of origination and extinction of species within marine taxa are correlated because heritable traits influence these two processes in similar ways. He also found that taxa with high background rates suffered disproportionate losses during the major Phanerozoic extinction events; thus there has been an episodic weeding out of taxa with high rates of origination and extinction through the history of animal life. There is also evidence that speciation, i.e., number of branches along a lineage, is correlated with total divergence along that lineage (Webster et al. 2003, Pagel et al. 2006). If this is correct, then the preferential survival of taxa that experience less speciation and divergence along their branches means that the molecular "clock" should slow down after each major biotic crisis. That means that the use of rates of divergence calculated from later, internal nodes will overestimate the amount of actual time elapsed since earlier nodes (Fig. 3). This, among many other systematic effects, could explain the deep Precambrian divergence times produced by some analyses.

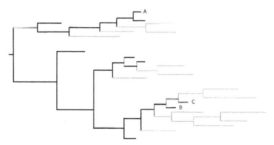

Fig. 3. Illustration of how the tendency of rapidly evolving species to go extinct will cause a change in the tempo of the molecular "clock" with time. Light shaded branches represent extinct lineages. Branch length represent evolutionary distance, not time. The evolutionary distance between taxa A and B is much larger than that predicted based on the distance between B and C because the latter is based on more slowly evolving species that survived extinction events whereas the former includes internal branches along which evolution was more rapid.

The Life of Lomagundi

There are a handful of reports of megascopic fossils much older than the Ediacaran (Han and Runnegar 1992, Shixing and Huineng 1995, Bengtson et al. 2007), but the fossil nature and taxonomic affinity of many of the are disputed (Lamb et al. 2007). All such finds are subject to intense scrutiny because they are so rare and there is no continuous record connecting them to the Ediacaran. It is clear that a diverse, benthic fauna did not leave a fossil record in Paleoproterozoic, but could much simpler, planktonic or sessile organisms have existed? Could there have been an early event that triggered the appearance of metazoan lineages, which did not become larger, more taphonomically cooperative animals until the oxygenation of the entire ocean ca. 580 Ma?

The biosphere experienced a geochemically unique period between the oxygenation of the atmosphere 2400–2300 Ma and the buildup of sulfur in the oceans (as sulfide and sulfate) due to the oxygenation. The residence time of sulfate in the modern oceans is about 10 Myr (Schlesinger 1997) and would have been considerably longer if sulfate weathering rates were lower under a low oxygen Proterozoic atmosphere. If the Canfield ocean model is grossly correct, banded iron formation deposition and the persistence of ferrous iron in the deep ocean requires low sulfide concentrations until 1800 Ma (pyrite being extremely insoluble). During this interval, marine sulfur cycling may have played a minor role in setting the over-all redox balance of the oceans and atmospheric oxygen content; instead ferrous iron and organic carbon and their oxidation/respiration may have been important.

Following atmospheric oxidation the $\delta^{13}C$ of carbonates rose to +8 parts per mil and may have reached 18 parts per mil in isolated basins, during an extraordinary 200 Myr period known as the Lomagundi event (Aharon 2005). Over this same interval, sulfate concentrations in the (shallow) ocean increased dramatically (Bekker et al. 2006), major banded iron formations disappeared and the isotopic fractionation $\delta^{56}Fe$ of sedimentary iron changed from principally negative values, perhaps reflective of ferrihydroxide precipitation, to positive values, indicating formation of pyrite, until about 1800 Ma (Rouxel et al. 2005). Because this occurred *after* the evidence for oxygenation of the atmosphere and intense glaciation ca. 2400–2300 Ma (Bau et al. 1999) it is unlikely to be directly related to either oxygenation (Des Marais et al. 1992) or a hypothetical "snowball Earth" episode (Kirschvink et al. 2000, Bekker et al. 2004).

Carbonate $\delta^{13}C$ reflects the relative burial fluxes of organic (light) carbon to carbonate and either productivity and/or burial efficiency must have increased eight-fold (Aharon 2005), or carbonate deposition decreased by a similar amount, because a marked change in the isotopic discrimination between organic and inorganic carbon is not found in the available data (Aharon 2005). However, these other explanations are not supported by the sediment record, which shows abundant carbonates and scarce organic deposits (Aharon 2005). An alternative scenario is that the heavy carbon reflects a larger isotopic gradient between the shallow oceans (where the carbonate record formed) and light carbon in the deep ocean, perhaps associated with the development of a large suspended or dissolved organic matter pool. Such a gradient may also have existed during the late Precambrian (Jiang et al. 2007).

This gradient could have been maintained by inefficient mixing in a near-stratified ocean, a more effective biological "pump" of organic matter into deep waters, or less efficient remineralization of that organic matter. The last could have occurred because of a paucity of electron acceptors for anaerobic metabolisms in the deep ocean during this period: sulfur was scavenged (as pyrite) by a stochiometric excess of ferrous iron and the remaining iron itself would have reacted rapidly with oxygen before the latter could be involved in the remineralization of organic carbon, producing insoluble iron hydroxides:

$$Fe^{2+} + \frac{1}{4}O_2 + \frac{5}{2}H_2O \rightarrow Fe(OH)_3 + 2H^+$$

that precipitated to the seafloor (Walker 1987). To produce the Lomagundi isotopic anomaly, the deep oceans would have had to sequester at least 0.02 gigatonnes of organic carbon per year for at least ~100 Myr, an amount more than 1000 times larger than the estimated reservoir of organic matter in the modern oceans. (The organic matter concentration would be ~1 g/l, comparable to chemostat cell cultures!) This is consistent with predictions based on modeling of the Proterozoic carbon cycle (Rothman et al. 2003), but it is unclear how such organic matter would have been maintained in the presence of anaerobic decomposition processes, particularly methanogenesis. This would have introduced light carbon back into the atmosphere and surface oceans, countering the heavy isotope signal. Perhaps organic carbon was removed by sedimentation on the sea floor and eventual subduction.

Could oxygen have risen to levels sufficient for multicellular eukaryotic metabolism during this period? The residence time of oxygen in the atmosphere with respect to modern burial rates is ~2 Myr (Schlesinger 1997). More modest amounts of oxygen could have accumulated quickly. The last (youngest) of the early Paleoproterozoic banded iron formations—the Hotazel formation of the Transvaal Supergroup in the Kalahari, include massive Mn formations that are consistent with mixing of oxygenated waters with anoxic (but not sulfidic) Fe-Mn rich deep waters (Schneiderhan et al. 2006). Because Mn will precipitate out only after most Fe is removed, the presence of the Mn formations suggests that well-oxygenated conditions were reached, at least transiently and locally. Such oxygenation could have driven by the sequestration of organic matter in the deep oceans as seems to be recorded in the Lomagundi event, however, if ferrous iron was present, the equivalent oxidizing power may have been diverted to ferric iron and then to carbonates during microbially-mediated sedimentary diagenesis. The absence of banded iron formations during this interval may mean that abundant ferrous iron was not present, or that oxygen levels in the surface ocean were held low. However, during a transition when sulfide and iron began to be competitive in some ocean basins, perhaps around 2000 Ma, oxygen levels may have briefly risen to a (uncertain) level capable of supporting some sort of multicellular eukaryotic life, perhaps in the estuarine environments discussed above.

The body plans of these hypothetical organisms may not have resembled anything found amongst the Ediacaran or Phanerozoic fauna and they would not necessarily have represented any of the extant lineages of Metazoa. We may safely assume that the absence of bioturbated sediments and flourishing of stromatolites through most of the Proterozoic rule out any extensive burrowing or scavenging lifestyle. Genome sequencing of animals from the major basal lineages has shown that most of the major gene families associated with a multicellularity and development were present in the last common ancestor and that gene duplication, diversification, and loss in existing families has been an important, if not the dominant process along each individual lineages (Nichols et al. 2006, Putnam et al. 2007, Baguna et al. 2008). That could mean that we can say very little about the ecology and physiology of individual basal lineages soon after the divergence, e.g. any sort of adaptive radiation is not detectable after 600 Myr of separate evolution. That is to say, the ancestors of sponges, jellyfish, and worms may have had completely different forms and lifecycles. For example, it was recently discovered that a parasitic myxozoan worm is actually a member of the Cnidaria (Jimenez-Guri et al. 2007). In an estuarine environment where riparian or tidal currents were strong, but there was insufficient oxygen to support swimming motility, a benthic or even sessile lifestyle would have been advantageous. Burrowing would have presumably been discouraged by the rapid disappearance of oxygen below the sediment-water interface. Filter feeding would have been a natural means to earn a living in the currents.

Once sulfur had accumulated (as sulfate and sulfide) in the Proterozoic ocean, many of the habitats of these organisms would have been polluted by sulfide and efficient reaction of sulfide with the atmosphere would have lowered oxygen levels both in surface waters and the atmosphere. The antecedents to animals would have found themselves restricted to the oxygenated refugia described above, perhaps going extinct, or, surviving their trial by sulfide, then liberated after more than a billion years had elapsed to re-invade the world again.

References

Aharon, P. 2005. Redox stratification and anoxia of the early Precambrian oceans: implications for carbon isotope excursions and oxidation. Precambrian Research 137: 207–222.

Andolfatto, P. 2007. Hitchhiking effects of recurrent beneficial amino acid substitutions in the *Drosophila melanogaster* genome. Genome Research 17: 1755–1762.

Arnold, G.L. and A.D. Anbar, J. Barling, and T.W. Lyons. 2004. Molybdenum isotope evidence for widespread anoxia in mid-Proterozoic oceans. Science 304: 87–90.

Baguna, J. and P. Martinez, J. Paps, and M. Riutort. 2008. Back in time: a new systematic proposal for the Bilateria. Philosophical Transations of the Royal Society of London. Series B Biological Sciences 363: 1481–1491.

Bau, M. and R.L. Romer, V. Luders, and N.J. Beukes. 1999. Pb, O, and C isotopes in silicified Moodraai dolomite (Transvaal Supergroup South Africa): implications for the composition of Paleoproerozoic seawater and 'dating' the increase of oxygen in the Precambrian atmosphere. Earth and Planetary Science Letters 174: 43–57.

Begun, D.J. and A.K. Holloway, K. Stevens, L.W. Hillier, Y.-P. Poh, M.W. Hahn, P. Nista, C.D. Jones, A.D. Kern, C.N. Dewey, L. Pachter, E. Myers, and C.H. Langley. 2007. Population genomics: whole-genome analysis of polymorphism and divergence in *Drosophila simulans*. PLos Biology 5: e310.

Bekker, A. and H. Holland, P.L. Wang, D. Rumble III, H.J. Stein, J.L. Hannah, L.L. Coetzee, and N.J. Beuke. 2004. Dating the rise of atmospheric oxygen. Nature 427: 117–120.

Bekker, A. and J.A. Karhu, and A.J. Kaufman. 2006. Carbon isotope record for the onset of the Lomagundi carbon isotope excursion in the Great Lakes area, North America. 148: 145–180.

Bengtson, S. and B. Rasmussen, and Krape. 2007. The Paleoproterozoic megascopic Stirling biota. Paleobiology 33: 351–381.

Blair, J.E. and S.B. Hedges. 2005. Molecular clocks do not support the Cambrian explosion. Molecular Biology and Evolution 22: 387–390.

Brocks, J.J. and G.D. Love, R.E. Summons, A.H. Knoll, G.A. Logan, and S.A. Bowden. 2005. Biomarker evidence for green and purple sulphur bacteria in a stratified Palaeoproterozoic sea. Nature 437: 866–870.

Broeker, W.S. and T.-H. Peng. 1982. Tracers in the Sea. Eldigio Press, Palisades, New York.

Butterfield, N.J. 2007. Macroevolution and macroecology through deep time. Journal of Paleontology 50: 41–55.

Canfield, D.E. 1998. A new model for Proterozoic ocean chemistry. Nature 396: 450–453.

Canfield, D.E. and R. Raiswell. 1999. The evolution of the sulfur cycle. American Journal of Science 299: 697–723.

Canfield, D.E. and S.W. Poulton, and G.M. Narbonne. 2007. Late Neoproterozoic deep-ocean oxygenation and the rise of animal life. Science 315: 92–95.

Conway-Morris, S. 2004. Life's Solution: Inevitable Humans in a Lonely Universe.Cambridge University Press, Cambridge.

Des Marais, D.J. and H. Strauss, R.E. Summons, and J.M. Hayes. 1992. Carbon isotope evidence for the stepwise oxidation of the Proterozoic environment. Nature 359: 605–609.

Droser, M.L. and S. Jensen, and J.G. Gehling. 2002. Trace fossils and substrates of the terminal Proterozoic-Cambrian transition: Implications for the record of early bilaterians and sediment mixing. Proc. Natl. Acad. Sci. USA 99: 12572–12576.

Dzik, J. 2003. Anatomical information content in the Ediacaran fossils and their possible zoological affinities. Integrative and Comparative Biology 43: 114–126.

Gaidos, E. and T. Dubuc, M. Dunford, P. McAndrew, J. Padilla-Gamino, B. Studer, K. Weersing, and S. Stanley. 2007. The Precambrian emergence of animal life: a geobiological perspective. Geobiology 5: 351–373.

Gould, S.J. 1989. Wonderful Life: The Burgess Shale and the Nature of History. W.W. Norton & Co., New York.

Grieshaber, M.K. and S. Volkel. 1998. Animal adaptations for tolerance and exploitations of poisonous sulfide. Annu. Rev. Physiology 60: 33–53.

Hagadorn, J.W. 2002. Chengjiang: early record of the Cambrian Explosion. D. J. Bottjer. Exceptional Fossil Preservation: A Unique View on the Evolution of Marine Life. New York, Columbia University Press 35–60.

Han, T.M. and B. Runnegar. 1992. Megascopic eukaryotic algae from the 2.1-billion-year-old Negaunee iron-formation, Michigan. Science 257: 232–235.

Hurtgen, M.T. and M.A. Arthur, and G.P. Halverson. 2005. Neoproterozoic sulfur isotopes, the evolution of microbial sulfur species, and the burial efficiency of sulfide as sedimentary pyrite. Geology 33: 41–44.

Jiang, G. and A.J. Kaufman, N. Christie-Black, S. Zhang, and H. Wu. 2007. Carbon isotope variability across the Ediacaran Yangtze platform in South China: implications for a large surface-to-deep ocean $\delta^{13}C$ gradient. Earth and Planetary Science Letters 261: 303–320.

Jimenez-Guri, E. and H. Philippe, B. Okamura and P.W.H. Holland. 2007. *Buddenbrockia* is a cnidarian worm. Science 317: 116–118.

Kah, L.C. and T.W. Lyons, and T.D. Frank. 2004. Low marine sulphate and protracted oxygenation of the Proterozoic biosphere. Nature 431: 834–837.

Kasting, J.F. 1987. Theoretical constrains on oxygen and carbon dioxide concentrations in the Precambrian atmosphere. Precambrian Research 34: 305–329.

Kaufman, A.J. 2005. The calibration of Ediacaran time. Science 308: 59–60.

Kennedy, M. and M. Droser, L.M. Mayer, D. Pevear, and D. Mrofka. 2006. Late Precambrian oxygenation; inception of the clay mineral factory. Science 311: 1446–1449.

Kimura, M. 1968. Evolutionary rate at the molecular level. Nature 217: 624–626.

King, N. 2004. The unicellular ancestry of animal development. Developmental Cell 7: 313–325.

Kirschvink, J.L. and E.J. Gaidos, L.E. Bertani, N.J. Beukes, J. Gutzmer, L.N. Maepa, and R.E. Steinberger. 2000. Paleoproterozoic Snowball Earth: extreme climatic and geochemical global change and its biological consequences. Proc. Natl. Acad. Sci. USA 97: 1400–1405.

Klein, C. 2005. Some Precambrian banded iron-formations (BIFs) from around the world: Their age, geologic setting, mineralogy, metamorphism, geochemistry, and origin. American Mineralogist 90: 1473–1499.

Kump, L.R. and A. Pavlov, and M.A. Arthur. 2005. Massive release of hydrogen sulfide to the surface ocean and atmosphere during intervals of oceanic anoxia. Geology 33: 397–400.

Lamb, D.M. and S.M. Awramik, and S. Zhu. 2007. Paleoproterozoic compression-like structures from the Changzhougou Formation, China: Eukaryotes or clasts? Precambrian Research 154: 236–247.

Lenton, T.M. and A.J. Watson. 2004. Biotic enhancement of weathering, atmospheric oxygen and carbon dioxide in the Neoproterozoic. Geophysical Research Letters 31: L05202.

Lindstrom, M. 1995. The environment of the Early Cambrian Chengjiang fauna. International Cambrian Explosion Symposium, Programme and Abstracts.

Logan, G.A. and J.M. Hayes, G.B. Hieshima, and R.E. Summons. 1995. Terminal Proterozoic reorganization of biogeochemical cycles. Nature 376: 53–56.

Mangum, C. and W. Van Winkle. 1973. Responses of aquatic invertebrates to declining oxygen conditions. American Zoologist 13: 529–541.

McFadden, K.A. and J. Huang, X. Chu, G.Q. Jiang, A.J. Kaufman, C. Zhou, X. Yuan, and S. Xiao. 2008. Pulsed oxidation and biological evolution in the Ediacaran Doushantuo Formation. Proc. Natl. Acad. Sci. USA 105: 3197–3202.

Millero, F.J. 2005. Chemical Oceanography. CRC, Boca Raton, FL.

Narbonne, G.M. 2005. The Ediacara biota: Neoproterozoic origin of animals and their ecosystems. Annual Reviews of Earth and Planetary Sciences 33: 421–442.

Nichols, S.A. and W. Dirks, J.S. Pearse, and N. King. 2006. Early evolution of animal cell signaling and adhesion genes. Proc. Natl. Acad. Sci. USA 103: 12451–12456.

Ohta, T. 2005. The nearly neutral theory is dead. The current significance and standing of neutral and nearly neutral theories. BioEssays 18: 673–677.

Pagel, M. and C. Venditti, and A. Meade. 2006. Large punctuational contribution of speciation to evolutionary divergence at the molecular level. Science 314: 119–121.

Poulton, S.W. and P.W. Fralick, and D.E. Canfield. 2004. The transition to a sulphidic ocean ~1.84 billion years ago. Nature 431: 173–177.

Putnam, N.H. and M. Srivastava, U. Hellsten, B. Dirks, J. Chapman, A. Salamov, A. Terry, H. Shapiro, E. Lindquist, V.V. Kapitonov, J. Jurka, G. Genikhovich, I.V. Grigoriev, S. M. Lucas, R.E. Steele, J.R. Finnerty, U. Technau, M.Q. Martindale, and D.S. Rokhsar. 2007. Sea anemone genome reveals ancestral eumetazoan gene repertoire and genomic organization. Science 317: 86–94.

Rothman, D. and J. Hayes, and R. Summons. 2003. Dynamics of the Neoproterozoic carbon cycle. Proc. Natl. Acad. Sci. USA 100: 8124–3129.

Rouxel, O.J. and A. Bekker, and K.J. Edwards. 2005. Iron isotope constraints on the Archean and Paleoproterozoic ocean redox state. Science 307: 1088–1091.

Runnegar, B. 1982. A molecular-clock date for the origin of the animal phyla. Lethaia 15: 199–205.

Schlesinger, W.H. 1997. Biogeochemistry: An Analysis of Global Change. Academic Press, San Diego.

Schneiderhan, E.A. and J. Gutzmer, H. Strauss, K. Mezger, and N.J. Beukes. 2006. The chemostratigraphy of a Paleoproterozoic MnF-BIF succession—the Voelwater Subgroup of the Transvaal Supergroup in Griqualand West, South Africa. South African Journal of Geology 109: 63–80.

Schroder, S. and J. Grotzinger. 2007. Evidence for anoxia at the Ediacaran-Cambrain boundary: the record of redox-sensitive trace elements and rare earth elements in Oman. Journal of the Geological Society 164: 175–187.

Seilacher, A. 1989. Vendozoa: organismic construction in the Proterozoic biosphere. Lethaia 17: 229–239.

Shen, Y. and A. Knoll, and M. Walter 2003. Evidence for low sulphate and anoxia in a mid-Proterozoic marine basin. Nature 423: 632–635.

Shixing, Z. and C. Huineng. 1995. Megascopic multicellualr organisms from the 1700-million-year-old Tuanshanzi Formation in the Jixian area, North China. Science 270: 620–622.

Slack, J.F. and T. Grenne, A. Bekker, O.J. Rouxel, and P.A. Lindberg. 2007. Suboxic deep seawater in the late Paleoproterozoic: evidence from hematitic chert and iron formation related to seafloor-hydrotehrmal sulfide deposits, central Arizona, USA. Earth and Planetary Science Letters 255: 243–256.

Stanley, S. 2007. An analysis of the history of marine animal diversity. Paleobiology 33(4s): 1–55.

Thamdrup, B. and K. Finster, J.W. Hansen, and F. Bak 1993. Bacterial disproportionation of S^0 and coupled to chemical reduction of iron or manganese. Applied and Environmental Microbiology 59: 101–108.

Walker, J.C.G. 1987. Was the Archean biosphere upside down? Nature 329: 710–712.

Webster, A.J. and R. J.H. Payne, and M. Pagel. 2003. Molecular phylogenies link rates of evolution and speciation. Science 301: 478.

Wray, G.A. and J.S. Levinton, and L.H. Shapiro. 1996. Molecular evidence for deep Precambrian divergences among metazoan phyla. Science 274: 568–573.

Zopfi, J. and T.G. Ferdelman, B.B. Jorgensen, A. Teske, and B. Thamdrup. 2001. Influence of water column dynamics on sulfide oxidation and other biogeochemical processes in the chemocline Marine Chemistry 74: 29–51.

Chapter 16

Redefining Stem Cells and Assembling Germ Plasm: Key Transitions in the Evolution of the Germ Line

John Srouji[1] and *Cassandra Extavour*[2]*

Introduction

A discussion of "key transitions" in the evolution of animals often invokes mental images of large-scale morphological or behavioural changes: the fin-to-limb transition, avian beak shape changes, the transition from simply holding objects to using them purposefully as tools. These types of changes clearly occurred in evolution and had great adaptive value. Other types of changes, however, have also occurred in the morphologies and behaviours of single cells and cell lineages. A complete understanding of many "key transitions" involving new structures and new cell types must therefore incorporate the molecular genetic basis for the novel or modified cell behaviours that can lead to novel structures.

For example, the fin-to-limb transition cannot be considered without first considering the origin of paired fins. The appearance of these appendages clearly predated their adaptive transformation into terrestrial limbs. At the

[1]Department of Molecular and Cellular Biology, Harvard University, 16 Divinity Avenue, Cambridge MA 02138, USA.
E-mail: jrsrouji@fas.harvard.edu
[2]Department of Organismic and Evolutionary Biology, Harvard University, 16 Divinity, Avenue, Cambridge MA 02138, USA.
E-mail: extavour@oeb.harvard.edu

level of the completed morphological product, the fossil record shows that midline unpaired appendages (fins) were present before paired appendages (Zhang and Hou 2004). At the level of behaviour of the individual cells that participate in the development of these structures, gene expression studies suggest that bilaterally paired groups of cells adopted a developmental program hitherto used by unpaired midline appendage anlagen (Freitas et al. 2006). In other words, cells that had previously held a given developmental capacity were able to expand their biological potential and perform a new function, thus giving rise to a new morphology. In this case, the bilateral anlagen may have co-opted a developmental program that was already in use by other cells in the organism.

A second example illustrates that some features require us to consider the emergence of an apparently new cell type. In this case, rather than cells adopting entire genetic regulatory programs already in use by other cells in the embryo, cells may instead have acquired novel combinations or modifications of genetic regulatory programs, allowing them to perform functions that are new to the embryo. Years of evolutionary analysis of avian beaks have shown that beak shape is an important target of adaptive morphological changes (Darwin 1859, Grant et al. 1976, Grant 1999, Schluter 2000). One important level of analysis has been aimed at the gene regulatory differences that are present in birds with differently shaped upper beaks (Abzhanov et al. 2004, Abzhanov et al. 2006). A second, related level of analysis is possible, however, which focuses on the very existence of these beak regions: the analysis of the neural crest (Douarin and Kalcheim 2009). Neural crest cells give rise to the upper beak pattern itself, and are responsible for most vertebrate craniofacial variation (Noden 1975, Hu et al. 2003). From their origins in the dorsal neural tube, neural crest cells migrate away from the neural tube in an anterior to posterior progression, finally coming to rest at several different places in the embryo. Both during and after this migration, these cells differentiate into an enormous diversity of cell types, including neurons, glia, pigmented cells, cartilage, and bone. The first neural crest cells would thus literally have gone where no cell had gone before.

The neural crest is unique to vertebrates, yet cell lineages with some, but not all, neural crest properties have been identified in non-vertebrate chordates (Jeffery et al. 2004, Ota et al. 2007, Jeffery et al. 2008). The evolution of the "true" neural crest may therefore have involved the acquisition of neural crest-specific characters (including migration in different directions and the potential to give rise to multiple terminal cell fates) that pre-existed in an either fundamentally migratory but not pluripotent, or pluripotent but stationary, cell lineage (Wada 2001, Kee et al. 2007, Sauka-Spengler et al. 2007, Donoghue et al. 2008, Jeffery et al. 2008).

This chapter will examine the evolution of a cell type that, not unlike the neural crest, seems to have arisen at a definable branch point within the Metazoa, and was more likely to have arisen by modification of a pre-existing cell type than to have appeared entirely *de novo*: the dedicated germ line.

The novelty of the bilaterian germ line

The cells of the germ line are those uniquely responsible for undergoing gametogenesis during adult reproductive life. The eggs and sperm that germ cells produce ensure both organismal reproduction and species continuity. All sexually reproducing organisms, including plants, specify germ cells at some stage of reproductive life. However, bilaterians are the only clade with a single, embryonically specified cell lineage that is capable only of producing gametes. Many non-bilaterians have pluripotent stem cell-like lineages, which can both undergo gametogenesis and produce a variety of differentiated somatic cell types (discussed further below). In contrast, bilaterians generally have a uniquely specified germ line that is established once during embryogenesis and cannot self-renew. While germ line stem cells (GLSCs), the self-renewing precursors to gametes, exist in some animals, embryonic primordial germ cells (PGCs) are not a naturally self-renewing population before they commit to gametogenesis. Instead, similar to somatic cell types, a limited, species-specific number of germ cells is generated during embryogenesis. These cells generally cannot be replaced if they are lost during embryogenesis (but see Takamura et al. 2002, Modrell et al. submitted), and only they can take on the job of creating gametes in the sexually mature animal.

The emergence of the dedicated germ line during embryogenesis thus represents a key transition in animal evolution: the transition from a state in which any cell could contribute to the next generation, to one where that potential was restricted to a tiny number of cells in the organism. In other words, the issue of the evolution of the germ line is also that of the evolution of the true soma, cells that contribute to the body plan but whose genetic material is barred from being transmitted to future generations.

The Dedicated Germ Line as an Outcome of Multicellularity

One of the consequences of the evolution of stable multicellularity was a fundamental shift in reproductive strategy. In unicellular species, each individual is responsible for its own reproduction, and passes on its genes to every one of its descendants. In contrast, in multicellular species, the majority of an organism's cells will not have their particular gene complement transmitted to the next generation: only a small fraction of the organism's cells, the germ cells, will have that privilege. Similarly, every single somatic function that was formerly performed by the unicellular organism (for example, motility, nutrient intake, or waste excretion) then became the responsibility of only a subset of cells in the multicellular organism. Thus, while the physical phenomenon of multicellularity would have required the acquisition of cell adhesion and cell signalling molecules (King and Carroll 2001, Nichols et al. 2006, Abedin and King 2008, Newman and Bhat 2009), the adaptive value of multicellularity is based on the principle of division of labor among cells with theoretically equivalent genetic potential (Willensdorfer 2008). Such equivalent genetic potential, however, is not guaranteed. We must therefore consider what happens in the case that genotypes are not identical, or do not produce identical phenotypes, in cells of a multicellular organism.

Heterogeneity and cellular competition in the germ line

Although we usually think of all cells in an organism having identical genomes, in reality somatic mutation at any point after first cleavage can result in a genetically mosaic individual. When cells are not genetically identical, and only some of them can contribute to making gametes, we must consider the impact of cellular competition. Genetic heterogeneity in a cellular population has been shown to lead to competition and natural selection (Keller 1999), in the same way that adult organisms and populations of organisms are subject to selection (Darwin 1859, Wallace 1885). Many scholars have therefore reasoned that because unchecked competition between aggregated cells poses a threat to multicellularity, a dedicated germ line is required for true multicellularity (see for example Buss 1987, Michod and Roze 2001). Higher-order levels of cooperation, such as insect colonies and some mutualisms, have been proposed to represent "organisms" with no need for a germ-soma distinction to ensure organismal cohesion (Queller and Strassmann 2009). However, in a single multicellular organism with a dedicated germ line, reproduction is a "non-exchangeable benefit" (Michod and Roze 2001), because no somatic cell can produce gametes. In this situation, "cooperation" in the form of maintaining multicellularity, is less costly than abandoning the aggregate (Michod 2005, Michod 2007).

The complexity of the cooperation/conflict problem posed by multicellularity increases, however, even when the gametogenic lineage is restricted, if the germ line itself is genetically mosaic. The most extreme case of such mosaicism occurs in plants, where individual, spatially separated stem cell lineages produce "multiple germ lines." In plants, a stem cell lineage called the shoot apical meristem (SAM) is produced at the tips of the aerial parts of the plant (Sharma et al. 2003). The SAM undergoes self-renewing divisions to give rise to gametogenic cells, and also produces various somatic cell types during the continuous growth of the plant (Dickinson and Grant-Downton 2009, Stahl and Simon 2009). With the production of every new reproductive organ (flower), germ cells must be established *de novo* from each SAM. The high levels of somatic mosacism, combined with the longevity of many plants, means that individual SAMs can differ genetically, resulting in germ lines of multiple genotypes being produced by a single plant (Whitham and Slobodchikoff 1981, Kleckowski 1986, Schultz and Scofield 2009).

The plant mechanism of germ line determination might seem to resemble that seen in some basal bilaterians, where a pluripotent stem cell lineage gives rise to both somatic cells and gametes throughout adult life (discussed below). However, this is only superficially similar to the production of gametes in plants. Whereas basal bilaterian gametogenic stem cells arise from a single founder population during embryogenesis, the individual SAMs of a plant are not all clonally related. Thus, in the plant case, the opportunities for a single plant to produce genetically heterogeneous gametes are greater than those for an individual cnidarian or acoel flatworm. In other words, an important determinant of the possible heterogeneity of the gametic population is whether the germ line arises from a small, early-determined clone, or whether it is polyclonal and derived late in development. Germline specification mechanisms are therefore highly relevant to the selective pressures and cellular competition that influence the germ line during pre-gametic development.

Once the animal germ line is specified, not all PGCs may have the opportunity to contribute to the gametic population in the next generation. Competition between genetically mosaic somatic cells is a well-documented phenomenon (García-Bellido et al. 1973, Morata and Ripoll 1975, Moreno et al. 2002, Moreno and Basler 2004, Oliver et al. 2004, Oertel et al. 2006). Similarly, when genetic mosaicism is induced in embryonic PGCs, they compete to enter the germ line (Extavour and García-Bellido 2001). Standing genetic variation has also been clearly shown to result in natural selection acting on the germ line in colonial ascidians. In these animals, embryos produce tadpole larvae that join genetically heterogeneous colonies. The larvae mature into individual zooids, which participate in the colony by sharing a common test (outer covering) and by undergoing a degree of internal anatomical fusion to share common nutritive, excretory, and reproductive functions (Milkman 1967). Each zooid has a population of putative stem cells that are established during embryogenesis (Brown et al. 2009) and circulate in the hemolymph of the entire colony. These cells give rise to gametes and thus act as shared, circulating germ line progenitors for the colony (Berrill 1941, Mukai and Watanabe 1976, Sabbadin and Zaniolo 1979). Elegant molecular lineage experiments have shown that the germ line from one zooid can "invade" a colony, effectively outcompeting the germ cells contributed by other colony members (Stoner and Weissman 1996, Stoner et al. 1999, Laird et al. 2005). These observations support the hypothesis put forward by Buss (1982), who suggested that genetically chimaeric animals should possess a self/non-self recognition system in order to prevent germ line takeover by closely related genomes (de Tomaso et al. 2004, 2005) .

Investigation of the developmental origin of the germ line in individual zooids has shown that gametogenic stem cells circulating in the colony can be distinguished from circulating somatic stem cells based on the expression of many of the same genes expressed by germ cells in solitary (non-colonial) animals (Sunanaga et al. 2006, Sunanaga et al. 2007, Sunanaga et al. 2008, Rosner et al. 2009). Recent work on the embryonic origin of these cells has further suggested that the "germ line" contributed by each zooid to the competing gamete precursor pool, is specified as a small, early-derived lineage of cells close to the beginning of embryogenesis. These cells may go on to contribute to gametes years after their initial specification (Brown et al. 2009).

Finally, even in cases where cells are genetically identical, stochastic differences in gene expression and cellular metabolism levels are expected to produce heterogeneous phenotypes, which may be subject to natural selection (discussed in Khare and Shaulsky 2006). Thus germ cells, which can be considered a special subset of stem cells, are themselves fundamental units of selection, and their genomes are critical loci of evolutionary change (Weissman 2000).

Evolutionary Implications of Developmental Timing and Molecular Mechanism of Germ Line Specification

We have established that the developmental origin of the germ line is an important event with implications that are not just reproductive, but also evolutionary. Two key aspects of this developmental origin are the timing of germ line specification and the molecular mechanisms directing it.

Timing of germ cell specification

The timing of germ cell specification during embryogenesis affects the degree of gametic, and subsequent somatic, mosaicism because every round of cell division provides an opportunity for mutation (Drost and Lee 1998). DNA replication errors can result in hereditary mutations, and increased mitotic rates can lead to increased mutation rates (Sweasy et al. 2006) and accelerated aging (Ban and Kai 2009). In most animals, once PGCs have been specified in the embryo, they divide little or not at all until gametogenesis begins. Even after gamete production starts, the fact that more mitoses are required to produce sperm than eggs has been suggested to cause increased mutation rates, and therefore higher rates of evolution, in genes transmitted by males (reviewed by Ellegren 2007). Haldane's studies of female carriers of haemophilia led to the first suggestion of male-biased evolution over 70 years ago (Haldane 1935, Haldane 1947), but these ideas were not subjected to further theoretical or empirical studies until decades later (Miyata et al. 1987). In mammals, the mutation rate of male-transmitted genes is higher than that of female-transmitted genes, and is correlated with the higher total number of cell divisions required to produce spermatids as opposed to oocytes (Shimmin et al. 1993, Chang et al. 1994, but see Sandstedt and Tucker 2005). In contrast, some studies in *Drosophila* have provided evidence for an absence of male-biased mutation (Bauer and Aquadro 1997), while others suggest that some weak male-biased mutation does occur (Bachtrog 2008).

The strength of the bias appears to be greater overall in mammals than in Drosophilids, which may be related to the different timing of germ cell specification in these groups. Most biased mutation studies have calculated the number of cell divisions from "zygote" to gamete, without taking into consideration when during zygotic development the germline is created. Mammalian germ cells are specified after many rounds of embryonic cell division, and they continue to divide extensively between specification and gametogenesis, giving those cells an increased opportunity to collect mutations generated during early mitoses (Chiquoine 1954, Zamboni and Merchant 1973, Clark and Eddy 1975, Tam and Snow 1981, Ginsburg et al. 1990). Drosophilid germ cells, on the other hand, are specified after fewer mitotic divisions, and do not divide between specification and gametogenesis (Sonnenblick 1941, Turner and Mahowald 1976, Foe and Alberts 1983, Campos-Ortega and Hartenstein 1985, Hay et al. 1988). The developmental timing of germ cell specification must thus be addressed not only by developmental biologists, but also in any consideration of metazoan evolution.

Molecular mechanisms of germ cell specification

The developmental mechanisms used by metazoans to specify PGCs early in embryogenesis can be grouped into two broad categories. Previous literature has referred to these two types as epigenesis and preformation (Nieuwkoop and Sutasurya 1981, Extavour and Akam 2003). We will avoid using these terms in this chapter, due to the varied and often confusing use of these terms in the biological literature (Haig 2004, Callebaut 2008), and particularly in light of the recent rise of "epigenetics" (in the sense of non-genomically encoded, heritable gene expression phenotype changes (Ko and McLaren 2006, Hayashi and Surani 2009). Instead, we will refer to inherited germ line determinants (formerly "preformation") as the "inheritance mode," and to inductive signal-dependent PGC specification(formerly "epigenesis") as the "inductive mode."

Under the inductive mode, signals produced from one cell population induce another cell population to adopt germ cell fate (reviewed by Extavour and Akam 2003). The signals are necessary and sufficient for germ cell formation. When PGCs are specified in this way, the specification event takes place relatively late in development. By this time, the cells that become PGCs have already undergone several rounds of cell division, and therefore of putative somatic mutation and selection. Under this scenario, gametic populations are expected to be relatively heterogeneous, as the founder population of PGCs may be relatively large. In animals, the heterogeneity of the gametic founder population never reaches the extremes seen in plants (discussed above), but is still considerable. In mice, for example, the PGC founder population of at least 150 cells has divided mitotically at least 22 times before becoming committed to gametogenesis (summarized by Drost and Lee 1998).

Experimental evidence for the inductive mode in wild-type development, in the form of experimental embryology and genetic knockdown experiments, comes from work on mice (Tam and Zhou 1996, Lawson et al. 1999, Ying et al. 2000, Ying and Zhao 2001, de Sousa Lopes et al. 2004, Ohinata et al. 2009) and salamanders (Humphrey 1929, McCosh 1930, Nieuwkoop 1947). There is also experimental evidence that some animals can employ this mode under abnormal conditions where the endogenous germ line has been removed (Takamura et al. 2002, Modrell et al. submitted). An additional wealth of cytological, histological, embryological, electron microscopy, and lineage tracing data from all major metazoan clades collectively suggests that the inductive mode is the most common mode of germ cell specifications in bilaterians (Extavour and Akam 2003). Outside of the bilaterians, germ line development and gene expression patterns are consistent with the inductive mode of germ cell specification (reviewed by Extavour and Akam 2003, see for example Extavour et al. 2005). In these groups, germ cells are formed either by a dedicated germ line stem cell (GLSC) population, or by a pluripotent stem cell population that can produce both gametes and various somatic cell types. No data currently support the existence of an inheritance mode of PGC specification in bilaterian outgroups.

The inheritance mode is characterized by the maternal provision of germ cell determinants (reveiwed by Extavour and Akam 2003). Before the end of oogenesis, or shortly after fertilization, the determinants are asymmetrically localized to a region of the ooplasm (the germ plasm) that will be inherited by the future PGCs. In some cases, the PGCs may incorporate this germ plasm after initial rounds of syncitial cleavage (see classical descriptions of Metschnikoff 1866, Huettner 1923). In other cases, germ plasm is inherited directly through successive rounds of asymmetric, holoblastic cleavage until all of the germ plasm is contained within one or a few cells, which are the first PGCs (see for example Browne et al. 2005, Extavour 2005). When PGCs are specified by inheritance, their separation from the soma takes place very early in embryogenesis, sometimes as early as second cleavage (Grbic et al. 1998, Donnell et al. 2004, Zhurov et al. 2004) and not later than sixth cleavage (Nishida 1987, Fujimura and Takamura 2000). The germ line in these cases is thus subject to limited somatic selection, and the resulting gametic population is less likely to be genetically heterogeneous. At the same time, fixation of mutations may be more likely than in the inductive mode, since the small founder population of PGCs means that early-occurring mutations will appear in all or most of the gametes (Drost and Lee 1998).

Because this is the mode of germ cell specification displayed by the genetic laboratory model organisms *Drosophila melanogaster* (fruit fly), *Caenorhabditis elegans* (nematode), *Danio rerio* (zebrafish), and *Xenopus laevis* (frog), it has been examined in great molecular genetic detail (reviewed by Raz 2003, Zhou and King 2004, Hayashi et al. 2007, Strome and Lehmann 2007). Largely due to the enormous amounts of molecular genetic data concerning this mode of PGC specification, it was until recently widely assumed that the inheritance mode represented an ancestral animal developmental mechanism (Wolpert et al. 2002). However, as mentioned above, the examination of PGC specification mode distribution across a broad phylogenetic range suggests that the inheritance mode may have evolved convergently several times during metazoan evolution (Extavour and Akam 2003). This hypothesis now forms part of many synthetic treatments of developmental biology (Gilbert and Singer 2006, Wolpert et al. 2007).

Two Major Transitions in the Evolution of the Germ Line

Of all of the evolutionary steps that must have taken place along the road to the spectrum of mechanisms used by extant animals to specify their germline, two stand out as being of particular interest. The first is the evolution of a lineage-restricted stem cell population (GLSCs) from pluripotent stem cells. The second is the repeated, convergent evolution of an inheritance mode of PGC specification, from an ancestral inductive mode. The following sections deal with each of these two critical transitions, and propose testable hypotheses for further investigation.

From Stem Cell to Germ Cell

Comparative and phylogenetic analyses of PGC specification modes across the metazoans have led to the hypothesis that germ cells have their evolutionary origins in a pluripotent stem cell population that was present in the last common bilaterian ancestor (Extavour 2007a, b).

Similar hypotheses have been proposed by researchers examining the *in vivo* and *in vitro* similarities between germ cells, somatic stem cells, and cultured stem cells of extant animals (Sanchez Alvarado and Kang 2005, Agata et al. 2006). Nonetheless, it is still unclear how the molecular genetic program controlling pluripotent stem cell identity could have been modified to yield a gametogenic stem cell program. To this end, we will first review the relationships, potentials and conversions between these various cell types. A note regarding terminology is in order here: the term "stem cell" is not always consistently defined in developmental biology research, clinical research or the popular media (Shostak 2006, Lander 2009). For the purposes of this chapter, a "stem cell" simply refers to a cell that is self-renewing. Namely, upon mitotic division, one daughter differentiates, and the second daughter does not differentiate, but rather engages in another self-renewing mitosis. The common second criterion of pluripotency will not be applied, so as to be able to include unipotent germ line stem cells (GLSCs). In these cells, mitoses are self renewing, but their differentiated daughters are all of the same cell type, that is, gametes or gametogonia.

Dedicated Germ Cells from Pluripotent Stem Cells

Germ cells have sometimes been called the "ultimate" or "mother of all" stem cells because although their immediate differentiated products are of a single cell type, their final products (oocytes, the founder cell of all animal embryogenesis) ultimately give rise to all cell types of an organism (Donovan 1998, Spradling and Zheng 2007, Cinalli et al. 2008, Rangan et al. 2008, Rangan et al. 2009). From such a perspective, germ cells seem "more pluripotent" than some naturally occurring or laboratory-developed (somatic) stem cell populations. However, in order to detect a pattern in the wide variety of observed germ cell/stem cell relationships, it is more useful to think of the production of lineage-restricted germ cells or GLSCs, as a true differentiation event. In other words, deriving either GLSCs or gametes from somatic stem cells will be considered a reduction in pluripotency (Fig. 1).

Transitions from stem cell to germ cell in normal development

This transition occurs naturally during adult reproductive life in extant members of bilaterian outgroups. Cnidarians and sponges have pluripotent stem cells that give rise both to differentiated somatic cells of various types, and to male and female gametes (Fig. 1A) (Müller 2006). Within Bilateria but branching basally to protostomes and deuterostomes are the acoel flatworms (Hejnol et al. 2009b), also of interest when considering bilaterian germ cell origin because they are a sister group to all other

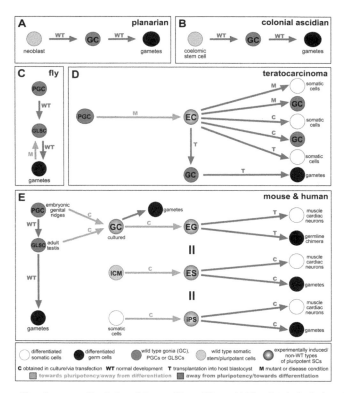

Fig. 1. Stem cell and germ cell relationships and transitions. (A) In platyhelminths, pluripotent stem cells called neoblasts can differentiate into gonia (GC), which generate gametes. **(B)** In colonial ascidians, pluripotent stem cells circulate in the haemolymph, and may differentiate into gonia (GC), which generate gametes. The existence of lineage-restricted germline stem cells in colonial ascidians remains a possibility under active investigation (Brown et al. 2009, Kawamura and Sunanaga 2010). **(C)** In *Drosophila*, embryonically specified primordial germ cells (PGCs) differentiate into unipotent germline stem cells (GLSCs), which generate gametes. However, GLSC daughters that have begun the differentiation process into gametes can be induced to revert to a GLSC fate (Niki and Mahowald 2003). **(D)** In teratocarcinomas, misguided embryonic PGCs are thought to convert to an "embryonal carcinoma" (EC) state. When transplanted into a host blastocyst, EC descendants can populate both the soma and the germline, and generate functional gametes. In situ, EC cells can differentiate into cells with both somatic and germ cell characteristics. In *in vitro* culture, EC cells be induced to differentiate into both somatic cell types and germ cells. **(E)** In normally developing mouse and human embryos, embryonically specified PGCs colonize the embryonic genital ridges, differentiate either into gonia (in females) or into germline stem cells (in males), and subsequently generate gametes. PGCs from the embryonic genital ridges or GLSCs from the adult male testis, can be cultured to form embryonic germ (EG) cells. EG cells can contribute to both somatic and germline cells when transplanted into host blastocysts. Similarly, embryonic stem (ES) cells derived from culturing cells of the inner cell mass (ICM), and differentiated somatic cells converted into induced pluripotent stem (iPS) cells by treatment with variations on the "Yamanaka factors," (Takahashi and Yamanaka 2006, Okita et al. 2007, Takahashi et al. 2007)can both be induced to adopt differentiated somatic and germ cell fate characteristics *in vitro*.

Color image of this figure appears in the color plate section at the end of the book.

bilaterians. Work on acoel gametogenesis suggests that inheritance does not play a role in embryonic germ cell specification (Falleni and Gremigni 1990, Gschwentner et al. 2001, Hejnol and Martindale 2008).

Although the Platyhelminthes, excluding Acoelomorpha, appear to be firmly nested within protosome bilaterians (Dunn et al. 2008, Hejnol et al. 2009a), their embryogenesis is very similar to that described for acoels (Boyer et al. 1998, Henry et al. 2000). Data on platyhelminth germ cell specification may therefore be informative for understanding early bilaterian strategies for PGC and stem cell function. Flatworm germ cells are derived from pluripotent stem cells called neoblasts (Fig. 1A) (Baguñà 1981). These cells can also give rise to several types of somatic cells. Although germ cells and neoblasts have been the subject of much recent work in these animals, there is still no clear evidence supporting the existence of an exclusively gametogenic subpopulation of neoblasts (Reddien et al. 2005, Guo et al. 2006). In contrast, colonial ascidians (discussed above) display multiple stem cell types, one of which is GLSCs (Fig. 1B).

Transitions from stem cell to germ cell in disease conditions

The transition from stem cell to gamete can also occur *in vivo* as a result of specific genetic or epigenetic alterations. Teratocarcinomas and teratomas are two types of germ cell neoplasms, thus called because they are derived from misplaced or ectopically occurring PGCs (Fig. 1D). Teratomas are generally benign, non-invasive, and do not recur after surgical removal (Stevens and Little 1954, Stevens 1967, Heerema-McKenney et al. 2005). By contrast, teratocarcinomas are malignant, can be invasive and recurring, and are composed of a combination of teratomas and embryonal carcinoma (EC) cells. EC cells, in turn, are thought to be the misregulated products of ectopic or misguided embryonic PGCs (Fig. 1D) (Kleinsmith and Pierce 1964, Martin and Evans 1974, Martin 1975, Graham 1977, Hoei-Hansen et al. 2006). These two types of germ cell tumor can give rise both to germ cells and to several types of somatic cell, resulting in tumors that may contain differentiated hair, bone or teeth (Kleinsmith and Pierce 1964, Kahan and Ephrussi 1970). The demonstrated pluripotency of these tumors is attributed to the EC cells they contain. The higher differentiation potential of EC cells, as compared to cells of non-differentiating tumors, may be attributable to their germ cell origin.

It has been suggested that a "germ cell state" of reprogramming "back to" pluripotency exists, and must be traversed by differentiated cells in order to regain pluripotency (Hayashi and Surani 2009). This hypothesis is consistent with the observation that EC cells of germ cell neoplasms are able to differentiate into many cell types; in other words, they are demonstrably pluripotent. When cultured and subsequently transplanted into blastocysts, EC cells can contribute both to somatic cells and to the germline of the hosts (Mintz and Illmensee 1975, Stewart and Mintz 1981, Stewart and Mintz 1982). In summary, germ cell neoplasms provide an example of PGCs, or possibly even lineage-restricted GLSCs, that first increase in pluripotency to become EC cells, and then lose pluripotency as they move towards either GLSC, gametic, or somatic cell fate.

Transitions from stem cell to germ cell in vitro

The various methods of inducing laboratory-cultured stem cells to differentiate as gametogenic cells again demonstrate the close relationship between these cell types (Fig. 1E). We will consider three main categories of laboratory-derived stem cells. (1) Embryonic Germ (EG) cells are obtained by culturing PGCs from embryonic genital ridges (Matsui et al. 1992, Resnick et al. 1992, Rohwedel et al. 1996, Shamblott et al. 1998, Kanatsu-Shinohara et al. 2004, Kerr et al. 2006). Although cultured spermatogonial cells from adult testes are called SSCs or maGSs in the literature, in this chapter we will consider them within this first category, since they are also stem cells derived from cells that had achieved germline specification through normal development. (2) Embryonic Stem (ES) cells are obtained by culturing cells of the inner cell mass (ICM) of blastocyst stage embryos (Evans and Kaufman 1981, Martin 1981, Thomson et al. 1998). (3) Induced Pluripotent Stem (iPS) cells are obtained by transfecting differentiated somatic cells with an appropriate combination of transcription factors (Takahashi and Yamanaka 2006, Blelloch et al. 2007, Takahashi et al. 2007, reviewed by Amabile and Meissner 2009).

Once in culture, all of these stem cell types can be induced to "reduce pluripotency" by differentiating either as somatic cells, or as germ cells (reviewed by Marques-Mari et al. 2009). EG cells show germline transmission when injected into donor blastocysts, indicating that they can give rise to functional gametogenic cells (Kanatsu-Shinohara et al. 2003, Guan et al. 2006). ES cells are capable of adopting many germ cell characteristics (Kee et al. 2006, Clark et al. 2004). They can produce "oocytes" that are competent to enter meiosis, form follicle-like structures, and parthenogenetically produce blastocyst-like masses (Hübner et al. 2003). They can upregulate germ cell-specific genes and undergo spermatogenesis (Toyooka et al. 2003). Finally, iPS cells can produce cells that upregulate germ cell-specific genes and contribute to the germline of a host blastocyst (Okita et al. 2007, Park et al. 2009). For all of these laboratory-cultured stem cells, however, fully functional germ cells have never been achieved without passage through an embryonic system (host blastocyst), suggesting that additional factors are required to drive true germ cell differentiation *in vitro* (Niwa 2007).

In summary, pluripotent stem cells have the capacity to move away from pluripotency towards a special, restricted stem cell identity as a GLSC or germ cell. They can do this as a part of normal development, as in the case of flatworms or colonial ascidians, in disease conditions such as teratocarcinomas, and under culture conditions. Both in the latter case and in wild type development, upregulation of germ cell-specific genes such as *vasa* accompanies the transition.

Pluripotent Stem Cells from Dedicated Germ Cells

A number of lines of evidence suggest that not only can pluripotent stem cells reduce pluripotency to acquire germ cell or GLSC fate, but also that the reverse is true: somatic cells, GLSCs, PGCs, or even germ cells that have already begun to enter gametogenesis, can "backtrack" to acquire or increase pluripotency. Fruit fly germ cells that have already begun to differentate into oocytes can be induced to dedifferentiate and revert to a germ line stem cell identity (Fig. 1C) (Niki and Mahowald 2003). The most notable *in vitro* examples of this are cultured PGCs or GLSCs that give rise to embryonic germ (EG) cells, cultured ICM cells that give rise to ES cells, and iPS cells.

PGCs from the embryonic genital ridges, or GLSCs from adult testes, can acquire pluripotency if cultured with leukemia inhibitory factor (LIF), fibroblast growth factor (bFGF), and the mitogen or survival factor kit ligand (KL) (Dolci et al. 1991, Godin et al. 1991, Matsui et al. 1991). For ICM cells to make the transition to ES cells, they must be cultured with a feeder layer providing at least LIF and bone morphogenetic proteins (BMPs 2/4). However, the specific family and levels of BMP are critical, since addition of appropriate concentrations of BMP4 and/or BMP8b can cause failure to acquire pluripotency, and adoption of the more restricted fate of gametes (reviewed by Zhang and Li 2005). Finally, differentiated somatic cells of many kinds can be induced to acquire pluripotency by transfection or other delivery (Blelloch et al. 2007, Okita et al. 2008, Stadtfeld et al. 2008) of the so-called "Yamanaka factors," comprising a cocktail of transcription factors that promote pluripotency (Yamanaka et al. 2006, Takahashi et al. 2007, Yamanaka 2007).

Factors that are not directly dependent on transcriptional regulation have also been shown to influence the stem cell/germ cell decision. Extended mitosis of cultured cells inhibits germ cell differentiation, and is associated with a tendency towards continued self-renewal (Kimura et al. 2003). From an evolutionary point of view, this is consistent with positive selection acting on cells that have undergone fewer mutagenic events (mitoses) as founders of the germ line.

Increased pluripotency by vertebrate stem cells is often accompanied or regulated by the transcription factors Oct4 and Nanog (Chambers et al. 2003, Mitsui et al. 2003, Hatano et al. 2005, Lavial et al. 2007). These genes clearly play an important role in vertebrate stem cell biology. However, they are linage-restricted genes (unique to deuterostomes) (Booth and Holland 2004, Odintsova 2009). As a result, they cannot play a part in any evolutionary scenario concerning the transition, in a bilaterian ancestor, from pluripotent stem cell to dedicated PGC or GLSC.

Modifying Stem Cell Programs: From Generic to Germ Line-Specific

Many experiments aimed at achieving a molecular definition of "stemness" have suggested that the transcriptional regulatory landscape is largely similar between vertebrate stem cells and germ cells (Burns and Zon 2002, Ivanova et al. 2002, Evsikov and Solter 2003, Fortunel et al. 2003, Ramalho-Santos and Willenbring 2007, Sun et al. 2007). However, other studies have succeeded in identifying clear transcriptional and proteomic profile differences between GLSCs and ES cells (Sperger et al. 2003, Fujino et al. 2006, Kurosaki et al. 2007). Moreover, the upregulation of germline-specific genes in more "generic" stem cell types is correlated with a transition from somatic stem cell to GLSC *in vivo* (Sunanaga et al. 2006, Sunanaga et al. 2007), and has been shown to induce GC characteristics in somatic stem cells *in vitro* (Lavial et al. 2009). At the level of specific genes, several conserved molecules are expressed in both germ cells and all types of stem cells. These include Piwi family proteins, Tudor family proteins, and *PL10* gene products. Genes specifically upregulated in dedicated germlines, but downregulated in or absent from pluripotent stem cell types include the *vasa* family members, and possibly *nanos* (reviewed by Ewen-Campen et al. 2010).

In summary, the extensive molecular signatures and functional potential of germ cells and stem cells suggest a shared evolutionary origin for these cell types. A dedicated germ line is likely to have appeared before the divergence of Bilateria, but after the advent of animal multicellularity. This in turn suggests that germ cell-specific components would have been added to a pre-existing pluripotency network. An ancestral pluripotency network would have been responsive to BMP signalling. It would also have included members of Piwi and Tudor class proteins, which are conserved components of both germ plasm and stem cell cytoplasm across the metazoans (reviewed by Ewen-Campen et al. 2010). *PL10*, the ancestor of the *vasa* gene family, is expressed in both germ line and stem cell populations in extant metazoans (Mochizuki et al. 2001). It is therefore likely that *PL10*, rather than a true *vasa* family member, participated in an ancestral "germ cell regulatory module".

From Inductive Signalling to Germ Plasm Inheritance

The transition from "generic" stem cell to germ line stem cell would, by parsimony, have happened once in the lineage leading to the Bilateria. The ancestral mechanism of germ cell fate acquisition was likely to have been inductive signalling, possibly via BMP response as outlined above. Subsequently, however, an inheritance mechanism would have evolved independently in multiple branches of both protostomes and deuterostomes. Given that it has arisen in at least ten metazoan phyla (Extavour and Akam 2003, Extavour 2007a), some researchers have addressed the possible selective advantages of evolving inheritance mechanisms (Johnson et al. 2003, Crother et al. 2007); these issues will not be discussed here. Instead, the following section will consider molecular developmental mechanisms that could explain such convergence, given what we know about germ cell specification and development in extant groups.

The key to achieving the inheritance mode of germ line specification is the germ plasm. This is a special cytoplasm whose components are deposited during oogenesis. Thus, the developing progeny are dependent upon the fidelity of the mother's genome for proper germ cell formation. The components are then assembled (asymmetrically localized) into a special region of cytoplasm either before the end of oogenesis or immediately following fertilization. This specialized cytoplasm may have a diffuse appearance of loosely localized molecular components, often called nuage or simply germ plasm (coined by Weismann 1892, see for example Noda and Kanai 1977). Germ plasm can also display a compact appearance in the form of a discrete (though not membrane-bounded) organelle(s), which may be called germinal granules, Balbiani bodies, or oosomes (first described by Balbiani 1864, see for example Strome and Wood 1982, Gutzeit 1985). In the inheritance mode, germ plasm components are retained at the end of oogenesis and inherited by the embryo. Animals with an inductive mode of germ cell specification show similar cytoplasmic inclusions in germ cells once they acquire their fate, and during gametogenesis. However, in this mode, germ plasm components are cleared from the ooplasm before the end of oogenesis, leaving PGCs in the developing embryo to assemble their germ plasm *de novo*.

Heterochrony of Germ Plasm Component Expression or Localization

A comparison of asymmetrical germ plasm localization in different groups with the inheritance mode reveals that the molecular mechanisms necessary to assemble germ plasm are variable. Microtubule-driven localisation is critical for germ plasm assembly in *X. laevis* (Ressom and

Dixon 1988, Robb et al. 1996). In *C. elegans*, however, an actin-based mechanism controls the cytoplasmic flows that localize germ plasm components (called P granules) (Hird and White 1993, Hird 1996, Hird et al. 1996). In *D. rerio* both actinomyosin function and the mictrotubule cytoskeleton are implicated in germ plasm assembly (Pelegri et al. 1999, Knaut et al. 2000, Urven et al. 2006). Similarly, both microtubules and actin filaments have been shown to play a role in fruit fly germ plasm assembly (Erdelyi et al. 1995, Pokrywka and Stephenson 1995, Lantz et al. 1999, Jankovics et al. 2002, Zimyanin et al. 2008). The mechanisms driving germ plasm assembly in chickens are unknown (Tsunekawa et al. 2000).

The timing of localization also differs between groups. In fruit flies and zebrafish, germ plasm assembly begins before the end of oogenesis (Illmensee et al. 1976, Olsen et al. 1997), but in *C. elegans* asymmetric localisation of P granules takes place following fertilization (Strome and Wood 1982). *X. laevis* employs a two-step assembly mechanism, where some components are localized during oogenesis, and others following fertilization (reviewed by King et al. 2005).

In contrast to these differing mechanisms of assembly, the molecules that make up the germ plasm are remarkably conserved. The *vasa*, *tudor*, and *piwi* gene families discussed above, as well as *nanos* and *staufen* homologues, have been identified in all metazoans studied, and their gene products are germ plasm components. Strikingly, many of these gene products are not only localized to embryonic germ plasm and PGCs, but are also expressed during and required for gametogenesis. In the inheritance mode, PGCs therefore acquire these gene products directly from the ooplasm. In animals that use the inductive mode, however, these genes must be upregulated *de novo* in PGCs. An evolutionary switch from induction to inheritance could therefore be explained if a mechanism for stabilization, retention, and localization of germ plasm molecules during oogenesis were to evolve in some lineages. In other words, a heterochronic shift in the timing of regulation and/or localization of germ plasm genes could explain the repeated convergent evolution of the inheritance mode in metazoans (Extavour 2007a, Extavour 2007b).

Evidence for such a heterochronic regulatory change is provided by analysis of the evolution and expression of *vasa* and *PL10* homologues from an anthozoan cnidarian (Extavour et al. 2005). In the sea anemone *Nematostella vectensis*, the *vasa* locus has undergone a gene duplication event after diverging from the *PL10* ancestor. During embryogenesis, *PL10* and one of the *vasa* duplicates are not expressed during late oogenesis, and show zygotic expression in presumptive germ cells. The later-diverging paralogue, however, displays both maternal and zygotic expression. This suggests that following duplication the second *vasa* locus evolved regulatory mechanisms that allowed its prolonged expression during oogenesis, such that the transcript was available for cytoplasmic inheritance by cells in the early embryo.

Germ Plasm Nucleators

Conceptually, the easiest way to achieve germ plasm assembly at a given time would be to have a single molecule that was itself necessary and sufficient to nucleate germ plasm components. Under the inheritance model, cytoplasmic germ cell determinants are necessary and sufficient for specifying PGCs. Indeed, when germ plasm is removed or damaged by physical means, germ cell formation is either disrupted or eliminated (see for example Hegner 1908, Geigy 1931, Buehr and Blackler 1970). Conversely, when transplanted wholesale to ectopic locations, germ plasm can be sufficient to autonomously specify ectopic germ cells (Illmensee and Mahowald 1974, Okada et al. 1974, Illmensee and Mahowald 1976). The prediction for a necessary and sufficient "germ plasm nucleator" molecule would therefore be that it, too, would be able to drive PGC formation ectopically, and impede PGC formation when removed. Two metazoan genes are known whose products possess these properties.

Drosophila oskar *provides a solution to the localization problem*

Oskar (*osk*) was first identified in a screen for maternal effect genes on the third chromosome in *Drosophila melanogaster* (Lehmann and Nüsslein-Volhard 1986). *osk* mRNA accumulation and translation are localized to the posterior cytoplasm during oogenesis (Ephrussi et al. 1991, Kim-Ha et al. 1991, Kim-Ha et al. 1995). Its localization requires microtubules and the plus end-directed motor protein kinesin (Lehmann and Nüsslein-Volhard 1986, Brendza et al. 2000, Zimyanin et al. 2008). Localized translation of Osk protein is achieved both by activating translation of posteriorly localized *osk* transcripts and by inhibiting translation of unlocalized transcripts (Kim-Ha et al. 1995, Micklem et al. 2000, Chekulaeva et al. 2006, Klattenhoff et al. 2007, Klattenhoff and Theurkauf 2008). Osk thus displays the germ plasm localization pattern we would expect of a "germ plasm nucleator".

Functional studies have also shown that the genetic and biochemical properties of *osk* satisfy the necessity and sufficiency requirements for a germ plasm nucleator. Loss-of-function *oskar* mutants do not form germ cells (Lehmann and Nüsslein-Volhard 1986). Conversely, *osk* gene products can autonomously recruit germ plasm components, resulting in ectopic germ cells that are capable of functional gametogenesis (Ephrussi et al. 1991, Ephrussi and Lehmann 1992, Smith et al. 1992). However, *osk's* ability to ectopically assemble germ plasm depends on the presence of other germ plasm factors: ectopic *osk* cannot produce ectopic PGCs in *vas* or *tud* mutants (Ephrussi and Lehmann 1992). Accordingly, Osk protein has been shown to directly interact with Vasa protein, and with Staufen protein, another germ plasm component (Breitwieser et al. 1996). Osk also recruits *nanos* mRNA (Ephrussi et al. 1991, Smith et al. 1992, Kim-Ha et al. 1995), a third component of germ plasm that is also needed for posterior and abdominal patterning (Nüsslein-Volhard et al. 1987, Lehmann and Nusslein-Volhard 1991). These data are consistent with a model where the role of Osk is to recruit germ plasm components rather than to induce PGC fate directly.

Zebrafish bucky ball *as a germ plasm nucleator*

Until recently, *osk* was the only gene known to be both necessary and sufficient for germ plasm nucleation, and therefore for germ cell formation. However, a recent zebrafish screen for mutants affecting anterior-posterior polarity (Dosch et al. 2004) uncovered the gene *bucky ball* (*buc*), whose phenotype is strikingly similar to that of *oskar*'s. Like *oskar* transcripts during fly oogenesis, during zebrafish oogenesis *buc* transcripts are localised to the vegetal pole, where germ plasm begins to accumulate. Following fertilization, germ plasm becomes localised to the early cleavage furrows of the zebrafish embryos (Olsen et al. 1997, Yoon et al. 1997), as does Buc protein (Marlow and Mullins 2008). *buc* loss-of-function mutants do not form germ cells (Marlow and Mullins 2008), and germ plasm components, including transcripts of the *vas*, *dazl*, and *nos* genes, are not localized correctly (Bontems et al. 2009). *buc* appears to be not just sufficient, but also necessary for germ cell formation, as ectopic expression of *buc* in early embryonic cells that would normally not give rise to germ cells results in supernumerary PGCs. These ectopic cells are derived from the cells containing ectopic *buc*, and they localize germ plasm components and migrate to the gonad along with wild type PGCs (Bontems et al. 2009).

Commonalities Between Germ Plasm Nucleators

Given the data available for fruit fly *osk* and zebrafish *buc*, one might predict that in other groups using the inheritance mode of germ plasm specification, *osk* and *buc* homologues would provide the key to early germ plasm assembly. However, both of these genes appear to be recent evolutionary novelties within Diptera and Vertebrata, respectively. Moreover, both of these genes encode novel proteins with neither identifiable functional domains nor predictable secondary structure, making it difficult to understand how they are able to perform analogous biological functions. In the case of *oskar*, sequence and functional comparisons even within closely related species suggest that this gene is not only a relatively recent dipteran novelty, but is also evolving rapidly within the Diptera.

Functional comparison of dipteran oskar homologues

oskar homologues are identifiable in all 12 Drosophilid genomes and in three mosquito genomes (Fig. 2). Although a "consensus" Drosophilid sequence can be deduced, much of the protein is variable at the amino

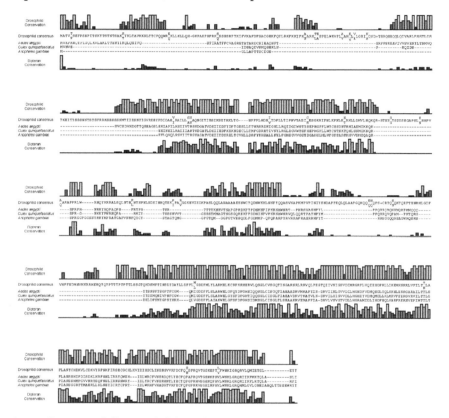

Fig. 2. Alignment of dipteran Osk homologs. An amino acid MAFFT (Katoh and Toh 2008) alignment of twelve Drosophilid and three probable mosquito *osk* homologs (identified via BLAST) has been condensed to the modal consensus Drosophilid *osk* sequence (top line) and the three mosquito proteins (bottom lines). Where two residues in a column are found to be most and equally frequent in the Drosophilid consensus sequence, both are presented; a "+" denotes when more than two residues are found to be equally frequent. Physico-chemical conservation (generated in Jalview v2.4 (Waterhouse et al. 2009)) among the twelve Drosophilid *osk* proteins is shown in the histogram above the alignment, and among all fifteen Dipteran proteins in the histogram below the alignment. Highly conserved positions are lighter and taller in the histogram, whereas poorly conserved positions are darker and shorter. Sequences (and accession numbers in parentheses) aligned are from *Drosophila melanogaster* (NP_731295), *D. sechellia* (XP_002031969), *D. simulans* (XP_002104196), *D. yakuba* (XP_002096875), *D. erecta* (XP_001980894), *D. ananassae* (XP_001953297), *D. persimilis* (XP_002017385), *D. pseudoobscura pseudoobscura* (XP_001359508), *D. virilis* (Q24741), *D. mojavensis* (XP_002000116), *D. grimshawi* (XP_001994345), *D. immigrans* (ABH12272), *Aedes aegypti* (XP_001656415), *Culex quinquefasciatus* (XP_001848641), and *Anopheles gambiae* (XP_313289).

acid level even when comparing species that are thought to have diverged as little as 12 million years ago (Tamura et al. 2004). Two large regions, an ~100 amino acid N-terminal domain, and a larger ~210 amino acid C-terminal domain, show a relatively high level of physico-chemical conservation within Drosophilids. Comparison with three probable mosquito homologues shows that these two domains are the areas of highest physico-chemical conservation across all compared dipterans. However, the extent of similarity decreases when the mosquito sequences are added to the analysis. Furthermore, the mosquito sequences are highly diverged from the Drosophilid sequences even in these regions. Finally, outside of these regions, the mosquito sequences are essentially not alignable with the Drosophilid sequences, due to significant insertions in the Drosophilid lineages.

Only two *oskar* homologues besides *D. melanogaster osk* have been assessed functionally. Cytoplasm from the posterior (pole plasm) of the early *D. immigrans* embryo can induce ectopic, functional germ cells when introduced into *D. melanogaster* embryos (Mahowald et al. 1976). This *D. immigrans* pole plasm potential appears to be exclusively due to the *D. immigrans oskar* homologue (*immosk*), which can rescue both the posterior patterning and germ cell formation defects of *D. melanogaster oskar* loss of function mutants (Jones and Macdonald 2007). However, the morphology of *D. immigrans* germ plasm is very distinct from that of *D. melanogaster* (Mahowald 1962, Mahowald 1968), suggesting that *immosk* interacts with its germ plasm binding partners differently in the two species. However, the known binding partners of Osk, including Vasa, are very highly conserved across the metazoans at the amino acid level (Fig. 3) (Mochizuki et al. 2001, Extavour et al. 2005), much more so than the Osk proteins. This provides an interpretation for the observation that when *immosk* replaces endogenous *osk* in *D. melanogaster* embryos, the morphology of the resulting germ plasm matches that of *D. immigrans*, rather than that of *D. melanogaster* (Jones and Macdonald 2007). We hypothesize that despite high conservation of most germ plasm components, germ plasm morphology differs between these two species because the amino acid changes between the two Osk proteins affect specific molecular interactions with their conserved binding partners. Since diverging from their last common ancestor 30-40 million years ago (Spicer 1988, Russo et al. 1995, Remsen and O'Grady 2002), the evolutionary changes in *D. melanogaster* and *D. immigrans osk* have been sufficient to change Osk protein's specific physical interactions with germ plasm components. However, the changes have not been sufficient to disrupt these interactions altogether, since both Osk proteins can provide posterior patterning (presumably via *nanos* mRNA localization) and germ plasm assembly.

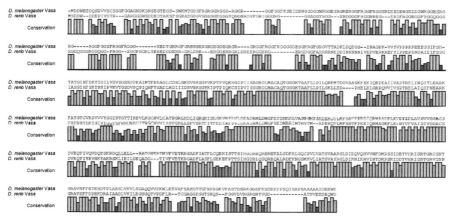

Fig. 3. Vasa proteins show clear homology between *D. melanogaster* and *D. rerio.* Needleman-Wunsch global pairwise alignment of *Drosophila melanogaster* and *Danio rerio* Vasa proteins is presented with physico-chemical conservation (generated in Jalview v2.4 (Waterhouse et al. 2009)) at each residue in the histogram below the alignment. Histogram interpretation is as in Figure 2. Sequences (and accession numbers in parentheses) aligned are from *Drosophila melanogaster* (P25158) and *Danio rerio* (A1L1Z2).

A similar divergence time separates *D. melanogaster* from *D. virilis* (Remsen and O'Grady 2002, Tamura et al. 2004). Similar to *immosk*, *D. virilis oskar* (*virosk*) is able to rescue the posterior and abdominal patterning defects exhibited by *D. melanogaster osk* loss of function mutants (Webster et al. 1994). However, *virosk* cannot maintain *osk* mRNA levels in the germ plasm, and cannot rescue the germ cell formation phenotype of *D. melanogaster osk* mutants. Furthermore, when introduced into wild type *D. melanogaster* (with two functioning copies of endogenous *osk*), *virosk* disrupts endogenous *osk* localization, and induces dominant maternal-effect lethality. These observations indicate that significant changes in *osk* sequence and function occurred in the 30-60 million years separating *D. virilis* and *D. melanogaster* from their last common ancestor (Remsen and O'Grady 2002, Tamura et al. 2004). Specifically, an Osk protein ancestral to those in *D. virilis* and *D. melanogaster* may have had the ability to localize *nanos* mRNA, thus ensuring posterior and abdominal patterning, but not the ability to bind and nucleate other germ plasm components. This hypothesis is consistent with the absence of pole cells (PGCs arising early in embryogenesis) reported for some lower dipterans (Rohr et al. 1999), which may either lack an *osk* homologue altogether, or lack one with the ability to nucleate germ plasm.

Functional similarity between buc *and* osk *cannot be explained by homology*

More data on the function of other dipteran *osk* homologues are clearly needed to broaden our understanding of a gene that has played a key role in the evolution of the inheritance mode of germ cell specification. *bucky ball*, the zebrafish "solution" to the inheritance mode problem seems to be a functional analogue of *D. melanogster oskar*. However, there is no evidence that *buc* and *osk* are homologues in the classical sense of common descent from an shared ancestral genomic sequence (Remane 1952). Although comparing the two sequences may appear to show some physico-chemical similarity in a central domain (Fig. 4), these predictions are in fact artefacts of a pairwise comparison between two sequences that cannot confidently be aligned.

The conundrum of these two non-homologous proteins that interact with homologous binding partners to achieve an analogous biological function may be explained by similarities at a level higher than amino acid sequence. The observed molecular interactions of Osk with Vasa and Staufen proteins, and the hypothesized interactions of Buc with the zebrafish homologues of Osk and Vasa, will be governed by the specific biochemical properties of Osk and Buc, including tertiary structure. If three-dimensional structure is similar for both Osk and Buc, conserved molecular interactions with other (homologous) germ plasm components could explain their analogous roles in germ cell specification. Such a finding would mean that the convergent evolution driving acquisition of the inheritance mode in flies and fish was due to selection acting on the level of protein structure, rather than at the nucleotide or amino acid sequence level.

There is evidence from other systems that selective pressures acting on tertiary structure can result in highly similar biological functions in the absence of demonstrable genetic homology. The relationship between the bacterial protein MreB and eukaryotic Actin proteins provides one such example. These genes are domain-specific: *actin* genes are specific to eukaryotes, while *mreB* homologues are found in several bacterial species (Doi et al. 1988, Jones et al. 2001). Primary sequence alignment between Actin and MreB homologues (Jones et al. 2001) does not reveal greater amino acid identities, or longer regions of amino acid similarity, than an Osk and Buc alignment (Fig. 4). At the amino acid level, there is little identity, and there are very limited regions of physico-chemical similarity (Jones et al. 2001). However, structural analysis of MreB and Actin has shown that there is extensive tertiary similarity between these two proteins, such that their crystal structure models are nearly identical (van den Ent et al. 2001). MreB has also been shown to polymerize into filaments close to the cell cortex, similar to Actin (van den Ent et al. 2001).

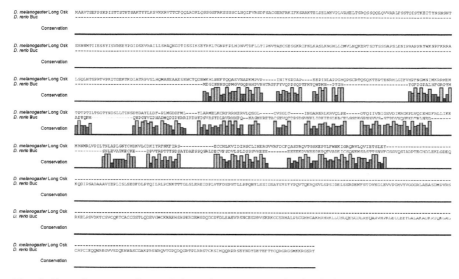

Fig. 4. *D. melanogaster* Long Osk and *Danio rerio* bucky ball are not homologues.
Needleman-Wunsch global pairwise alignment of *Drosophila melanogaster* Long Osk and
Danio rerio Buc proteins is presented with physico-chemical conservation (generated in
Jalview v2.4 (Waterhouse et al. 2009)) at each residue in the histogram below the alignment.
Histogram interpretation is as in Figure 2. Sequences (and accession numbers in parentheses)
aligned are from *Drosophila melanogaster* (NP_731295) and *Danio rerio* (XP_688879).

Evolving Early Germ Plasm Assembly: Selective Pressure on Tertiary Structure

While phylogenetic analysis suggests that *mreB* and *actin* genes are not
true genetic homologues, many have argued that their striking structural
similarities argue for cryptic homology (Kabsch and Holmes 1995, Egelman
2001, van den Ent et al. 2001, Doolittle and York 2002). Under this scenario,
selective pressures would have maintained tertiary structure, and therefore
biochemical function, by permitting extensive amino acid changes as long
as 3D structure was sufficiently unchanged. The accumulated amino acid
changes are predicted to be so great that the true homology relationships
of the descendant proteins in extant groups are obscured. New approaches
to understanding how evolution acts on protein structure and function
may be needed to shed light on this problem (Halabi et al. 2009).

While we cannot definitively rule out the possibility that *bucky ball*
and *oskar* are cryptic evolutionary homologues, their primary structures
provide no evidence in favor of such a hypothesis. This leads us to suggest
an intriguing scenario to explain the convergent evolution of maternal or
early embryonic germ plasm assembly, and therefore of the inheritance
mode of germ cell specification. We hypothesize that the driving force
behind this convergence was selective pressure on the tertiary structure

of germ plasm nucleators, such that they were able to bind pre-existing, conserved germ plasm factors. The genes encoding these factors, such as *vasa*, *nanos*, and *staufen*, are clearly ancient in metazoans and predate the evolution of the inheritance mode.

Summary

The germ cell lineage of extant metazoans is hypothesized to share ancestry with other pluripotent stem cell populations. An ancestral stem cell lineage with the capacity to give rise to both somatic and gametogenic cells, would have undergone modifications to its genetic regulatory program such that it acquired the specialization characteristic of the germ line. These modifications included changes to previously generic pluripotency regulators, such as *PL10* family genes and *piwi* class genes. *PL10* family modifications resulted in the *vasa* family of genes. To test this hypothesis, assessment of the physico-chemical properties and functional abilities with respect to gametogenesis and PGC development should be carried out on bilaterian *vasa* homologues and *PL10* homologues from bilaterian outgroups.

In the evolution of multicellularity, all divisions of labor in the form of specialization of cell types, must have been the result of cell-cell signalling (Pires-daSilva and Sommer 2003). This is consistent with the hypothesis that the inductive signalling mode of germ cell specification is ancestral to the metazoans. Early germ line genetic regulatory networks may have been responsive to signalling from BMP family members. Indeed, germ cell and stem cell function in extant animals can be regulated by BMP family members even in animals using an inheritance mode to specify germ cells (Chen and McKearin 2003, Kawase et al. 2004, Pan et al. 2007, Guo and Wang 2009, Rhiner et al. 2009, Wilkinson et al. 2009), or in cultured cells undergoing an *in vitro* switch from pluripotency to immortality or from pluripotency to gamete production (reviewed by Zhang and Li 2005). Further studies of the possible involvement of BMPs in specifying germ cells in animals with no evidence for early germ plasm will help provide support for or against this hypothesis. Similarly, testing the responsiveness of germ cells to BMP signalling in animals that use the inheritance mode will inform our understanding of the evolution of signals that induce germ cell formation.

A transition from an ancestral inductive mode to a comparatively derived inheritance mode occurred in multiple lineages of the bilaterian radiation. Few similarities are observed in the molecular mechanisms used by different animals to achieve early germ cell specification. *Drosophila oskar* and zebrafish *bucky ball* gene products are the only examples to date of molecules that are both necessary and sufficient for germ cell formation.

The lack of homology between these two genes stands in contrast to the high conservation of the germ plasm factors with which they interact. We therefore hypothesize that selective pressure at the level of tertiary structure of germ plasm nucleator molecules was the driving force behind the convergent evolution of the inheritance mode in these two clades. Comparative structural, biochemical, and functional studies of these two proteins will be needed to evaluate this hypothesis.

Acknowledgements

This work was supported by funding from the National Science Foundation (NSF IOS-0817678).

References

Abedin, M. and N. King. 2008. The premetazoan ancestry of cadherins. Science 319: 946–948.

Abzhanov, A. and M. Protas, B.R. Grant, P.R. Grant, and C.J. Tabin. 2004. Bmp4 and morphological variation of beaks in Darwin's finches. Science 305: 1462–1465.

Abzhanov, A. and W.P. Kuo, C. Hartmann, B.R. Grant, P.R. Grant, and C.J. Tabin. 2006. The calmodulin pathway and evolution of elongated beak morphology in Darwin's finches. Nature 442: 563–567.

Agata, K. and E. Nakajima, N. Funayama, N. Shibata, Y. Saito, and Y. Umesono. 2006. Two different evolutionary origins of stem cell systems and their molecular basis. Sem. Cell Dev. Biol. 17: 503–509.

Amabile, G. and A. Meissner. 2009. Induced pluripotent stem cells: current progress and potential for regenerative medicine. Trends Mol. Med. 15: 59–68.

Bachtrog, D. 2008. Evidence for male-driven evolution in *Drosophila*. Mol. Biol. Evol. 25: 617–619.

Baguñà, J. 1981. Planarian neoblasts. Nature 290: 14–15.

Balbiani, M. 1864. Sur la constitution du germe dans l'oeuf animal avant la fecondation. Comptes Rendus de l Academie des Sciences. Serie III, Sciences de la Vie 58: 584–588.

Ban, N. and M. Kai. 2009. Implication of replicative stress-related stem cell ageing in radiation-induced murine leukaemia. Br. J. Cancer 101: 363–371.

Bauer, V.L. and C.F. Aquadro. 1997. Rates of DNA sequence evolution are not sex-biased in *Drosophila melanogaster* and *D. simulans*. Mol. Biol. Evol. 14: 1252–1257.

Berrill, N.J. 1941. The development of the bud in *Botryllus*. Biol. Bull. 80: 169–184.

Blelloch, R. and M. Venere, J. Yen, and M. Ramalho-Santos. 2007. Generation of induced pluripotent stem cells in the absence of drug selection. Cell Stem Cell 1: 245–247.

Bontems, F. and A. Stein, F. Marlow, J. Lyautey, T. Gupta, M.C. Mullins, and R. Dosch. 2009. Bucky Ball Organizes Germ Plasm Assembly in Zebrafish. Curr. Biol. 19: 414–422.

Booth, H.A.F. and P.W.H. Holland. 2004. Eleven daughters of NANOG. Genomics 84: 229–238.

Boyer, B.C. and J.J. Henry, and M.Q. Martindale. 1998. The cell lineage of a polyclad turbellarian embryo reveals close similarity to coelomate spiralians. Dev. Biol. 204: 111–123.

Breitwieser, W. and F.-H. Markussen, H. Horstmann, and A. Ephrussi. 1996. Oskar protein interaction with Vasa represents an essential step in polar granule assembly. Genes & Development 10: 2179–2188.

Brendza, R.P. and L.R. Serbus, J.B. Duffy, and W.M. Saxton. 2000. A function for kinesin I in the posterior transport of *oskar* mRNA and Staufen protein. Science 289: 2120–2122.

Brown, F.D. and S. Tiozzo, M.M. Roux, K. Ishizuka, B.J. Swalla, and A.W. De Tomaso. 2009. Early lineage specification of long-lived germline precursors in the colonial ascidian *Botryllus schlosseri*. Development 136: 3485–3494.

Browne, W.E. and A.L. Price, M. Gerberding, and N.H. Patel. 2005. Stages of embryonic development in the amphipod crustacean, *Parhyale hawaiensis*. Genesis 42: 124–149.

Buehr, M. and A.W. Blackler. 1970. Sterility and partial sterility in the South African clawed toad following the pricking of the egg. J. Embryol. Exp. Morphol. 23: 375–384.

Burns, C.E. and L.I. Zon. 2002. Portrait of a stem cell. Dev. Cell. 3: 612–613.

Buss, L.W. 1982. Somatic cell parasitism and the evolution of somatic tissue compatibility. Proc. Natl. Acad. Sci. USA 79: 5337–5341.

Buss, L.W. 1987. The Evolution of Individuality. Princeton University Press, Princeton.

Callebaut, M. 2008. Historical evolution of preformistic versus neoformistic (epigenetic) thinking in embryology. Belg. J. Zool. 138: 20–35.

Campos-Ortega, J.A. and V. Hartenstein 1985. The Embryonic Development of *Drosophila melanogaster*. Springer-Verlag, Jena.

Chambers, I. and D. Colby, M. Robertson, J. Nichols, S. Lee, S. Tweedie, and A. Smith. 2003. Functional expression cloning of Nanog, a pluripotency sustaining factor in embryonic stem cells. Cell 113: 643–655.

Chang, B.H. and L.C. Shimmin, S.K. Shyue, D. Hewett-Emmett, and W.H. Li. 1994. Weak male-driven molecular evolution in rodents. Proc. Natl. Acad. Sci. USA 91: 827–831.

Chekulaeva, M. and M.W. Hentze, and A. Ephrussi. 2006. Bruno acts as a dual repressor of *oskar* translation, promoting mRNA oligomerization and formation of silencing particles. Cell 124: 521–533.

Chen, D. and D. McKearin. 2003. Dpp signaling silences *bam* transcription directly to establish asymmetric divisions of germline stem cells. Curr. Biol. 13: 1786–1791.

Chiquoine, A.D. 1954. The identification, origin, and migration of the primordial germ cells in the mouse embryo. Anat. Rec. 2: 135–146.

Cinalli, R.M. and P. Rangan, and R. Lehmann. 2008. Germ cells are forever. Cell 132: 559–562.

Clark, J.M. and E.M. Eddy. 1975. Fine structural observations on the origin and associations of primordial germ cells of the mouse. Dev. Biol. 47: 136–155.

Crother, B.I. and M.E. White, and A.D. Johnson. 2007. Inferring developmental constraint and constraint release: Primordial germ cell determination mechanisms as examples. J. Theor. Biol. 248: 322–330.

Darwin, C. 1859. On the Origin of Species by Means of Natural Selection. Grant Richards, London.

de Sousa Lopes, S.M. and B.A. Roelen, R.M. Monteiro, R. Emmens, H.Y. Lin, E. Li, K.A. Lawson, and C.L. Mummery. 2004. BMP signaling mediated by ALK2 in the visceral endoderm is necessary for the generation of primordial germ cells in the mouse embryo. Genes Dev. 18: 1838–1849.

Dickinson, H.G. and R. Grant-Downton. 2009. Bridging the generation gap: flowering plant gametophytes and animal germlines reveal unexpected similarities. Biol. Rev. Camb. Philos. Soc. 84: 589–615.

Doi, M. and M. Wachi, F. Ishino, S. Tomioka, M. Ito, Y. Sakagami, A. Suzuki, and M. Matsuhashi. 1988. Determinations of the DNA sequence of the mreB gene and of the gene products of the mre region that function in formation of the rod shape of *Escherichia coli* cells. J. Bacteriol. 170: 4619–4624.

Dolci, S. and D.E. Williams, M.K. Ernst, J.L. Resnick, C.I. Brannan, L.F. Lock, S.D. Lyman, H.S. Boswell, and P.J. Donovan. 1991. Requirement for mast cell growth factor for primordial germ cell survival in culture. Nature 352: 809–811.

Donnell, D.M. and L.S. Corley, G. Chen, and M.R. Strand. 2004. Caste determination in a polyembryonic wasp involves inheritance of germ cells. Proc. Natl. Acad. Sci. USA 101: 10095–10100.

Donoghue, P.C. and A. Graham, and R.N. Kelsh. 2008. The origin and evolution of the neural crest. Bioessays 30: 530–541.

Donovan, P.J. 1998. The germ cell—the mother of all stem cells. Int. J. Dev. Biol. 42: 1043–1050.

Doolittle, R.F. and A.L. York. 2002. Bacterial actins? An evolutionary perspective. BioEssays 24: 293–296.

Dosch, R. and D.S. Wagner, K.A. Mintzer, G. Runke, A.P. Wiemelt, and M.C. Mullins. 2004. Maternal control of vertebrate development before the midblastula transition: mutants from the zebrafish I. Dev. Cell. 6: 771–780.

Douarin, N.L. and C. Kalcheim. 2009. The Neural Crest. Cambridge University Press, Cambridge.

Drost, J.B. and W.R. Lee. 1998. The developmental basis for germline mosaicism in mouse and *Drosophila melanogaster*. Genetica 102–103: 421–443.

Dunn, C.W. and A. Hejnol, D.Q. Matus, K. Pang, W.E. Browne, S.A. Smith, E. Seaver, G.W. Rouse, M. Obst, G.D. Edgecombe, M.V. Sorensen, S.H. Haddock, A. Schmidt-Rhaesa, A. Okusu, R.M. Kristensen, W.C. Wheeler, M.Q. Martindale, and G. Giribet. 2008. Broad phylogenomic sampling improves resolution of the animal tree of life. Nature 452: 745–749.

Egelman, E.H. 2001. Molecular evolution: actin's long lost relative found. Curr. Biol. 11: R1022–1024.

Ellegren, H. 2007. Characteristics, causes and evolutionary consequences of male-biased mutation. Proc. R. Soc. Lond. B. Biol. Sci. 274: 1–10.

Ephrussi, A. and L.K. Dickinson, and R. Lehmann. 1991. Oskar organizes the germ plasm and directs localization of the posterior determinant *nanos*. Cell 66: 37–50.

Ephrussi, A. and R. Lehmann. 1992. Induction of germ cell formation by *oskar*. Nature 358: 387–392.

Erdelyi, M. and A.M. Michon, A. Guichet, J.B. Glotzer, and A. Ephrussi. 1995. Requirement for *Drosophila* cytoplasmic tropomyosin in *oskar* mRNA localization. Nature 377: 524–527.

Evans, M.J. and M.H. Kaufman. 1981. Establishment in culture of pluripotential cells from mouse embryos. Nature 292: 154–156.

Evsikov, A.V. and D. Solter. 2003. Comment on " 'Stemness': transcriptional profiling of embryonic and adult stem cells" and "a stem cell molecular signature". Science 302: 393.

Ewen-Campen, B. and E.E. Schwager, and C. Extavour. 2009. The molecular machinery of germ line specification. Mol. Reprod. Dev. 77: 3–18.

Extavour, C. and A. García-Bellido. 2001. Germ cell selection in genetic mosaics in *Drosophila melanogaster*. Proc. Natl. Acad. Sci. USA 98: 11341–11346.

Extavour, C. and M.E. Akam. 2003. Mechanisms of germ cell specification across the metazoans: epigenesis and preformation. Development 130: 5869–5884.

Extavour, C. and K. Pang, D.Q. Matus, and M.Q. Martindale. 2005. *vasa* and *nanos* expression patterns in a sea anemone and the evolution of bilaterian germ cell specification mechanisms. Evol. Dev. 7: 201–215.

Extavour, C.G. 2005. The fate of isolated blastomeres with respect to germ cell formation in the amphipod crustacean *Parhyale hawaiensis*. Dev. Biol. 277: 387–402.

Extavour, C. 2007a. Evolution of the bilaterian germ line: lineage origin and modulation of specification mechanisms. Integr. Comp. Biol. 47: 770–785.

Extavour, C. Urbisexuality: The evolution of bilaterian germ cell specification and reproductive systems. pp. 317–338. *In*: A. Minelli, and G. Fusco. [eds.] 2007b. Evolving Pathways: Key Themes in Evolutionary Developmental Biology. Cambridge University Press, Cambridge.

Falleni, A. and V. Gremigni. 1990. Ultrastructural study of oogenesis in the acoel turbellarian *Convoluta*. Tissue Cell 22: 301–310.

Foe, V.E. and B.M. Alberts. 1983. Studies of nuclear and cytoplasmic behaviour during the five mitotic cycles that precede gastrulation in *Drosophila* embryogenesis. J. Cell Sci. 61: 31–70.

Fortunel, N.O. and H.H. Otu, H.H. Ng, J. Chen, X. Mu, T. Chevassut, X. Li, M. Joseph, C. Bailey, J.A. Hatzfeld, A. Hatzfeld, F. Usta, V.B. Vega, P.M. Long, T.A. Libermann, and B. Lim. 2003. Comment on " 'Stemness': transcriptional profiling of embryonic and adult stem cells" and "a stem cell molecular signature". Science 302: 393.

Freitas, R. and G. Zhang, and M.J. Cohn. 2006. Evidence that mechanisms of fin development evolved in the midline of early vertebrates. Nature 442: 1033–1037.

Fujimura, M. and K. Takamura. 2000. Characterization of an ascidian DEAD-box gene, *Ci-DEAD1*: specific expression in the germ cells and its mRNA localization in the posterior-most blastomeres in early embryos. Dev. Genes Evol. 210: 64–72.

Fujino, R.S. and Y. Ishikawa, K. Tanaka, M. Kanatsu-Shinohara, K. Tamura, H. Kogo, T. Shinohara, and T. Hara. 2006. Capillary morphogenesis gene (CMG)-1 is among the genes differentially expressed in mouse male germ line stem cells and embryonic stem cells. Mol. Reprod. Dev. 73: 955–966.

García-Bellido, A. and P. Ripoll, and G. Morata. 1973. Developmental compartmentalisation of the wing disc of *Drosophila*. Nature New Biology 245: 251–253.

Geigy, R. 1931. Action de l'ultra-violet sur le pôle germinal dans l'oef de *Drosophila melanogaster*. Rev. Suisse Zool. 38: 187–288.

Gilbert, S.F. and S.R. Singer 2006. Developmental Biology. Sinauer Associates, Inc., Sunderland, MA.

Ginsburg, M. and M.H.L. Snow, and A. McLaren. 1990. Primordial germ cells in the mouse embryo during gastrulation. Development pp. 521–528.

Godin, I. and R. Deed, J. Cooke, K. Zsebo, M. Dexter, and C.C. Wylie. 1991. Effects of the *steel* gene product on mouse primordial germ cells in culture. Nature 352: 807–809.

Graham, C.F. Teratocarcinoma cells and normal mouse embryogenesis. pp. 315–394. *In*: Sherman M.I. [ed.] 1977. Concepts in Mammalian Embryogenesis. MIT Monograph, Cambridge.

Grant, P.R. and B.R. Grant, J.N. Smith, I.J. Abbott, and L.K. Abbott. 1976. Darwin's finches: population variation and natural selection. Proc. Natl. Acad. Sci. USA 73: 257–261.

Grant, P.R. 1999. Ecology and Evolution of Darwin's Finches. Princeton University Press, Princeton, NJ.

Grbic, M. and L.M. Nagy, and M.R. Strand. 1998. Development of polyembryonic insects: a major departure from typical insect embryogenesis. Dev. Genes Evol. 208: 69–81.

Gschwentner, R. and P. Ladurner, K. Nimeth, and R. Rieger. 2001. Stem cells in a basal bilaterian. S-phase and mitotic cells in *Convolutriloba longifissura* (Acoela, Platyhelminthes). Cell Tiss. Res. 304: 401–408.

Guan, K. and K. Nayernia, L.S. Maier, S. Wagner, R. Dressel, J.H. Lee, J. Nolte, F. Wolf, M. Li, W. Engel, and G. Hasenfuss. 2006. Pluripotency of spermatogonial stem cells from adult mouse testis. Nature 440: 1199–1203.

Guo, T. and A.H. Peters, and P.A. Newmark. 2006. A Bruno-like gene is required for stem cell maintenance in planarians. Dev. Cell. 11: 159–169.

Guo, Z. and Z. Wang. 2009. The glypican Dally is required in the niche for the maintenance of germline stem cells and short-range BMP signaling in the *Drosophila* ovary. Development 136: 3627–3635.

Gutzeit, H.O. 1985. Oosome formation during in vitro oogenesis in *Bradysia tritici* (syn. *Sciara ocellaris*). Roux' Arch. Dev. Biol. 194: 404–410.

Haig, D. 2004. The (dual) origin of epigenetics. Cold Spring Harbor Symp. Quant. Biol. 69: 67–70.

Halabi, N. and O. Rivoire, S. Leibler, and R. Ranganathan. 2009. Protein sectors: evolutionary units of three-dimensional structure. Cell 138: 774–786.

Haldane, J.B.S. 1935. The rate of spontanous mutation of a human gene. J. Genet. 31: 317–326.

Haldane, J.B.S. 1947. The mutation rate of the gene for haemophilia and its segregation ratios in males and females. Ann. Eugenics 13: 262–271.

Hatano, S.Y. and M. Tada, H. Kimura, S. Yamaguchi, T. Kono, T. Nakano, H. Suemori, N. Nakatsuji, and T. Tada. 2005. Pluripotential competence of cells associated with Nanog activity. Mech. Dev. 122: 67–79.

Hay, B. and L. Ackerman, S. Barbel, L.Y. Jan, and Y.N. Jan. 1988. Identification of a component of *Drosophila* polar granules. Development 103: 625–640.

Hayashi, K. and S.M. de Sousa Lopes, and M.A. Surani. 2007. Germ cell specification in mice. Science 316: 394–396.

Hayashi, K. and M.A. Surani. 2009. Resetting the epigenome beyond pluripotency in the germline. Cell Stem Cell 4: 493–498.

Heerema-McKenney, A. and M.R. Harrison, B. Bratton, J. Farrell, and C. Zaloudek. 2005. Congenital teratoma: a clinicopathologic study of 22 fetal and neonatal tumors. Am. J. Surg. Pathol. 29: 29–38.

Hegner, R.W. 1908. Effects of removing the germ-cell determinants from the eggs of some chrysomelid beetles. Preliminary report. Biol. Bull. 16: 19–26.

Hejnol, A. and M.Q. Martindale. 2008. Acoel development supports a simple planula-like urbilaterian. Philos. Trans. R. Soc. Lond. B. Biol. Sci. 363: 1493–1501.

Hejnol, A. and M. Obst, A. Stamatakis, M. Ott, G.W. Rouse, G.D. Edgecombe, P. Martinez, J. Baguna, X. Bailly, U. Jondelius, M. Wiens, W.E. Muller, E. Seaver, W.C. Wheeler, M.Q. Martindale, G. Giribet, and C.W. Dunn. 2009a. Assessing the root of bilaterian animals with scalable phylogenomic methods. Proc. R. Soc. Lond. B. Biol. Sci. 276: 4261–4270.

Hejnol, A. and M. Obst, A. Stamatakis, M. Ott, G.W. Rouse, G.D. Edgecombe, P. Martinez, J. Baguna, X. Bailly, U. Jondelius, M. Wiens, W.E. Muller, E. Seaver, W.C. Wheeler, M.Q. Martindale, G. Giribet, and C.W. Dunn. 2009b. Assessing the root of bilaterian animals with scalable phylogenomic methods. Proc. Biol. Sci. 276: 4261–4270.

Henry, J.Q. and M.Q. Martindale, and B.C. Boyer. 2000. The unique developmental program of the acoel flatworm, *Neochildia fusca*. Dev. Biol. 220: 285–295.

Hird, S.N. and J.G. White. 1993. Cortical and cytoplasmic flow polarity in early embryonic cells of *Caenorhabditis elegans*. J. Cell Biol. 121: 1343–1355.

Hird, S. 1996. Cortical actin movements during the first cell cycle of the *Caenorhabditis elegans* embryo. J. Cell Sci. 109: 525–533.

Hird, S.N. and J.E. Paulsen, and S. Strome. 1996. Segregation of germ granules in living *Caenorhabditis elegans* embryos: cell-type-specific mechanisms for cytoplasmic localisation. Development 122: 1303–1312.

Hoei-Hansen, C.E. and A. Sehested, M. Juhler, Y.-F.C. Lau, N.E. Skakkebaek, H. Laursen, and E. Rajpert-de Meyts. 2006. New evidence for the origin of intracranial germ cell tumours from primordial germ cells: expression of pluripotency and cell differentiation markers. J. Pathol. 209: 25–33.

Hu, D. and R.S. Marcucio, and J.A. Helms. 2003. A zone of frontonasal ectoderm regulates patterning and growth in the face. Development 130: 1749–1758.

Hübner, K. and G. Fuhrmann, L.K. Christenson, J. Kehler, R. Reinbold, R. De La Fuente, J. Wood, J.F. Strauss, 3rd, M. Boiani, and H.R. Scholer. 2003. Derivation of oocytes from mouse embryonic stem cells. Science 300: 1251–1256.

Huettner, A.F. 1923. The origin of the germ cells in *Drosophila melanogaster*. J. Morphol. 2: 385–422.

Humphrey, R.R. 1929. The early position of the primordial germ cells in Urodeles: evidence from experimental studies. Anat. Rec. 42: 301–314.

Illmensee, K. and A.P. Mahowald. 1974. Transplantation of posterior polar plasm in *Drosophila*. Induction of germ cells at the anterior pole of the egg. Proc. Natl. Acad. Sci. USA 4: 1016–1020.

Illmensee, K. and A.P. Mahowald. 1976. The autonomous function of germ plasm in a somatic region of the *Drosophila* egg. Exp. Cell Res. 97: 127–140.

Illmensee, K. and A.P. Mahowald, and M.R. Loomis. 1976. The ontogeny of germ plasm during oogenesis in *Drosophila*. Dev. Biol. 49: 40–65.

Ivanova, N.B. and J.T. Dimos, C. Schaniel, J.A. Hackney, K.A. Moore, and I.R. Lemischka. 2002. A stem cell molecular signature. Science 298: 601–604.

Jankovics, F. and R. Sinka, T. Lukacsovich, and M. Erdelyi. 2002. MOESIN crosslinks actin and cell membrane in Drosophila oocytes and is required for OSKAR anchoring. Curr. Biol. 12: 2060–2065.

Jeffery, W.R. and A.G. Strickler, and Y. Yamamoto. 2004. Migratory neural crest-like cells form body pigmentation in a urochordate embryo. Nature 431: 696–699.

Jeffery, W.R. and T. Chiba, F.R. Krajka, C. Deyts, N. Satoh, and J.S. Joly. 2008. Trunk lateral cells are neural crest-like cells in the ascidian *Ciona intestinalis*: insights into the ancestry and evolution of the neural crest. Dev. Biol. 324: 152–160.

Johnson, A.D. and M. Drum, R.F. Bachvarova, T. Masi, M.E. White, and B.I. Crother. 2003. Evolution of predetermined germ cells in vertebrate embryos: implications for macroevolution. Evol. Dev. 5: 414–431.

Jones, L.J. and R. Carballido-López, and J. Errington. 2001. Control of cell shape in bacteria: helical, actin-like filaments in *Bacillus subtilis*. Cell 104: 913–922.

Jones, J.R. and P.M. Macdonald. 2007. Oskar controls morphology of polar granules and nuclear bodies in *Drosophila*. Development 134: 233–236.

Kabsch, W. and K.C. Holmes. 1995. The actin fold. FASEB J. 9: 167–174.

Kahan, B.W. and B. Ephrussi. 1970. Developmental potentialities of clonal in vitro cultures of mouse testicular teratoma. J. Nat. Cancer Inst. 44: 1015–1036.

Kanatsu-Shinohara, M. and N. Ogonuki, K. Inoue, A. Ogura, S. Toyokuni, T. Honjo, and T. Shinohara. 2003. Allogeneic offspring produced by male germ line stem cell transplantation into infertile mouse testis. Biol. Reprod. 68: 167–173.

Kanatsu-Shinohara, M. and K. Inoue, J. Lee, M. Yoshimoto, N. Ogonuki, H. Miki, S. Baba, T. Kato, Y. Kazuki, S. Toyokuni, M. Toyoshima, O. Niwa, M. Oshimura, T. Heike, T. Nakahata, F. Ishino, A. Ogura, and T. Shinohara. 2004. Generation of pluripotent stem cells from neonatal mouse testis. Cell 119: 1001–1012.

Katoh, K. and H. Toh. 2008. Recent developments in the MAFFT multiple sequence alignment program. Brief. Bioinform. Q: 286–298.

Kawamura, K. and T. Sunanaga. 2010. Hemoblasts in colonial tunicates: are they stem cells or tissue-restricted progenitor cells? Dev. Growth Differ. 52: 69–76.

Kawase, E. and M.D. Wong, B.C. Ding, and T. Xie. 2004. Gbb/Bmp signaling is essential for maintaining germline stem cells and for repressing *bam* transcription in the *Drosophila* testis. Development 131: 1365–1375.

Kee, Y. and B.J. Hwang, P.W. Sternberg, and M. Bronner-Fraser. 2007. Evolutionary conservation of cell migration genes: from nematode neurons to vertebrate neural crest. Genes Dev. 21: 391–396.

Keller, L. (Ed.), 1999. Levels of Selection in Evolution. Princeton University Press, Princeton, NJ.

Kerr, C.L. and M.J. Shamblott, and J.D. Gearhart. 2006. Pluripotent stem cells from germ cells. Methods Enzymol. 419: 400–426.

Khare, A. and G. Shaulsky. 2006. First among equals: competition between genetically identical cells. Nat. Rev. Genet. 7: 577–583.

Kim-Ha, J. and J.L. Smith, and P.M. Macdonald. 1991. *oskar* mRNA is localized to the posterior pole of the *Drosophila* oocyte. Cell 66: 23–35.

Kim-Ha, J. and K. Kerr, and P.M. Macdonald. 1995. Translational regulation of *oskar* mRNA by bruno, an ovarian RNA-binding protein, is essential. Cell 81: 403–412.

Kimura, T. and A. Suzuki, Y. Fujita, K. Yomogida, H. Lomeli, N. Asada, M. Ikeuchi, A. Nagy, T.W. Mak, and T. Nakano. 2003. Conditional loss of PTEN leads to testicular teratoma and enhances embryonic germ cell production. Development 130: 1691–1700.

King, N. and S.B. Carroll. 2001. A receptor tyrosine kinase from choanoflagellates: molecular insights into early animal evolution. Proc. Natl. Acad. Sci. USA 98: 15032–15037.

King, M.L. and T.J. Messitt, and K.L. Mowry. 2005. Putting RNAs in the right place at the right time: RNA localization in the frog oocyte. Biol. Cell 97: 19–33.

Klattenhoff, C. and D.P. Bratu, N. McGinnis-Schultz, B.S. Koppetsch, H.A. Cook, and W.E. Theurkauf. 2007. *Drosophila* rasiRNA pathway mutations disrupt embryonic axis specification through activation of an ATR/Chk2 DNA damage response. Developmental Cell 12: 45–55.

Klattenhoff, C. and W. Theurkauf. 2008. Biogenesis and germline functions of piRNAs. Development 135: 3–9.

Kleckowski, E.J. Mutation, apical meristems and developmental selection in plants. pp. 79–83. *In*: J.P. Gustafson. [eds.] 1986. Genetics, Molecules and Evolution. Plenum, New York.

Kleinsmith, L.J. and G.B. Pierce, Jr. 1964. Multipotentiality of single embryonal carcinoma cells. Cancer Res. 24: 1544–1551.

Knaut, H. and F. Pelegri, K. Bohmann, H. Schwarz, and C. Nusslein-Volhard. 2000. Zebrafish *vasa* RNA but not its protein is a component of the germ plasm and segregates asymmetrically before germline specification. J. Cell Biol. 149: 875–888.

Ko, M.S. and A. McLaren. 2006. Epigenetics of germ cells, stem cells, and early embryos. Dev. Cell. 10: 161–166.

Kurosaki, H. and Y. Kazuki, M. Hiratsuka, T. Inoue, Y. Matsui, C.C. Wang, M. Kanatsu-Shinohara, T. Shinohara, T. Toda, and M. Oshimura. 2007. A comparison study in the proteomic signatures of multipotent germline stem cells, embryonic stem cells, and germline stem cells. Biochem. Biophys. Res. Comm. 353: 259–267.

Lander, A. 2009. The 'stem cell' concept: is it holding us back? J. Biol. 8: 70.

Lantz, V.A. and S.E. Clemens, and K.G. Miller. 1999. The actin cytoskeleton is required for maintenance of posterior pole plasm components in the *Drosophila* embryo. Mech. Dev. 85: 111–122.

Lavial, F. and H. Acloque, F. Bertocchini, D.J. Macleod, S. Boast, E. Bachelard, G. Montillet, S. Thenot, H.M. Sang, C.D. Stern, J. Samarut, and B. Pain. 2007. The Oct4 homologue PouV and Nanog regulate pluripotency in chicken embryonic stem cells. Development 134: 3549–3563.

Lavial, F. and H. Acloque, E. Bachelard, M.A. Nieto, J. Samarut, and B. Pain. 2009. Ectopic expression of Cvh (Chicken Vasa homologue) mediates the reprogramming of chicken embryonic stem cells to a germ cell fate. Dev. Biol. 330: 73–82.

Lawson, K.A. and N.R. Dunn, B.A. Roelen, L.M. Zeinstra, A.M. Davis, C.V. Wright, J.P. Korving, and B.L. Hogan. 1999. *Bmp4* is required for the generation of primordial germ cells in the mouse embryo. Genes Dev. 13: 424–436.

Lehmann, R. and C. Nüsslein-Volhard. 1986. Abdominal segmentation, pole cell formation, and embryonic polarity require the localized activity of *oskar*, a maternal gene in *Drosophila*. Cell 47: 144–152.

Lehmann, R. and C. Nusslein-Volhard. 1991. The maternal gene *nanos* has a central role in posterior pattern formation of the *Drosophila* embryo. Development 112: 679–691.

Mahowald, A.P. 1962. Fine structure of pole cells and polar granules in *Drosophila melanogaster*. J. Exp. Zool. pp. 201–215.

Mahowald, A.P. 1968. Polar granules of *Drosophila*. II. Ultrastructural changes during early embryogenesis. J. Exp. Zool. 167: 237–261.

Mahowald, A.P. and K. Illmensee, and F.R. Turner. 1976. Interspecific transplantation of polar plasm between *Drosophila* embryos. J. Cell Biol. 70: 358–373.

Marlow, F.L. and M.C. Mullins. 2008. *Bucky ball* functions in Balbiani body assembly and animal-vegetal polarity in the oocyte and follicle cell layer in zebrafish. Dev. Biol. 321: 40–50.

Marques-Mari, A.I. and O. Lacham-Kaplan, J.V. Medrano, A. Pellicer, and C. Simón. 2009. Differentiation of germ cells and gametes from stem cells. Hum. Reprod. Update 15: 379–390.

Martin, G.R. and M.J. Evans. 1974. The morphology and growth of a pluripotent teratocarcinoma cell line and its derivatives in tissue culture. Cell 2: 163–172.

Martin, G.R. 1975. Teratocarcinomas as a model system for the study of embryogenesis and neoplasia. Cell 5: 229–243.

Martin, G.R. 1981. Isolation of a pluripotent cell line from early mouse embryos cultured in medium conditioned by teratocarcinoma stem cells. Proc. Natl. Acad. Sci. USA 78: 7634–7638.

Matsui, Y. and D. Toksoz, S. Nishikawa, D. Williams, K. Zsebo, and B.L. Hogan. 1991. Effect of Steel factor and leukaemia inhibitory factor on murine primordial germ cells in culture. Nature 353: 750–752.

Matsui, Y. and K. Zsebo, and B.L. Hogan. 1992. Derivation of pluripotential embryonic stem cells from murine primordial germ cells in culture. Cell 70: 841–847.

McCosh, G.K. 1930. The origin of the germ cells in *Amblystoma maculatum*. J. Morphol. 50: 569–611.

Metschnikoff, E. 1866. Embryologische Studien an Insekten. Zeit. f. wiss. Zool. 16: 389–500.

Michod, R.E. and D. Roze. 2001. Cooperation and conflict in the evolution of multicellularity. Heredity 86: 1–7.

Michod, R. 2005. On the transfer of fitness from the cell to the multicellular organism. Biol. Philos. 20: 967–987.

Michod, R.E. 2007. Evolution of individuality during the transition from unicellular to multicellular life. Proc. Natl. Acad. Sci. USA 104 Suppl 1: 8613–8618.

Micklem, D.R. and J. Adams, S. Grünert, and D. St Johnston. 2000. Distinct roles of two conserved Staufen domains in *oskar* mRNA localization and translation. EMBO J. 19: 1366–1377.

Milkman, R. 1967. Genetics and developmental studies on *Botryllus schlosseri*. Biol. Bull. 132: 229–243.

Mintz, B. and K. Illmensee. 1975. Normal genetically mosaic mice produced from malignant teratocarcinoma cells. Proc. Natl. Acad. Sci. USA 72: 3585–3589.

Mitsui, K. and Y. Tokuzawa, H. Itoh, K. Segawa, M. Murakami, K. Takahashi, M. Maruyama, M. Maeda, and S. Yamanaka. 2003. The homeoprotein Nanog is required for maintenance of pluripotency in mouse epiblast and ES cells. Cell 113: 631–642.

Miyata, T. and H. Hayashida, K. Kuma, K. Mitsuyasu, and T. Yasunaga. 1987. Male-driven molecular evolution: a model and nucleotide sequence analysis. Cold Spring Harb. Symp. Quant. Biol. 52: 863–867.

Mochizuki, K. and C. Nishimiya-Fujisawa, and T. Fujisawa. 2001. Universal occurrence of the *vasa*-related genes among metazoans and their germline expression in *Hydra*. Dev. Genes Evol. 211: 299–308.

Modrell, M. and A.L. Price, J. Havemann, C. Extavour, M. Gerberding, and N.H. Patel. submitted. Germline replacement following ablation of the primordial germ cells in *Parhyale hawaiensis*. Science. (submitted).

Morata, G. and P. Ripoll. 1975. Minutes: Mutants of *Drosophila* autonomously affecting cell division rate. Dev. Biol. pp. 211–221.

Moreno, E. and K. Basler, and G. Morata. 2002. Cells compete for *decapentaplegic* survival factor to prevent apoptosis in *Drosophila* wing development. Nature 416: 755–759.

Moreno, E. and K. Basler. 2004. dMyc transforms cells into super-competitors. Cell 117: 117–129.

Mukai, H. and H. Watanabe. 1976. Studies on the formation of germ cells in a compound ascidian, *Botryllus primigenus* Oka. J. Morphol. 148: 337–362.

Müller, W.E. 2006. The stem cell concept in sponges (Porifera): Metazoan traits. Sem. Cell Dev. Biol. 17: 481–491.

Newman, S.A. and R. Bhat. 2009. Dynamical patterning modules: a "pattern language" for development and evolution of multicellular form. Int. J. Dev. Biol. 53: 693–705.

Nichols, S.A. and W. Dirks, J.S. Pearse, and N. King. 2006. Early evolution of animal cell signaling and adhesion genes. Proc. Natl. Acad. Sci. USA 103: 12451–12456.

Nieuwkoop, P.D. 1947. Experimental observations on the origin and determination of the germ cells, and on the development of the lateral plates and germ ridges in the urodeles. Arch. Neerl. Zool. 8: 1–205.

Nieuwkoop, P.D. and L.A. Sutasurya. 1981. Primordial Germ Cells in the Invertebrates: From Epigenesis to Preformation. Cambridge University Press, Cambridge.

Niki, Y. and A.P. Mahowald. 2003. Ovarian cystocytes can repopulate the embryonic germ line and produce functional gametes. Proc. Natl. Acad. Sci. USA 100: 14042–14045.

Nishida, H. 1987. Cell lineage analysis in ascidian embryos by intracellular injection of a tracer enzyme. III. Up to the tissue restricted stage. Dev. Biol. 121: 526–541.

Niwa, H. 2007. How is pluripotency determined and maintained? Development 134: 635–646.

Noda, K. and C. Kanai. 1977. An ultrastructural observation of *Pelmatohydra robusta* at sexual and asexual stages, with a special reference to "germinal plasm". J. Ultrastruct. Res. 61: 284–294.

Noden, D.M. 1975. An analysis of migratory behavior of avian cephalic neural crest cells. Dev. Biol. 42: 106–130.

Nüsslein-Volhard, C. and H.G. Frohnhöfer, and Ruth Lehmann. 1987. Determination of anteroposterior polarity in *Drosophila*. Science 238: 1675–1681.

Odintsova, N.A. 2009. Stem cells of marine invertebrates: regulation of proliferation and differentiation processes in vitro. Tsitologiia 51: 367–372.

Oertel, M. and A. Menthena, M.D. Dabeva, and D.A. Shafritz. 2006. Cell competition leads to a high level of normal liver reconstitution by transplanted fetal liver stem/progenitor cells. Gastroenterology 130: 507–520.

Ohinata, Y. and H. Ohta, M. Shigeta, K. Yamanaka, T. Wakayama, and M. Saitou. 2009. A signaling principle for the specification of the germ cell lineage in mice. Cell 137: 571–584.

Okada, M. and I.A. Kleinman, and H.A. Schneiderman. 1974. Restoration of fertility in sterilized *Drosophila* eggs by transplantation of polar cytoplasm. Dev. Biol. 37: 43–54.

Okita, K. and T. Ichisaka, and S. Yamanaka. 2007. Generation of germline-competent induced pluripotent stem cells. Nature 448: 313–317.

Okita, K. and M. Nakagawa, H. Hyenjong, T. Ichisaka, and S. Yamanaka. 2008. Generation of mouse induced pluripotent stem cells without viral vectors. Science 322: 949–953.

Oliver, E.R. and T.L. Saunders, S.A. Tarlé, and T. Glaser. 2004. Ribosomal protein L24 defect in belly spot and tail (Bst), a mouse Minute. Development 131: 3907–3920.

Olsen, C.E. and R. Aasland, and A. Fjose. 1997. A *vasa*-like gene in zebrafish identifies putative primordial germ cells. Mech. Dev. 66: 95–105.

Ota, K.G. and S. Kuraku, and S. Kuratani. 2007. Hagfish embryology with reference to the evolution of the neural crest. Nature 446: 672–675.

Pan, L. and S. Chen, C. Weng, G. Call, D. Zhu, H. Tang, N. Zhang, and T. Xie. 2007. Stem cell aging is controlled both intrinsically and extrinsically in the *Drosophila* ovary. Cell Stem Cell 1: 458–469.

Pelegri, F. and H. Knaut, H.M. Maischein, S. Schulte-Merker, and C. Nüsslein-Volhard. 1999. A mutation in the zebrafish maternal-effect gene *nebel* affects furrow formation and *vasa* RNA localization. Curr. Biol. 9: 1431–1440.

Pires-daSilva, A. and R.J. Sommer. 2003. The evolution of signalling pathways in animal development. Nat. Rev. Genet. 4: 39–49.

Pokrywka, N.J. and E.C. Stephenson. 1995. Microtubules are a general component of mRNA localization systems in *Drosophila* oocytes. Dev. Biol. 167: 363–370.

Queller, D.C. and J.E. Strassmann. 2009. Beyond society: the evolution of organismality. Phil. Trans. Roy. Soc. London B. 364: 3143–3155.

Ramalho-Santos, M. and H. Willenbring. 2007. On the origin of the term "stem cell". Cell Stem Cell 1: 35–38.

Rangan, P. and M. DeGennaro, and R. Lehmann. 2008. Regulating gene expression in the *Drosophila* germ line. Cold Spring Harb. Symp. Quant. Biol. 73: 1–8.

Rangan, P. and M. DeGennaro, K. Jaime-Bustamante, R.-X. Coux, R.G. Martinho, and R. Lehmann. 2009. Temporal and spatial control of germ-plasm RNAs. Curr. Biol. 19: 72–77.

Raz, E. 2003. Primordial germ-cell development: the zebrafish perspective. Nat. Rev. Genet. 4: 690–700.

Reddien, P.W. and N.J. Oviedo, J.R. Jennings, J.C. Jenkin, and A. Sánchez Alvarado. 2005. SMEDWI-2 is a PIWI-like protein that regulates planarian stem cells. Science 310: 1327–1330.

Remane, A. 1952. Die Grundlagen des natürlichen Systems, der vergleichenden Anatomie und der Phylogenetik. Akademische Verlagsgesellschaft, Berlin.

Remsen, J. and P. O'Grady. 2002. Phylogeny of Drosophilinae (Diptera: Drosophilidae), with comments on combined analysis and character support. Mol. Phylogenet. Evol. 24: 249–264.

Resnick, J.L. and L.S. Bixler, L. Cheng, and P.J. Donovan. 1992. Long-term proliferation of mouse primordial germ cells in culture. Nature 359: 550–551.

Ressom, R.E. and K.E. Dixon. 1988. Relocation and reorganization of germ plasm in *Xenopus* embryos after fertilization. Development 103: 507–518.

Rhiner, C. and B. Díaz, M. Portela, J.F. Poyatos, I. Fernández-Ruiz, J.M. López-Gay, O. Gerlitz, and E. Moreno. 2009. Persistent competition among stem cells and their daughters in the *Drosophila* ovary germline niche. Development 136: 995–1006.

Robb, D.L. and J. Heasman, J. Raats, and C. Wylie. 1996. A Kinesin-like protein is required for germ plasm aggregation in *Xenopus*. Cell pp. 823–831.

Rohr, K.B. and D. Tautz, and K. Sander. 1999. Segmentation gene expression in the mothmidge *Clogmia albipunctata* (Diptera, Psychodidae) and other primitive dipterans. Dev. Genes Evol. 209: 145–154.

Rohwedel, J. and U. Sehlmeyer, J. Shan, A. Meister, and A.M. Wobus. 1996. Primordial germ cell-derived mouse embryonic germ (EG) cells in vitro resemble undifferentiated stem cells with respect to differentiation capacity and cell cycle distribution. Cell Biol. Int. 20: 579–587.

Rosner, A. and E. Moiseeva, Y. Rinkevich, Z. Lapidot, and B. Rinkevich. 2009. Vasa and the germ line lineage in a colonial urochordate. Dev. Biol. 331: 113–128.

Russo, C.A. and N. Takezaki, and M. Nei. 1995. Molecular phylogeny and divergence times of drosophilid species. Mol. Biol. Evol. 12: 391–404.

Sabbadin, A. and G. Zaniolo. 1979. Sexual differentiation and germ cell transfer in the colonial ascidian *Botryllus schlosseri*. J. Exp. Zool. 207: 289–304.

Sanchez Alvarado, A. and H. Kang. 2005. Multicellularity, stem cells, and the neoblasts of the planarian *Schmidtea mediterranea*. Exp. Cell Res. 306: 299–308.

Sandstedt, S.A. and P.K. Tucker. 2005. Male-driven evolution in closely related species of the mouse genus *Mus*. J. Mol. Evol. 61: 138–144.

Sauka-Spengler, T. and D. Meulemans, M. Jones, and M. Bronner-Fraser. 2007. Ancient evolutionary origin of the neural crest gene regulatory network. Dev. Cell 13: 405–420.

Schluter, D. 2000. The Ecology of Adaptive Radiation. Oxford University Press, Oxford.

Schultz, S.T. and D.G. Scofield. 2009. Mutation accumulation in real branches: fitness assays for genomic deleterious mutation rate and effect in large-statured plants. Am. Nat. 174: 163–175.

Shamblott, M.J. and J. Axelman, S. Wang, E.M. Bugg, J.W. Littlefield, P.J. Donovan, P.D. Blumenthal, G.R. Huggins, and J.D. Gearhart. 1998. Derivation of pluripotent stem cells from cultured human primordial germ cells. Proc. Natl. Acad. Sci. USA 95: 13726–13731.

Sharma, V.K. and C. Carles, and J.C. Fletcher. 2003. Maintenance of stem cell populations in plants. Proc. Natl. Acad. Sci. USA 100 Suppl 1: 11823–11829.

Shimmin, L.C. and B.H. Chang, and W.H. Li. 1993. Male-driven evolution of DNA sequences. Nature 362: 745–747.

Shostak, S. 2006. (Re)defining stem cells. BioEssays 28: 301–308.

Smith, J.L. and J.E. Wilson, and P.M. Macdonald. 1992. Overexpression of oskar directs ectopic activation of nanos and presumptive pole cell formation in *Drosophila* embryos. Cell 70: 849–859.

Sonnenblick, B.P. 1941. Germ cell movements and sex differentiation of the gonads in the *Drosophila* embryo. Proc. Natl. Acad. Sci. USA 27: 484–489.

Sperger, J.M. and X. Chen, J.S. Draper, J.E. Antosiewicz, C.H. Chon, S.B. Jones, J.D. Brooks, P.W. Andrews, P.O. Brown, and J.A. Thomson. 2003. Gene expression patterns in human embryonic stem cells and human pluripotent germ cell tumors. Proc. Natl. Acad. Sci. USA 100: 13350–13355.

Spicer, G.S. 1988. Molecular evolution among some *Drosophila* species groups as indicated by two-dimensional electrophoresis. J. Mol. Evol. 27: 250–260.

Spradling, A.C. and Y. Zheng. 2007. Developmental biology. The mother of all stem cells? Science 315: 469–470.

Stadtfeld, M. and M. Nagaya, J. Utikal, G. Weir, and K. Hochedlinger. 2008. Induced pluripotent stem cells generated without viral integration. Science 322: 945–949.

Stahl, Y. and R. Simon. 2009. Plant primary meristems: shared functions and regulatory mechanisms. Curr. Opin. Plant. Biol. 13: 53–58.

Stevens, L.C. and C.C. Little. 1954. Spontaneous testicular teratomas in an inbred strain of mice. Proc. Natl. Acad. Sci. USA 40: 1080–1087.

Stevens, L.C. 1967. Origin of testicular teratomas from primordial germ cells in mice. J. Natl. Cancer Inst. 38: 549–552.

Stewart, T.A. and B. Mintz. 1981. Successive generations of mice produced from an established culture line of euploid teratocarcinoma cells. Proc. Natl. Acad. Sci. USA 78: 6314–6318.

Stewart, T.A. and B. Mintz. 1982. Recurrent germ-line transmission of the teratocarcinoma genome from the METT-1 culture line to progeny in vivo. J. Exp. Zool. 224: 465–469.

Stoner, D.S. and I.L. Weissman. 1996. Somatic and germ cell parasitism in a colonial ascidian: possible role for a highly polymorphic allorecognition system. Proc. Natl. Acad. Sci. USA 93: 15254–15259.

Stoner, D.S. and B. Rinkevich, and I.L. Weissman. 1999. Heritable germ and somatic cell lineage competitions in chimeric colonial protochordates. Proc. Natl. Acad. Sci. USA 96: 9148–9153.

Strome, S. and W.B. Wood. 1982. Immunofluorescence visualization of germ-line-specific cytoplasmic granules in embryos, larvae, and adults of *Caenorhabditis elegans*. Proc. Natl. Acad. Sci. USA 79: 1558–1562.

Strome, S. and R. Lehmann. 2007. Germ versus soma decisions: lessons from flies and worms. Science 316: 392–393.

Sun, Y. and H. Li, Y. Liu, S. Shin, M.P. Mattson, M.S. Rao, and M. Zhan. 2007. Cross-species transcriptional profiles establish a functional portrait of embryonic stem cells. Genomics 89: 22–35.

Sunanaga, T. and Y. Saito, and K. Kawamura. 2006. Postembryonic epigenesis of Vasa-positive germ cells from aggregated hemoblasts in the colonial ascidian, *Botryllus primigenus*. Dev. Growth and Differ. 48: 87–100.

Sunanaga, T. and A. Watanabe, and K. Kawamura. 2007. Involvement of *vasa* homolog in germline recruitment from coelomic stem cells in budding tunicates. Dev. Genes Evol. 217: 1–11.

Sunanaga, T. and M. Satoh, and K. Kawamura. 2008. The role of Nanos homologue in gametogenesis and blastogenesis with special reference to male germ cell formation in the colonial ascidian, *Botryllus primigenus*. Dev. Biol. 324: 31–40.

Sweasy, J.B. and J.M. Lauper, and K.A. Eckert. 2006. DNA polymerases and human diseases. Rad. Res. 166: 693–714.

Takahashi, K. and S. Yamanaka. 2006. Induction of pluripotent stem cells from mouse embryonic and adult fibroblast cultures by defined factors. Cell 126: 663–676.

Takahashi, K. and K. Tanabe, M. Ohnuki, M. Narita, T. Ichisaka, K. Tomoda, and S. Yamanaka. 2007. Induction of pluripotent stem cells from adult human fibroblasts by defined factors. Cell 131: 861–872.

Takamura, K. and M. Fujimura, and Y. Yamaguchi. 2002. Primordial germ cells originate from the endodermal strand cells in the ascidian *Ciona intestinalis*. Dev. Genes Evol. 212: 11–18.

Tam, P.P.L. and M.H.L. Snow. 1981. Proliferation and migration of primordial germ cells during compensatory growth in mouse embryos. J. Emb. Exp. Morph. 64: 133–147.

Tam, P.P. and S.X. Zhou. 1996. The allocation of epiblast cells to ectodermal and germ-line lineages is influenced by the position of the cells in the gastrulating mouse embryo. Dev. Biol. 178: 124–132.

Tamura, K. and S. Subramanian, and S. Kumar. 2004. Temporal patterns of fruit fly (*Drosophila*) evolution revealed by mutation clocks. Mol. Biol. Evol. 21: 36–44.

Thomson, J.A. and J. Itskovitz-Eldor, S.S. Shapiro, M.A. Waknitz, J.J. Swiergiel, V.S. Marshall, and J.M. Jones. 1998. Embryonic stem cell lines derived from human blastocysts. Science 282: 1145–1147.

Toyooka, Y. and N. Tsunekawa, R. Akasu, and T. Noce. 2003. Embryonic stem cells can form germ cells in vitro. Proc. Natl. Acad. Sci. USA 100: 11457–11462.

Tsunekawa, N. and M. Naito, Y. Sakai, T. Nishida, and T. Noce. 2000. Isolation of chicken *vasa* homolog gene and tracing the origin of primordial germ cells. Development 127: 2741–2750.

Turner, F.R. and A.P. Mahowald. 1976. Scanning Electron Microscopy of *Drosophila* Embryogenesis. 1. The Structure of the Egg Envelopes and the Formation of the Cellular Blastoderm. Dev. Biol. pp. 95–108.

Urven, L.E. and T. Yabe, and F. Pelegri. 2006. A role for non-muscle myosin II function in furrow maturation in the early zebrafish embryo. J. Cell Sci. 119: 4342–4352.

van den Ent, F. and L.A. Amos, and J. Löwe. 2001. Prokaryotic origin of the actin cytoskeleton. Nature 413: 39–44.

Wada, H. 2001. Origin and evolution of the neural crest: a hypothetical reconstruction of its evolutionary history. Dev. Growth Differ. 43: 509–520.

Wallace, A.R. 1885. On the law which has regulated the introduction of new species. Ann. Mag. Nat. Hist. 16: 184–196.

Waterhouse, A.M. and J.B. Procter, D.M. Martin, M. Clamp, and G.J. Barton. 2009. Jalview Version 2—a multiple sequence alignment editor and analysis workbench. Bioinformatics 25: 1189–1191.

Webster, P.J. and J. Suen, and P.M. Macdonald. 1994. *Drosophila virilis oskar* transgenes direct body patterning but not pole cell formation or maintenance of mRNA localization in *D. melanogaster*. Development 120: 2027–2037.

Weismann, A. 1892. The Germ-Plasm: A Theory of Heredity. Walter Scott, Ltd., London.

Weissman, I.L. 2000. Stem cells: units of development, units of regeneration, and units in evolution. Cell 100: 157–168.

Whitham, T.G. and C.N. Slobodchikoff. 1981. Evolution by individuals, plant-herbivore interactions, and mosaics of genetic variability: the adaptive significance of somatic mutations in plants. Oecologia 49: 287–292.

Wilkinson, R.N. and C. Pouget, M. Gering, A.J. Russell, S.G. Davies, D. Kimelman, and R. Patient. 2009. Hedgehog and Bmp polarize hematopoietic stem cell emergence in the zebrafish dorsal aorta. Dev. Cell 16: 909–916.

Willensdorfer, M. 2008. Organism size promotes the evolution of specialized cells in multicellular digital organisms. J. Evol. Biol. 21: 104–110.

Wolpert, L. and R. Beddington, T. Jessell, P. Lawrence, E. Meyerowitz, and J. Smith. 2002. Principles of Development. Oxford University Press, Oxford.

Wolpert, L. and T. Jessell, P. Lawrence, E. Meyerowitz, E. Robertson, and J. Smith 2007. Principles of Development. Oxford University Press Ltd., Bath.

Yamanaka, Y. and A. Ralston, R.O. Stephenson, and J. Rossant. 2006. Cell and molecular regulation of the mouse blastocyst. Dev. Dyn. 235: 2301–2314.

Yamanaka, S. 2007. Strategies and new developments in the generation of patient-specific pluripotent stem cells. Cell Stem Cell 1: 39–49.

Ying, Y. and X.M. Liu, A. Marble, K.A. Lawson, and G.Q. Zhao. 2000. Requirement of *Bmp8b* for the generation of primordial germ cells in the mouse. Mole. Endocrinol. 14: 1053–1063.

Ying, Y. and G.Q. Zhao. 2001. Cooperation of endoderm-derived BMP2 and extraembryonic ectoderm-derived BMP4 in primordial germ cell generation in the mouse. Dev. Biol. 232: 484–492.

Yoon, C. and K. Kawakami, and N. Hopkins. 1997. Zebrafish *vasa* homologue RNA is localized to the cleavage planes of 2- and 4-cell-stage embryos and is expressed in the primordial germ cells. Development 124: 3157–3166.

Zamboni, L. and H. Merchant. 1973. The fine morphology of mouse primordial germ cells in extragonadal locations. Am. J. Anat. 137: 299–335.

Zhang, X.G. and X.G. Hou. 2004. Evidence for a single median fin-fold and tail in the Lower Cambrian vertebrate, *Haikouichthys ercaicunensis*. J. Evol. Biol. 17: 1162–1166.

Zhang, J. and L. Li. 2005. BMP signaling and stem cell regulation. Dev. Biol. 284: 1–11.

Zhou, Y. and M.L. King. 2004. Sending RNAs into the future: RNA localization and germ cell fate. IUBMB Life 56: 19–27.

Zhurov, V. and T. Terzin, and M. Grbic. 2004. Early blastomere determines embryo proliferation and caste fate in a polyembryonic wasp. Nature 432: 764–769.

Zimyanin, V.L. and K. Belaya, J. Pecreaux, M.J. Gilchrist, A. Clark, I. Davis, and D. St Johnston. 2008. *In vivo* imaging of *oskar* mRNA transport reveals the mechanism of posterior localization. Cell 134: 843–853.

Questions and Discussion

Section 1. Tangled Roots in the Animal Tree of Life

Chapter 1. Vazquez et al., *Putting animals in their place within a context of eukaryotic innovations*

1.1. How does taxon sampling affect the conclusions made in this chapter? Can a "reasonably designed" taxonomic sampling answer the questions posed in this chapter?

1.2 Take the tree in Fig. 1 and map on it the characters discussed in the text that define the major groups in the tree. For instance, foraminifera are defined by granular reticulopodia.

1.3. Compare the relationships in Fig. 1 with the Tree of Life Web Project topology for these taxa (http//tolweb.org). Are there differences in topologies? What nodes in the ToL at this level have been considered to be controversial?

1.4. Discuss the idea that the molecular results for relationships of Cercozoa, Amoebozoa and Rhizaria do not agree with morphological synapomorphies. What might be the reasons for the lack of discernable morphological innovations for these clades?

1.5. What are the reasons for the authors' statement that microbial lineages defy many of the long-held assumptions about eukaryotes? Discuss the dynamics of mitotic mechanisms in the taxa described in this chapter. How does the mitotic apparatus and process affect our view of the early origin of animals?

Chapter 2. Kocot et al., *Elucidating animal phylogeny: advances in knowledge and forthcoming challenges*

2.1. Discuss the lack of resolution at the base of the tree in Fig. 1 of this chapter. What is the cause of this lack of resolution according to the authors? In the same context the authors of this chapter suggest that "the branching order of the three major lineages of Arthropoda, the relationships within Lophotrochozoa, and the placement of several so-called minor groups such as Chaetognatha." Discuss these problems too in light of recent molecular studies.

2.2. Discuss the utility of the 18S rDNA gene in phylogenetics. What are its advantages and what are its disadvantages? How will next generation sequencing approaches (like 454 sequencing) affect the current understanding of animal phylogeny?

2.3. For each of the major bilaterian clades describe the information that allows phylogeneticists to lay claim to the existence of these major clades. Are there any clades supported entirely by molecular evidence?

2.4. Make a list of the major groups of bilaterians that have been reclassified or reorganized as a result of recent molecular work. In addition to the list of groups, write a statement of how the reclassification or reorganization affects our understanding of morphological change.

Chapter 3. Lavrov, *Key transitions in animal evolution: a mitochondrial DNA perspective*

3.1. List the deviations that the non-bilaterian animal mtDNA genomes have with bilaterian mt DNA genomes. What does "bialaterian like" mtDNA mean?

3.2. Are the phylogenetic inferences made using mtDNA sequences different from the inferences made with gene order of the mt DNA genome? Why have gene order studies been delayed in this context? How will this problem of including gene order studies be overcome?

3.3. What critical taxa are missing in the assessment of the tree of life at the base of metazoa? Why?

3.4. Discuss the utility of mitochondrial genome size and gene content (not gene order) in phylogenetic analysis of animals.

3.5. Can we associate any steps in mitochondrial DNA evolution with the major branches of animals?

3.6. How does the tree from mtDNA analyses in this chapter (Fig. 4) differ from the consensus tree presented in the previous chapter? What are the reasons for any discrepancies?

Chapter 4.   Deutsch, *Pending issues in development and phylogeny of arthropods*

4.1. What are the major kinds of homology discussed in this chapter. Why is homology so important a concept in reconstructing phylogeny of the arthropods? How does the anatomical concept of homology relate to DNA and protein sequences?

4.2. What are the basic anatomical characteristics that Deutsch argues are evidence for Arthropoda? From the studies discussed in the previous chapters does the molecular evidence agree with this inference?

4.3. Critically assess the monophyly or polyphyly of the Myriapodia?

4.4. Where does the mandible come from? How is it used in the ordering of the Arthropoda?

4.5. Why are Hexapods terrestrial crustaceans? What is the collembolan controversy and how does it impact the systematics of the Hexapoda.

4.6. According to the great biologist Ernst Myer, eyes evolved in the animal lineage more than twenty times ? How many times does Deutsch think eyes evolved in arthropods? What is the evidence he presents for this conclusion. Critically assess his conclusion.

Section 2. The Earliest Animals: From Genes to Transitions

Chapter 5.   Nickel, *The pre-nervous system and beyond—poriferan milestones in the early evolution of the metazoan nervous system*

5.1. Even though porifera lack a nervous system, they are often suggested as a critical system to study for a complete understanding of the nervous system. Why?

5.2. Why are contractions and possible locomotion in sponges so critical to our understanding of the organism? What is our phylogenetic understanding of these two important aspects of sponge biology?

5.3. What are integrative systems? What are the specific integrative systems discussed in the context of sponges?

5.4. Would it be too terribly surprising to discover action potential in sponges? What other groups demonstrate he existence of action potentials other than those with nervous systems?

5.5. Is it surprising that 36 members of gene families involved in the post synaptic scaffold are found in sponges? Pick one of the gene families that is present (voltage gated ion channels, ionotropic and metabotropic glutamate receptors) and discuss the role of these genes in animals with nervous systems and then place the presence of this gene family in sponges in context.

5.6. What is LCAM? What kind of nervous system did it have?

Chapter 6. Galliot et al., *A key innovation in evolution, the emergence of neurogenesis: cellular and molecular cues from cnidarian nervous systems*

6.1. How does the phylogenetic scenario discussed by Galliot et al. compare to those in other chapters in this book?

6.2. What are nematocysts? How do they fit into the nervous system of cnidarians? What is the common progenitor hypothesis for neamtocysts and neurons?

6.3. Describe the anatomic organization of at least one cnidarian nervous system. Compare this to a similar characterization from one of your colleagues using this book for a cnidarian in a different phylogenetic class. Would you characterize the differences you see as "complex and highly variable"?

6.4. What are the behaviors that exist in cnidarians that have influenced nervous system structure an function?

6.5. As Galliot et al ask "Can we trace back in cnidarians an ancestral neurogenic circuitry?"

6.6. Why would Galliot et al describe the cnidarian nervous system as paradigmatic?

Chapter 7. Boero and Piraino, *From cnidaria to "higher" metazoa in one step*

7.1. What phylogenetic placement of Cnidaria does Boero prefer? How does this conflict with the placement of Cnidaria in Chapters 2,3 and 6? How might the evolutionary scenarios discussed in this chapter be altered if the Cnidaria-Bilateria sister relationship is false?

7.2. How does the inverted cone relate to the generation of new taxa (i.e. in speciation)? Now consider what you answered for the first part of this question in the context of this quote from this chapter:

"It has often been said that higher taxa are the product of our way of thinking and that species are the only "real" evolutionary units. But if higher taxa are monophyletic, then speciation events, as seen in Fig. 2, do have different bearings on the descendants of the newly formed species, according to the developmental changes that generated them. "

7.3. What is paedomorphosis? What is peromorphoisis? How does Boero think these processes are involved in the cone metaphore?

7.4. What is the "one step" that Boero refers to in the title of this chapter? Is it really one step? Does it have to be one step?

7.5. Assess the statements that Cnidaria are the "basis of metazoan evolution" and "most of the features that are considered as milstones of metazoan evolution evolved in the Cnidaria".

Chapter 8.   Jacobs et al., *Basal metazoan sensory evolution*

8.1. Using the definition of animal promulgated by Jacobs et al. in the first sentence of their paper, take the major groups of animals Bilatiera, Cnidaria, Ctenophora, Porifera and Placozoa and discuss how well each fits with the definition.

8.2. Discuss what is meant by "differential persistence and modification of an ancestral condition" and contrast this with "wholesale "cooptation" of genes". How do these two possible scenarios for the evolution of structures in the metazoa stack up as explanation for the conflicting patterns of gene expression and morphology observed across the metazoan tree.

8.3. List the characteristics that sense organs in Metazoa have in common. Discuss these characteristics in the context of gene regulation in bilaterians.

8.4. What is the significance of sharing developmental regulatory genes across developmental and organ systems?

8.5. What do the kidney have to do with sensory systems? What about limb development?

8.6. Critically assess the evidence for the homology of medusan sensory structures with Bilaterian sensory structures. What other scenarios are possible for the similarity of these structures?

8.7. Characterize what is meant by motor coordination of sponges. Do sponges have sensory systems? Explain why or why not.

8.8. How do the authors tie in choanocyte structure into their ideas about sensory structure? In light of the various tree topologies that have been discussed in earlier chapters of this book, how do the authors ideas about sensory structure homology hold up?

Chapter 9. Ball et al., *Cnidarian gene expression patterns and the origins of bilaterality—are cnidarians reading the same game plan as "higher" animals?*

9.1. What are the two major textbook characteristics discussed in this chapter that separate Cnidaria from Bilateria? Assess the validity of these characteristics as differentiating Cnidaria from Bilatieria. Are there possibly others?

9.2. Does oral-aboral correspond to anterior-posterior and, if so, what is the correspondence? How does the question "What is a Hox gene" relate to the first part of this question?

9.3. Which came first Hox genes, Hox system or axial patterning? Explain your answer.

9.4. What is the significance of the absence of miRNA's in sponges and their presence in Cnidaria? What about the absence of mir-10 genes in acoels?

9.5. Cnidaria have a series of developmental genes that Bilateria also have for axial patterning. Make a list of these major classes of genes. Now in separate columns describe the role of the genes in Cnidaria in one column and in Bilateria in another column. What are the implications of these comparisons for animal evolution and the statement that Cnidarians "might reflect a separate experiment (or perhaps a separate set of experiments) in axial patterning".

Chapter 10. Oakley and Plachetzki, *Key transitions during animal phototransduction evolution: Co-duplication as a null hypothesis for the evolutionary origins of novel traits*

10.1. What are the two major processes that the authors suggest are important in evolutionary transitions. How do these relate to the key transitions in the evolution of the bilateria?

10.2. Discuss the role of duplication in the major transitions in the bilateria. What is the role of divergence in combination with duplication?

10.3. Explain the difference between co-duplication and co-option. How can we differentiate co-duplication from co-option?

10.4. Describe phototransducing cascades. Discuss the evolution of rods and cones in the context of these cascades. How have trees aided in the discussion of phototransducing cascades?

10.5. Describe the role of co-option in the evolution of phototransducing networks.

10.6. Describe the process by which G-proteins have been involved in the evolution of phototransduction. How has tree thinking aided in the elucidation of the role of G-proteins in phototransduction evolution?

Chapter 11.   Kappen, *Vertebrate Hox genes and specializations in mammals*

11.1. How can altered Hox gene function lead to divergence of anatomical structures?

11.2. How does Hox gene function affect the evolution of mammalian female reproductive tract?

11.3. Describe the role of Hox genes in the placenta . How might the role of Hox genes in the placenta be used to shed light on the transitions made to mammals?

11.4. Describe the patterns of placenta specific regulation of hox genes. How well do these patterns support the statement that "subtle differences in the activity of Hox gene enhancers may explain differences in expression patterns that are associated with phenotypic differences in the features of specific structures."

11.5. Evaluate the predictions that Kappen makes in the summary of her paper. How likely are we to evaluate these predictions? How much will examining the predictions tell us about the evolution of novelty in organisms with Hox genes?

Chapter 12. Pearse and Voigt, *Field biology of placozoans* **(Trichoplax):** *distribution, diversity, biotic interactions*

12.1. Explain why Placozoa would be considered a diploblast.

12.2. What is unique about the collecting procedure for Placozoa. Is this procedure biased in the kinds of organisms it will obtain?

12.3. Discuss the biogeography of Placozoa. How might this biogeography have affected the evolution of this organism?

12.4. Is Placozoa a single species in a single phylum?

12.5. Outline the life-history of placozoans. Use a flow diagram if needed to make this clear. Where are the gaps in our knowledge of the life history? How do these gaps affect our understanding of the evolution of this organism.

Chapter 13. Schierwater et al., **Trichoplax** *and Placozoa: one of the crucial keys to understanding metazoan evolution*

13.1. Compare and contrast the following hypotheses that concern the origin of the Metazoa and the placement of the Placozoa in the animal tree of life—gastraea, blastaea, placula hypothesis.

13.2. For five taxa (Bilateria, Ctenophora, Cnidaria, Porifera and Placozoa) how many possible bifurcating trees are there that describe the relationships of the five taxa (assuming Monosiga is an outgroup). Consult Table 1, Fig. 1 and some of the references mentioned in the references and estimate how many "viable" hypotheses exist for the relationships of these taxa. Compare the possible number with the number of hypotheses that exist in the literature.

13.3. Critically assess the claim of the authors in this chapter that Placozoa is "no longer a phylum of one".

13.4. Discuss the claim of the authors in this chapter that Placozoa are an excellent model system for defining the basal state of metazoans.

13.5. Why is quality data like morphology, functional morphology, molecular morphology and gene gain or loss data unambiguously showing a basal position for Placozoa, while DNA or protein sequence data create conflicting scenarios?

13.6. If the phylum Placozoa harbors dozens of species and possibly several genera, families or even orders, why is it that the phylum is yet officially monophylectic?

Chapter 14. Blackstone, *A food's-eye view of animal transitions*

14.1. Discuss what Blackstone refers to as two levels of selection in multicellular animals. How doe these two levels interact with each other ? Are their interactions antagonistic?

14.2. What is an interior to exterior metabolic gradient and why would it be important in the evolution of animals?

14.3. Blackstone mentions several times the "urmetazoan". What is he referring to? How do we know anything about the urmetazoan ?

14.4. What is ROS and why is it important in the context of animal evolution? Why are antioxidants important in this context?

14.5. Outline the three transitions in feeding that Blackstone articulates. Give at least two evolutionary events or parts of the transitions that Blackstone feels are important for the transitions.

Chapter 15. Gaidos, *Lost in transition: the biogeochemical context of animal origins*

15.1. Why is the study of the chemistry of the Precambrian environment important? List three possible constraints and the impact they might have on organismal evolution.

15.2. Where might animals have first thrived in the Proterozoic world? Where would the highest dissolved oxygen concentrations be found and why would this be an important fact to discover?

15.3. How old are the Metazoa? What methods have been used to estimate this age? How does Gaidos reconcile the deep evolutionary divergence inferred with molecular clocks with the fossil record?

15.4. What is Lomagundi and why does Gaidos think it is so important? How might it affect the life cycles of animals?

Chapter 16. Srouji and Extavour, *Redefining stem cells and assembling germ plasm: key transitions in the evolution of the germ line*

16.1. Why do the authors suggest that:

A complete understanding of many "key transitions" involving new structures and new cell types must therefore incorporate the molecular genetic basis for the novel or modified cell behaviors that can lead to novel structures. Give at least two reasons.

16.2. What are the novelties of the bilaterian germ line?

16.3. What molecular mechanisms are involved in the specification of germ cells? How is timing of events in the development of germ cells involved?

16.4. Why have germ cells been called the "mother" of all stem cells? How are disease states involved in the transition of germ cells to stem cells?

16.5. What kind of events are involved in modifying stem cell lineages?

16.6. What is the role of heterochrony in germ plasm development?

Index

Color Plate Section

Chapter 1

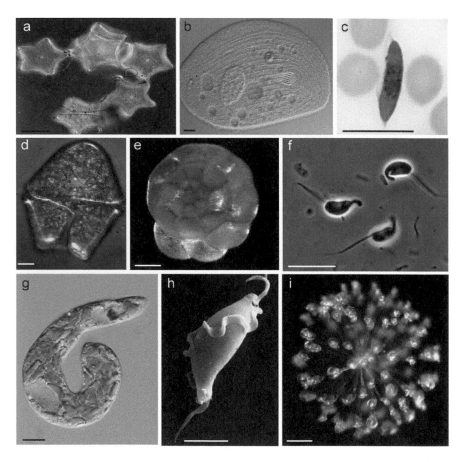

Fig. 2. Representative taxa from major eukaryotic lineages discussed. Stramenopiles: **(a)** Triceratium pentacrinus, a diatom; Alveolates (b-d): **(b)** Trithigmostoma cucullulus, a ciliate, **(c)** Plasmodium falciparum, an apicomplexan, and **(d)** Akashiwo sanguinea, a dinoflagellate; **(e)** Ammonia, a foraminiferan; Excavata: **(f)** Rhynchomonas nasuta, a kinetoplastid and **(g)** Euglena mutabilis, a euglenid; **(h)** Tritrichomonas muris, a parabasalid; Opisthokonta: **(i)** Conochilus, a colonial rotifer. Scale bars: b-d,f-h = 10 micrometers; a, e = 100 micrometers; i = 200 micrometers. All images are provided by micro*scope (http: //starcentral.mbl.edu/microscope/portal.php) except c, which is used under PLoS Biology's Open Access rules and can be found in Lacroix et al. (2005), credit Dr. Mae Melvin.

Chapter 2

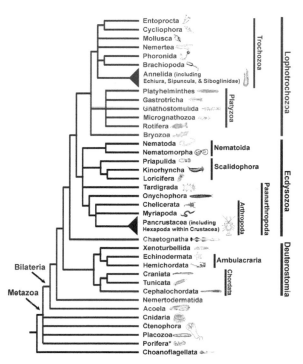

Fig. 1. Current understanding of animal phylogeny. This tree is conservatively drawn to illustrate major concepts inferred from recent studies as described in the text. The tree is color-coded: 'basal metazoans' are purple, deuterostomes are red, ecdysozoans are blue, and lohpotrochozoans are green. The monophyly of Porifera has been questioned, but for the sake of simplicity it is drawn as a single lineage here (see text).

Chapter 3

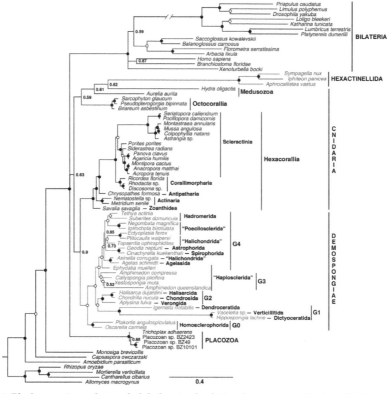

Fig. 4. Phylogenetic analysis of global animal relationships using mitochondrial sequence data. Posterior majority-rule consensus tree obtained from the analysis of 2,539 aligned amino acid positions under the CAT+F+Γ model in the PhyloBayes program is shown. We ran four independent chains for ~60,000 generations and sampled every 10th tree after the first 1000 burnin cycles. The convergence among the chains was monitored with the maxdiff statistics and the analysis was terminated after maxdiff became less than 0.3. The number/circle at each node represents the Bayesian posterior probability. Posterior probability = 1 is indicated by a filled circle; that > 0.95 – by an open circle; smaller posterior probabilities are shown as numbers. Amino acid sequences for non-bilaterian animals were derived from the GenBank files listed in table 1. Amino-acid sequences for *Cantharellus cibarius* mtDNA were downloaded from http://megasun.bch.umontreal.ca/People/lang/FMGP/proteins. html; those for *Capsaspora owczarzaki* mtDNA were provided by Franz Lang (Université de Montréal). Other sequences were derived from the GenBank files: *Arbacia lixula* NC_001770, *Balanoglossus carnosus* NC_001887, *Branchiostoma floridae* NC_000834, *Drosophila yakuba* NC_001322, *Florometra serratissima* NC_001878, *Homo sapiens* NC_001807, *Katharina tunicata* NC_001636, *Limulus polyphemus* NC_003057, *Loligo bleekeri* NC_002507, *Lumbricus terrestris* NC_001673, *Platynereis dumerilii* NC_000931, *Priapulus caudatus* NC_008557, *Saccoglossus kowalevskii* NC_007438, *Xenoturbella bocki* NC_008556, *Amoebidium parasiticum* AF538042-AF538052, *Monosiga brevicollis* NC_004309, *Allomyces macrogynus* NC_001715, *Mortierella verticillata* NC_006838, *Rhizopus oryzae* NC_006836.

Color image of this figure appears in the color plate section at the end of the book.

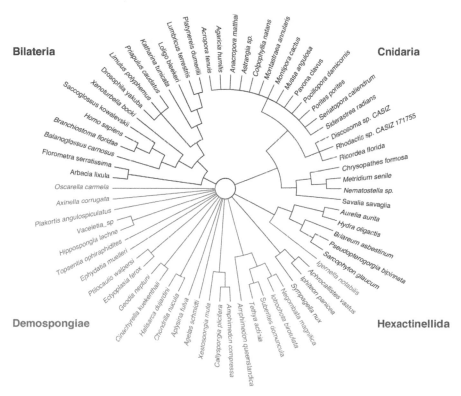

Fig. 5. Global animal relationships based on mitochondrial gene order data. Strict consensus tree is shown from the Maximum Parsimony analysis on Multistate Encodings (Boore et al. 1995, Wang et al. 2002). Gene orders were encoded as described previously (Lavrov and Lang 2005a) and analyzed using heuristic search with 100 random addition replicates in PAUP*4.0b10 and TBR branch swapping option.

Chapter 4

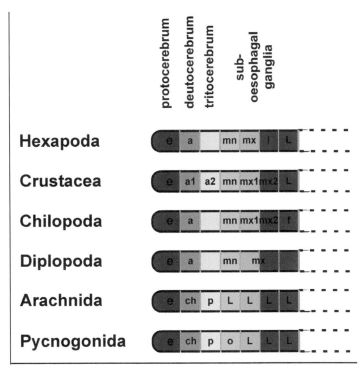

Fig. 3. Homology of anterior segments in arthropods.

Hexapoda: No appendage corresponding to the tritocerebral segment. Crustacea: In many species, the first legs of the thorax (or pereion) are transformed in a pair of maxillipedes, i.e. maxilla-like appendages. Chilopoda: No appendage corresponding to the tritocerebral segment; the first appendage of the trunk is transformed in a pair of forcipules (poison claws). Diplopoda: No appendage corresponding to the tritocerebral segment; the first and second pairs of maxillae are fused; no appendage in the first trunk segment. Arachnida: The deuto-cerebral segment bears a pair of chelicerae, the tritocerebral segment a pair of pedipalps; Arachnids have four pairs of walking legs. Pycnogonida: The first pair of legs is transformed in a pair of ovigers. Taking into account the ovigers as modified legs, pycnogonids have more legs than arachnids.

a: antennae; a1: first pair of antennae; a2: second pair of antennae; ch: chelicerae; e: eyes; f: forcipules; l: labium; L: legs; mn: mandibles; mx: maxillae; mx1: first pair of maxillae; mx2: second pair of maxillae; o: ovigers; p: pedipalps.

Chapter 5

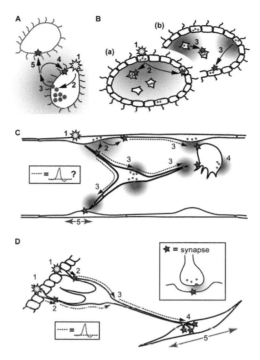

Fig. 4. Explanatory scheme presenting evolutionary levels as integral parts of the 'paracrine-to-electrochemical-dominance transition' hypothesis. For details refer to the text. Diffusion gradients **A.** Level of auto- and paracrine signalling in unicellular eukaryotes. (1) External stimulus followed by (2) a intracellular signal leading to (3) the release of a chemical messenger mediating (4) autocrine and (5) paracrine reception. **B.** Hypothetical level of the LCAM: (1) a sensed extracorporeal stimulus results in (2) the release of a paracrine messenger into the extracellular matrix (blue gradient). This diffusing signal reaches receptive cells which act as receptor-effectors *sensu* Grundfest including (3) secretion of secondary chemical signals (integration) or the same chemical substance (positive feedback). These secondary signal gradients (orange gradients) differ from the primary signal gradient and eventually reach other cells. **C.** Hypothetical intermediate level, eventually present in Demospongiae: again (1) an external stimulus is received by epithelial cells, followed by (2) the release of a chemical messenger (blue gradient). This signal is received by neighboring epithelial cells and mesenchymal multipolar cells. It locally triggers the opening of ligand-gated ion channel in the receptor cell, which (3) results in a potential change within the whole cell or a spreading local potential change over the cell body (action potential). The incurrent of specific ions (e.g. Ca^{2+}) and/or secondary messenger systems locally trigger the secondary release of extracellular messengers (orange gradients). During the proposed evolutionary transition, the secretory complexes are compartmentalized within the cells. The secondary signal is received by other cells and triggers secondary effects like (4) secretion, cell locomotion and (5) contraction of epithelial cells. **D.** Level of a simple nervous system representing the simple reflex-arc sensu Parker, from (1) the incoming external signal at epithelial receptor cells over (2) synapses to a nerve cell, generating excitatory or inhibitory postsynaptic potentials. This neuron (3) might integrate signals from separate sensory cells and generate an outgoing signal at (4) a synapse towards a muscle cell, representing a highly differentiated contractile effector cell (5).

Chapter 6

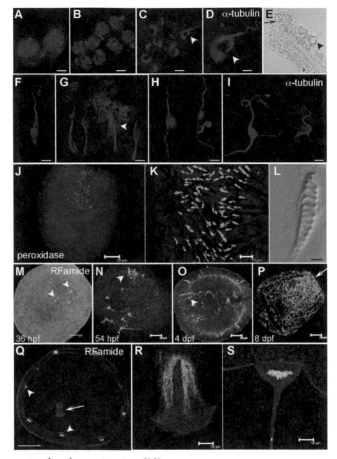

Fig. 2. The neuronal and nematocyte cell lineages.
A-E) The *Hydra* nematocyte lineage : interstitial stem cells **(A)** synchronously divide, providing nematoblasts **(B)** that differentiate a typical capsule (nematocyst) as observed in mature nematocytes **(C,D)**. Arrowheads : discharged nematocyst. **E)** Bright-field view of a tentacle with nematocytes either undischarged (arrow) or discharged (arrowhead) embedded in large epithelial battery cells. F-I) *Hydra* neurons detected after maceration with alpha-tubulin antibody (red) and DAPI (blue): sensory **(F, G)**, bipolar **(H)** or multipolar **(I)** also named ganglionic cells. J-L) *Nematostella* spirocytes detected in larva thanks to their peroxidase activity (green). Spirocytes are anthozoan mechanosensory cells involved in adhesion to prey and non-prey (Kass-Simon and Scappaticci 2002). M-P) Neurogenesis in the developing *Nematostella* : Early neurons expressing the neuropeptide RFamide (arrowheads) appear in the endodermal layer **(M)**, then migrate to the ectoderm **(N)** where they form a net **(O)**. In the newly metamorphosed polyp **(P)**, the density of RF-amide neurons is higher in the oral region (arrow). Q-S) *Clytia* medusa showing a high density of RF-amide neurons (red) in the manubrium (Q arrow, R) and in the tentacle bulbs (Q arrowheads, S). Blue : DAPI staining ; green : endogenous GFP. Scale bars: 2 µm (A,B,D,L), 5 µm (E-I, C), 10 µm (K), 50 µm (J, M-P, R-S), 500 µm (Q).

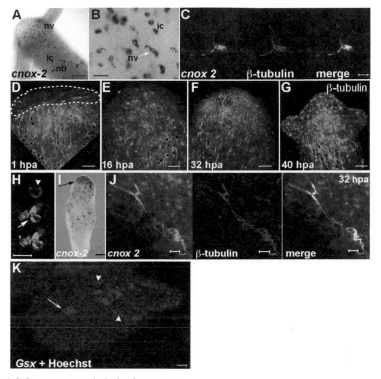

Fig. 4. Adult neurogenesis in hydrozoans
A-C) Neurons (arrow) located at the base of the head region between the tentacle roots express the ParaHox *Gsx* orthologue, *cnox-2*. Along the body column, *cnox-2*+ cells are interstitial cells (ic) and proliferating nematoblasts (nb). **D-G)** *De novo* neurogenesis during *Hydra* head regeneration. Following bisection the tip of the head-regenerating half is depleted of neurons (D, outline), progressively repopulated with neuronal progenitors (U) and mature neurons (V) detected with anti beta-tubulin (green). The apical nervous system at 40 hpa is still less dense than in adult polyps (W). hpa : hours post-amputation. **H-J)** During head regeneration, *cnox-2* is expressed in progenitors and differentiating neurons in the presumptive head region (arrow) shown here at 32 hpa. I) Apical BrdU labeled cells (red) expressing *cnox-2* (green). J) Apical neurons co-expressing *cnox-2* transcripts (red) and beta-tubulin (green). **K)** In the *Clytia* medusa, the proliferative (arrow) and differentiating (arrowheads) zones of the tentacle bulbs express the ParaHox gene *Gsx*, detected here with FastRed (red). Scale bars: 200 µm (A), 10 µm (B, I, J), 20 µm (K), 50 µm (D-G).

Chapter 10

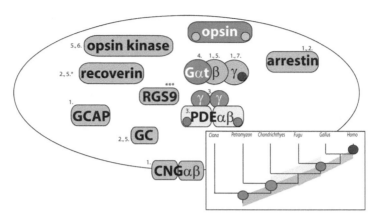

Fig. 5. The timing of origin of components of rod phototransduction. Phototransduction genes are illustrated as symbols that are colored according to when a gene duplication event led to paralogs expressed separately in rods and cones. Red are mammal-specific gene duplications, orange are tetrapod-specific, yellow duplicated between the common ancestor of *Petromyzon*+mammals and *Fugu*+mammals, green are vertebrate-specific (assuming monophyly of hagfish+*Petromyzon*), grey are genes duplicated between the ancestor of *Ciona*+mammals and *Fugu*+mammals. Genes colored in grey are targets for study in *Petromyzon*, genes colored in yellow are targets for study in cartilaginous fishes. References for gene trees are indicated with numbers next to each gene family's symbol: 1. (Nordström et al. 2004); 2. (Hisatomi and Tokunaga 2002); 3. (Muradov et al. 2007); 4. (Muradov et al. 2008); 5. (Larhammar et al. 2009); 6. (Yasutaka et al. 2006); 7. (Chen et al. 2007). Small orange circles in PDE indicate timing of origin of alpha and beta PDE, both expressed within rods. Small pink circles in opsin indicates that multiple recent gene duplications have occurred within vertebrates, usually having specialized for different wavelength sensitivity in cones. *Recoverin is not known to have rod and cone-specific expression. **Mammal-specific duplication present and expressed in various tissues.*** RGS9 is expressed in both rods and cones, but at different levels. The gene is unique to vertebrates. Its sister gene family is RGS11, which is expressed in retinal neurons. Our unpublished phylogenetic analyses indicates that RGS9 and RGS11 are sister paralog, and that the purple urchin has a single copy that is sister to the RGRS9/11 clade. Agnathans or condrycthyes have not been investigated. Therefore, RGS9 and RGS11 duplicated within deuterostomes, before the common ancestor of teleosts and humans.

Chapter 11

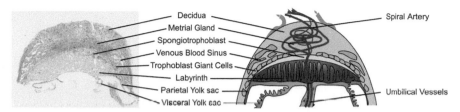

Fig. 2. Mouse placenta at midgestation.

Legend: Left panel shows a section from paraffin embedded mouse placenta isolated at gestation day 10.5, stained with the Periodic acid Schiff method, counterstained with methyl green.

Chapter 13

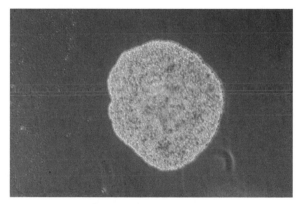

Fig. 2A. Photograph of a live *Trichoplax adhaerens*. This specimen belongs to the so-called "Grell clone" and is about 2 mm in diameter.

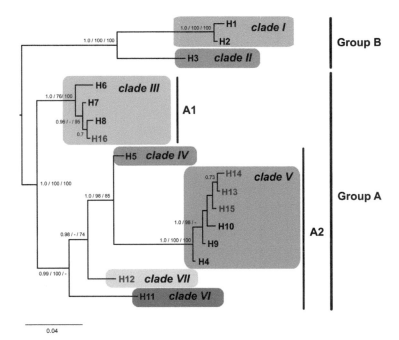

Fig. 3. 16S haplotype cladogram of all known placozoan lineages. The cladogram shows a distinctive hierarchical arrangement independent of the tree-building algorithm applied. Haplotype numbers (H) refer to strains listed in Table 1. Numbers beside nodes are from left to right: Baysian posterior probabilities, Maximum likelihood and Maximum Parsimony bootstrap support. Values below 70% are marked with '-'. Two main groups ('A' and 'B') are found within the Placozoa probably representing higher taxonomic units. Within group 'A' two subgroups ('A1' and 'A2') are clearly distinguishable. Red labeling marks formerly undescribed haplotypes. (from Eitel and Schierwater 2010).

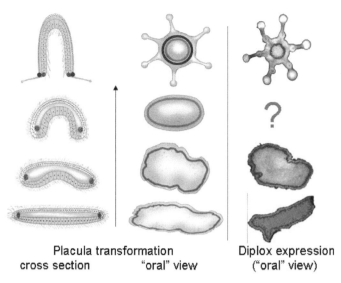

Placula transformation
cross section   "oral" view

Diplox expression
("oral" view)

Fig. 4. Modern Placula Hypothesis. Modern interpretation and modification of the placula hypothesis of metazoan origins. A non-symmetric and axis-lacking Placula bauplan transforms into a typical symmetric metazoan bauplan with a defined oral–aboral body axis. The Placula transforms from a primitive disk consisting of an upper and a lower epithelium (lower row) to a form with an external feeding cavity between its lower epithelium and the substrate (2nd row from bottom). For this the Placula lifts up the center of its body, as naturally seen in feeding *Trichoplax*. If this process is continued, the external feeding cavity increases (cross section, 3rd row) while at the same time the outer body shape changes from irregular to more circular (see oral views). Eventually the process results in a bauplan where the formerly upper epithelium of the Placula remains "outside" (and forms the ectoderm) and the formerly lower epithelium becomes "inside" (and forms the entoderm, upper row). This stage represents the basic bauplan of Cnidaria and Porifera. Three of the four transformation stages have living counterparts in form of resting *Trichoplax*, feeding *Trichoplax*, and cnidarian polyps and medusae (right column). From a developmental genetics point of view a single regulatory gene would be required to control separation between the lower and upper epithelium (three lower rows).We find this realized in *Trichoplax* in the form of the putative Proto-Hox/Para-Hox gene, *Trox-2*. More than one regulatory gene would be required to organize new head structures

originating from the ectoderm-entoderm boundary of the oral pole (upper row) in Cnidaria. Indeed (and quite intriguing), two cnidarian orthologs of the *Trox-2* gene, *Cnox-1* and *Cnox-3*, show exactly these hypothesized expression patterns (Diplox expression upper row; for simplicity only the ring for *Cnox-1* expression is shown). (from Schierwater et al. 2009a).

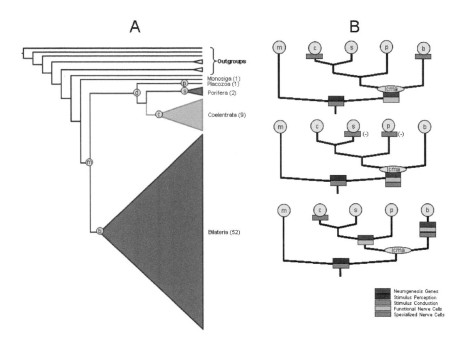

Fig. 5A. Phylogenetic tree with relationships within Bilateria, Coelenterata, and Porifera collapsed. Numbers in parentheses refer to number of species in each of these groups. To establish tree support measures we ran both maximum likelihood (ML) and maximum parsimony (MP) analyses. In the reporting of tree statistics we give the ML value first and the MP value second. Tree support statistics for the nodes marked by circles with letters inside are: (B) Bilateria 100/100, (C) Coelenterata100/82, (S) Porifera 100/100, (D) Diploblasta 100/99, (M) Metazoa 100/63; (P) Placozoa is a single taxon. Within the Bilateria - Deuterostomia 100/100, Protostomia 100/100. (from Schierwater et al. 2009a).

Fig. 5B. Phylogenetic scenarios for the evolution of nerve cells mapped onto the Diploblast-Bilateria Sister hypothesis. Five potential characters (represented by colored boxes in the figure) important in the evolution of nerve cells are mapped onto the Diploblast-Bilateria Sister. Most qualities of a nerve cell seem to have been present already in the last common metazoan ancestor (lcma in light blue). In the top figure we present the most parsimonious explanation for the evolution of these five characters (6 steps). Only the specialization of multi-functional proto-nerve cells into uni-functional nerve cells would have occurred in parallel in Bilateria and Coelenterata in the above scenario. The middle scenario is similar to the top only instead of hypothesizing independent gain of specialized nerve cells it hypothesizes independent loss of specialized nerve cells (7 steps). The bottom tree shows a highly unlikely scenario where the number of steps is nearly twice the top scenario (from Schierwater et al. 2009b).

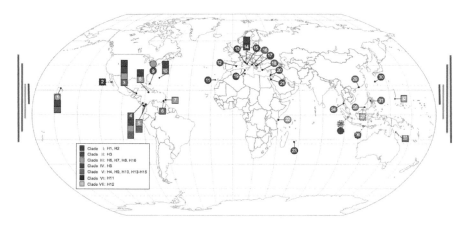

Fig. 6. Worldwide distribution of genetically characterized placozoan specimens. Aquarium samples (A.s.) with presumed origin are labeled with dashed lines. Note that several numbers combine multiple sampling sites. **1.** Oahu, Hawaii (US), **2.** Southern California (A.s., US), **3.** Caribbean coast of Belize, **4.** Caribbean coast of Panama, **5.** Pacific coast of Panama, **6.** Cubagua Island/Margarita Island (Venezuela), **7.** Grenada, **8.** Discovery Bay (Jamaica), **9.** Bahamas, **10.** Bermuda (GB), **11.** Tenerife, Canary Islands (Spain), **12.** Majorca, Balearic Islands (Spain), **13.** Castiglioncello (Italy), **14.** Orbetello Lagoon (Italy), **15.** San Felice Circeo (Italy), **16.** Otranto (Italy), **17.** Kateríni and Ormos Panagias (Greece), **18.** Bay of Turunç (Turkey), **19.** Gulf of Hammamet and near Zarzis (Tunisia), **20.** Caesarea (Israel), **21.** Elat (Israel), **23.** Mombasa (Kenya), **23.** Réunion (France), **24.** Laem Pakarang (Thailand), **25.** 'Indonesia' (A.s.), **26.** Bali (A.s), **27.** 'Indo-Pacific' (A.s.), **28.** Kota Kinabalu, Sabah (Malaysia), **29.** Hong Kong (China), **30.** Okinawa, Ryukyu Islands (Japan), **32.** Boracay (Philippines), **32.** Guam (US), **33.** Lizard Island, (NE Australia). (from Eitel and Schierwater 2010).

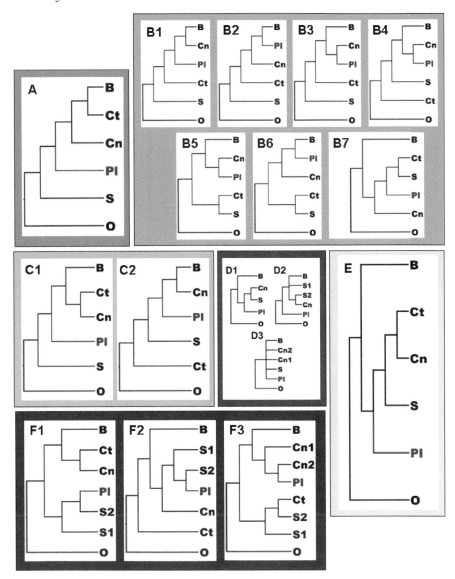

Fig. 7. An overview of published intra-relationships of the four diploblastic groups (Placozoa, Porifera, Cnidaria, Ctenophora) and their inter-relationship to the Bilateria. Shown are a few examples for each of the five character groups defined in Table 1: morphology **(A)**, ribosomal DNA **(B)**, nuclear encoded protein sequences **(C)**, mitochondrial encoded protein sequences **(D)** and combined data sources **(E)**. Placozoans have been placed at nearly every possible relationship to the other four groups, even within Porifera (F1, F2) and within Cnidaria (F3). Hence, a consensus on the phylogenetic placement of the Placozoa is still missing.

O = outgroup(s), S = Porifera, Pl = Placozoa, Cn = Cnidaria, Ct = Ctenophora, B = Bilateria.

Chapter 14

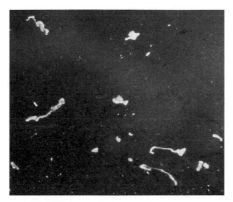

Fig. 5. *Trichoplax* individuals gathering in an area where ephemeral algae is growing in an aquarium. Larger individuals are several millimeters long. While placozoans lack several key features of the stem bilaterians, they may nonetheless exhibit some parallels in life history—rapid development and reproduction in resource-rich areas, followed by dispersal and colonization of other such areas. In the competition between placozoans and stem bilaterians, the head and mouth of the latter were likely decisive advantages.

Chapter 16

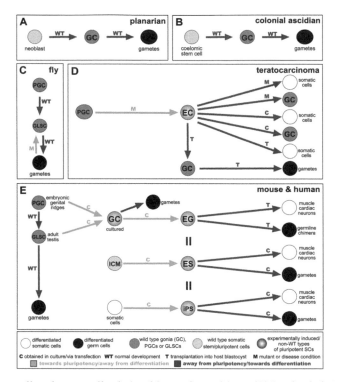

Fig. 1. Stem cell and germ cell relationships and transitions. (A) In platyhelminths, pluripotent stem cells called neoblasts can differentiate into gonia (GC), which generate gametes. **(B)** In colonial ascidians, pluripotent stem cells circulate in the haemolymph, and may differentiate into gonia (GC), which generate gametes. The existence of lineage-restricted germline stem cells in colonial ascidians remains a possibility under active investigation (Brown et al. 2009, Kawamura and Sunanaga 2010). **(C)** In *Drosophila*, embryonically specified primordial germ cells (PGCs) differentiate into unipotent germline stem cells (GLSCs), which generate gametes. However, GLSC daughters that have begun the differentiation process into gametes can be induced to revert to a GLSC fate (Niki and Mahowald 2003). **(D)** In teratocarcinomas, misguided embryonic PGCs are thought to convert to an "embryonal carcinoma" (EC) state. When transplanted into a host blastocyst, EC descendants can populate both the soma and the germline, and generate functional gametes. In situ, EC cells can differentiate into cells with both somatic and germ cell characteristics. In *in vitro* culture, EC cells be induced to differentiate into both somatic cell types and germ cells. **(E)** In normally developing mouse and human embryos, embryonically specified PGCs colonize the embryonic genital ridges, differentiate either into gonia (in females) or into germline stem cells (in males), and subsequently generate gametes. PGCs from the embryonic genital ridges or GLSCs from the adult male testis, can be cultured to form embryonic germ (EG) cells. EG cells can contribute to both somatic and germline cells when transplanted into host blastocysts. Similarly, embryonic stem (ES) cells derived from culturing cells of the inner cell mass (ICM), and differentiated somatic cells converted into induced pluripotent stem (iPS) cells by treatment with variations on the "Yamanaka factors," (Takahashi and Yamanaka 2006, Okita et al. 2007, Takahashi et al. 2007) can both be induced to adopt differentiated somatic and germ cell fate characteristics *in vitro*.